3rd International Conference on Biological Sciences and Biotechnology (ICBSB 2017)

Exploration and Conservation of Natural Resources for Biotechnology Development

IOP Conference Series: Earth and Environmental Science
Volume 130

Medan, Indonesia
23 – 24 August 2017

ISBN: 978-1-5108-6009-4
ISSN: 1755-1307

Printed from e-media with permission by:

Curran Associates, Inc.
57 Morehouse Lane
Red Hook, NY 12571

proceedings.com

Some format issues inherent in the e-media version may also appear in this print version.

Printed by Curran Associates, Inc. (2018)

Institute of Physics
Dirac House, Temple Back
Bristol BS1 6BE UK

Phone: 44 1 17 929 7481
Fax: 44 1 17 920 0979

techtracking@iop.org

Additional copies of this publication are available from:

Curran Associates, Inc.
57 Morehouse Lane
Red Hook, NY 12571 USA
Phone: 845-758-0400
Fax: 845-758-2633
Email: curran@proceedings.com
Web: www.proceedings.com

Table of contents

Volume 130

The 3rd International Conference on Biological Sciences and Biotechnology
23–24 August 2017, Medan, Indonesia

Preface

Papers

On behalf of the organizing committee of the International Conference and Department of Biology, University of Sumatera Utara, we would like to express our warmest welcome for all our participants and speakers to the 3rd International Conference on Biological Sciences and Biotechnology. The conference itself was held in Medan, August 23rd 2017, with the theme "Exploration and Conservation of Natural Resources for Biotechnology Development".

With over 40 selected manuscripts presented in the conference from authors across the globe, this proceeding aimed to serve as a platform to exchange information of the research methods, results, and experiences from various field of biological science and biotechnology all over the world, including microbiology, plant and animal biosciences, biodiversity and environmental sciences, as well as biopharmacy and biomedical science. In addition to that, we hope that this proceeding could promote research opportunities for the participants and speakers both from Indonesia and worldwide.

In this occasion, the organizing committee would express our greatest gratitude for the manuscripts from our four keynote speakers in the conference: Prof. Alison Roberts and Prof. Steven T. Gregory from University of Rhode Island, USA, Prof. Shigeyuki Baba from University of Ryukyus, Japan, and Prof. Sutiman B. Sumitro from Brawijaya University, Indonesia, to participate in this seminar proceeding. The organizing committee would also like to thank all our editors for their precious time and patience to read and revise the manuscripts in this proceeding, as well as to IOP Publishing for their warm service to publish this volume of proceeding.

Thank you very much and we are looking forward for your next participation on the next conference.

Warmest regards,

Prof. Dr. Dwi Suryanto, M.Sc.
Chairman of the 3rd ICBSB 2017

IOP Conf. Series: Earth and Environmental Science **130** (2018) 011002 doi:10.1088/1755-1315/130/1/011002

ACKNOWLEDGEMENT

The committee of the 3rd International Conferece on Biological Sciences and Biotechnology would like to express our greatest gratitude to University of Sumatera Utara, Konsorsium Biologi Indonesia (KOBI), and PT Olympus Indonesia for sponsoring part of the expenses during the seminar, as well as University of Rhode Island (USA), University of Ryukyus (Japan), and Brawijaya University (Indonesia) for providing us noble keynote speakers for the conference. In addition, we would also like to gratefully thank all our participants and oral speakers, as well as all the efforts and contributions that everyone has made for the seminar, which we definitely cannot mention one by one, that the conference would not be as successful without your participation. We are looking forward for your participation in the next ICBSB in 2019!

Warmest Regards,

Chairman

IOP Conf. Series: Earth and Environmental Science **130** (2018) 011003 doi:10.1088/1755-1315/130/1/011003

Organizer
Department of Biology, University of Sumatera Utara

Honorary Board

Prof. Dr. Runtung, SH, Mhum
Rector of University of Sumatera Utara

Dr. Kerista Sebayang, MSi
Dean of Faculty of Mathematics and Natural Sciences, University of Sumatera Utara

International Board
Prof. Alison Roberts, Ph.D
University of Rhode Island, USA

Scientific Committee

Prof. Ing. Peter Valasek, Ph.D
Czech University of Life Sciences Prague, Czech

Prof. Dr. Syafruddin Ilyas, M.Biomed
University of Sumatera Utara, Indonesia

Prof. Dr. Sutiman Bambang Sumitro
Brawijaya University, Indonesia

Prof. Dr. Dwi Suryanto, M.Sc
University of Sumatera Utara, Indonesia

Prof. Erman Munir, M.Sc
University of Sumatera Utara, Indonesia

Prof. Syamsuardi, M.Sc
University of Sumatera Utara, Indonesia

Saharman Gea, Ph.D
University of Sumatera Utara, Indonesia

Dr. Nursahara Pasaribu, M.Sc
University of Sumatera Utara, Indonesia

Organizing Committee
Chairman
Prof. Dr. Dwi Suryanto, M.Sc

Secretary
Dr. It Jamilah, M.Sc
Dra. Nunuk Priyani, M.Sc
Dr. Erni Jumilawaty, M.Si

Treasury
Dr. Isnaini Nurwahyuni, M.Sc

Technical Program Committee

Dr. Saleha Hannum, M.Si
Dr. Liana Dwi Sri Hastuti, M.Si
Dr. Suci Rahayu, M.Si
Dr. Kiki Nurtjahja, M.Sc
Dr. Etti Sartina Siregar, M.Si
Dr. Masitta Tanjung, M.Si
Dr. Miswar Budi Mulya, M.Si
Drs. Nursal, M.Si
Dra. Denny Supriharti, M.Sc
Dra. Mayang Sari Yeanny, M.Si

Dr. Hesti Wahyuningsih, M.Si
Dr. Elimasni, M.Si
Dr. Salomo Hutahaean, M.Si
Dr. Alief Aththorick, M.Si
Dr. Yurnaliza, M.Si
Dr. Kaniwa Berliani, M.Si
Dra. Emita Sabri, M.Si
Drs. Riyanto Sinaga, M.Si
Doni Aldo Siahaan, M.Si

IOP Conf. Series: Earth and Environmental Science **130** (2018) 011004 doi:10.1088/1755-1315/130/1/011004

Peer review statement

All papers published in this volume of *IOP Conference Series: Earth and Environmental Science* have been peer reviewed through processes administered by the proceedings Editors. Reviews were conducted by expert referees to the professional and scientific standards expected of a proceedings journal published by IOP Publishing.

Cellulose microfibril structure: inspirations from plant diversity

A W Roberts

Department of Biological Sciences, University of Rhode Island, 120 Flagg Road, Kingston RI 02881 USA

Email: aroberts@uri.edu

Abstract. Cellulose microfibrils are synthesized at the plasma membrane by cellulose synthase catalytic subunits that associate to form cellulose synthesis complexes. Variation in the organization of these complexes underlies the variation in cellulose microfibril structure among diverse organisms. However, little is known about how the catalytic subunits interact to form complexes with different morphologies. We are using an evolutionary approach to investigate the roles of different catalytic subunit isoforms in organisms that have rosette-type cellulose synthesis complexes.

1. Diversity in cellulose microfibril structure and biosynthesis

Cellulose is an important component of many commercial products and a carbon-neutral energy source. It is produced by diverse organisms in the form of microfibrils, which consist of bundles of 1,4-β-glucan chains stabilized by inter-chain hydrogen bonds. These microfibrils are synthesized by cellulose synthase catalytic subunits that associate in the plasma membrane to form Cellulose Synthesis Complexes (CSC). The number and arrangement of catalytic subunits varies among different taxa, and this variation in CSC organization has been correlated with variation in cellulose microfibril structure [1]. For example, linear CSCs in bacteria [2], brown algae [3], dinoflagellates [4], and red algae [5, 6] produce ribbon-like cellulose microfibrils. In some chlorophycean green algae, large CSCs produce microfibrils containing more than 100 glucan chains [7]. The CSCs of land plant and their close charophycean green algal relative are hexagonal structures known as rosettes ([8], which contain 18 catalytic subunits and therefore produce cellulose microfibrils consisting of 18 glucan chains [9]. These rosette CSCs are composed of specialized cellulose synthases designated CESAs [10, 11] and it has been proposed that domains unique to CESAs participate in CSC assembly [12]. The CESAs diversified early in the evolution of the seed plant lineage to form a medium-sized family whose members are functionally-specialized [13]. Although nearly all commercial cellulose is derived from organisms that have CESAs and rosette CSCs, we know little about the evolutionary origin of CESA proteins or the selective pressures that drove their functional specialization.

2. CESAs functional specialization in Arabidopsis

The specialized functions of the ten members of the Arabidopsis CESA family have been well characterized. First, it has been shown that different CESAs are responsible for primary and secondary cell wall deposition. Specifically, the cellulose microfibrils in secondary cell walls are synthesized by AtCESA4, -7 and -8 [14] and those in in the primary cell walls are synthesized by AtCESA1, -3, and members of the 6-like group [15, 16]. However, promoter swap experiment have shown that at least some primary AtCESAs can substitute for secondary AtCESAs and vice versa, i.e. *AtCESA3pro::AtCESA7* partially rescues *atcesa3* and *AtCESA8pro::AtCESA1* partially rescues *atcesa8* [17]. This is consistent with a role for sub-functionalization in the evolution of seed plant *CESA*s. A role for neo-functionalization is reflected in the shared phenotype of *atcesa4*, *atcesa7,* and *atcesa8* null mutants [18] and the inability to complement these mutants by expressing one of the other secondary *AtCESA*s with the promoter for the missing isoform [19]. This, combined with similar studies of the primary AtCESAs [15, 16], suggests a requirement for three functionally distinct, non-interchangeable CESA isoforms for both primary and secondary cell wall biosynthesis. In addition, there is some functional redundancy among the 6-like members of the *AtCESA* family, *AtCESA2*, *-5*, *-6* and *-9* [16].

3. *Physcomitrella patens* as a model for investigating CESA and CSC evolution

Physcomitrella patens represents the mosses, a lineage of non-vascular plants that diverged from the seed plant lineage over 440 million years ago [20]. The life cycle of *P. patens* [21] includes a haploid phase consisting of a system of protonemal filaments and leafy gametophores (Fig. 1), and a diploid phase that consists of short-stalked sporangia. Although true vascular tissue is lacking, the leaf midribs of *P. patens* contain supportive stereid cells with thick secondary cell walls and thin-walled hydroids that function in transport [22]. The ability to make targeted changes in the *P. patens* genome by homologous recombination [23] along with a sequenced genome [24, 25] has led to the extensive use of this organism to study the evolution of developmental mechanisms in plants [26, 27].

Figure 1. Protonemal filament (top) and leafy gametophores (bottom) of *P. patens*.

Like other land plants, *P. patens* has rosette CSCs [9, 28]. The genome of *P. patens* contains seven *CESA* genes and phylogenetic analysis has shown that the moss and seed plant *CESA* families evolved independently from a single common ancestral gene [19, 29-31]. Thus, the diversification and functional specialization of *CESA*s followed independent paths in the two lineages [28]. We have investigated *CESA* functional specialization in *P. patens* using a targeted gene knockout approach in an effort to better understand the selective pressures that drove the evolution of *CESA* families.

4. CESA functional specialization in *P. patens*

Knock out of individual *PpCESA*s revealed that *PpCESA5* is the only family member with an obvious single-knockout morphological phenotype. The *ppcesa5*KO lines are distinguished by their inability to produce leafy gametophores (Fig. 2). Gametophore buds develop normally through the 4-cell stage, but these cells expand and divide abnormally, producing irregular cell clumps instead of an organized leaf-generating meristem [32]. The defects in cell division and cell expansion are consistent with a role for PpCESA5 in biosynthesis of the primary cell wall.

Figure 2. Phenotype of *ppcesa5*KO with irregular clumps of tissue (arrows, left), in place of wild type leafy gametophores (right).

The production of double knockouts revealed phenotypes for *ppcesa3/8*KO, *ppcesa6/7*KO and *ppcesa4/10*KO. In *ppcesa3/8*KO and *ppcesa6/7*KO lines, the leaf midribs showed reduced birefringence and staining with the cellulose binding dye Pontamine Falst Scarlet 4B (S4B; Fig. 3), consistent with cellulose deficiency [33]. Labeling of sections with the carbohydrate binding module 3A indicated that the secondary cell walls of the stereid cells in the leaf midribs were cellulose deficient [33]. Quantitative analysis of S4B fluorescence revealed a slight deficiency in midrib cellulose deposition in *ppcesa8*KO. However, the *ppcesa*3KOs were indistinguishable from the wild type, indicating that *PpCESA3* and *PpCESA8* are partially redundant. The *ppcesa3/8*KO phenotype could not be rescued by expressing *PpCESA6, PpCESA7, PpCESA4* or *PpCESA10* under control of the *PpCESA8* promoter, indicating that the evolutionary history of the PpCESA family included neofunctionalization. However, we were able to rescue the *ppcesa3/8*KO by expressing *PpCESA5* under control of the *PpCESA8* promoter. This indicates that evolution of the PpCESA family also involved subfunctionalization. The *ppcesa4/10*KO lines have defects in protonemal growth (unpublished) and are currently being investigated.

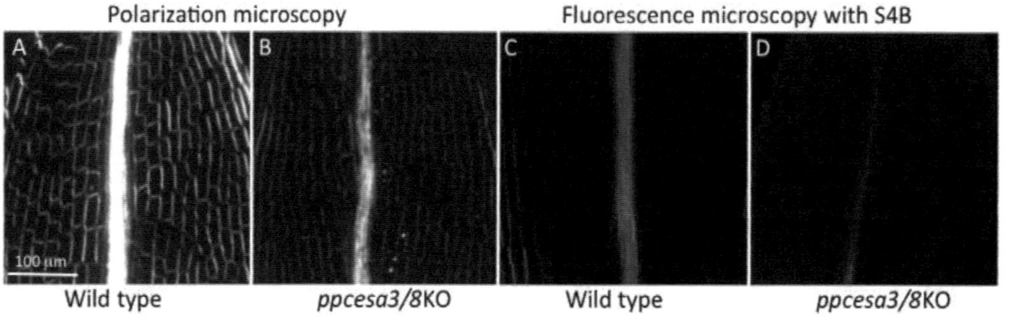

Figure 3. Polarization and fluorescence microscopy of leaves from wild type and *ppcesa3/8*KO *P. patens* showing cellulose deficiency in midribs.

5. Conclusion

Phylogenetic analysis has confirmed that the *CESA* families of seed plants and mosses diversified independently from a single common ancestor [19, 29-31]. However, there are similarities in functional specialization of the *PpCESA* and *AtCESA* families that are attributable to convergent evolution. First, the CESAs that deposit secondary cell walls (*PpCESA3, -6* and *-7*) and primary cell walls (*PpCESA5*) are functionally differentiated [32, 33]. Second, subfunctionalization has occurred as indicated by the ability of *PpCESA5*, which is normally functions in primary cell wall deposition, to rescue the *ppcesa3/8*KO secondary cell wall phenotype when expressed with the *PpCESA8* promoter. Third, neofunctionalization is indicated by the shared phenotype of *ppcesa3/8*KO *and ppcesa6/7*KO and the inability of *PpCESA8pro::PpCESA7* to complement the *ppcesa3/8*KO phenotype. We are currently investigating whether the secondary PpCESAs physically interact to form a CSC. Finally, there is some functional redundancy in the PpCESA family since a strong secondary cell wall phenotype is observed only when *PpCESA3* and *-8* or *PpCESA6* and *-7* are knocked out together [33].

Acknowledgment

Work reviewed here was supported by the National Science Foundation under Award IOS-1257047 and the Center for LignoCellulose Structure and Formation, an Energy Frontier Research Center funded by the U.S. Department of Energy, Office of Science, Office of Basic Energy Sciences under Award DE-SC0001090.

References

[1] Tsekos I (1999) The sites of cellulose synthesis in algae: Diversity and evolution of cellulose-synthesizing enzyme complexes *J Phycol* **35** 635-55

[2] Brown R M, Jr., Willison J H M and Richardson C L (1976) Cellulose biosynthesis in *Acetobacter xylinum*: Visualization of the site of synthesis and direct measurement of the *in vivo* process *Proc. Natl. Acad. Sci. USA* **73** 4565-9

[3] Peng H B and Jaffe L F (1976) Cell-wall formation in Pelvetia embryos. A freeze-fracture study *Planta* **133** 57-71

[4] Sekida S, Horiguchi T and Okuda K (2004) Development of thecal plates and pellicle inthe dinoflagellate *Scrippsiella hexapraecingula* (Peridiniales, Dinophyceae) elucidated

by changes in stainability of the associated membranes *European Journal of Phycology* **39** 105-14

[5] Tsekos I and Reiss H-D (1992) Occurrence of the putative microfibril-synthesizing complexes (linear terminal complexes) in the plasma membrane ofth epiphytic marine red alga *Erythrocladia subintegra* Rosenv. *Protoplasma* **169** 57-67

[6] Tsekos I, Reiss H D and Schnepf E (1993) Cell-wall structure and supramolecular organization of the plasma membrane of marine red algae visualized by freeze-fracture *Acta Botanica Neerlandica* **92** 119–32

[7] Itoh T (1990) Cellulose synthesizing complexes in some giant marine algae *J. Cell Sci.* **95** 309-19

[8] Mueller S C and Brown R M, Jr. (1980) Evidence for an intramembrane component associated with a cellulose microfibril-synthesizing complex in higher plants *J Cell Biol* **84** 315-26

[9] Nixon B T, Mansouri K, Singh A, Du J, Davis J K, Lee J G, Slabaugh E, Vandavasi V G, O'Neill H, Roberts E M, Roberts A W, Yingling Y G and Haigler C H (2016) Comparative structural and computational analysis supports eighteen cellulose synthases in the plant cellulose synthesis complex *Sci Rep* **6** 28696

[10] Pear J R, Kawagoe Y, Schreckengost W E, Delmer D P and Stalker D M (1996) Higher plants contain homologs of the bacterial *celA* genes encoding the catalytic subunit of cellulose synthase *Proc Natl Acad Sci U S A* **93** 12637-42

[11] Kimura S, Laosinchai W, Itoh T, Cui X, Linder C R and Brown R M, Jr. (1999) Immunogold labeling of rosette terminal cellulose-synthesizing complexes in the vascular plant *Vigna angularis Plant Cell* **11** 2075-85

[12] Delmer D P (1999) Cellulose biosynthesis: Exciting times for a difficult field of study *Annu Rev Plant Physiol Plant Mol Biol* **50** 245-76

[13] Somerville C (2006) Cellulose synthesis in higher plants *Annu Rev Cell Dev Biol* **22** 53-78

[14] Taylor N G, Howells R M, Huttly A K, Vickers K and Turner S R (2003) Interactions among three distinct CesA proteins essential for cellulose synthesis *Proc Natl Acad Sci U S A* **100** 1450-5

[15] Desprez T, Juraniec M, Crowell E F, Jouy H, Pochylova Z, Parcy F, Hofte H, Gonneau M and Vernhettes S (2007) Organization of cellulose synthase complexes involved in primary cell wall synthesis in *Arabidopsis thaliana Proc Natl Acad Sci U S A* **104** 15572-7

[16] Persson S, Paredez A, Carroll A, Palsdottir H, Doblin M, Poindexter P, Khitrov N, Auer M and Somerville C R (2007) Genetic evidence for three unique components in primary cell-wall cellulose synthase complexes in Arabidopsis *Proc Natl Acad Sci U S A* **104** 15566-71

[17] Carroll A, Mansoori N, Li S, Lei L, Vernhettes S, Visser R G, Somerville C, Gu Y and Trindade L M (2012) Complexes with mixed primary and secondary cellulose synthases are functional in Arabidopsis plants *Plant Physiol* **160** 726-37

[18] Taylor N G, Laurie S and Turner S R (2000) Multiple cellulose synthase catalytic subunits are required for cellulose synthesis in Arabidopsis *Plant Cell* **12** 2529-39

[19] Kumar M, Atanassov I and Turner S (2016) Functional analysis of cellulose synthase CESA protein class-specificity *Plant Physiol* **173** 970-83

[20] Pires N D and Dolan L (2012) Morphological evolution in land plants: new designs with old genes *Philos Trans R Soc Lond B Biol Sci* **367** 508-18

[21] Cove D J and Knight C D (1993) The moss *Physcomitrella patens*, a model system with potential for the study of plant reproduction *Plant Cell* **5** 1483-8

[22] Xu B, Ohtani M, Yamaguchi M, Toyooka K, Wakazaki M, Sato M, Kubo M, Nakano Y, Sano R, Hiwatashi Y, Murata T, Kurata T, Yoneda A, Kato K, Hasebe M and Demura T (2014) Contribution of NAC transcription factors to plant adaptation to land *Science* **343** 1505-8

[23] Schaefer D G and Zryd J P (1997) Efficient gene targeting in the moss *Physcomitrella patens Plant J* **11** 1195-206

[24] Zimmer A D, Lang D, Buchta K, Rombauts S, Nishiyama T, Hasebe M, Van de Peer Y, Rensing S A and Reski R (2013) Reannotation and extended community resources for the genome of the non-seed plant *Physcomitrella patens* provide insights into the evolution of plant gene structures and functions *BMC Genomics* **14** 498

[25] Rensing S A, Lang D, Zimmer A D, Terry A, Salamov A, Shapiro H, Nishiyama T, Perroud P F, Lindquist E A, Kamisugi Y, Tanahashi T, Sakakibara K, Fujita T, Oishi K, Shin I T, Kuroki Y, Toyoda A, Suzuki Y, Hashimoto S, Yamaguchi K, Sugano S, Kohara Y, Fujiyama A, Anterola A, Aoki S, Ashton N, Barbazuk W B, Barker E, Bennetzen J L, Blankenship R, Cho S H, Dutcher S K, Estelle M, Fawcett J A, Gundlach H, Hanada K, Heyl A, Hicks K A, Hughes J, Lohr M, Mayer K, Melkozernov A, Murata T, Nelson D R, Pils B, Prigge M, Reiss B, Renner T, Rombauts S, Rushton P J, Sanderfoot A, Schween G, Shiu S H, Stueber K, Theodoulou F L, Tu H, Van de Peer Y, Verrier P J, Waters E, Wood A, Yang L, Cove D, Cuming A C, Hasebe M, Lucas S, Mishler B D, Reski R, Grigoriev I V, Quatrano R S and Boore J L (2008) The *Physcomitrella* genome reveals evolutionary insights into the conquest of land by plants *Science* **319** 64-9

[26] Cove D, Benzanilla M, Harries P and Quatrano R (2006) Mosses as model systems for the study of metabolism and development *Annu Rev Plant Biol* **57** 497-520

[27] Prigge M J and Bezanilla M (2010) Evolutionary crossroads in developmental biology: *Physcomitrella patens Development* **137** 3535-43

[28] Roberts A W, Roberts E M and Haigler C H (2012) Moss cell walls: structure and biosynthesis *Front Plant Sci* **3** 166

[29] Roberts A W and Bushoven J T (2007) The cellulose synthase (*CESA*) gene superfamily of the moss *Physcomitrella patens Plant Mol Biol* **63** 207-19

[30] Yin Y, Huang J and Xu Y (2009) The cellulose synthase superfamily in fully sequenced plants and algae *BMC Plant Biol* **9** 99

[31] Harholt J, Sørensen I, Fangel J, Roberts A, Willats W G T, Scheller H V, Petersen B L, Banks J A and Ulvskov P (2012) The glycosyltransferase repertoire of the spikemoss *Selaginella moellendorffii* and a comparative study of its cell wall *PLoS One* **7** e35846

[32] Goss C A, Brockmann D J, Bushoven J T and Roberts A W (2012) A *CELLULOSE SYNTHASE* (*CESA*) gene essential for gametophore morphogenesis in the moss *Physcomitrella patens Planta* **235** 1355-67

[33] Norris J H, Li X, Huang S, Van de Meene A M L, Tran M L, Killeavy E, Chaves A M, Mallon B, Mercure D, Tan H-T, Burton R A, Doblin M S, Kim S H and Roberts A W (2017) Functional specialization of cellulose synthase isoforms in a moss shows parallels with seed plants *Plant Physiol* **175** 210-22

Thermus Thermophilus as a Model System for the Study of Ribosomal Antibiotic Resistance

Steven T. Gregory

Department of Cell and Molecular Biology, College of the Environment and Life Science
The University of Rhode Island
Kingston, RI 02881 USA
and
Department of Molecular Biology, Cell Biology and Biochemistry, Brown University
Providence, RI 02912 USA

Email: stgregory@uri.edu

Abstract. Ribosomes are the intracellular ribonucleoprotein machines responsible for the translation of mRNA sequence into protein sequence. As an essential cell component, the ribosome is the target of numerous antibiotics that bind to critical functional sites to impair protein synthesis. Mutations causing resistance to antibiotics arise in antibiotic binding sites, and an understanding of the basis of resistance will be an essential component of efforts to develop new antibiotics by rational drug design. We have identified a number of antibiotic-resistance mutations in ribosomal genes of the thermophilic bacterium *Thermus thermophilus*. This species offers two primary advantages for examining the structural basis of antibiotic-resistance, in particular, its potential for genetic manipulation and the suitability of its ribosomes for analysis by X-ray crystallography. Mutations we have identified in this organism are in many instances identical to those found in other bacterial species, including important pathogens, a result of the extreme conservation of ribosome functional sites. Here I summarize the advantages of this organism as a model system to study antibiotic-resistance mechanisms at the molecular level.

1. Introduction: *Thermus thermophilus* as a Model Organism

Members of the *Thermus* genus are all extremely thermophilic bacteria belonging to the *Deinococcus-Thermus* phylum, one of the most deeply branching clades in the bacterial domain [1]. The type species of the genus *Thermus* was discovered in 1969 and named *Thermus aquaticus* [2]. Although *T. thermophilus* was discovered a year earlier, its taxonomic affiliation was not properly described until 1974 [3]. *T. thermophilus* has become one of the most widely used thermophiles due to several distinct advantages it provides as a model organism [4]. Several species have been found to be naturally transformable with genomic or plasmid DNA [5]. Together with the development of plasmid vectors and antibiotic-resistance genes functional at high temperatures, this provides the potential for genetic manipulations including the construction of gene knockouts [reviewed in 4].

T. thermophilus is easily cultivated, growing aerobically with an optimum growth temperature between 65 and 72 °C, depending on the strain. These temperatures are within the usable range for agar as a solidifying agent in culture media, and cultivation of *T. thermophilus* requires no specialized laboratory equipment. It forms colonies on agar plates, allowing genetic selections and colony purification. Our laboratory has employed three different strains of *T. thermophilus*. Strains HB8 (the type strain of the species) and HB27 were isolated from terrestrial hot springs in Japan [3], while strain IB-21 was isolated from a submarine hot spring off the coast of Iceland in 1986 [6]. The genome sequence of HB27 has been published [7], while the HB8 sequence is available through public databases. We have successfully isolated antibiotic-resistant mutants of each of these strains. *T. thermophilus* is sensitive to a wide range of antibiotics, allowing the isolation of mutants with phenotypes corresponding to those found among many important bacterial pathogens. We have also noted that genetic crosses by transformation between these strains occurs readily, attesting to their close relationship and allowing mutations to be transferred to various backgrounds.

Importantly, ribosomes from *T. thermophilus* form crystals that diffract X-rays to high resolution, allowing their three-dimensional structure to be determined [reviewed in 8]. Figure 1 shows the structure of the *T. thermophilus* 70S ribosome. Further, structures of antibiotic-ribosome complexes have been determined, thereby establishing the precise mode of ribosome-antibiotic interaction [for instance, 9, 10]. Importantly, these results are consistent with similar structural findings using mesophilic ribosomes [11, 12], thus validating the use of this thermophile as a model system. Such results aid in predicting possible mechanisms of resistance. More importantly, they provide a framework for the interpretation of structures of mutant ribosomes determined experimentally.

2. Antibiotic-resistance Mutations in Ribosomes of *T. thermophilus*.

For genetic studies of ribosomes, *T. thermophilus* has the attribute of having only two copies of each rRNA gene [7]. Deletion of one of these to produce strains having single rRNA gene copies allows the isolation of recessive mutations. One potential concern regarding the use of this species to investigate antibiotic-resistance mutations is the very thermophilic nature that is responsible for the facility with which its ribosomes form ordered crystals. As a result, some laboratories will choose to acquire high-resolution structural data using ribosomes from pathogenic species [for example, 13].

Several factors combine to argue in favor of *T. thermophilus* as a model system, however. First and foremost, antibiotics bind very specifically to their binding sites and do so with ribosomes from diverse clades within bacterial phylogeny. These binding sites are highly conserved in sequence; this conservation is due to the functionally critical nature of these sites, which is why antibiotics bind to them in the first place. Thus, the conserved binding of antibiotics to ribosomes from across the bacterial domain provides a strong case that ribosome sequence conservation reflects an underlying conservation of three-dimensional structure. This conclusion is supported by structures of ribosomes from several sources. Finally, as I will describe below, we have found that similar or identical mutations produce similar or identical phenotypes, which in turn implies that they produce similar or identical changes in ribosome structure to prevent binding. Thus, it can be argued that information from structural studies of antibiotic-resistant thermophilic ribosomes is sufficiently robust to interpret resistance mutations found in mesophilic bacterial pathogens.

2.1 Aminoglycoside-resistance Mutations.

Aminoglycosides, including streptomycin, kanamycin, paromomycin, neomycin, hygromycin B, and many others, act by binding at or near the decoding site on the 30S subunit, and in doing so, induce translational errors [reviewed in 14]. While paromomycin (and the structurally related drugs kanamycin and neomycin) and hygromycin B bind at distinct sites within 16S rRNA helix 44, streptomycin has a more complex binding site that consists of two 16S rRNA pseudoknots and 30S subunit ribosomal protein

S12. Consistent with these drugs binding to non-overlapping sites, mutations conferring resistance to these two antibiotic groups arise at distinct places in the ribosome structure. Streptomycin-resistance mutations in *T. thermophilus* result in amino acid substitutions in S12 [15, 16] or in base substitutions in the central pseudoknot of 16S rRNA [16]. Resistance to kanamycin is conferred by an A1408G substitution while hygromycin B resistance is conferred by base substitutions at the U1406-U1495 base pair [17].

X-ray crystal structures of 30S subunits bearing base substitutions in the central pseudoknot of 16S rRNA show clear changes in position due to disruption of the U13-A915-U20 base triple or the Type II G21-A914 base pair [18]. However, streptomycin contacts the central pseudoknot via interactions with the sugar-phosphate backbone, and makes no direct interaction with rRNA bases, as shown in Figure 2. In crystal structures of mutant ribosomes, the backbone trajectory remains very close to that of wild-type ribosomes, defying a straightforward explanation for resistance. Indeed, these mutant ribosomes are able to bind streptomycin at high concentrations. One possible interpretation of the structural data is that disruption of pseudoknot base-base hydrogen bonding increases backbone conformational dynamics, thereby reducing streptomycin binding affinity.

An interesting variation on the streptomycin-resistance phenotype is streptomycin dependence. Streptomycin-dependent mutants are incapable of growth in the absence of the drug. Given the capacity of streptomycin and other aminoglycosides to induce miscoding, it is perhaps not surprising that streptomycin-resistance mutations generally cause the ribosome to become hyper-accurate. Streptomycin dependence has been described as having an extreme form of hyperaccuracy [14]. Dependence mutations can arise either as amino acid substitutions in ribosomal protein S12, or as base substitutions in 16S rRNA. Crystal structures of the former are quite revealing, as they show a severe distortion of the 16S rRNA helix 44 decoding site, such that the ribosome's ability to monitor the geometry of codon recognition would be grievously if not critically impaired [19]. However, either binding of streptomycin or the inclusion of a suppressing mutation were found to restore the decoding site structure to a near-native conformation. In light of these observations, the notion of streptomycin dependence as a form of extreme hyperaccuracy would appear to be an oversimplification.

2.2 Thiostrepton-resistance Mutations.

Thiostrepton and related peptide antibiotics bind to a cleft between 50S ribosomal protein L11 and its 23S rRNA binding site. Its mode of action is to block the EF-G-dependent translocation step of protein synthesis [14]. *T. thermophilus* is exquisitely sensitive to thiostrepton, and mutants are readily selected with both amino acid substitutions in L11 and base substitutions in 23S rRNA occurring [20]. Some of these are located directly in the thiostrepton binding site, while several others occur at the hinge between the two domains of L11. Curiously, each of the L11 amino acid substitutions occurs at a proline residue. As prolines constrain polypeptide conformation, such substitutions may act by increasing the flexibility of various parts of L11 as they impinge upon the thiostrepton binding site. Thus, changes in conformational dynamics within L11 may potentially impact the binding affinity of thiostrepton to mutant ribosomes. In contrast, the sites of base substitutions in 23S rRNA, A1067 or A1095, are located at contact sites for thiostrepton, allowing a straightforward explanation for resistance.

2.3 Capreomycin-resistance Mutations.

Capreomycin and viomycin are examples of tuberactinomycin antibiotics [14] that bind to the interface between the 30S and 50S subunits, and in so doing, interfere with intersubunit rotations that occur during the translocation step of protein synthesis [21]. In our initial selections for capreomycin-resistant mutants of *T. thermophilus*, we identified only base substitutions in helix 44 of 16S rRNA, specifically A1408G, C1409G, and G1491A, mutations previously associated with aminoglycoside resistance [16]. This finding is consistent with aminoglycosides and tuberactinomycins having overlapping binding sites. In a

IOP Conf. Series: Earth and Environmental Science **130** (2018) 012002 doi:10.1088/1755-1315/130/1/012002

subsequent study, we identified mutations affecting 23S rRNA helix 69, including the base substitutions A1913U and mU1915G and the single base deletion ΔmU1915 [22]. These two rRNA structures, despite being on different subunits, interact directly with one another as part of an intersubunit bridge and together constitute the tuberactinomycin binding site [21]. Distortion of the drug binding site seems to be sufficient to cause resistance, although firmer conclusions await high-resolution structures of mutant ribosomes.

2.4 Chloramphenicol and Macrolides.

The peptidyltransferase active site (also known as the peptidyltransferase center, or PTC) consists exclusively of nucleotides of 23S rRNA in the 50S ribosomal subunit. The PTC is the most frequently targeted functional site of the ribosome, and multiple structurally unrelated classes of antibiotics bind to it [14]. Among these are macrolides (such as erythromycin, tylosin, and azithromycin, among many others), lincosamides (including lincomycin and clindamycin), streptogramins (virginiamycin, pristinamycin), chloramphenicol, hygromycin A, and sparsomycin.

Among the very first *T. thermophilus* antibiotic-resistant mutants we identified was an erythromycin-resistant mutant having the single base substitution A2058G of 23S rRNA [23]. Erythromycin is a macrolide that interferes with polypeptide chain elongation into the exit channel [14]. This mutation has been identified in a number of organisms, and is usually found to confer cross-resistance to lincosamides and streptogramin B antibiotics. Chloramphenicol by contrast is a true peptidyltransferase inhibitor, binding directly to the active site on the 50S ribosomal subunit and inhibiting the catalytic step of peptide bond formation. Selection for chloramphenicol-resistant mutants of *T. thermophilus* produced the largest variety of base substitutions, including A2030G, G2061A, A2062G, G2447A, C2452U, A2453G, U2500C, A2503G, U2504G, U2504C, U2504A, and G2505A, consistent with the highly complex three-dimensional organization of the PTC [17]. Because numerous antibiotics bind to the PTC, mutations conferring resistance to one antibiotic often confer resistance to one or more other antibiotics. This was indeed found for mutants in our study, as revealed by cross-resistance tests. Furthermore, base substitutions in the PTC occur at functionally critical residues, and can have significant to severe effects on growth and peptidyltransferase catalytic activity [24].

Given the large number of antibiotics that bind to the PTC, and the correspondingly large number of potential resistance mutations that can arise within it, one interesting question is how combinations of multiple mutations might impact resistance phenotypes. For instance, selection on multiple antibiotics, either sequentially or in combination, could conceivably give rise to multiple mutations in the PTC. Complex cross-resistance patterns could emerge, which is information that could prove of benefit in designing new drug combination therapies. How such mutation combinations might phenotypically interact is completely unknown and worthy of future investigation.

3. Conclusions

T. thermophilus in many ways provides an ideal model system for investigating antibiotic-resistance mechanisms. Its facile genetics combined with the capacity of its ribosomes to crystallize allow for detailed investigations into the precise mechanism by which mutations in the ribosome interfere with antibiotic action. Given that mutations in this organism are most often identical to those found in pathogenic species, *T. thermophilus* provides an excellent model for establishing basic principles that are applicable to medically important organisms. While the solution of pathogen ribosome structures may yield insights into specific organisms, the benefit of such insights would need to outweigh the potential hazards of growing large-scale cultures of antibiotic-resistant pathogens for ribosome purification. Given the remarkable conservation of ribosome structure, particularly in the functional centers that comprise antibiotic binding sites, the use of a model system as a proxy for pathogenic organisms may continue to prove advantageous.

Acknowledgements
The work described here represents the combined effort of a very large number of people working in multiple laboratories, without whose contributions this work would not have been possible. I am especially grateful to Albert E. Dahlberg and Gerwald Jogl, whose collaborations have been invaluable.

Funding/Financial Disclosure
The work described here was supported by grants from the US National Institutes of Health GM19756 to Albert E. Dahlberg and GM094157 to Steven T. Gregory and Gerwald Jogl.

References

[1] Williams RA, Smith KE, Welch SG, Micallef J, Sharp RJ (1995) DNA relatedness of *Thermus* strains, description of *Thermus brockianus* sp. nov., and proposal to reestablish *Thermus thermophilus* (Oshima and Imahori). Int J Syst Bacteriol 45: 495–499.
[2] Brock TD, Freeze H (1969) *Thermus aquaticus* gen. n. and sp. n., a nonsporulating extreme thermophile. J Bacteriol 98: 289–297.
[3] Oshima T, Imahori K (1974) Description of *Thermus thermophilus* (Yoshida and Oshima) comb. nov., a nonsporulating thermophilic bacterium from a Japanese thermal spa. Int J Syst Bacteriol 24: 102–112.
[4] Cava F, Hidalgo A, Berenguer J (2009) *Thermus thermophilus* as biological model. Extremophiles 13: 213–231. (doi: 10.1007/s00792-009-0226-6)
[5] Koyama Y, Hoshino T, Tomizuka N, Furukawa K (1986) Genetic transformation of the extreme thermophile *Thermus thermophilus* and of other *Thermus* spp. J Bacteriol 166: 338–340.
[6] Henne A, Brüggemann H, Raasch C, Wiezer A, Hartsch T, et al. (2004) The genome sequence of the extreme thermophile *Thermus thermophilus*. Nat Biotechnol 22: 547–553. (doi: 10.1038/nbt956)
[7] Kristjansson JK, Hreggvidsson GO, Alfredsson GA (1986) Isolation of halotolerant *Thermus* spp. from submarine hot springs in Iceland. Appl Environ Microbiol 52: 1313–1316.
[8] Schmeing TM, Ramakrishnan V (2009) What recent ribosome structures have revealed about the mechanism of translation. Nature 461: 1234–1242. (doi: 10.1038/nature08403)
[9] Carter AP, Clemons WM, Brodersen DE, Morgan-Warren RJ, Wimberly BT, Ramakrishnan V (2000) Functional insights from the structure of the 30S ribosomal subunit and its interactions with antibiotics. Nature 407: 340–348. (doi: 10.1038/35030019)
[10] Bulkley D, Innis CA, Blaha G, Steitz TA (2010) Revisiting the structures of several antibiotics bound to the bacterial ribosome. Proc Natl Acad Sci USA 107: 17158–17163. (doi: 10.1073/pnas.1008685107)
[11] Schlünzen F, Zarivach R, Harms J, Bashan A, Tocilj A, Albrecht R, Yonath A, Franceschi F (2001) Structural basis for the interaction of antibiotics with the peptidyl transferase centre in eubacteria. Nature 413: 814–821. (doi: 10.1038/35101544)
[12] Dunkle JA, Xiong L, Mankin AS, Cate JHD (2010) Structures of the *Escherichia coli* ribosome with antibiotics bound near the peptidyl transferase center explain spectra of drug action. Proc Natl Acad Sci USA 107: 17152–17157. (doi: 10.1073/pnas.1007988107)
[13] Hentschel J, Burnside C, Mignot I, Leibundgut M, Boehringer D, Ban N (2017) The complete structure of the *Mycobacterium smegmatis* 70S ribosome. Cell Rep 20: 149-160. (doi: 10.1016/j.celrep.2017.06.029)
[14] Gale EF, Cundliffe E, Reynolds PE, Richmond MH, Waring MJ (1981) The molecular basis of antibiotic action, p. 402–547. John Wiley & Sons, London, United Kingdom.

[15] Gregory ST, Cate JH, Dahlberg AE (2001) Streptomycin-resistant and streptomycin-dependent mutants of the extreme thermophile *Thermus thermophilus*. J Mol Biol 309: 333–338. (doi: 10.1006/jmbi.2001.4676)

[16] Gregory ST, Dahlberg AE (2009) Genetic and structural analysis of base substitutions in the central pseudoknot of *Thermus thermophilus* 16S ribosomal RNA. RNA 15: 215–223. (doi: 10.1261/rna.1374809)

[17] Gregory ST, Carr JF, Rodriguez-Correa D, Dahlberg AE (2005) Mutational analysis of 16S and 23S rRNA genes of *Thermus thermophilus*. J Bacteriol 187: 4804–4812. (doi:10.1128/JB.187.14.4804-4812.2005)

[18] Demirci H, Murphy FV, Murphy EL, Connetti JL, Dahlberg AE, Jogl G, Gregory ST (2014) Structural analysis of base substitutions in *Thermus thermophilus* 16S rRNA conferring streptomycin resistance. Antimicrob Agents Chemother 58: 4308–4317. (doi: 10.1128/AAC.02857-14)

[19] Demirci H, Wang L, Murphy FV, Murphy EL, Carr JF, Blanchard SC, Jogl G, Dahlberg AE, Gregory ST (2013) The central role of protein S12 in organizing the structure of the decoding site of the ribosome. RNA 19: 1791–1801. (doi: 10.1261/rna.040030.113)

[20] Cameron DM, Thompson J, Gregory ST, March PE, Dahlberg AE (2004) Thiostrepton-resistant mutants of *Thermus thermophilus*. Nucleic Acids Res 32: 3220–3227. (doi: 10.1093/nar/gkh644)

[21] Stanley RE, Blaha G, Grodzicki RL, Strickler MD, Steitz TA (2010) The structures of the anti-tuberculosis antibiotics viomycin and capreomycin bound to the 70S ribosome. Nat Struct Mol Biol 17: 289–293. (doi: 10.1038/nsmb.1755)

[22] Monshupanee T, Gregory ST, Douthwaite S, Chungjatupornchai W, Dahlberg AE (2008) Mutations in conserved helix 69 of 23S rRNA of *Thermus thermophilus* that affect capreomycin resistance but not posttranscriptional modifications. J Bacteriol 190: 7754–7761. (doi: 10.1128/JB.00984-08)

[23] Gregory ST, Cate JH, Dahlberg AE (2001) Spontaneous erythromycin resistance mutation in a 23S rRNA gene, *rrlA*, of the extreme thermophile *Thermus thermophilus* IB-21. J Bacteriol 183: 4382–4385. (doi: 10.1128/JB.183.14.4382-4385.2001)

[24] Rodriguez-Correa D, Dahlberg AE (2008) Kinetic and thermodynamic studies of peptidyltransferase in ribosomes from the extreme thermophile *Thermus thermophilus*. RNA 14: 2314-8. (doi: 10.1261/rna.1146008)

[25] Voorhees RM, Weixlbaumer A, Loakes D, Kelley AC, Ramakrishnan V (2009) Insights into substrate stabilization from snapshots of the peptidyltransferase center of the intact 70S ribosome. Nat Struct Mol Biol 16: 528- (doi: 10.1038/nsmb.1577)

[26] Demirci H, Murphy F, Murphy E, Gregory ST, Dahlberg AE, Jogl G. 2013. A structural basis for streptomycin-induced misreading of the genetic code. Nat Commun 4: 1355-1358. (doi: 10.1038/ncomms2346)

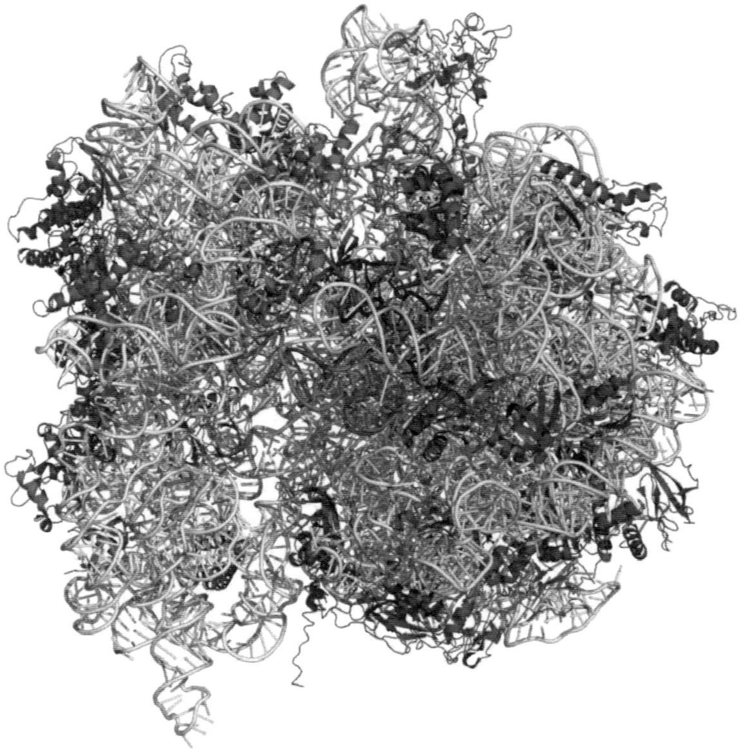

Figure 1. The crystal structure of the *T. thermophilus* 70S ribosome (pdb entries 2wdk and 2wdl; ref 25). The 30S subunit is to the left, with 16S rRNA colored palecyan and small subunit proteins colored skyblue. The 50S subunit is to the right, with 23S rRNA and 5S rRNA colored grey and large subunit proteins colored purple. Three tRNAs are located at the subunit interface, with the A-site tRNA colored green, P-site tRNA colored red, and E-site tRNA colored orange. These tRNAs mark the functional centers of the ribosome, also the binding sites for numerous antibiotics.

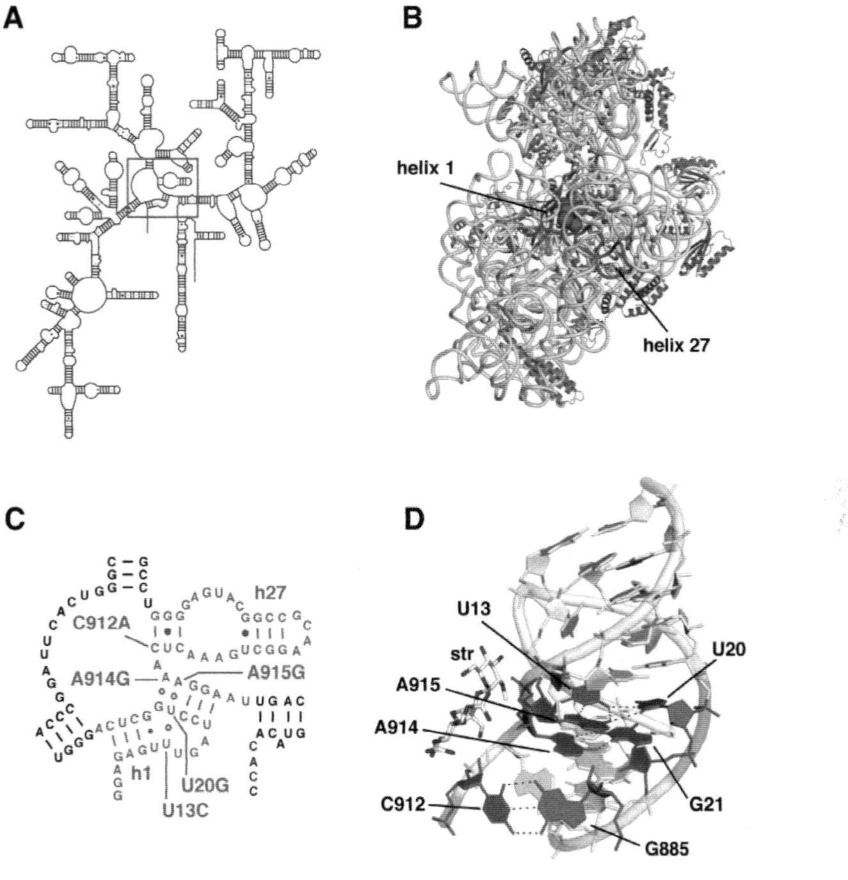

Figure 2. Streptomycin-resistance mutations in 16S rRNA, an example of antibiotic-resistance mutations found in *T. thermophilus*. (A) Secondary structure model of 16S rRNA, with the central pseudoknot region enclosed by a blue box. (B) Crystal structure fo the 30S ribosomal subunit, with 16S rRNA helices 1 and 27 shown in blue, and streptomycin shown as blue spheres (pdb entry 4DR3; ref 26). (C) Sites of single base substitutions in the secondary structure model of 16S rRNA. (D) Close-up view of the central pseudoknot, showing hydrogen bonding interactions affected by resistance mutations in blue. Note that streptomycin makes contacts with the pseudoknot exclusively via the backbone and makes no direct contacts with bases.

Herbal Medicine, Radical Scavenger and Metal Detoxification: Bioinorganic, Complexity and Nano Science Perspectives

Sutiman B. Sumitro[1*] and Sukmaningsih Alit[2]

1. Laboratory of Cell, Molecular and Nano Biology, Brawijaya University
2. Department of Biology, Udayana University

*Email: sutiman@ub.ac.id

Abstract. Developing Complexity Science and Nano Biological perspective giving the ideas of interfacing between modern physical and biological sciences for more comprehensive understanding of life. The study of bioinorganic is a trans-disciplinary, and will initiate the way to more comprehensive and better understanding life. We can talk about energy generation, motive forces and energy transfer at the level of macromolecules. We can then develop understanding biological behavior on nano size biological materials and its higher order using modern physics as well as thermodynamic law. This is a necessity to ovoid partial understanding of life that are not match with holism. In animal tissues, the accumulation or overwhelmed production of free radicals can damage cells and are believed to accelerate the progression of cancer, cardiovascular disease, and age-related diseases. Thus a guarded balance of radical species is imperative. *Edward Kosower* [1] proposed an idea of biradical in an aromatic organic compounds. Each of which having unpaired electrons. The magnetic force of this compound used for making agregation based on their magnetic characters. Bioinorganic low molecular weight complex compounds composing herbal medicine can bind toxic metals. This low molecular weight complex molecules then easily excerted the metals from the body, removing them from their either intracellular or extracellular existences. This bioinorganic chelation potential is now inspiring a new therapeutic strategies.

Keywords: Bioinorganic-complex compound-scavenger-radicals-intoxication

1. Bioinorganic and Holism Perspectives

Biological system is complex structure with specific function dedicated to perform normal ordered organizational system. Most people are not even conscious that biologists are using what is called the mechanical view of reality when they think and talk. In fact many biological scientists including Medical Science, they still practiced reductionism, reducing things into their parts and examining the parts to understand what made them tick. They reduced life to cells, molecules while the concepts are remaining within Newtonian. In the early 20th Century, the certainty of Newton's mechanics was undermined by quantum mechanics and the Uncertainty Principle developed by Werner Heisenberg . This is the new challenges to the old Newtonian view of reality. So some biological scientists began abandoning the Newtonian worldview, while most of them and the ordinary people held on to it.

Developing Complexity Science and Nano Biological perspective are giving the ideas of interfacing between modern physical and biological sciences for more comprehensive understanding of life [2,3]. We can then develop understanding biological behavior on nano size biological materials and its higher order using modern physics as well as thermodynamic law.

In the second half of the 20th century coordination chemistry developed which broadened understanding of the organic chemical bond to transition metal ions called as bioinorganic complex compounds [4,5]. The study is commonly called as Bioinorganic Chemistry talking about biochemical function of "inorganic elements". The consequence of the works is interdisciplinary research, regarding to synthesis and analysis of "model systems". The advantage of Bioinorganic Chemistry is that we can learn any biochemical process from nature in more holism and original fashion overcoming partial-analytic-reductive concepts which are normally developed from a test tube in the laboratory.

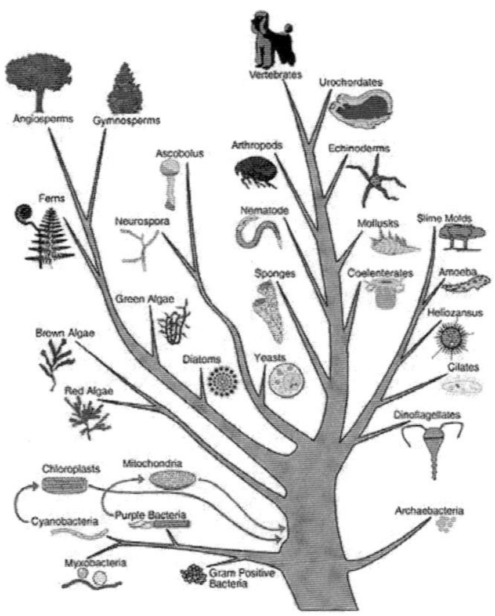

Figure 1. Evolution of organism along with bioinorganic complex compounds related to energy generation

The study is very much concerned on control and the use of metal ions in biochemical processes which are optimized system by evolution (Figure 1), producing efficient system of collection, conversion and storage of energy. Bioinorganic chemistry is important for the realization of the implications of electron-transfer proteins, substrate bindings and activation, atom and group transfer chemistry, as well as metal properties in biological chemistry. Many of them are naturally occurred in biological processes, such as respiration and photosynthesis [6]. The bioinorganic are metalloproteinase such as Superoxide Dismutase (SOD), Catalase within mitochondria as well as Hemoglobin in red blood cells and chlorophyll in leaf cells; they are containing complex organic compounds with metal as central coordinator atom (Figure 2).

The process system is in mode of moderate conditions processes stand-in by metal proteins. The main field of research is biological relevant complexes namely biochemistry with metal as center of

IOP Conf. Series: Earth and Environmental Science **130** (2018) 012003 doi:10.1088/1755-1315/130/1/012003

coordination chemistry [7]. This will include bio-mineralization namely biochemistry and solid state (materials) chemistry, enzyme which are not dealing with analysis but also synthesis and characterization of model systems by means of computer software. Bioinorganic is a multidisciplinary study and play important roles in understanding regulating combustion, polymerization, and chemical energy transfer processes in physiology. Therefore the characterization of metal ions in biological systems and in particular those bound to proteins has flourished.

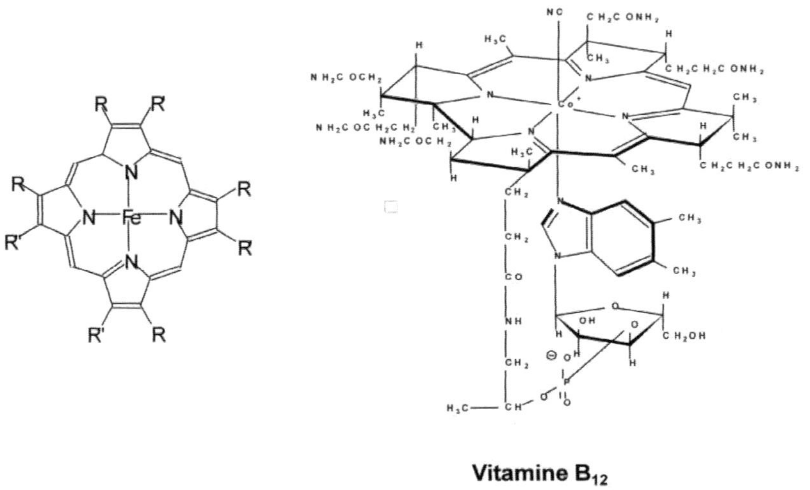

Figure 2. Bioinorganic, low molecular weight complex with metal as coordinator

The study of bioinorganic is a trans-disciplinary, and will initiate the way to more comprehensive and better understanding life [8]. We can talk about energy generation, motive forces and energy transfer at the level of macromolecules.

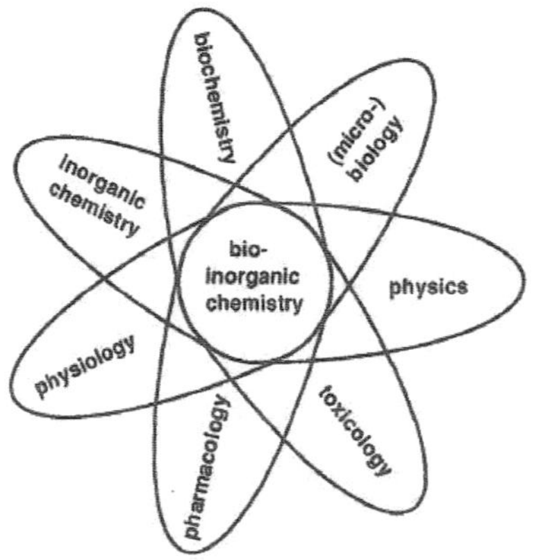

Figure 3. Bioinorganic is a multidisciplinary study

We can then develop understanding biological behavior on nano size biological materials and its higher order using modern physics as well as thermodynamic law. This is a necessity to ovoid partial understanding of life that is not match with holism.

2. Life Process, Nano Size Material, Energy, Bioinorganic and Free Radical Scavenging System

Biological system is complex structure. All sub cellular material such as proteins, DNA, RNA and membranous materials are complex molecular structures which have specific function dedicated to perform ordered organizational system. They are all having functional structure size within 1 to 100 nm. Their behaviors when proceeding life process within the cell are not merely can be thought following Newtonian but also demonstrating electrical and magnetic phenomena. Their intracellular activities are showing their smartness which are following modern physics thoughts.

Life process even though there is a unified basic pattern of life, and we can be more precise and say that this pattern is a network pattern, these networks are not structures — at least most of them — they are functional networks [9]. The energy generation is occurred along with electron transitions amongst atoms which unpaired electron has called as radical (Figure 4). This can be a species of atom or group of atoms (molecule) as gas• that has at least one unpaired electron and is therefore unstable and highly reactive, having rate const. > 10^8 M^{-1} sec^{-1}.

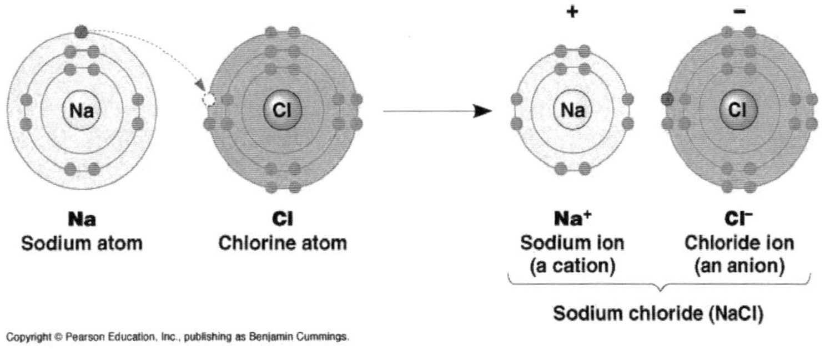

Figure 4. Transitional electron. Radical is atom with unpaired electron

Edward Kosower [1] proposed an idea of biradical in an aromatic organic compounds, each of which having unpaired electrons. The magnetic force of this compound used for making agregation based on their magnetic characters (Figure 5).

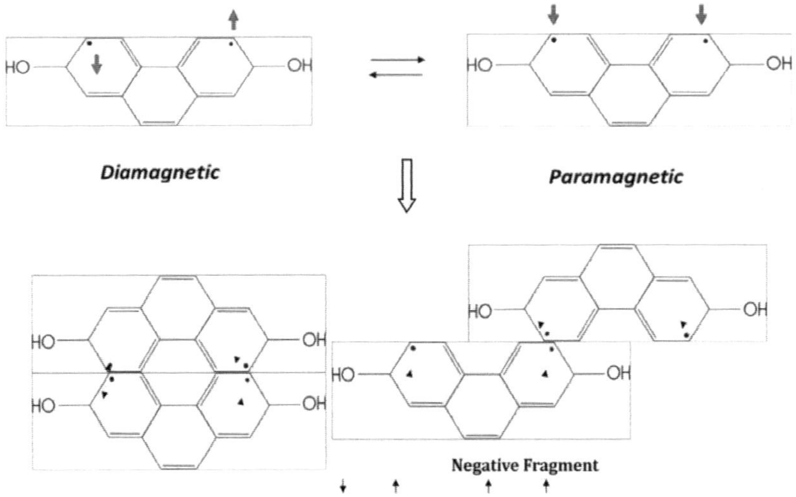

Figure 5. **Biradical,** two atoms with unpaired electron in a compound.This will initiate different type of molecular aggregation

Metals can play a further role in the design of anion receptors, not as binding centers, but rather as architectural elements. Complex molecules such as DNA, Hemoglobin, chlorophyll, SOD and Catalase plays an important role in scavenging free radicals in life [10]. These compounds have a molecule that binds to the metal. These Metal ions form hydroxyl radicals when it reacts with H_2O_2 through the Haber Weiss or Fenton reaction [1].

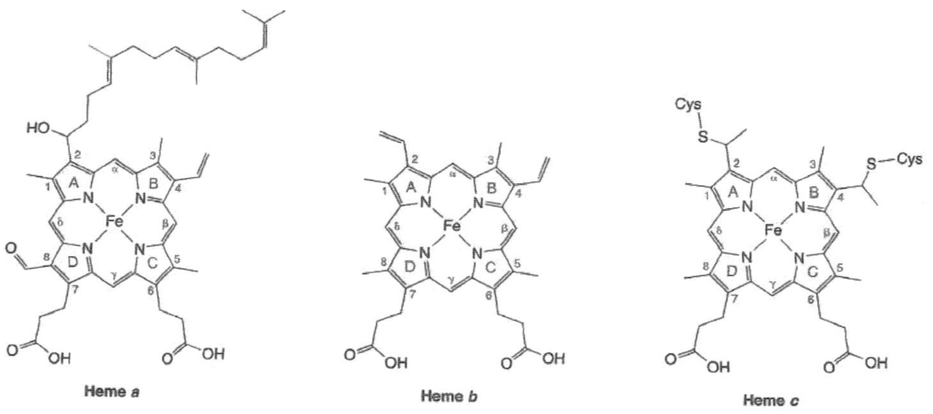

Figure 6. TETRAPYRROLES. Partially unsaturated tetradentate, macrocyclic ligands. Stable, rigid, planar or nearly planar ring system. Deprotonated forms bind metal ions tightly and size selectively extensive conjugation leads to very intense colors and potentially to redox activity

IOP Conf. Series: Earth and Environmental Science **130** (2018) 012003 doi:10.1088/1755-1315/130/1/012003

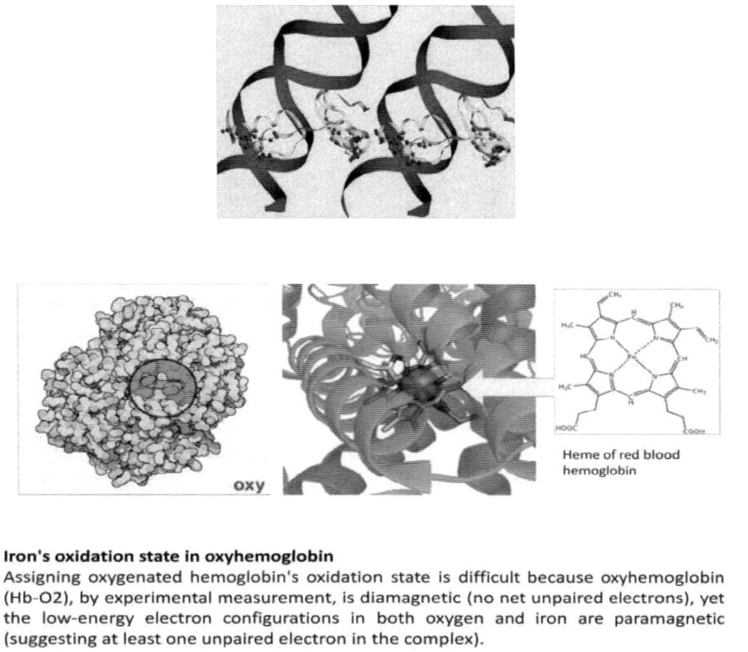

[10]

Figure 7. Bioinorganic bind with DNA (upper) and red blood globins (lower)

The role of radicals in living systems is dual. On the one hand radicals are vital for normal biological functions but on the other hand they are responsible for a variety of deleterious biological processes. In animal tissues, the accumulation or overwhelmed production of free radicals can damage cells and are believed to accelerate the progression of cancer, cardiovascular disease, and age-related diseases. Thus a guarded balance of radical species is imperative. Metal ions have a pronounced tendency to interact with molecules to give structures of varying complexity, which are described by classical coordination chemistry [11].

Bioinorganic compound, namely porphyrin will exhibit differently when they interact with various globint under diffrent expopsed pressure as seen in figure 8. Porphyrins are the conjugate acids of ligands that bind metals to form complexes. The metal ions such as Fe, Cu and Mg usually have a charge of 2^+ or 3^+. A porphyrin without a metal-ion in its cavity is a “free base”. In Hemoglobin and Myoglobin they contain iron-containing porphyrins which are called hemes are found extensively in nature. Hemoglobin and Myoglobin are two O_2-binding proteins that contain iron porphyrins. Various Cytochromes and scavenging proteins such as Super Oxide Dismutase and Catalase are also Hemoproteins.

IOP Conf. Series: Earth and Environmental Science **130** (2018) 012003 doi:10.1088/1755-1315/130/1/012003

$$K_{Mb} = \frac{[Mb(O_2)]}{[Mb][O_2]}$$

$$K_{Hb} = \frac{[Hb(O_2)_4]}{[Hb][O_2]^{2.8}}$$

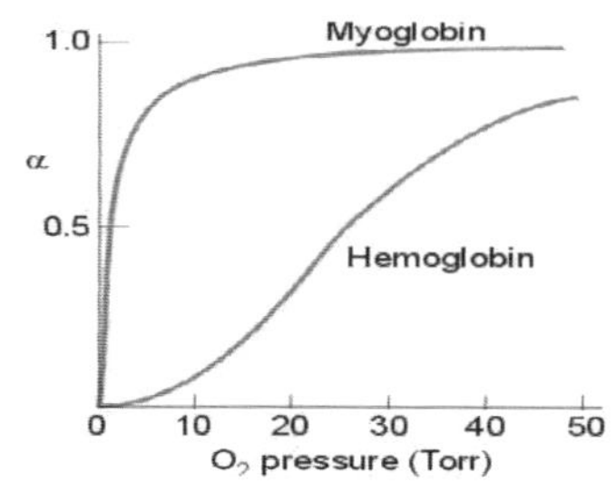

Figure 8. The globin behave in diferent ways when it bound with different bioinorganic compounds

3. Bioinorganic on Herbal Studies

Herbal medicine use plant or animal materials for medicinal purposes. The study of herbal medicine is often talking about extracted mixing complex compounds. Most of them are remain with their native structures and in the form of complex compounds. Bioinorganic studies using EPR on herbal medicine is shown on figure 7 where extracted fruit flesh and seeds enable to reduce radical of cigarette smoke when observed by EPR Spectroscopic.

Bioinorganic low molecular weight complex compounds composing herbal medicine can bind toxic metals. This low molecular weight complex is molecules then easily excreted the metals from the body, removing them from their either intracellular or extracellular existences. This bioinorganic chelation potential is now inspiring a new therapeutic strategies [12]. The following figure 7 is a theoretical concept of mercury chelation by complex phenanthrene and nicotine contained in tobacco smoke.

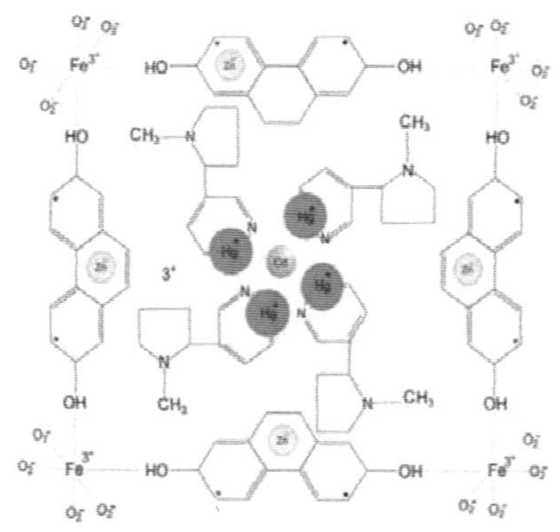

Figure 9. The nanostructures of complex molecule characterized by *intra spin-spin interaction of the conjugated di-HO•-Phenanthrenedienyl* with very mobile electron in their surface.

The mobile electron on surface nanostructure of complex molecule generated by *intra spin-spin interaction of the conjugated di-HO•-Phenanthrenedienyl* can exhibit electron transition to shade free radicals as shown by their change in their magnetic resonance (Figure 10).

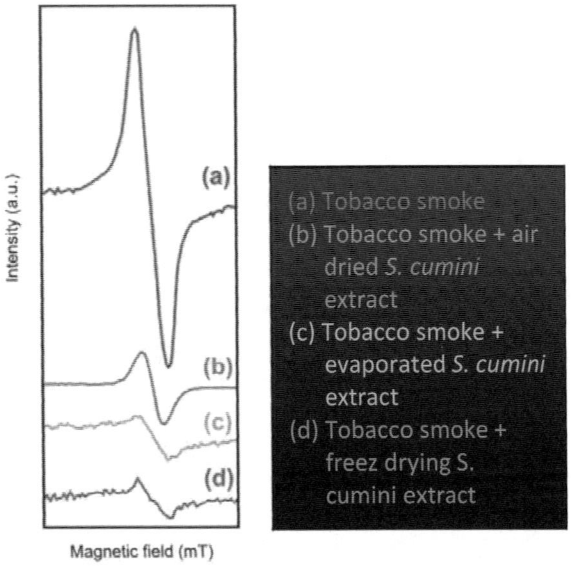

Figure 10. Short-lived free radical observed under EPR Spectroscopy showing free radical shading of Tobacco smoke streamed pass through filter coated by seed extract of *S. cumini*.

4. Instrumentations

The research methodologies to study characteristic of bioinorganic complex compounds of herbal medicinal materials are dealing with analytical techniques aimed to study their natural structures, metal chelating properties, reconstructing their tree dimensional formation as well as evaluating their potential as radical scavenger. Their crystals properties usually showed by the light retardation or birefringence under polarizing microscope, and more specifically observed using X-ray diffraction spectroscopy (XRD). The particles size is measured using Particle Analyzer and observed under Scanning Electron Microscope. The metal components confirm by XRF and SEM. Meanwhile, the LC-MS test can be used to observe the existence of antioxidant compounds with chelating potential to bind transitional metals such as Fe, Cu and Mg forming bioinorganic complex compounds. The paramagnetic properties is observed using Electron Spin Resonance Spectroscopy (ESR). Here is the list of instrumentation needed for bioinorganic characterization:

- Diffraction methods for 3d structure: Problem: crystallization of proteins and Complex structures, "high" resolution (ca. 0.2 nm), no identification of hydrogen atoms
- NMR (Nuclear Magnetic Resonance) for local structure and dynamical properties of species
- Scanning Electron Microscopy (3d structure with medium resolution)
- ESR (Electron Spin Resonance Spectroscopy or EPR (Electron Paramagnetic Resonance) can be used to observe electronic properties of species containing unpaired electrons
- Mössbauer spectroscopy used for identification of species with quadrupol moment
- Optical spectroscopy observing color, birefringence, light retardation
- X-ray absorption techniques for local structure
- SQUID for characterization of magnetic materials
- Cyclovoltammetry for characterization of electron transfer

IOP Conf. Series: Earth and Environmental Science **130** (2018) 012003 doi:10.1088/1755-1315/130/1/012003

Acknowledgment

We are very grateful thank to Dr. Gretha Zahar for her valuable suggestions and contribution in developing our understanding on Physical Organic Chemistry. She also kindly let us use her unpublished work namely *the nanostructures of complex molecule* (Figure 9) for this publication.

References

[1] Kosower EM (1967) An introduction to physical organic chemistry. John Wiley, University of California. USA.
[2] Luisi PL (2006) The Emergence of Life. From Chemical Origins to Synthetic Biology. Cambridge UniversityPress.(http://www.chemiebiologie.unisiegen.de/ac/hjd/lehre/master/bioinorganic_handout.pdf)
[3] Pisani F (2007) Network as a unifying pattern of life involving different process at different levels: An Interview with Fritjof Capra. International Journal of Communication1. Feature 5-25. El Paris-Le Monde-Reforma.
[4] Gray, HB, EI Stiefel, J Selverstone Valentine, I Bertini Eds (2006) Biological Inorganic Chemistry: Structure and Reactivity. University Science Books.
[5] Kienle L (2017) Basics of Bioinorganic Chemistry, part 1. Max-Planck-Institut für Festkörperforschun Stuttgart.
[6] Bertini I (ed) (2009) Organic and Bio-inorganic Chemistry. Encyclopedia of Life Support Systems (Vol 1). EOLSS Publishers / UNESCO, Eolss Publishers Co. Ltd., United Kingdom. www.eolss.net.
[7] Brend AH (ed) (2000) Relativistic Effects in Heavy-Element Chemistry and Physics. John Wiley.
[8] Rehder D (2008) Introduction to Bioinorganic Chemistry, Lecture Notes, University of Lund.
[9] Capra F (2007) Complexity and Life in Reframing Complexity: Perspective from the North and South vol. 1 (Capra F., A. Juarrero, P. Sotolongo and J. van Uden eds.). ISCE Publishing, Mansfield, MA. USA.
[10] Albert, B, JD Watson, D Bray, M Raff (2008) Molecular Biology of The Cell. Garland Science, Publishing Co.
[11] Sumitro, SBS, Widyarti, S Permana (2016) Cell Biology. Department of Biology, Brawijaya University.
[12] Swaran JS. and V Pachauri (2010) Chelation in metal Intoxication. Int. J. Environ. Public Health, (7): 2745-2788; doi: 10.3390/ijerph7072745.

Correlation between homocysteine and dyslipidemia in ischaemic stroke patients with and without hypertension

Cut Aria Arina[1], Darwin Amir[2], Yahwardiah Siregar[1], Rosita J. Sembiring[1]

[1]Medical Faculty, University of Sumatera Utara, Jl. dr. Mansur no. 5 Kampus USU Medan 20155
[2]Medical Faculty, University Andalas, Padang, West Sumatera

[*]Email: ariaarina2010@gmail.com

Abstract. Almost 80% of strokes are ischaemic and stroke is the third most common cause of death in developed countries, . The treatment of stroke still limited, the best approach to reduce mortality and morbidity is primary prevention through modification of acquired risk factors. Hypertension and dyslipidemia are one of the major risk factor for stroke while homocysteine is a less well-documented risk factor. The purpose of this study was to know the correlation between homocysteine and dyslipidemia in ischaemic stroke patients with and without hypertension. This study is a cross sectional study; the sample were taken consecutively. All sample matched with inclusion and exclusion criteria, demography data and blood sample were taken. Demography data was analyzed using descriptive statistic, to analyze the relation, we used Chi-Square test. p value <0,05 was significant. Of the 100 patients, were divided into two groups, with hypertension, and without hypertension, hyperhomocysteinemia was found in 62 patients (59 patients had mild hyperhomocysteinemia, three patients had moderate hyperhomocysteinemia) and dyslipidemia was found in 60 patients. There is a significant relation between homocysteine and dyslipidemia in ischaemic stroke patients with hypertension, p value = 0,009. A significant correlation between homocysteine and dyslipidemia might be because both of them have an important role in the acceleration of the atherosclerotic formation by activation platelet and thrombus, but we still need further study to get more explanation about the relation.

Keywords: atherosclerotic disease, blood pressure, risk factors of stroke

1. Introduction

Almost 80% of strokes are ischaemic, and the remainder is haemorrhagic [1], stroke is the third most common cause of death in developed countries [1,2]. The treatment of stroke still limited, the best approach to reduce mortality and morbidity is primary prevention through modification of acquired risk factors [3,4].

Hyperhomocysteinemia is a *less well-documented modifiable risk factor* [5], that correlated with vascular disease, including stroke. [6,7,8] An increasing total plasma homocysteine is an independent risk factor for atherosclerosis and increased the risk of ischaemic stroke about two and half times. Epidemiologic studies showed that a moderate increasing plasma homocysteine level (>10 µmol/L) could increase the risk of stroke. [9] Hyperhomocysteinemia can caused endothelial dysfunction that correlated with activation platelet and thrombus [10].

Data from cohort and case control studies showed a positive correlation between total concentration of homocysteine and the risk of stroke. [11,12] An increased plasma serum level about five μmol/L will increase the risk of stroke about 50% [11], while decreased plasma serum level about three μmol/L will decrease the risk of stroke about 24%. [13]

Dyslipidemia has an important role as the cause of disability and mortality. Mechanism dyslipidemia as the cause of ischaemic stroke still unclear, its has been reported that there is a relation between cholesterol total, LDL, and HDL with the risk of ischaemic stroke. [14]

Disorders of lipid metabolism such as elevated serum levels of total cholesterol, triglyceride, low density lipoprotein, high density lipoprotein, apolipoprotein A have been considered to
There is a study that has proved that cholesterol total, HDL, LDL, has relation with acute ischaemic stroke, but the role in pathogenesis of ischaemic stroke still unclear. Others studies reported that cholesterol total, LDL, HDL related with the risk factor of ischaemic stroke. [15]

Another study reported a decreased relation between lipid and stroke. From 296 stroke patients and control, found that cholesterol total, trygliserida, and HDL had connection with the risk of ischaemic stroke, even higher total ratio of cholesterol/ HDL increased the risk. [14]
The purpose of this study was to know the correlation between homocysteine and dyslipidemia in ischaemic stroke patients.

2. Materials and Methods

This study is a cross sectional study; the sample was taken consecutively. The study was conducted with the approval of the ethic committee of the institutions involved and informed consent was obtained from all subjects or their relatives (for comatose patients) for the inclusion of the study. Inclusion criteria included first were the presence of stroke at the present hospital admission. The neurological deficits were confirmed in all cases by computerized tomography (CT) scan.
The homocysteine was classified into two categories: (1) normal and (2) hyperhomocysteinemia. Dyslipidemia was classified into two categories: (1) normal, (2) dyslipidemia.
All sample matched with inclusion and exclusion criteria, demography data and others data were taken. Demography data was analyzed using descriptive statistic, to analyze the relation, we used Chi-Square test. P value <0,05 was significant.

3. Results and Discussion

Of the 100 patients with ischaemic stroke patients were divided into two groups, with, and without hypertension. A total of 63 men had an ischaemic stroke, while 37 women had an ischaemic stroke, 35 patients at the age below 50 y.o (24 male, 11 female), 65 patients at the age above 50 y.o (39 male, 26 female), hyperhomocysteinemia was found in 62 patients, dyslipidemia was found in 60 patients, smoking was found in 45 patients, diabetes mellitus was found in 64 patients.

Table 1. Baseline Characteristic of the Ischaemic Stroke Patients

Subjects		Male	Female
Age	< 50y.o	24	11
	> 50y.o	39	26
Hyperhomocysteine	Yes	41	21
	No	22	16
Dyslipidemia	Yes	40	20
	No	23	17
Smoking	Yes	43	2
	No	20	35
Diabetes mellitus	Yes	42	22
	No	21	15

Table 2. Relation between Homocysteinemia and Dyslipidemia

Ischaemic Stroke	Hyperhomocysteinemia	Dyslipidemia	p = 0,009
With Hypertension	47	46	
Without hypertension	15	14	

There is a significant relation between homocysteine and dyslipidemia, p=0,009 (p<0,005).

This study found that, among 100 patients with ischaemic stroke, hyperhomocysteinemia was found in 62 patients (59 patients had mild hyperhomocysteinemia, three patients had moderate hyperhomocysteinemia) and dyslipidemia was found in 60 patients. Many epidemiologic studies found that increased level of homocysteine is an independent risk factor for atherosclerosis and thrombosis artery. There are many advere effects of homocysteine, most of them are related to endothelial dysfunction which is related to platelet activation and thrombus formation. [10]

Newest study has explain the relation between lipid and ischaemic stroke, and has shown that ischaemic stroke and the size of carotid atheroma can be reduced by lipid lowering agents. Its has been reported that the asscociation between cholesterol total, LDL, HDL with the incidence of ischaemic stroke. [14]

Tian et al. 2014 found that cholesterol total, LDL, HDL related with the incidence of acute ischaemic stroke [14]. Kurth et al. 2007 and Shahar et al. 2003 reported that cholestreol total, LDL, HDL related with risk factor of ischaemic stroke [15,16]. Shahar et al. 2003 and Bowman et al. 2003, reported a decreased relation between lipid and stroke [16,17].

Atherosclerotic is the cause of death for more than half death in west countries, [16] its cause the most serious morbidity and mortality [17]. It is a slow progressive artery disease, produced by the thickness of tunica intima that caused by fibrous accumulation that gradually will narrowing the vessel lumen and caused thrombus [16].

Atherosclerosis risk factors act at several points on this pathogenic pathway, such as the plaques generally cause clinical manifestations by producing flow-limiting stenoses that lead to tissue ischaemia, or by provoking thrombi that can interrupt blood flow locally or embolize and lodge in distal arteries. [18,19]

There are several limitations of our study. First the study enrolled hospital based stroke patients rather than patients from a community based general population, secondly, this study did not collect data on others several major risk factors that may influence the homocysteine level.

4. Conclusion

A significant correlation between hyperhomocysteinemia and dyslipidemia might be because both of them have an important role in the acceleration of the atherosclerotic formation by activation platelet and thrombus, that caused atherosclerotic vascular disease. Atherosclerosis is the main cause of death for more than half in west countries. This is a slow progressive artery disease, with the thickness of tunica intima that happens because of gradually accumulation of fibrosa that make the lumen smaller and become thrombus. Lumen narrowing as the effects of the plaque accumulation caused stiffness of the artery wall that can cause ischaemic. [16]

Acknowledgements
The authors gratefully acknowledge that the present research is supported by Ministry of Research and Technology and Higher Education Republic of Indonesia. The support is under the research grant

TALENTA Universitas Sumatera Utara of The Year 2016 Contract Number: 1/UN.5.2.3.1/PPM/SP/2016

References

[1] Caplan LR, (2009) Caplan's Stroke: A Clinical Approach. 3rd ed. Butterworth-Heineman. Boston.
[2] Goldstein LB, Bushnell, CD, Braun, LT, Chaturvedi, S, Creager, MA, Culebras, A, et al. (2011) Guidelines for the Primary Prevention of Stroke. A Guideline for Healthcare Professionals From the American Heart Association / American Stroke Association. Stroke, vol. 42, pp. 517-84.
[3] Casas JP, Bautista, LE, Smeeth, L, Sharma, P, Hingorani, AD, (2005) Homocysteine and stroke: evidence on a causal link from mendelian randomisation. The Lancet, vol. 365, no. 9455, pp. 224-32.
[4] Sacco RL (2000), Pathogenesis, Classification and Epidemiology of Cerebrovascular Disease In: Rowland L.P. (ed). Merrit's Neurology. 10th ed. pp. 217-29. Lippincott Williams & Wilkins. Philadelphia.
[5] Goldtein LB, Adam, R, Albert, MJ, Appel, LJ, Brass, LM., Bushnell, CD, et al. (2006) Primary Prevention of Ischemic Stroke: A Guideline from the American Heart Association / American Stroke Association Council. Stroke, vol. 37, pp. 1583-633.
[6] Rahman A, Gupta, RD, Quraishi, FA, Saha, UK, Amin, N (2013), Relationship Between Homocysteine and Ischaemic Stroke. J. Medicine, vol. 14, pp. 47-51
[7] Men X, Li, J, Zhang, B, Zhang, L, Li, H, Lu, Z (2013), Homocysteine and C-Reactive Protein Associated with Progression and Prognosis of Intracranial Branch Atheromatous Disease. PloS ONE, vol, 8, no.9, pp. 1-7.
[8] Murakami H, Lemitsu, M, Sanada, K, Gando, Y, Ohmori, Y, Kawakami, R, et al. (2011) Association among objectively measured physical activity, fasting plasma homocysteine concentration, and MTHFR C677T genotype. Eur J Appl Physiol, vol 111, pp. 2997-3005.
[9] Palvonic AM, Pekmezovic, T, Obrenovic, R, Novakovic, I, Tomic, G, Mijajlovic, M (2011) Increased total homocysteine level is associated with clinical status and severity of white matter changes in symptomatic patients with subcortical small vessel disease. Clinical Neurology and Neurosurgery, vol. 113, pp. 711-5.
[10] Voetsch B, Loscalzo, J (2004) Genetic Determinants of Artrial Thrombosis. Arterioscler Thromb Vasc Biol, vol. 24, pp. 216-29.
[11] Low HQ, Chen, CPLH., Kasiman, K, Thalamuthu, A, Ng, SS, Foo, JN, et al. (2011) A Comprehensive Association Analysis of Homocysteine Metabolic Pathway Genes in Singaporean Chinese with Ischemic Stroke. PLoS ONE, vol, 6, no.9, pp, 1-6.
[12] Narayan SK, Firbank, MJ, Saxby, BK, Stansby, G., Hansrani, M, O'Brien, JT, et al. (2011) Elevated Plasma Homocysteine Is Associated with Increased Brain Atrophy Rates in Older Subjects with Mild Hypertension. Demen Geriatr Cogn Disord, vol. 31, pp. 341-8.
[13] Holmes MV, Newcombe, V, Hubacek, JA, Safat, R, Ricketts, SL, Cooper, J, et al. (2011) Effect modification by population dietary folate on the association between MTHFR genotype, homocysteine, and stroke risk: a meta-analysis of genetic studies and randomized trials. Lancet, vol. 378, pp. 584-94.
[14] Tian X, Tao ZT, MeiY, Huan Z, Qing LW, Yan K, et al. (2014) Dyslipidemia and Outcome in Patients with Acute Ischemic Stroke. Biomedical and Environmental Sciences Journal; 27(2): 106-110.
[15] Kurth T, Everett BM, Buring JE, Kase CS, Ridker PM, Gaziano JM, (2007) Lipid Levels and The Risk of Ischaemic Stroke in Women. Neurology. 68(8):556-562.

[16] Shahar E, Chambles LE, Rosamond WD, Boland LL, Ballantyne CM, McGovern PG, Sharrett AR (2003) Plasma Lipid Profile and Incident Ischemic Stroke The Atherosclerosis Risk in Communities (ARIC) Study. Stroke.34: 623-631.

[17] Bowman TS, Sesso HD, Ma J, Kurth T, KaseCS, Stampher MJ, Gaziano, JM (2003) Cholesterol and The Risk of Ischaemic Stroke. Stroke. 34:2930-2934.

[18] Silbernagel S (2007) Heart and Circulation: Atherosclerosis. In: Silbernagel, S., Lang F. Color Atlas of Pathophysiology. Thieme Syuat Gard. New York. pp. 236-239.

[19] Hossmann KA, Heiss WD (2010) Section 1 Chapter 1: Etiology, Pathophysiology, and Imaging. Neuropathology and Pathophysiology of Stroke. In: Brainin M, Heiss W.D (editors). Textbook of Stroke Medicine. New York: Cambridge University Press.p.1-77.

[20] Bilic i, Dzamonja G, Lusic I, Matijaca M, Caljkusic K. (2009) Risk Factor and Outcome Differences Between Ischaemic and Hemorrhagic Stroke. Acta Clin Croat (Original Scientific Paper). 48: 399-403.

[21] Libby P, Ridker PM, Hansson GK (2011) Progress and challenges in translating the biology of atherosclerosis. Nature. 473: 317-323.

Contrast induced nephropathy in hypertensive patients after elective percutaneous coronary intervention

Cut Aryfa Andra[1*], Andi Khairul[1], Cut Aria Arina[2], Zulfikri Mukhtar[1], Isfanuddin Nyak Kaoy[1]

[1]Department of Cardiology, University of Sumatera Utara, Jl. dr. Mansur Kampus USU Medan 20155
[2]Department of Neurology, University of Sumatera Utara, Jl. dr. Mansur Kampus USU Medan 20155

[*]Email: andra1711@gmail.com

Abstract. Contrast induced nephropathy (CIN) is the third lead cause of hospital acquired renal failure and was associated with significant morbidity and mortality. We hypothesized that hypertension is an independent risk factor for the development of CIN in patients undergoing elective percutaneous coronary intervention (PCI). The case-control method was used, 138 patients scheduled for elective PCI. We measured serum creatinine at baseline and after 24 hours of the procedure. CIN was defined as arising in serum creatinine of at least 44 μmol/l (0,5 mg/dl) or 25% rise from baseline. All patients received low osmolality nonionic contrast during PCI. Hypertension was defined as self-reported a history of treated or untreated diagnosed high blood pressure. One hundred thirty-eight patients (74,6%) were male, and 35 patients (25,4%) were female. Among the 138 patients, 86 (62,3%) were hypertensive patients whereas 52 (37,7%) were nonhypertensive patients. There was no difference in baseline serum creatinine levels and the amount of contrast media in patient with and without CIN. CIN developed in 42 patients, 39 patients (92,9%) were hypertensive compared to 3 patients (7,1%) without hypertension with p value < 0,05. (Odds ratio 16,8, 95% CI 4.542 - 62,412). This study showed that hypertension was a risk factor for the development of CIN

Keywords: complications, contrast media, coronary angiography, risk factors, nephropathy

1. Introduction

Contrast media (CM) was used increasingly in diagnostic and interventional procedure. This result in the rising incidence of iatrogenic renal function impairment caused by exposure to CM, a condition known as CIN. Radiographic CM is responsible for 11% of cases of hospital-acquired renal insufficiency, the third most common cause of renal failure after impaired renal perfusion and the use of nephrotoxic medications. Among all procedures utilizing CM for diagnostic or therapeutic purposes, coronary angiography and percutaneous coronary interventions (PCI) was associated with the highest rates of CIN [1]. CIN is a serious clinical problem associated with increased morbidity and mortality [2,3]. It is the third most common cause of hospital-acquired acute renal failure, after surgery and hypotension [4].

The definition of CIN includes three necessary components: an absolute or relative increase in serum creatinine compared to the baseline values, a temporal relationship between the rise in serum creatinine and exposure to a contrast agent, and the exclusion of alternative explanations for renal impairment [1]. CIN is typically defined in the recent literature as an increase in serum creatinine (SCr) occurring within the first 24 h after contrast exposure and peaking up to 5 days afterwards and is usually defined as an increase in SCr concentration of ≥ 0.5 or 1.0 mg/dl or relative increase of ≥ 25 or 50% over baseline [5,6]. Development of CIN has important effects on patient outcome and may contribute to other post-PCI complication including death, myocardial infarction and stroke [5]. The consequences of CIN are prolonged hospital stay, increased nonrenal morbidity and increased cardiovascular risk and mortality [7].

Arterial hypertension is a major cardiovascular risk factor and an important cause of chronic renal dysfunction. Early studies documented the presence of arteriolosclerosis and tubulointerstitial changes in nearly 100% of kidneys from patients with hypertension. Patients with hypertension would constitute a large and important risk group because of the high prevalence of the disorder in the general population as well as among patients undergoing coronary angiography [8]. The aim of the present report is to evaluate the impact of hypertension on the incidence of contrast nephropathy in patients undergoing percutaneous coronary intervention.

2. Methods

This study is case contol study, the sample was taken consecutively. The written informed consent was obtained from all subjects who were admitted to Haji Adam Malik Hospital in North Sumatera, Medan, Indonesia during the study. Inclusion criteria were patients with documented SCr taken within one week before elective PCI and age >18 years old. We excluded patients who had end stage renal disease and required dialysis, patients who had recent contrast exposure before elective PCI.

Patients underwent elective PCI according to the current guidelines. All patients received low osmolality nonionic contrast during PCI. CIN in our study was defined as an increase in serum creatinine of at least 44 µmol/l (0.5 mg/dl) or 25% rise from baseline recorded 24 hours before and after elective PCI. Hypertension was defined as self-reported a history of treated or untreated diagnosed high blood pressure.

Patients were assigned to received hydration with 0.9% saline (isotonic) with an infusion rate of 1 ml/kg body weight/hour. After the procedure, hydration was continued until 12 hours.

All sample matched with inclusion and exclusion criteria, demography data and others data were taken. Demography data were expressed as mean ± standard deviation or median and range. We used Chi-Square test (nominal data) and Students t-test (numerical data) to univariate analysis. Logistic regression analysis was used to multivariate analysis to test the hypothesis that hypertension is an independent risk factor for development CIN. P value < 0.05 was significant.

3. Results and Discussion

Total 138 patients included in this study. One hundred thirty-eight patients (74.6%) were male, and 35 patients (25.4%) were female. Among the 138 patients, 86 (62.3%) were hypertensive patients whereas 52 (37.7%) were nonhypertensive patients. Table 1 showed the differences in the baseline clinical characteristic between patients with CIN and without CIN. There was no difference in gender, age, history of smoking, hyperlipidemia and family history, whereas there were differences in body mass index, hypertension and diabetes mellitus with the p value <0.05.

Table 1. Baseline Cinical Characteristic

	Variable	Total	CIN (-) N=96	CIN (+) N=42	p value
1	Men (%)	103 (74.6%)	71 (74.0%)	32 (76.2%)	0.782
2	Age (years)	56.7 ± 8.9	56,2 ± 9,6	58.5 ± 7.1	0.120
3	Body Mass Index (kg/m^2)	25.7 ± 3.3	25,3 ± 3,1	28.6 ± 2.8	<0.001*
4	Hypertension (%)	86 (62.3%)	47 (49.0%)	39 (92.9%)	<0.001*
5	Diabetes Mellitus (%)	31 (22.5%)	16 (16.7%)	15 (35.7%)	0.014*
6	Smoker	87 (63.0%)	61 (63.5%)	26 (61.9%)	0.855
7	Hyperlipidemia	33 (23.9%)	25 (26.0%)	8 (19.0%)	0.375
8	Family History	13 (9.5%)	12 (12.6%)	1 (2.4%)	0.065

Table 2 showed the differences in the baseline laboratory values. There was no difference in baseline serum creatinine levels and the amount of contrast media in patient with and without CIN, whereas there were differences in GFR<60 and leucocyte level (p value<0.05).

Table 2. Baseline Laboratory Values

Variable	**Total**	**CIN (-) N=96**	**CIN (+) N=42**	**P value**
GFR ≤60	51 (37.0%)	41 (42.7%)	10 (23.8%)	0.034*
Hemoglobin	13.0 ± 1.7	13.0 ± 1.6	13.1 ± 1.7	0.703
Leucocyte	8000 (3080 – 19600)	8505 (3080 – 19600)	7570 (4400 – 17490)	0.003
Hematocrit	40.0 ± 5.1	40.1 ± 5.2	39.7 ± 4.9	0.685
Ureum	30.0 (14.0 – 105.0)	30 (14 – 105)	29 (16 – 76.3)	0.749
Creatinin	1.1 (0.6 – 2.8)	1,1 (0,6 – 2,8)	1.0 (0.6 – 2.3)	0.232
Blood glucose	112 (63.0 – 364.0)	110 (63 – 364)	116.5 (81 – 364)	0.103
SGOT	26 (12 – 292)	26 (12 – 292)	27.5 (12 – 90)	0.976
SGPT	24 (6 – 128)	25 (6 – 128)	21 (6 – 71)	0.676
Contrast volume	200 (50 – 400)	200 (50 – 400)	200 (75 – 300)	0.080

Table 3 showed the prevalence of CIN in hypertension and non-hypertension patients. CIN was developed in 42 patients, 39 patients (92.9%) were hypertensive compared to 3 patients (7.1%) without hypertension with p value < 0.05. Logistic regression analysis was used to multivariate analysis to test the hypothesis that hypertension is an independent risk factor for development CIN and the result showed that hypertension was a risk factor for the development of CIN with p value <0.001, Odds ratio 16.836 and 95% CI 4.542-62.412.

Table 3. CIN in patients with and without hypertension

	CIN (+)	CIN (-)
Hypertension	39 (92.9%)	47 (49.0%)
Non Hypertension	3 (7.1%)	49 (51%)

IOP Conf. Series: Earth and Environmental Science **130** (2018) 012006 doi:10.1088/1755-1315/130/1/012006

Tabel 4. Multivariate analysis

Variable	P value	OR	95% CI
Hypertension	<0.001	16.836	4.542 – 62.412
Diabetes Mellitus	0.024	3.186	1.168 – 8.695

The main result of the present study is that hypertension is an independent risk factor for the development of contrast nephropathy. A recently published analysis reported an incidence of contrast nephropathy of 13% among patients with glomerular filtration rate above 60 ml/min/1.73 m^2. Nevertheless, hypertension was also an independent predictor of contrast nephropathy in this study [9]

The subtle renal injury could be a reason for increased risk of contrast nephropathy in hypertensive patients. Nearly all patients with hypertension have pathological changes on renal biopsy [8]. Cowley et al. proposed reduced renal medullary flow with local tissue ischemia as a cause of hypertension. The same mechanism could be responsible for increased incidence of contrast nephropathy [10].

Previous studies raised the debate whether or not only patients with preexisting renal dysfunction should receive hydration before exposure to radiographic contrast agents [11]. Our result indicates that hydration should not be restricted to patients with established kidney disease. Especially patients with hypertension should be carefully prepared before being exposed to radiographic contrast media.

4. Conclusions

Hypertension is an independent risk factor for the development of CIN. Care must be taken before radiographic contrast agents are used in hypertensive patients.

Acknowledgments

The authors gratefully acknowledge that the present research is supported by Ministry of Research and Technology and Higher Education Republic of Indonesia. The support is under the research grant BP-PTN USU of Year 2017.

References

[1] Mehran R, Nikolsky E (2006) Contrast induced nephropathy: definition, epidemiology, and patients at risk. *Kidney International*. 69:S11-S15.

[2] Toprak O, Cirit M (2006) Risk factor for contrast induced nephropathy. *Kidney Blood Press Res*. 29:84-93.

[3] Marenzi G, Assanelli E, Campodonico J, Lauri G, Marana I, Demetrio M, et al. (2009) Contrast volume during primary percutaneous coronary intervention and subsequent contrast induced nephropathy and mortality. *Ann Intern Med*. 150:170-7.

[4] Gleeson TG, Bulugahapitiya S (2004) Contast induced nephropathy. *AJR*.183:1673-89.

[5] Brinker JA, Davidson CJ, Laskey W. (2005) Preventing in hospital cardiac and renal complications in high risk PCI patients. *Eur Heart J Supp* .7(Suppl G):G13-G24.

[6] McCullough PA (2008) Contast induced acute kidney injury. *J Am Coll Cardiol*. 51:1419-28.

[7] Sterner G, Nyman U. Contrast medium induced nephropathy: aspects on incidence, consequences, risk factors and prevention. *Libyan J Med*. Downloaded from: www.ljm.org.ly

[8] Conen D, Buerkle G, Perruchoud AP, Buettner HJ, Mueller C (2006) Hypertension is an independent risk factor for contrast nephropathy after percutaneous coronary intervention. *International Journal Of Cardiology*.110:237241.

[9] Dangas G, Iakovou I, Nikolsky E, et al. (2005) Contrast induced nephropathy after percutaneous coronary intervention in relation to chronic kidney disease and hemodynamic variables. Am J Cardiol. 95:13-9.

IOP Conf. Series: Earth and Environmental Science **130** (2018) 012006 doi:10.1088/1755-1315/130/1/012006

[10] Cowley Jr AW, Mattson DL, Lu S, Roman RJ (1995) The renal medulla and hypertension. Hypertension. 25:663-73.
[11] Barrett BJ, Parfrey PS (2006) Preventing nephropathy induced by contrast medium. *N Eng J Med*. 354:379-86.

An Ability of Endophytic Bacteria from Nutgrass (*Cyperus rotundus*) from Lafau Beach of North Nias in Producing Indole Acetic Acid and in Solubilizing Phosphate

Atriani Zega, Dwi Suryanto*, Yurnaliza
Department of Biology, Faculty of Mathematics and Natural Sciences, Universitas Sumatera Utara, Jln. Bioteknologi No. 1, Medan 20155, Indonesia

*Email:dwisuryanto@usu.ac.id

Abstract: Endophytic bacteria have taken much attention for their potency to promote plant growth. This study was aimed to isolate endophytic bacteria from nutgrass (*Cyperus rotundus*) and to examine their potency in producing indole acetic acid (IAA) and in solubilizing phosphate. Isolation of endophytic bacteria was done by slicing and sterilizing root, stem, and leaf sample surface with alcohol 70% and sodium hypochlorite 2%, followed by incubation of the sliced samples in nutrient agar medium. Morphological characterization and simple biochemical tests were performed on bacterial isolates. All bacterial isolates were examined for their ability to produce indole acetic acid and to solubilize phosphate. Three isolates (AZ_5, AZ_{12} and AZ_6) out of fifteen indicated the ability to produce indole acetic acid and to solubilize phosphate. IAA producing test using spectrophotometry method showed that AZ_5, AZ_{12},and AZ_6 produce more IAA with concentration of 49,91, 48,18, and 44,45 ppm, respectively. Phosphate solubilizing test using Pikovskaya agar medium showed that the three isolates were able to solubilize phosphate with index of 6.27, 3,31, and 3.41 respectively.

Keywords: *Cyperus rotundus*, plant growth promoting bacteria, IAA production, phosphate solubilizing

1. Introduction

Endophytes are a group of bacteria that live in the plant tissue as neutral or symbiotic to the host plant. Strobel and Daisy [8] explained that endophytic bacteria can live in plants during a certain period of their life cycle. The role of endophytic bacteria is allegedly supported by their role in supplying nutrients for the plant growth either by phosphate solubilization, nitrogen fixation [1], or phytohormones such as indole-3-acetic acid (IAA), abscisic acid, gibberellic acid, and jasmonic acid [2,3,4,5]. Another benefit of endophytic bacteria that support plant growth is by protecting plants from pathogen infection through competition for habitat and nutrients [6,7].

Indole acetic acid plays a role in cell enlargement, as well as rapid change of gene expression, which causes the cells in the area of extension to produce new proteins as a constituent of the cell walls that affect plant development. Patten and Glick [12] reported that the endophytic bacteria that produce IAA stimulate the growth of the host rooting system.

It was reported that endophytic bacteria from paddy rice are capable of producing hormones IAA [11]. Duangpaeng et al. [13] reported that thirty five endophytic bacteria on organic rice crops are capable of producing hormones IAA. M*ontañez* et al. [14] isolated twenty two endophytic bacteria from *Zea mays* L. that produces IAA. He showed that the bacteria contributed to increase in root and stem biomass of corn.

Phosphate is the macronutrients for plant growth. Many endophytes were known for their ability to solubilize phosphate to help plant growth [10]. Chen et al. [15] reported that endophytic bacteria from *Manihot esculenta* Crantz root are potential to solubilize phosphate.. Joe et al. [16] also reported that endophytic bacteria from plant *Phyllanthus amarus* Schum & Thonn was capable to solubilize phosphate.

In this study, isolation of endophytic bacteria of nutgrass (*Cyperus rotundus*) was done. Nutgrass is common grass species of tropical area and easily found in various regions of Indonesia. So far, the endophytic bacteria of nut grass with the ability both to produce IAA and solubilize phosphate have not been studied yet. The aims of this study were to know the potential of endophytic bacteria from nutgrass of Lafau Beach of North Nias to produce IAA and to solubilize phosphate.

2. Materials and Methods

2.1 Plant samples

Three nutgrass samples of *C. rotundus* were collected from three different locations in Lahewa, North Nias. The plant materials used in this study were root, stem and fresh leaf.

2.2 Isolation of endophytic bacteria

Samples of root, stem, and leafs were cut into 1-2 cm long. These pieces were rinsed with sterile water, surface-disinfected by soaking in 70% alcohol for 3 minutes, rinsed with sterile water, soaked in 1% NaOCl for 5 minutes, rinsed again with sterile water, soaked in 70% alcohol for another 1 minute, and put into sterile water. The sample was placed on a surface of filter paper aseptically, and put on the surface of the nutrient agar (NA) containing ketoconazole 0.3% (w/v).The samples were incubated at 37°C for three days. Colonies of bacteria that appeared were further purified using streak plate method and stored in a slant-butt in NA [17].

2.3 Morphological and biochemical characterization of endophytic bacteria

Bacterial isolates were morphologically characterized by observing their edges, color, elevation (appearance from the side), Gram staining, and motility test, as well as for their physiological characterization including starch and gelatin hydrolysis, triple sugar iron agar (TSIA) test and citrate utilization test [18].

2.4 Phosphate solubilization

The abilityof endophytic bacteria to solubilize phosphate was determined on Pikovskaya agar using spot inoculation method [19]. Each isolates were grown in Pikovskaya's medium and incubated at 35 ± 2°C for five days. Diameter of the colony as well as the clear zone was measured. Phosphate solubilization Index (PSI) was calculated as:

$$\mathrm{PSI} = \frac{\text{Diameter of clear zone}}{\text{Diameter of bacterial colony}} \qquad (1)[20]$$

IOP Conf. Series: Earth and Environmental Science **130** (2018) 012007 doi:10.1088/1755-1315/130/1/012007

2.5 Measurement of IAA

IAA production was determined using modified Salkowaski reagent as described by Gordon and Weber [21]. A 100 µL of 24 hours bacterial culture in Pikovskaya broth ($OD_{600} \approx 0,5$) was inoculated into 100 mL of nutrient broth medium in 250 mL flask containing 0.1 mg mL^{-1} L-tryptophan. The flask was incubated at 35 ± 2°C for 48 hours in a rotary shaker (120 rpm). The culture was centrifuged for 30 minutes at 10,000 rpm. Two mL of the supernatant was mixed with equal volume of Salkowski reagent (0.1 M $FeCl_3$ + 400 mL of concentrated sulfuric acid (H_2SO_4) + 580 mL of aquadest). The solution were mixed by shaking and allowed to stay at ambient temperature for 30 minutes until pink color appeared. IAA concentration was measured spectrophotometrically of 530 nm. Un-inoculated broth was used as control.

3. Results and Discussion

3.1 Bacterial isolation

A total of 21 endophytic bacterial isolates were found in root, stem and leaves of nut grass (Table 1). M*ontañez* et al. [14] isolated twenty two endophytic bacteria of IAA producer from corn.

Table 1. Number of endophytic bacterial isolates from root, stem, and leaf of nut grass

Root	Stem	Leaves	Total
9	5	7	21

The bacterial isolates from nutgrass showed to have different characters including colony shapes, edges, and colors, and cell morphology as well. Ten isolates were Gram-negative, and others were Gram positive (Table 2.). Biochemical and physiological traits of these isolates were also characterized. The isolates showed different Biochemical and physiological characteristics (Table 3.).

Table 2. Characterization of colony and cell morphology and properties of endophytic bacteria isolates of nutgrass

Isolate code	Colony Morphology				Cell morphology	
	Form	Edge	Elevation	Colour	Shape	Grams' staining
AZ_1	*Irregular*	*Undulate*	*Flat*	white	Diplo	+
AZ_2	*Irregular*	*Lobate*	*Flat*	white	Bacilli	-
AZ_3	*Circular*	*Entire*	*Flat*	white	Bacilli	-
AZ_4	*Circular*	*Entire*	*Raised*	White-yellowish	Diplobacilli	-
AZ_5	*Circular*	*Entire*	*Raised*	Off-white	Diplobacilli	+
AZ_6	*Circular*	*Undulate*	*Flat*	white	Cocci	-
AZ_7	*Irregular*	*Lobate*	*Umbonate*	White-yellowish	Bacilli	+
AZ_8	*Circular*	*Undulate*	*Flat*	Off-white	Bacilli	-
AZ_9	*Irregular*	*Undulate*	*Flat*	white	Bacilli	+
AZ_{10}	*Circular*	*Entire*	*Flat*	white	Bacilli	-
AZ_{11}	*Irregular*	*Undulate*	*Raised*	Off-white	Diplobacilli	-
AZ_{12}	*Circular*	*Undulate*	*Raised*	White-yellowish	Bacilli	-
AZ_{13}	*Circular*	*Undulate*	*Raised*	White-yellowish	Coccobacillus	-
AZ_{14}	*Circular*	*Entire*	*Raised*	White-yellowish	Bacilli	-
AZ_{15}	*Circular*	*Undulate*	*Raised*	White-yellowish	Bacilli	+

Table 3. Physiology and biochemistry characteristics of the endophytic bacterial isolates from nutgrass

Isolate code	Starch utilization	Gelatine hydrolysis	Citrate utilization	Hydrogen sulphide producing	Motility test
AZ_1	+	-	-	+	-
AZ_2	-	-	-	+	-
AZ_3	+	+	-	+	+
AZ_4	+	+	-	+	+
AZ_5	+	+	-	+	+
AZ_6	-	-	+	+	-
AZ_7	-	-	-	+	+
AZ_8	+	+	-	+	-
AZ_9	-	+	-	+	+
AZ_{10}	-	-	+	+	+
AZ_{11}	+	-	-	+	-
AZ_{12}	-	-	-	+	+
AZ_{13}	+	-	+	+	+
AZ_{14}	-	+	-	+	+
AZ_{15}	-	-	+	+	+

3.2 Assay of phosphate solubilization of endophytic bacteria

This test was performed to observe the ability of endophytic bacterial isolates in dissolving phosphate. Screening was done by growing the 24-hours pure culture of the isolates in Pikovskaya medium containing $Ca_3(PO_4)_2$ for seven days.

The ability of phosphate solubilization of the isolates was indicated by clear zone around the bacterial colonies (Figure 1). One out of fifteen bacterial isolates did not form a clear zone around the colony indicating unable to solubilizing phosphate (Table 4). It was observed that AZ_5 showed to have relatively high ability to solubilize phosphate with PSI of 6.27, while AZ_2, AZ_{10}, and AZ_{14} showed low ability with PSI 1-1.15. Interestingly, the isolates showed to form clear zones in different days, indicated that each endophytic bacterial isolate had different capability in solubilizing phosphate. The ability to solubilize phosphate could be due to the medium complexity with the presence of other minerals as a complement, which makes the solubilization process more complicated [23].

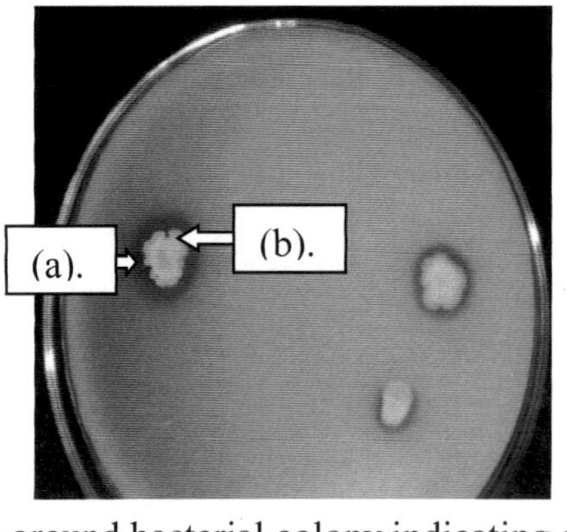

Figure 1. (a). Clear zone around bacterial colony indicating endophytic bacterial isolates in solubilizing phosphate, and (b). bacterial colony

Table 4. Phosphate solubilization index of endophytic bacterial isolates

Isolate code	PSI
AZ_1	1.33
AZ_2	1.15
AZ_3	1.83
AZ_4	3.75
AZ_5	6.27
AZ_6	3.41
AZ_7	1.36
AZ_8	1.16
AZ_9	1.71
AZ_{10}	1
AZ_{11}	2.35
AZ_{12}	3.31
AZ_{13}	2.75
AZ_{14}	1.15
AZ_{15}	3.13

3.3 Assay of indole acetic acid (IAA) production

All isolates showed to produce IAA. The production of IAA of the isolates seemed varied to some extent (Table 5). AZ_5 showed to have relatively high IAA production, while AZ_1 seemed to produce low IAA by 49.91 and 0.13 ppm, respectively (Table 5).

IAA is generally a type of plant hormone that is secreted by many bacterial symbionts regulating plant growth. Endophytic bacteria have a role in activating genes to colonize and to perform adaptation in plant tissues [24]. In this study, L-tryptophan was added in the culture as a precursor for IAA, although the production of IAA is also influenced by other precursor such as indol glycero phosphate [25].

Table 5. IAA production by endophytic bacteria isolates

Isolate code	Concentration (ppm)
AZ_1	7.18
AZ_2	8.09
AZ_3	14.45
AZ_4	18.09
AZ_5	49.91
AZ_6	44.45
AZ_7	19.00
AZ_8	12.64
AZ_9	31.73
AZ_{10}	19.91
AZ_{11}	18.09
AZ_{12}	48.18
AZ_{13}	24.45
AZ_{14}	8.09
AZ_{15}	13.54

4. Conclusion

Fiveteen endophytic bacteria were isolated from nutgrass (*Cyperus rotundus*). Eleven isolates were Gram-negative, and the rest were Gram-positive. Out of the fifteen isolates, AZ5

IOP Conf. Series: Earth and Environmental Science **130** (2018) 012007 doi:10.1088/1755-1315/130/1/012007

isolate showed to have relatively high IAA production and phosphate solubilization. This isolates might be potential to be used as biofertilizer.

Acknowledgements
The authors would like to express our greatest gratitude to North Nias District Government for funding this study.

References

[1] Marra LM, Soares CRFS, de Oliveira SM, Ferreira PAA, Soares BL, de Fráguas Carvalho R, de Lima JM, de Souza Moreira F M. 2012. Biological nitrogen fixation and phosphate solubilization by bacteria isolated from tropical soils. *Plant and Soil*, **57**, 289-307.

[2] Cohen AC, Bottini R, Piccoli P. 2008. *Azospirillum brasilense* sp. 245 produces ABA in chemically-defined culture mediumand increases ABA content in *Arabidopsis* plants. *Plant Growth Regulation*, **54**, 97-103.

[3] Gabriela F, Oscar M, Marḳa JI, Sergio A, Daniel A, Guillermina A. 2010. Endophytic bacteria improve seedling growth of sunflower under water stress, produce salicylic acid, and inhibit growth of pathogenic fungi. *Current Microbiology*, **61**, 485-493.

[4] Malfanova N, Kamilova F, Validov S, Shcherbakov A, Chebotar V, Tikhonovich I, Lugtenberg B. 2011. Characterization of *Bacillus subtilis* HC8, a novel plant-beneficial endophytic strain from giant hogweed. *Microbial Biotechnology*, **4**, 523-532.

[5] Patricia P, Claudia T, Ana C, Laura S, Paula C, Ricardo M, Rubén B. 2011. An endophytic bacterium isolated from roots of the halophyte *Prosopis strombulifera* produces ABA, IAA, gibberellins A1 and A3 and jasmonic acid in chemicallydefined culture medium. *Plant Growth Regulation*, 64, 207-210.

[6] Brooks D S, Gonzalez C F, Appel D N, Filer T H. 1994. Evaluation of endophytic bacteria as potential biocontrol agents for oak wilt. *Biological Control*, **4**, 373-381.

[7] Kuklinsky SJ, Araújo WL, Mendes R, Geraldi IO, Pizzirani-Kleiner AA, Azevedo JL. 2004. Isolation and characterization of soybean-associated bacteria and their potential for plant growth promotion. *EnvironmentalMicrobiology*, **6**, 1244-1251.

[8] Strobel dan Daisy (2003) Bioprospecting for Microbial Endophytes an Their Natural Products. Microbiol. and Mol. Biology, 67(4): 63-68.

[10] Anzuay MS, Frola O, Angelini JG, Luduena LM, Fabra A, Taurian T (2013) Genetic diversity of phosphate-solubilizing peanut (*Arachis hypogaea* L.) associated bacteria and mechanisms involved in this ability. *Symbiosis*, 60, 143-154.

[11] Yurnaliza MW, Siregar dan Priyani N (2011) Peran Bakteri Endofit Penghasil IAA (Indole Acetic Acid) Terseleksi Terhadap Pertumbuhan Tanaman Padi (*Oryza sativa* L.). Prosiding Seminar Nasional Biologi, 219-228.

[12] Patten CL and BR Glick (2002) Role of Pseudomonas putida Indole Acetic Acid in Development of The Host Plant Root System. *Appli Environ Microbiol*. 68:3795-3801.

[13] Duangpaeng P, Phetcharat S, Chanthapho N, Boonkantong N, Okuda (2012) The Study *Porcedia Engineering*, **32**,172-176.

IOP Conf. Series: Earth and Environmental Science **130** (2018) 012007 doi:10.1088/1755-1315/130/1/012007

[14] M*ontañez A,* Blanco AR, Barlocco C, Beracochea M, Sicardi M (2012) Characterization of cultivable putative endophytic plant growth promoting bacteria associated with maize cultivars (Zea mays L.) and their inoculation effects in vitro. *Applied Soil Ecology*, **58**, 21-28.

[15] Chen Y, Fan JB, Du L, Xu H, Zhang QH, He QY (2014) The application of phosphate solubilizing endophyte Pantoea dispersa triggers the microbial community in red acidic soil. *Applied Soil Ecology*, **84**, 235-244

[16] Joe MM, Devaraja S, Benson A, Tongmin S (2016) Isolation of phosphate solubilizing endophytic bacteria from Phyllanthus amarus Schum & Thonn: Evaluation of plant growthpromotion and antioxidant activity under salt stress. *Journal of Applied Research on Medicinal and Aromatic Plants*, ISSN: 2214-7861.

[17] Hallmann JG Berg (2006) *Spectrum and population dynamics of bacteri root endophytes*. Berlin, Heidelberg, Germany, Springer-Verlag. 15-31.

[18] Mac Faddin JF (1979) *Biochemical test for identification of medical bacteria*. Baltimore (US). The William and Wilkins Company, **1**(**5**), 2365-2370.

[19] Harley and LM Prescott (2003) *Microbiology*. 5th Edition. Mc Graw Hill, New York.

[20] Hidayati U 2014. *Potensi bakteri endofit asal tanaman karet sebagai pemacu pertumbuhan bibit batang bawah tanaman karet* (*Hevea brasiliensis* Müll. Arg.) [Disertation]. Bogor (ID): Bogor Institute of Agriculture.

[21] Gordon SA dan Weber RP (1951) Colorimetric Estimation of Indoleacetic Acid. Plant Physiol, 26(1):192–195.

[22] Yan Chen, Jian-Bo Fan, Lian Du, Huan Xu, Qi-Hai Zhang, Yuan-Qiu He (2014) The application of phosphate solubilizing endophyte Pantoea dispersa triggers the microbial *Applied Soil Ecology*, **84**, 235-244.

[23] Narsian V, Ahmed Abu, Samaha SM, Patel HH (2010) Rock phosphate dissolution by specific yeast. *Indian J. Microbiol*. **50** (1), 57-62.

[24] Ma Y, Prasad MNV, Rajkumar M, Freitas H (2011) Plant growth promoting rhizobacteria and endophytes accelerate phytoremediation of metalliferous soils. *Biotechnology Advances*, **29**, 248-258.

[25] Pollmann S, Düchting P, Weiler EW (2009) Tryptophan-dependent indole-3-acetic acid biosynthesis by 'IAA-synthase' proceeds via indole-3-acetamide. *Phytochemistry*, **70**, 523-531.

[26] Aryantha, INP, DP Lestari and NPD. Pangesti (2005) Potensi Isolat Bakteri Penghasil IAA dalam Peningkatan Pertumbuhan Kecambah Kacang Hijau pada Kondisi Hidroponik. *Jurnal Mikrobiologi Indonesia*. 9(2):43-46.

[27] Doty SL (2011) Nitrogen-fixing endophytic bacteria for improved plant growth. Maheswari DK, editor. *In Bacteria in Agrobiology*: Plant Growth Responses.Springer Berlin Heidelberg. 183-199.

[28] Feng K, Lu HM, Sheng HJ, Wang XL and Mao J (2006) Effect of organic ligands on biological availability of inorganic phosphorus in soils. *Pedosphere*. 14:85-92.

[29] Pikovskaya RI (1948) Mobilization of phosphorus in soil in connection with vital capacity of source microbial species. *Microbiologiya*,17, 362-370.

Optimization of the development of reproductive organs celepuk jawa (*otus angelinae*) owl which supplemented by turmeric powder

Tyas Rini Saraswati[1*], Enny Yusuf W.Yuniwarti[1], Silvana Tana[1]
[*]Biology Department. Faculty of Science and Mathematics, Diponegoro University, Semarang Indonesia.

*E-mail:tyasrinis63@gmail.com

Abstract. *Otus angelinae* is included as a protected animal because of its endangered existence. Whereas, it has many values such as for mice pest control. Therefore, this research aims to optimize the reproductive function of *Otus angelinae* by administering turmeric powder mixed in its feed. This study was held on a laboratory scale with two male and two female *Otus angelinae* three months of age. Each subject is divided into two groups: a control group and a treatment group which is treated with turmeric powder 108 mg/owl/day mixed in 30 g catfish/ day for a month. The parameter observed were the development of hierarchy follicles and the ovarium weight of female *Otus angelinae*, whereas the testis organs and testes weight were observed for the male. Both the female's and male's body weight, liver weight and the length of ductus reproduction were also observed. The data was analyzed descriptively. The results showed that the administration of turmeric powder can induce the development of ovarian follicles hierarchy and the length of ductus reproduction of female *Otus angelinae* and also induce the development of the testes and the length of ductus reproduction of male *Otus angelinae*. The addition of turmeric powder increased the liver weight of the female Otus angelinae, however it does not affect the body weight.

Keywords: follicle hierarchy, vittellogenin

1. Introduction

Owl is a bird of prey (carnivore) and a night animal (nocturnal). There are approximately 222 known species of owls, and they are spread around the world [1]. In Indonesia owl is a scarce and protected animal. The most popular type today is Celepuk jawa (*Otus angelinae*). *Otus angelinae* is the easiest type of owl to be tamed, and it also can be trained for various purposes. In the agricultural sector, it is often used as a pest control, specifically for mice and rat [2]. Using Owls to control rat is more effective than rat hunting or using rat poison. A pair of owls can protect 25 hectares of rice crops from rat. Within a year the owls can prey on 1300 rats. In terms of cost, rat control's cost by using owls is lower by 50 percent compared to by poison. Hence, owls breeding can become a lucrative business opportunity.

The low population of owls (*Otus angelinae*) is influenced by several factors. One of them is the low number of juvenile owls who are able to reach adult age. In the mating season, *Otus angelinae* females can produce as many as 6 eggs, but only 3 to 4 of them who are able to survive into adult owls. In addition, even though the oldest owlet usually becomes a powerful owl, the youngest owlet often has the weakest growth result. Thus, it grew less normal or dies.

Attempts to breed owls can be done by the optimization of the physiological condition of the owl's reproductive system. This can be done with turmeric powder supplementation. Previous study showed that the administration of turmeric powder is capable to accelerate the sexual maturity of female quail [3]. Turmeric powder supplementation can also improve the hierarchy of ovarian follicles in quail [4] and chickens [5], this is because of the active compound in turmeric that is hepatoprotective, which optimize the function of the liver in the synthesis vitellogenin (precursor yolk). Turmeric contains curcumin powder as much as 7.97%, which works to improve liver function [6]. The content of phytoestrogens in turmeric powder is 6.79% [6]. Phytoestrogens have estrogenic effects. It is able to make a bond with estrogen receptors, which leads to the increase of vitellogenin synthesis [7,8].Vitellogenin carried out through the bloodstream to the ovary, so it can improve follicular hierarchy and increase the number of eggs produced. The increased accumulation of the material forming the yolk inside the follicle also induces the increased embryo source of nutrients, so the embryo develops more optimal. Similarly, according to previous study by [3], a good development of embryo will also lead to a good physiology and immune system. Curcumin acts as an antioxidant, anti-inflammatory, antibacterial, and antiviral [9].

Based on those previous studies, the research on optimizing the reproduction of endangered species of Celepuk jawa (*Otus angelinae*) with turmeric powder could be used as an attempt to bioconservation.

2. Methods

The study was conducted in the laboratory of Biological Structure and Function of the Department of Animal Biology, Faculty of Science and Mathematics, University of Diponegoro. A total of 2 female Otus angelinae and 2 male *Otus angelinae* aged 3 months were divided into two experimental groups, namely the control group and the group treated turmeric powder 108 mg/owl/day, which were mixed in a feed. The feed is catfish meat weighing 15 g/day in the morning and evening. The treatment was given every day, for 1 month. Descriptive analysis was done on the development of ovarian follicles hierarchy of female *Otus angelinae*, while the testis organ was observed for the male. Both the female's and male's body weight, liver weight and the length of ductus reproduction were also observed.

IOP Conf. Series: Earth and Environmental Science **130** (2018) 012008 doi:10.1088/1755-1315/130/1/012008

3. Results and Discussion

The results of the research supplementation of turmeric powder on the male reproductive organs *Otus angelinae* showed the following results:

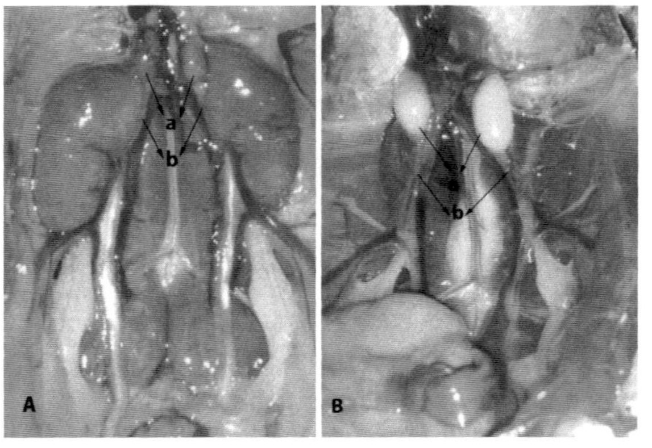

Figure 1.Testis *Otus angelinae* control (A) and which treated by turmeric powder (B). a.testis, b. vas deferens

The results of the research granting turmeric powder to the female reproductive organs *Otus angelinae* showed the following results

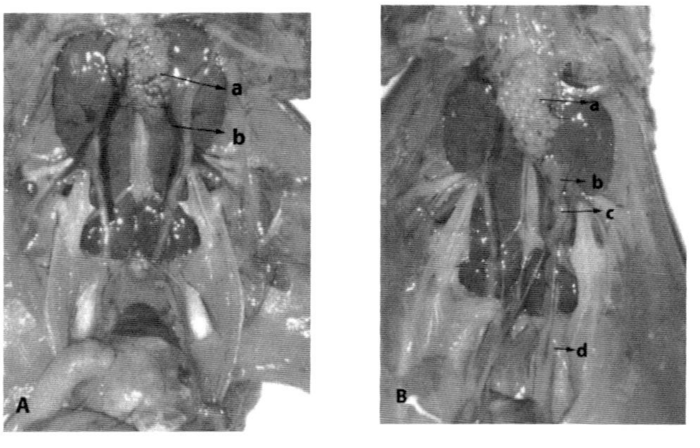

Figure 2. *Otus angelinae* reproduction system, (A) control and (B). treatment with turmeric powder a.ovarium, b. infundibulum, c. magnum, d. uterus

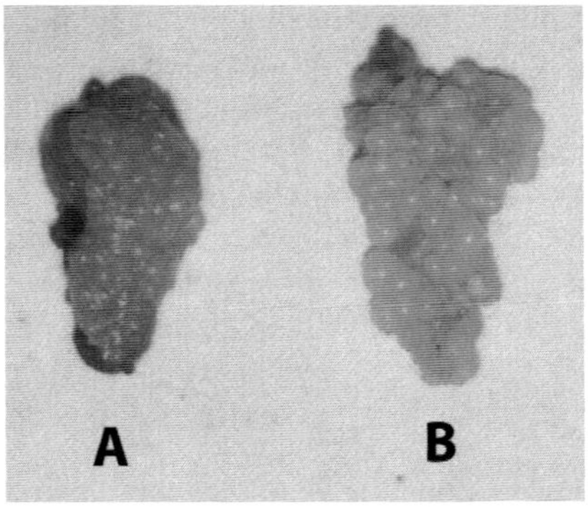

Figure 3. Follicle ovarium *Otus angelinae.* (A) control and (B) treatment by turmeric powder

The addition of turmeric powder does not increase the body weight, liver weight, the length of ductus reproduction, ovarium weight, testes weight *Otus angelinae*

Table 1: The measurement of body weight, liver, the length of ductus reproduction ovarium and testis of *Otus angelinae* male and female at control and which supplemented by turmeric powder

	Male		Female	
Parameter	Control	Treatment	Control	Treatment
Body weight (g)	9	9	9	9
Liver weight (g)	2.32	2.21	2.06	3.92
The length of ductus reproduction (cm)	3.2	5.6	6.5	7.5
Ovary weight (g)			0.13	0.24
Testis weight(g)	0.18	0.25		

K : Control

P : *Otus angelinae* which supplemented by turmeric powder

Based on Figure 1, male *Otus angelinae* has a pair of oval-shaped testes respectively located at the front of the upper lobes of the kidney. In untreated male *Otus angelinae*(control), aged three months, the testes weights 0.18 g. This is in accordance with the opinion which states that owl's testes during non-breeding season are very small. Turmeric powder supplementation affects the growth of the male reproductive organs. The weight of the testes in males *Otus angelinae* three months of age treated with turmeric powder for 1 month at 0.25 g, was approximately 1.5 times bigger than the control, and the color of the testis is yellow.

Testicular growth and reproductive organs occurs along with the development of sexual maturity. Sexual maturity is regulated by sex hormones. Gonadotropin-releasing hormone from the hypothalamus stimulate FSH (Follicle Stimulating Hormone) secretion. FSH stimulates cells to divide by mitosis of spermatogonia several times and ended with meiotic division to produce spermatids. FSH stimulates Sertoli cells to produce substances that can stimulate the spermiogenesis (spermatid differentiation into sperm). FSH and Sertoli cells are also stimulated by testosterone produced by the Leydig cells. Testosterone is a hormone that stimulates the growth and development of reproductive organs.

Besides hormonal factors, the growth of reproductive organs is also influenced by nutritional factors. Nutrition affects various aspects of reproduction, including maturity sexual and testicular growth. Effect of turmeric powder to the development of the testes is suspected because of the curcumin and β-carotene. The content of curcumin in turmeric powder is up to 7.97% [4]. Curcumin has a role as an antioxidant and anti-inflammatory [10,11]. Curcumin also plays a role in cell regeneration [12].

The analysis results of the content of β-carotene in turmeric powder amounted to 40.45 mg/100 mg. β-carotene is an antioxidant exogenous. Utilization of exogenous antioxidant compounds effectively is necessary to prevent the occurrence of oxidative stress. Exogenous antioxidant is a preventive defense system, where the system is the antioxidant works by cutting the chain reaction of free radical oxidation. β-carotene is a chain-breaking antioxidant. It is also a lipophilic which can play a role in sperm cell membrane to prevent lipid peroxidation. The study in rabbits showed that β-carotene can increase the male reproductive system and can support fertilization, increasing the motility and viability of spermatozoa [13]. Provision of β-carotene improves the quality of cement [14,15].

Based on Figure 2, the reproductive organs in female Otus angelinae consist of one ovarian that form hierarchy follicle. Hierarchy of follicles in the control (Figure 2A) shows undeveloped ovarian, which weights 0.13 g. Supplementation of turmeric powder is able to stimulate the growth of ovarian hierarchy, so that the weight of the ovaries reach to 0.25 g, although it has not seen any mature follicles. Female reproductive system depends on a complex hormonal mechanism. Gonadotropin hormone produced by the anterior pituitary consists of follicle stimulating hormone (FSH) and luteinizing hormone (LH). FSH hormones affect the growth of young follicles become mature follicles. Besides oocytes, in the follicle that is growing, there are some theca cells and granulosa cells. Furthermore, the FSH hormone also affects the secretion of estrogen and progesterone steroid that is produced by cells theca and granulosa cells, which are essential for the formation of the yolk. Through the blood stream estrogen is taken to liver. Liver has many estrogen receptors in the cytoplasm. Estrogen will further induce the synthesis vitellogenin (precursor yolk). The results showed that administration of turmeric powder in quail [3,4] in laying hens [5] can increase the biosynthesis vitellogenin in the liver, due to their content of phytoestrogens in powder turmeric as much as 6.79%, which has estrogen-like effect of inducing the formation of the yolk precursor [4].

The result in Table 1 showed that the addition of turmeric powder does not increase the body weight of both the female and male Otus angelinae. Based on the observation of the liver weight, the results showed that the liver of the *Otus angelinae* females treated with turmeric powder has the greatest weight, namely 3.92 g (Table 1). This indicates an increase in liver activity in the synthesis of the material that forming the yolk. Furthermore, the material of forming the yolk taken to liver to fill the follicle hierarchy as egg yolks, until mature follicles, while the addition of turmeric powder does not affect the liver weight of male *Otus angelinae* because they do not have vitelogenin. The addition of turmeric powder supplement could increase the length of the reproductive organ in female anf male *Otus angelinae*, this is proved that the reproduction organ grew sinergically with the development of both testes and ovarium to support the reproduction process. In conjunction with the development of the ovarium follicle hierarchy in female Otus angelinae and testes in male Otus angelinae, there was an increase in the ovarium and testes weight.

4. Conclusion

Based on the results of the study, it can be concluded that supplementation of turmeric powder can accelerate the development of reproductive organs of male and female *Otus angelinae*.

Acknowledgment
This article was part of the results of research funded by PNBP- RPP 2016. Our gratitude goes to the Directorate General of Higher Education and LPPM-Undip.

References

[1] Kitowski I (2013) Winter diet of the barn owl (*Tyto alba*) and the long-eared owl (*Asio otus*) in Eastern Poland. *North-Western Journal Of Zoology* **9** (1): 16-22
[2] Meyrom K, Motro Y, Leshem Y, Aviel S, Izhaki I, Argyle F and Charter M (2009) Nest-Box use by the Barn Olwl *Tyto alba* in a Biological Pest Control Program in the Beit She'an valley Israel. *Ardea* **97**(4) 463-467
[3] Saraswati TR and Tana S (2016) Effect Of Turmeric Powder Supplementation Upon The Age Of Sexual Maturity, Physical, And Chemical Quality Of The First Japanese Quail (*Coturnix Japonica*) Egg *Biosaintifika : Journal of Biology and Biology education* **8** (1).
[4] Saraswati TR, Manalu W, Ekastuti DR, Nastiti K (2013a) Increased Egg Production of Japanese quail (*Coturnix japonica*) by Improving Liver Funcion Through Turmeric Powder Supplementation. *International journal of poultry Science* **12**(10) 601-614.
[5] Saraswati TR, Manalu W, Ekastuti DR, Nastiti K (2014) Effect of Turmeric Powder to Estriol and Progesterone Hormon Profile of aying Hens During One Cycle of Ovulation. *International journal of Poultry Science* **13**(9) 504-509.
[6] Saraswati TR, Manalu W, Ekastuti DR, Nastiti K (2013b) The Role of Turmeric Powder in Lipid Metabolism and the Effect on The Quality of The First Quail's Egg. *The Journal of The Indonesian tropical Animal Agriculture* **38** (2):123-130.
[7] Levi L, Pekarski I, Gutman E, Fortina P, Hyslop T, Biran J, Levavi B, Lubzens E (2009) Licensee BioMed Central Ltd.http://www.biomedcentral.com/1471-2164/10/141.
[8] Ravindar PN, Babu KN, Sivaraman K (2007) Turmeric. The Genus Curcuma. CRC Press. London, New York.
[9] Aggarwal BB, Sundaram C, Malani N, Ichikawa H (2006) Curcumin: The Indian Solid Gold. SVNY-Aggarwal 16:3.
[10] Namratha K, Shenai P, Chatra.L, Rao PK, Veena KM, Prabhu RV (2013) Antioxidant and Anticancer Effect of Curcumin –A Review *J Contemp Med* **3**(2): 136-143.
[11] Servet A, Fatih T, Levent G, Akgul A, Erkan S, Ozkan, Ozalkan M.D (2016) The Anti-inflammatory and Antioxidant Effect of Curcumin in Middle Ear Infection **27**(5) 494-497.
[12] Sanivarapu R, Vallabhaneni V, and Verma V (2016) The Potential of Curcumin in Treatment of Spinal Cord Injury. *Neurology Research International*.
[13] Orazizadeh M, Khorsandi L, Absalan F, Hashemitabar M, Daneshi E (2014) Effect of beta-carotene on titanium oxide nanoparticles-induced testicular toxicity in mice. *J Assist Reprod Genet* **31**(5):561-8.
[14] Baghdadi HH, Kedwany FS, Yousef MI, Demerdash FM (2004) Role of alpha-tocopherol and beta-carotene in ameliorating the fenvalerate-induced changes in oxidative stress, hemato-biochemical parameters, and semen quality of male rats *J Environ Sci Health* **39**(3) 443-59.
[15] Shalaby F and Missiry MA (2000) Role of beta-carotene in ameliorating the cadmium-induced oxidative stress in rat brain and testis *J Biochem Mol Toxicol* **14**(5) 238-43.

Wheat bran soil inoculant of sumateran nematode-trapping fungi as biocontrol agents of the root-knot nematode *meloidogyne incognita* on deli tobacco (*nicotiana tabaccum* l) cv. deli 4

Liana Dwi Sri Hastuti[1*] and Jane Faull[2]

[1]Biology Department, University of Sumatera Utara, Jl. Bioteknologi no.1 Medan, Sumatera Utara, Indonesia.
[2]Biology Department, Birkbeck, University of London, Malet Street Bloomsbury, London WC1E 7HX

*Email: liana.hastuti@usu.ac.id

Abstract. A pot experiment was carried out to test the effectiveness of nematode-trapping fungi (NTF) isolated from Sumatera for controlling infection by the root-knot nematode (RKN) on Deli tobacco plant. Wheat bran soil containing 10^9 conidia of *Arthrobotrys. oligospora*, *Candellabrella musiformis* and *Dactylella eudermata* was added to the soil as a dry inoculum. Carbofuran was also applied as chemical agent and comparison treatment. Seedling tobacco (*Nicotiana tabacum* L.) cv. Deli 4 was inoculated with root knot (*Meloidogyne incognita* Chitwood.) seven days after the plant were transplanted to the pots. *A. oligospora*, *C. musiformis* and *D. eudermata* were found to be reliable as biocontrol agents, reducing the number of vermiform nematodes, swollen root, sausage shaped and galls in tobacco plant after 7, 15 and 30 days of infection with *M. incognita*. Treatment with NTF produced results that were comparable with Carbofuran® as a control agent in the reduction of the number of infections in tobacco plant caused by *M. incognita* in *Nicotiana tabacum* var. Deli 4. They also optimize the growth of the tobacco plants especially up to 15 days after infection.

Keywords: *A. oligospora*, *C. musiformis*, *D. eudermata*, *M.incognita,* Solid State Fermentation

1. Introduction

Recently, it has been reported that plant parasitic nematode infestations were spreading widely not only in West, Central, and East Java provinces, but also to the North Sumatera province [1]. They infect almost all species from the family of *Solanaceae* including tobacco and cause 32%-71% crop losses with a cost of approximately of 2 trillion rupiahs or about 149,835,200 USD [1].

Nicotiana tabacum var. Deli (Deli tobacco) is the preferred leaf used for wrapping cigars, with famous names like Black Sumatera. As one of the most economically important commodities in Indonesia, Deli tobacco also has a problem with RKN. Infection of RKN *Meloidogyne* species on tobacco plants leads to young plant mortality at the age of 30-45 days, and mortality can reach more

than 50% of the plant population per hectare. In addition, nematode attack can also cause growth retardation, decreased chlorophyll content, stunt growth and caused the death of the plant before it enters the generative phase [2]. Furthermore, RKN infestations can act as agents of predisposition for infection by other pathogens, tobacco infected by RKN will easily infected by *Ralstonia solanacearum* or *Phytophthora nicotianae* causing disease complexes.
The long term protection of an economically important plants such as tobacco and other crop plants as hosts against *Meloidogyne* spp. (*M. hapla*, *M. incognita*, *M.javanica* and *M. arenaria*) will be achieved more effectively when other management strategies are also applied. In addition to the application of antagonists, other techniques such as crop rotation, soil solarisation and use of nematicide should form an integrated pest management program.

Fungi as the biocontrol agent to control disease caused by plant nematode has only been reported in experiments under laboratory condition. Some fungi such as *Beauveria* sp., *Metarhizium*, *Paeciliomyces* sp, *Verticillium* sp, *Trichoderma harzianum* and some endophytic fungi have been known as entomopathogens with potential in nematode control. They have been tested as bio-insecticides for control of diseases caused by pests such as insect pathogen, also for plant parasitic nematode [3,4]. However, they have not been widely applied by farmers or growers in a farming area.
Fungi as the biocontrol agent to control disease caused by plant nematode has only been reported in experiments under laboratory condition. Some fungi such as *Beauvqria* sp., *Metarhizium*, *Paeciliomyces* sp, *Verticillium* sp, *Trichoderma harzianum* and some endophytic fungi have been known as entomopathogens with potential in nematode control.

There is very little research using NTF as biocontrol agents to control nematode pathogen such as RKN in Indonesia. As is well known, some fungi such as *Arthrobotrys* spp. (*Arthrobotrys dactyloides*, *A. brochopaga*, and *A. oligospora*) and *Monacrosporium* spp. (*Monacrosporium gephyropagum*, *M. cianopagum*, *M. doedycoides*), *Dactylaria* spp. and other *Orbiliaceae* have potential in the control nematode in the soil. [5] has been testing some media for production of inocula of nematophagous fungi *Dactylella* sp and *Duddingtonia flagrans* using ingredients available in Indonesia e.g. corn flour, cassava flour, rice flour and rice brand and CMA+ dung with some success. There has been research using such NTF to control plant parasitic nematodes in Indonesia; however, the experiments have only been carried out on laboratory scale.
The creation of fungal inoculum by using water as a suspension medium to create a conidial suspension for either injection or spray in laboratory or field conditions is a common method used by researchers and growers [6]. It must be considered, that if a bio-pesticide such as NTF inoculum is going to be used in large quantities for large-scale farming, mass production methods for industrial purposes. There are two current methods used for large scale biocontrol agent production, submerged fermentation (SMF) and solid-state fermentation (SSF) [7]. According to [8], NTF as biocontrol applied with SSF have an advantage over those produced by submerged fermentation: aerial fungal spores are more tolerant to UV radiation; have a higher spore stability; a higher spore resistance to drying; and higher spore germination rates for longer storage periods.

The objectives of this research are:
1. Use of wheat bran as a substrate for Indonesian NTF conidia production.
2. Testing efficacy of wheat bran inoculum as a biocontrol agent for *Meloidogyne incognita* and their effect on the growth of *Nicotiana tabacum* var. Deli 4.

2. Methods
2.1. Host Preparation
Seeds of Deli 4 tobacco were spread on moist white cloth, placed on glass lids and allowed to germinate. Three days old seedlings of Deli 4 tobaccos were removed to a squared tray containing sterile compost soil and incubated at 15-25°C in daylight. At 15 days of age tobacco plants were transplanted to the treatment pots containing 100g of soil and inoculum of NTF or Carbofuran. Soil around rhizosphere of host plant was injected with 1000 J2 of *M. incognita* seven days after receiving

15 days of seedling of Deli 4 tobacco. Deli tobacco plants in the treatments pots were maintained in the growth cabinet in a randomized pattern. The treatment pots were moved or rotated each two or three days regarding to obtain the same condition of light and temperature in the growth cabinet.

2.2. Inoculant Preparation

Three strains of NTF (*Arthrobotrys oligospora*, *Candelabrella musiformis*, *Dactylaria eudermata*) have been selected for further testing on the basis by proceeding experimental results reported in previous experiment. Wheat bran soil medium was used to make a dry inoculum. Each strain of NTF was cultured on Potato Dextrose Agar at 25^{0}C for 14 days. Five agar disks (5 mm in diameter) of each strain were added to 30 ml of wheat bran medium (1 unit wheat bran Bergin Fruit and Nut Company) : to 4 units compost in 50-ml Erlenmeyer flasks, and the cultures were then incubated for 14 days in darkness at 25°C [9].

Numbers of conidia were counted by a dilution series method where the 1 gram of wheat bran/compost taken from flask contain conidia was added into 9 ml of distilled water and number of conidia was count from the 10^{-4} dilution. A supernatant was obtained by pouring the suspension through sterile gauze into a sterile 15ml test tube. The conidial concentration was measured by the dilution method by counting the number of conidia in a 5 μL-drop 20 times under the microscope (100X) and multiplying the number counted by 100 to estimate the total number of conidia/ml. Each experimental pot containing 100g of sterile soil received 1g of wheat bran NTF inoculum. Inoculated pot soil was incubated in the growth cabinet (Bud Box Cabinet) 25°C for three days before receiving ornamental tobacco plants.

2.3. Chemical Agent Preparation

Carbofuran was used as a chemical control of nematodes as it is widely used in agriculture. The pesticide was diluted to a working concentration by dissolving 2 mg/10ml of solid Carbofuran in methanol to 200μg/ml or 2 mg/10ml liquid/suspension [10,11]. Carbofuran solution (10 ml) was poured into each pot, and controls received 10 ml of distilled water. The Carbofuran was poured into the pot (3 pots) together with fungal treatments three days before receiving a seedling tomato plant.

2.4. Root Knot Nematode Inoculant Preparation

Infected roots with galls were washed gently and cut into 1 cm section. Roots were vigorously shaken for 4 minutes in sodium hypochlorite solution (100 ml NaOCl 5-10%). Roots were poured and washed through a 250 μm sieve. Roots containing galls in the sieve were placed in a crystallizing dish, some petroleum jelly was spread on the borders of crystallizing dish to prevent the escape of nematodes and it was covered with a glass lid. The crystallizing dish containing nematodes was incubated at 25° C temperature room for 4-7 days. A piece of tomato root was examined under the microscope to confirm infection, and the infected tomato root was used as a source of inoculum. Mature nematode egg masses were selected, and three galls/egg mass (contain 400-600 per egg mass) were picked up with a pair of forceps. The egg mass was transferred into 15 ml tube containing sterile tap water. The 15 ml tube containing egg mass was left at room temperature to allow the juveniles (J2) to hatch. Most temperate species can be kept at 18-22°C and tropical and sub-tropical species at 25-30°C. Once hatched, the J2 was inoculated on to a previously transplanted tomato host. Preparation J2 of root-knot nematode used the combined technique described by Davies (Rothampstead Research) [12,13].

2.5. Soil Preparation

Soil was dried for four days and sieved, and after air drying and sieving, soil (100 grams per 8.5 cm pot) was sterilized with dry sterilization program (121°C at 1 atm. for 15 mins with a drying cycle) in 2000 ml beaker glass covered by alumunium foil. This stage was repeated once.

Under sterile conditions, sterile soil was transferred to 8.5 cm pots so that each pot contained 100 grams of sterile soil. The soil was then treated with Carbofuran and/or a conidial suspension of nematode-trapping fungi, and distilled water (10 ml) was used as a control.

There were three replicates of each treatment (Table 4.4). In order to avoid contamination, all treated pots were covered with para film on the top and bottom end of each pot. All treated pots were kept in the 25°C room for three days before receiving a tobacco seedling.

2.6. Harvesting and Root Assessment

The invasion and development of J2 stages, swollen root, sausage stage, pear-like (kidney-shape) of mature female [14] and galls was assessed at 7 and 15 days and 30 days.

3. Results

3.1. ASSESSMENT I: The effect of NTF and Carbofuran on the number of vermiform nematode and swollen roots.

On the first assessment, two stages *M. incognita* infection were observed as vermiform nematodes and swollen roots. Analysis of variance with Kruskal Wallis ANOVA on the data after seven days of infection with *M. incognita* showed that all treatments were statistically significant difference with the significant value less than 0.05 (Table 1).

Table 1. Analysis variance for the effect of NCA on the number of vermiform nematode and the swollen root of Deli tobacco after seven days infection.

Test Statistics[a,b]

	vermiform	swollen
Chi-Square	11.999	13.528
df	5	5
Asymp. Sig.	.035 Significant	.019 Significant

a. Kruskal Wallis Test

b. Grouping Variable: Nematode Control Agent

The tobacco plants treated by *C. musiformis* had significantly fewer vermiform nematode roots in comparison with the infected control (Fig. 3.1). Furthermore, all fungally treated plants showed lower numbers of vermiform nematode roots, followed by Carbofuran, than the positive control plants after seven days infected.

The fungal and Carbofuran treatments also had lower numbers of swollen roots than positive control plants. Tobacco plants treated by *D. eudermata* have the lowest number of swollen roots and positive control plants had the highest number of swollen roots. This is also support by analysis of variance by Kruskal Wallis ANOVA in Table 1. with a value $P<0.05$.

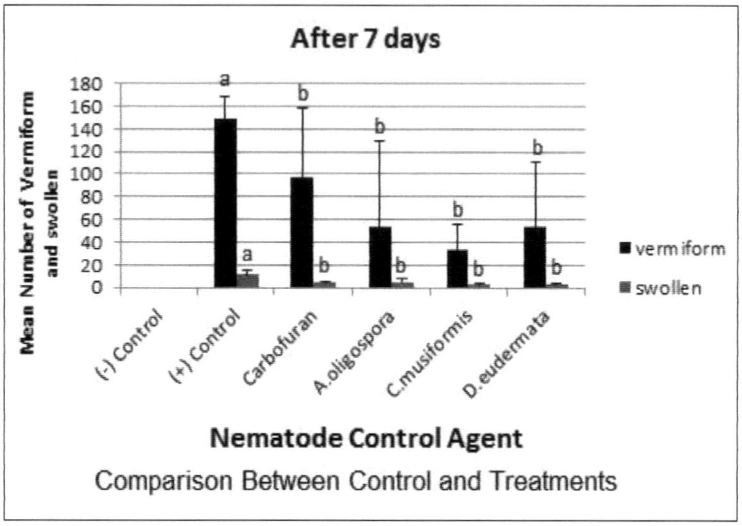

Figure 1. Effect of NCA on the number of vermiform nematodes and swollen root development by *M. incognita* on root Deli tobacco after seven days. Significantly different treatments are shown by an annotation letter above the bars.

3.2. ASSESSMENT II, The Effect of NCA on the number of infections 15 days after infected by Meloidogyne incognita

On the second assessment, four stages of infection were observed. There were a number of immature females and mature females, and a number of swollen roots and sausage shaped roots on the root of the positive control Deli tobacco treatment. Analysis of variance by Kruskal Wallis ANOVA show all treatments were statistically significant different with a significant value lower than 0.05 (Table 2).

Table 2. Analysis of variance for the effect of NCA on the number of immature female, mature female, swollen and sausage shaped of Deli tobacco after 15 day infection.

Test Statistics [a,b]

	Immature female	Mature female	Swollen	Sausage shaped
Chi-Square	10.568	12.598	12.547	13.840
df	5	5	5	5
Asymp. Sig.	.061 Significant	.027 Sig.	.028 Sig.	.017 Sig.

a. Kruskal Wallis Test

b. Grouping Variable: Nematode Control Agent

The effect of the addition of NCA on the roots of tobacco plant was significant in reducing the number of *M.incognita* infections after 15 days post inoculation. If compared with infective control plants, the numbers of infections were suppressed by the addition of chemical treatment and also fungal treatment (Figure 2). The lowest number of immature female was on tobacco plants treated by *A. oligospora*, while the maximum was on the infected control plants. The lowest number of mature females was found on tobacco plants treated by Carbofuran and *A. oligospora*. Tobacco plants treated by *D. eudermata* had the lowest number of swollen roots, and tobacco plants with the lowest number of sausage shaped roots were on tobacco treated by Carbofuran followed by *D. eudermata* treated plants (Figure 3).

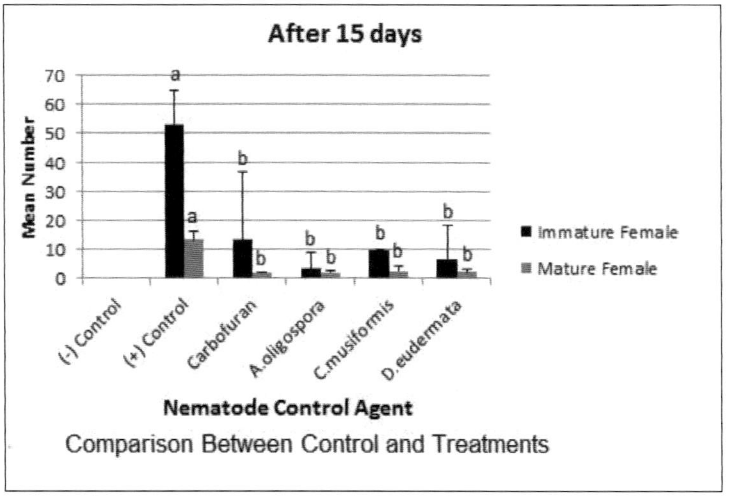

Figure 2. The effect of NCA on the number of immature female and mature female of *M. incognita* after 15 days. Significantly different treatments are shown by an annotation letter above the bars.

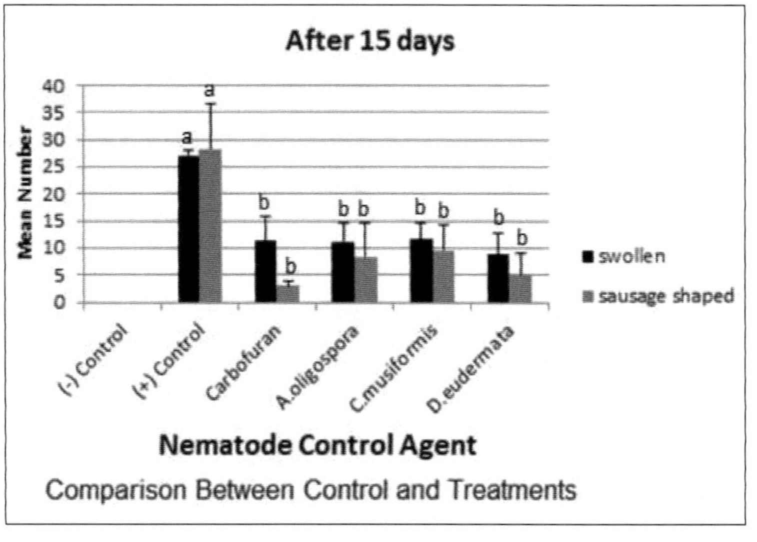

Figure 3. The effect of NCA on the number of swollen root and sausage shaped of *M. incognita* after 15 days. Significantly different treatments are shown by an annotation letter above the bars.

3.3. ASSESSMENT III, The Effect of NTF and Carbofuran on the number of infection after 30 days after infected by Meloidogyne incognita

On the third assessment (30 days after infection), four stages of infection were observed, these were immature females, mature females, sausage shaped roots of tobacco plants and galls. Analysis of variance by Kruskal Wallis ANOVA of data taken at 30 days post infection with nematodes, showed that all treatments were statistically significantly different in terms of the number of immature females, mature females, sausage shaped root and also galls (Table 3).

Table 3. Analysis of variance for the effect of NCA on the number of immature and mature female, sausage shaped and galls on Deli tobacco after 30 days of infection.

Test Statistics[a,b]

	immature	mature	Sausage shaped	Galls
Chi-Square	12.074	10.610	12.403	12.611
df	5	5	5	5
Asymp. Sig.	.034 Significant	.060 Sig.	.030 Sig.	.027 Sig.

a. Kruskal Wallis Test

b. Grouping Variable: Nematode Control Agent

Based on the data presented in Figure 4 the maximum number of immature, mature, sausage shaped roots, and galls were found on infected control plants. The lowest number of immature females was found on tobacco treated by *D. eudermata*, and there were no immature females found on roots after 30 days, following by the tobacco plants treated by *A. oligospora*. The lowest numbers of mature females were found on tobacco plants treated by Carbofuran followed by *C. musiformis* and *D. eudermata*. The lowest number of sausage shaped roots were found on tobacco plants treated by *D. eudermata*, and the lowest number of galls was on tobacco plants treated by *C. musiformis* (Figure 5).

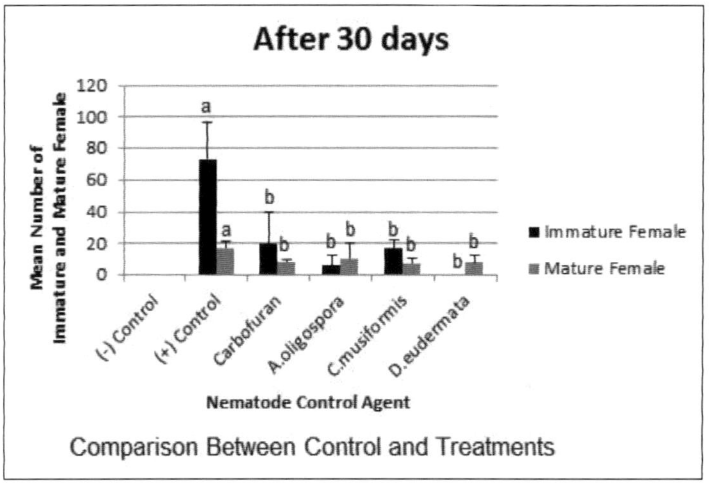

Figure 4. Effect of NCA on the number of immature female and mature female galls of *M. incognita* after 30 days infection. Significantly different treatments are shown by an annotation letter above the bars.

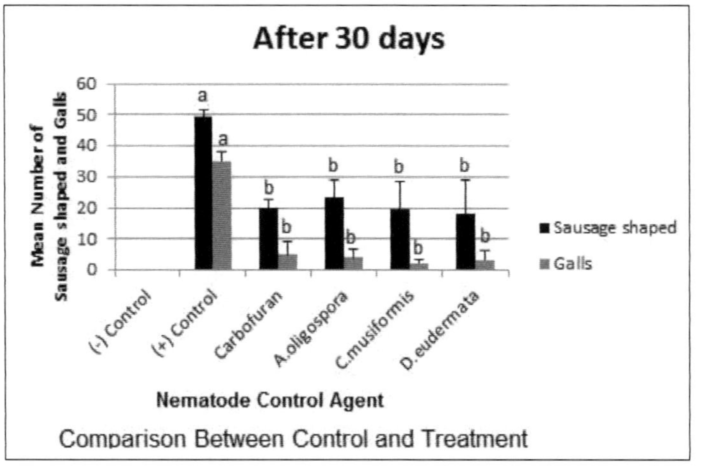

Figure 5. Effect of NCA on the number swollen and sausage shaped and galls development by *M. incognita* after 30 days. Significantly different treatments are shown by an annotation letter above the bars.

3.4. The effect of wheat bran inoculum of Nematode Control Agent to the Growth of Nicotiana tabacum var. Deli 4 over 30 days of infection.

The effect of NCA on the growth of Deli tobacco (*Nicotiana tabacum* var. Deli 4) is presented in the Fig. 3.6 below. Based on the line chart in Figure 6, either Carbofuran or fungal treatment plants showed an increase in the growth on the plant in terms of plant height, root depth, stem fresh weight, stem dry weight and root fresh weight parameter until the end of assessment. However, if compared with negative and positive control plants, only Carbofuran treated plants grew better in terms of all growth parameters.

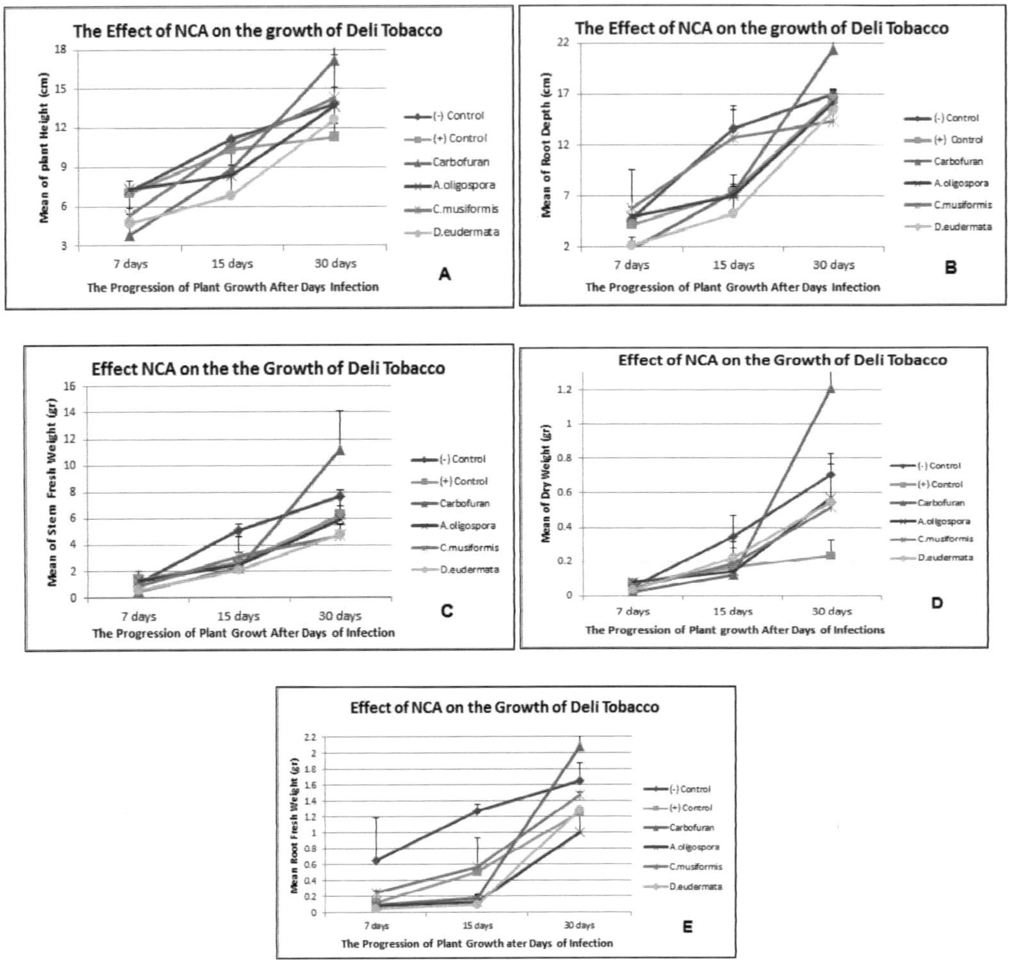

Figure 6. The effect of Nematode Control Agent on the growth progression of *Nicotiana tabacum* var. Deli 4 (Deli tobacco) 7, 15 and 30 days after infection *Meloidogyne incognita*.

All fungal treatments plants were better only in terms of the plant height parameter (Figure 6A) and stem dry weight parameter (Figure 6D) if compared with positive control plants. However, there is no different between fungal treatment plants with negative control plants in plant height parameter also stem dry weight parameter. *C. musiformis* and *D. eudermata* treated plants were better in terms of the root fresh weight parameter (Figure 6E), but there was no different between both fungal treatments plants with negative control plants.

4. Discussion

Analysis of variance to specify the effectiveness of Sumatran NTF in suppress the number of infections showed that all treatments were statistically significantly different throughout the experiment. Effects are first seen at 7 days after infection by J2 of *M. incognita*, and continue at 15 days and 30 days. *C. musiformis* is the most effective treatment in reducing the number of vermiform nematodes and *D. eudermata* is the most effective in reducing the number of swollen roots especially on the first assessment (7 days) after infection. As the nematode life cycle and infections continued, the number of immature and mature female, swollen and sausage shaped root were also reduced at the

IOP Conf. Series: Earth and Environmental Science **130** (2018) 012009 doi:10.1088/1755-1315/130/1/012009

second assessment (15 days) after infection and also the number of galls at the last assessment (30 days) after infection.

Analysis variance by Kruskal Wallis ANOVA also shows almost none of the treatments supported the growth of *Nicotiana tabacum* var. Deli 4 compared with the uninfected control plants. Even though the growth of Deli tobacco showed increases in all measured parameters; root depth, stem fresh weight, stem dry weight and root fresh weight until the end of assessment (Figure 6). A lack of plant growth stimulation when using NTF to control RKN has been previously reported. [15] attempted to reduce infection of *Meloidogyne mayaguensis* on tobacco *Nicotiana tabacum* L. var. Paraguay x Claro cv. *Mayiguana* using *Arthrobotrys oligospora*, *Arthrobotrys conoides*, *Arthrobotrys* sp. *Dactylaria sahelensis*, *Dactylaria* sp. and *Monacrosporium bembicodes* as BCAs. Numbers of infections were suppressed and growth of tobacco was improved, but most fungal treatments did not show any phyto-stimulant effects.

Similar research using wheat bran/sand inoculum (1:1 v/v) containing *Verticillium chlamydosporium* 1% w/w per treatment pot to control *Meloidogyne arenaria* on tomato plant was conducted by [16]. Research proved the inoculation of *V. chlamydosporium* in wheat bran/sand formula could reduce numbers of infections by *M. arenaria* but there was no stimulant growth effect on the growth of tomato plant.

Experiments on improving formulation and BCA dosage have had varied results. Based on the research done by [17], inoculation conidia of *A. dactyloides* in a gum-arabic pellets formula with several dosage levels of inoculum (0.2, 0.4, 0.6, 0.8 w/w) did not show a stimulant effect to the growth of tomato plant. The results showed that increasing the inoculum dosage application was not significant in increasing the plant height, root depth or stem fresh weight parameter. However, increasing inoculum dosage led to a significant increase in the dry weight of tomato plant until the end of the assessment.
The dosage of 1 g in 100 g growth medium (w/w) containing 10^9 conidia of NTF proved to work very well in increasing plant dry weight due to the formation of the hormone auxin in the plant host [18]. According to Ahemad and Kibret [18], auxin (IAA= Indole-3-acetic acid) will increase the content of organic and inorganic substances in the cells and at the same time will increase the dry weight and fresh weight of host plant.

Auxin has been shown to be present in high concentrations especially in the rhizosphere area when the microbial activities increase. The population of microbes and metabolic activity is increase in the presence of root exudates caused by RKN infestations [19]. However, the dosage used in this experiment could not stimulate formation of cytokinin hormone that plays a role in cell division, cell growth and leaf buds network so that high growth of plants was not significantly different [20].

In contrast, another work by Faria [20] using several dosage levels of BCA (0, 1×10^4, 5×10^4, 1×10^5, and 5×10^5 CFU/gr soil) of *L. lecanii* to control *M. incognita* on tomato plants, showed that $5\text{x}10^5$ CFU/gr soil (the highest dosage used) suppressed infection and also improved plant growth [21].

Further research is needed to establish the levels of dosage (w/w of inoculum/ growth medium) of inoculum needed to achieve biocontrol of NTF and also to better understand the effect of fungal treatment on the growth of host plant in terms of plant growth promotion. It is also important to determine the effect of inoculum in different forms, liquid versus solid state fermented, and differences in formulation. Applications and mass production of nematode-trapping fungi in SSF inoculum using local substrate available in Indonesia such as corn flour, cassava flour, rice flour mix with compost, sand or sow dust to commercial scale remains a challenge that must be solved. The work described in this chapter is a preliminary study which could be a reference for further research especially when using the biological control nematode such as *Meloidogyne* spp. applied in Indonesia.

5. Conclusion

Biological control using fungi is a sustainable alternative to dispose of conventional pesticide containing hazardous chemical compound. The use of Sumatran nematode-trapping fungi to control root-knot nematodes may be developed to protect the workers and environment from the pesticides. SSF inoculant such as wheat bran soil inoculant has some criteria as bio-control. They are easy to use and store, affordable, effective, and suitable solution to protect the crops especially in Sumatera Utara, Indonesia. Isolating indigenous strains of NTF from Sumatera Utara continued by pot trial and micro field trial experiment is need to be done to ensure the effectiveness Sumatran NTF of as biological control agents to control Plant Parasitic Nematode such as Root-Knot Nematode. Formulation of Sumatera Utara isolate should be adapted to ensure both good conservation of the Sumatran NTF and high virulence against root-knot nematode.

Acknowledgments

This study was supported, in part, by a Ristekdikti Grant from the Ministry of Research and Technology of Republic Indonesia. We thank Dr. K.G. Davies for providing *Meloidogyne incognita* as a root-knot nematode from Rothamstead Research and Dr. Jane Faull for generous and continue assistance.

References

[1] Mustika I (2005) Konsepsi dan Strategi Pengendalian Nematoda Parasit Tanaman Perkebunan di Indonesia. *Perspektif*, *4*(1), 20–32. Retrieved from http://perkebunan.litbang.deptan.go.id/upload.files/File/publikasi/perspektif/Perspektif_vol_4_No_1_2_IkaMustika.pdf

[2] Murthi RS, Lisnawita L, and Oemry S (2015) Potensi Bakteri Endofit dalam Meningkatkan Pertumbuhan Tanaman Tembakau yang Terinfeksi Nematoda Puru Akar (*Meloidogyne* spp.), *4*(1), 1881–1889.

[3] Purnomo H (2010) *Pengantar Pengendalian Hayati*. (S. Suryantoro, Ed.). Yogyakarta: C.V Andi. Retrieved from http://www.andipublisher.com

[4] Sunardi T, Nadrawati, MP, and Ginting S (2013) Eksplorasi Entomopathogen dan Patogenitasnya pada *Aphis craccivora* Koch. Laporan. Universitas Bengkulu, Bengkulu. Indonesia. 19p.

[5] Ahmad RZ (2002) Media Lokal Untuk Pertumbuhan Kapang Nematofagus Sebagai Sebuah Model (The Local Media For Growth Nematophagous Fungi As A Model). *Seminar Nasional Teknologi Peternakan Dan Veteriner*, 444–449.

[6] Askary TH (2015) Nematophagous fungi as biocontrol agents of Phytonematodes. In *Biocontrol Agents of Phytonematodes* (eds T.H. A, pp. 81–125). CABI.

[7] Nunez-Gaona O, Saucedo-Castoneda G, Alatorre-Rosas R, and Loera O (2010) Effect of moisture content and inoculum on the growth and conidia production by *Beauveria bassiana* on wheat bran. *Brazilian Archives of Biology and Technology*, *53*(4), 771–777.

[8] Brand D, Soccol CR, Sabu A, and Roussos S (2010) Production of fungal Biological control agents through solid state fermentation: a case study on *Paecilomyces lilacinus* against root-knot nematodes. *Mycological Control Using Paecilomyces*, *22*(1), 31–48.

[9] Kumari NS, and Sivakumar CV (2005) Integrated Management of root-knot nematode, *Meloidogyne incognita* infestation in Tomato and grapevine. *Commun Agric. Appl. Biol. Sci*, *70*(4), 909–14.

[10] Slaoui M, Ouhssine M, Berny E, and Elyachioui M (2006) Biodegradation of the carbofuran by a fungus isolated from treated soil. *African Journal of Biotechnology*, *6*(4), 419–423.

[12] Barker KR, and Weeks WW (1991) Relationships between Soil and Levels of *Meloidogyne incognita* and Tobacco Yield and Quality. Journal of Nematology, 23(1), 82–90.

[13] Roland N, Perry, Moens M, and Starr JM (2009) Root-knot Nematodes. London.

[14] Trigiano R (2007) Plant Pathology Concepts and Laboratory Exercises, Second Edition (2nd ed.). Boca Raton: CRC Press.

[15] Mateille T, & Ba A (1998) Potential effects of Sahelian nematophagous fungi against Meloidogyne mayaguensis on tobacco (*Nicotiana tabacum* L. var. Paraguay X Claro) [1997]. Annales Du Tabac. Section 2, 2(29), 61–72. Retrieved from http://www.versailles-grignon.inra.fr/

[16] De-Leij FAAM., and Kerry BR (1991) as a potential biological control agent for *Meloidogyne arenaria*. Revue Nématol., 14(1), 157–164.

[17] Sudirman (2010) Effects of Application Rates of Arthrobotrys dactyloides in Gum-Arabic Pellets on *Meloidogyne javanica* Penetration on Root, Growth, and Yield of Tomato. Retrieved from http://fp.unram.ac.id/data/2012/05/6Sudirman-Vol.3_No.2_Juli-2010.pdf

[18] Ahemad M, and Kibret M (2014) Mechanisms and applications of plant growth promoting rhizobacteria: Current perspective. Journal of King Saud University - Science, 26(1), 1–20.

[19] Trevisan S, Francioso O, Quaggiotti S, and Nardi S (2010) Humic substances biological activity at the plant-soil interface: from environmental aspects to molecular factors. Plant Signaling & Behavior, 5(6), 635–43.

[20] Davies PJ (2004). Plant Hormones. Biosynthesis, Signal Transduction, Action! (P. J. Davies, Ed.) (3rd ed.). Kluwer Academic.

[21] Faria AA (2013) Application Rate of *Purpureocillium lilacinum* Against Different Inoculum Density of Root-Knot Nematode (*Meloidogyne incognita*) of tomato. SHER-E-BANGLA AGRICULTURAL UNIVERSITY.

Polymorphism in beta fibrinogen -455 g/a gene was associated with diabetic in severe ischemic stroke patients

Kiking Ritarwan[1*], Alfansuri Kadri[1], Rosita Juwita Sembiring[2]

[1]Neurology Department, Faculty of Medicine, Universitas Sumatera Utara
[2]Clinical Pathology Department, Faculty of Medicine, Universitas Sumatera Utara

Email: kiking@usu.ac.id, kikingritarwan@gmail.com

Abstract. There is a association of polymorphism in the promoter region of the beta fibrinogen gene -455 G/A with enhancement plasma fibrinogen level. Diabetes mellitus is a risk factor for early neurologic deterioration in acute ischemic stroke. The prothrombotic fibrinogen protein is frequently elevated in patients with diabetes and may be association with poorer prognosis. This study evaluated the association of beta fibrinogen gene -455 G/A promoter polymorphism on modified Ranking Scale of Ischemic Stroke patients treated with diabetic and nondiabetic group. In a Cohort study design comprises 200 consecutive patients diabetic and a nondiabetic who, three months using completed a detailed outcome stroke. Of 200 samples genotype distribution were 27.1% for GG+GA and 0% for AA with diabetic and than 4.4% for GG+GA and 0.05% diabetic patients. Fibrinogen levels were higher in diabetic than nondiabetic group patients (307.7 + 106.3 vs 278 + 84 gr/dl, p=0.002). Fibrinogen level was found to be an independent predictor for diabetic patients. On Genotype GG+GA were associated wth diabetic and nondiabetic group patients. Modified Rankin Scale on day 90 were found associated with diabetic and nondiabetic patients.
Conclusion: Elevated fibrinogen level is dose-dependently associated with 90 days outcome severity stroke with diabetic following ischemic stroke

Keywords: beta fibrinogen promoter gene, diabetic-fibrinogen level, disease progression

1. Introduction

Diabetes Mellitus (DM) and hyperglycemia negatively affect the outcome and pathophysiology of acute ischemic stroke [1]. Stroke ranks third after *ischemic* heart disease and cancer as a cause of lost disability-adjusted life-years in high-income countries and as a cause of death worldwide [1]. The incidence of stroke varies among countries and increases exponentially with age. In Western societies, about 80% of strokes are caused by focal cerebral ischemia due to arterial occlusion, and the remaining 20% are caused by hemorrhages [2,3].

Ischemic brain injury is thought to result from cascade of events from energy depletion to cell death. Intermediate factors include an excess of extracellular excitatory amino acids, free-radical formation, and inflammation [3,4].

IOP Conf. Series: Earth and Environmental Science **130** (2018) 012010 doi:10.1088/1755-1315/130/1/012010

Acute stroke is typically characterized by sudden onset of a focal neurologic deficit, though some patients have a stepwise or gradual progression of symptomatic. *Common deficit* includes dysphasia, dysarthria, hemianopia, weakness, ataxia, sensory loss, and neglect. Symptom and signs are unilateral, and consciousness is generally normal or impaired only slightly, except in the case of some infarcts in the posterior circulation [4].

Prospective studies with large samples have suggested that the plasma fibrinogen level is an independent risk factor for coronary heart disease, diabetes mellitus or stroke [5]. Fibrinogen is a glycoprotein and associated with increased risk of arteriosclerosis disease in humans. G/A variability in the -455 locus of the beta fibrinogen promoter region, especially the carrier status of the A allele has previously been shown to be associated with elevated fibrinogen levels and to increase the risk of cardiovascular disease and ischemic stroke [3,6,7,8].

This study evaluated the association of beta fibrinogen gene -455 G/A promoter polymorphism on modified Rankin Scale of Ischemic Stroke patients treated with diabetic and nondiabetic group.

2. Materials and Methods

2.1 Subjects

All patients with acute ischemic stroke who were admitted to Adam Malik Hospital after Head CT scans were divided into two groups: diabetic and non diabetic patients for three months outcome. The study was approved by the ethics committee of Medical Faculty of Universitas Sumatera Utara. The study design was fully explained, and written information was offered to the patients; if they agreed to participate, the signed a written consent form.

2.2 Genetic Analysis

Genomic DNA was extracted from peripheral blood lymphocyte using a standard protocol. The polymerase chain reaction (PCR) primers for DNA –fragments in the promoter region of the fibrinogen gene -455 G/A polymorphism were 5'-GAACATTTTACCTTATGTGAATTAAGG-3' (forward primer) and 5'-GAAGCTCCAAGAAACCATCC-3' (reverse primer). PCR reaction was performed with Hae III Thermo in 50 micro liter reaction with 50 micrograms of genomic DNA, 200 ng of each appropriate primer and reverse primer, 200 micromol/L of each deoxynucleotide triphosphate, and 1U of Dynazyme II DNA Polymerase in 1 x reaction buffer (Finzymes OY). Samples were incubated for 5 minutes at 95° C, followed by 34 cycles of 1 minute at 95°C, 1 minute at 72°C. PCR Products (20 micro liters) were digested with 10 U of the HaeIII restriction enzyme (Promega Corp) and resolved in 2% agarose gel for determination of -455 G/A genotype. The amplification conditions were as follows: an initial denaturing step at 95° C for 7 second, followed by 35 amplification cycles of denaturation at 52°C for 45 second, 30 second at 72°C, 7 minute at 72° C and 7 minutes at 16°C [7]. Subsequent digestion with the restriction endonuclease HAEIII resulted in fragments of 181 base pair and 488 base pair for the more common genotype GG, 488 base pair and 669 base pair for genotype GA and 669 base pair for genotype AA [3,5,6]

2.3 Outcome Stroke

Prior knowledge of the expected outcome after stroke and its predictor is important for selection of appropriate instruments and analysis of clinical trials in the stroke field. In the clinical trials, the most widely used are modified Rankin Scale (MRS). The MRS is a global outcome rating scale ranging from 0 (no impairment) to 5 (bedridden, incontinent, requiring constant nursing care and attention) and 6 (fatal outcome). This study was divided scale one until two from MRS has good outcome and than scale three until 6 has bad outcome [3,9,10,11,13,15]

2.4 Data analysis

Data analysis include to determine the differences in plasma fibrinogen levels according to diabetic and non diabetic group used Paired T test and to determine changes in outcomes and plasma fibrinogen levels Chi-Square Test, Student T test and Regression analysis test. $P< 0.05$ was considered statistically significantly.

3. Results

Of 200 sample genotype distribution of the -455 G/A locus was 71 % for GG+ AA, 29% for G/A. At a young age (< 55 y.o) was 50.5%, male was 45.5% and Hyper fibrinogen 0 day was 55% and than hyper fibrinogen 90 days was 38%. Baseline characteristics of Patients gene -455 G/A beta fibrinogen are shown in table 1.

Table 1. Baseline Characteristic of the subjects (n=200)

Variable	n (%)
Age : < 55 years old	101 (50.5%)
>55 years old	99 (49.5%)
Sex (Male)	91 (45.5%)
Genotype GG + AA	142 (71%)
GA	58 (29%)
Fibrinogen 0 day	
Hipo fibrinogen (< 268.05)	90 (45%)
Hiper fibrinogen (> 268.05)	110 (55%)
Fibrinogen 90 day	
Hipo fibrinogen (< 272.7)	60 (30%)
Hiper fibrinogen (> 272.7)	76 (38%)

Table 2. Main characteristics of Patients in Diabetic and Nondiabetic Patients data According to gene -455 G/A beta fibrinogen

	DM (n=166)	Non DM (n=34)	*P*
Age	54.3 ± 12.6	63.5 ± 12.1	0.001
Sex (Male)	78 (85.7%)	13 (6.5%)	0.796
Fibrinogen 0	307.7 ± 106.3	278.5 ± 84	0.092
Fibrinogen 90	235.1 ± 94.8	214 ± 65.5	0.002
Genotype GG+GA	45 (27.1%)	155 (4.5%)	0.005
Genotype AA	0 (0)	2 (0.05)	0.367
BMI	34.2 ± 36	22.2 ± 2.1	0.001
MRS 0	64 (38.5%)	10 (29.4%)	0.994
MRS 90	32 (19.2%)	6 (17.6%)	0.003

There were 166 diabetic and 34 non diabetic patients. In DM Population, age (54.3 ± 12.6 yo vs 63.5 ± 12.1, p = 0.001), fibrinogen 90 (235.1 + 94.8 vs 214 ± 65.5%, p= 0.002), Genotype GG + GA (27.1% vs 4.5%, p = 0.005), BMI (34.2 ± 36 vs 22.2 ±2.1, p = 0.001), and outcome Modified Rankin Scale 90 days (19.2 % vs 17.6%, p = 0.003). There were significant differences in the genotype GG + GA but not significantly for allele A.

Table 3 below shows a further regression analysis in the matched population including: age, sex (male), Diabetes Mellitus, Fibrinogen and MRS for 0 days and 90 days and fibrinogen did not prove to be associated with Outcome MRS 0 days (p=0.355). Only fibrinogen 90 days proved to be *associated* with outcome MRS 90 days (p< 0.05). Body Mass Index proved tobe associated with early neurologic patients with outcome ischemic stroke 90 days (p< 0.005). On Regression analysis, DM (OR.1.10, 95% CI 0.91 – 1.25, p =0.210) and Body Mass Index (OR.1.07, 95% CI 1.05 – 1.08, p =0.001).

Table 3. A Logistic Regression model including potential factors associated with early neurologic patients with Outcome ischemic stroke 90 days.

	Odd's Ratio (95% Confidential Interval)	P
Age	0.726 (0.393 – 1.340)	0.304
Sex (Male)	1.170 (0.634 – 2.159)	0.878
DM	1.10 (0.91 – 1.25)	0.210
Fib0 MRS 0	1.010 (0.546 – 1.866)	0.355
Fib90 MRS 90	1.417 (0.752 – 2.670)	0,042
BMI	1.07 (1.05 – 1.08)	0.001

4. Discussion

In previous studies, fibrinogen has emerged as a risk factor for stroke, *ischemic* heart disease, Myocardial Infarction, venous thrombosis, and PAD [3,5,9,10,11] In this study, impaired fibrinogen levels from the first day of ischemic stroke patients treated with aspirin and 90 days after treatment.

At a young age (≤ 55 yo) was 50.5%, male was 45.5%. According Ritarwan et al found in the 136 sample taken, the genotype distribution of the -455 G/A locus was young age (≤ 55 years old) and significantly young and old age (p = 0,001) [3]. Imran found age for ischemic stroke 42.2 + 6.53 yo and p =0.076. This study found that the level of fibrinogen was significantly higher in ischemic stroke group than control (419. 2 mg/dl vs 351.1 mg/dl, p ≤ 0.000) [5]. In this study, fibrinogen 90 days were significantly from diabetic patients (p = 0.002).

In this cohort study, genotype distributions was GG + AA were 71 % and 29 % for GA. The -455 G/A fibrinogen polymorphism may aid in the development and progression of cerebral arteriosclerosis [10,14]. Nishiuma et al. [16] showed significant association (OR. 2.05; p = 0.05) between fibrinogen genotype and ischemic stroke. Martikainen et al. [10] showed The SAM Cohort, genotype distributions were 64.9% (GG). 31.8% (GA) and 3.35 (AA). Researchers from Korea, Liu et al. [17] found the average age based on beta-fibrinogen -455 G / A genotype GG was at 11.6 years, and GA genotype was 64.39 years and AA genotype was 65.09 ± 10.6 years. And found no significant differences between genotypes GG and GA and AA genotype (p> 0.05).

In this a study, based on forwarding stepwise a logistic regression model, the -455 G/A locus genotype showed a significant interaction fibrinogen level at modified Rankin Scale (MRS) day 90, where p=0.042. Identification of stroke risk factors based on the relative risk of age found was 1.417 times according to DM and plasma fibrinogen levels according to the MRS scale on day 90 was 1.10 times. Blake et al. [18] report prospective study, the distribution of the C148T beta fibrinogen polymorphism was similar among those subsequently developed cardiovascular events compared with those who did not. Specifically, the adjusted relative risk associated with the presence of T allele were 0.95 (95% CI 0.75-1.19) for all cases, 1.10 (0.83-1.45) for myocardial infarction, 0.93 (0.65-1.33) for stroke, and 0.71 (0.47-1.1) for venous thrombo-embolism

IOP Conf. Series: Earth and Environmental Science **130** (2018) 012010 doi:10.1088/1755-1315/130/1/012010

This is study to link diabetic hyper fibrinogenemia to a higher frequency of early neurologic deterioration after ischemic stroke. Epidemiological study has established that elevated fibrinogen levels are strongly and independently correlated with the risk of coronary disease (CAD) and peripheral arterial disease. Of note, elevated fibrinogen levels in our group patients with diabetic may be linked to antiplatelet resistance [3,10,12,17]

Conclusion

Elevated fibrinogen level is dose-dependently associated with 90 days outcome severity stroke with diabetic following ischemic stroke

References

1. Lee JS, Hong JM, Lee SE, Kang DR, Ovbiagele B, Demchuck AM, Lee JS. (2017) Association of fibrinogen level with early neurological deterioration among acute ischemic stroke patients with diabetes. BMC Neurology. 17:1-7
2. Hacke W, Kaste M, Bogousslavsky J, Brainin M, Gugging, M, Chamoro A, Less K, Leys D, Kwiencinski H, Toni D. (2003) European Stroke Initiative: Ischemic Stroke Prophylaxis and Treatment. EUSI. Heidelberg.
3. Ritarwan K, Triwidyawati, Lelo A (2016) Beta Fibrinogen gene -455 G/A polymorphisms as the determinant of Ischemic Stroke Outcome Severity in Response to Aspirin Treatment. International Journal of PharmTech Research. 9(4):147-152.
4. Feigin VL, Lawes, CM, Bennett DA, Anderson CS (2003) Stroke epidemiology: a review of population-based studies of incidence, prevalence, and case-fatality in the late 20th century, Lancet Neurol. 2:43-53.
5. Imran I, Lamsudin R, Idjradinata P, Achmad TH, Maskoen, A, Wibowo S, Harapan H (2015) Associated of beta fibrinogen promoter gene polymorphism (-148 C/T), hyperfibrinogenemia and ischemic stroke in young adult patients. The Egyptian Journal of Medical Human Genetics. 16:11-17.
6. Worp HB, Gijn JV (2007) Clinical Practice Acute Ishcemic Stroke. N. Eng J Med. 357:572-579.
7. Ahmed SH, Hu CJ, Paczynski R, Hsu, CY (2001) Pathophysiology of ischemic injury. In: Fischer, M. editor. Stroke Therapy. 2nd ed. Boston Butterworth- Heinemann. p25 – 49.
8. International Stroke Trial Collaborative Group (IST) (1997) a randomized trial of aspirin, subcutaneous heparin, both, or neither among 19.345 patients with acute ischemic stroke. Lancet. 349: 1569-1581.
9. Chinese Acute Stroke Trial (CAST) Collaborative Group (1997) randomized placebo-controlled trial of early aspirin use in 20.000 patients with acute ischemic stroke. Lancet 349: 1641-1649
10. Martikainen M, Pohjasvaara T, Mikkelsson J, Mantyla R, Kunnas T, Laippala P, Ilveskoski E, Kaste M, Karhunen PJ, Erkinjuntti T. (2003) Fibrinogen Gene Promoter -455 An Allele as a risk Factor for Lacunar Stroke. Stroke. 34:886-891.
11. Lee, S.H., Kim, M.K., Park, M.S., Choi, S.M., Kim, J.T., Kim, B.C., Cho, K.H (2008) Beta Fibrinogen gene -455 G/A Polymorphism in Korean Ischemic Stroke Patients. J. Clin. Neurol. 4(1):17-22.
12. Weimar, C., Kurth, T., Kraywinkel, K., Wagner, M., Busse, O, Habert, R.L., Dienner, H.C. (2002) Assessment of functioning and disability after ischemic stroke. Stroke. 33:2053-2059.
13. Qizilbash N, Jones L, Warlow C, Mann J (1991) Fibrinogen and lipid concentrations as risk factors for transient ischaemic attacks and minor ischaemic strokes. BMJ. 303:605-9
14. Wilhelmsen L, Svardsudd K, Korsan-Bengtsen K, Larsson B, Welin L, Tibblin G (1981) Fibrinogen as a risk factor for stroke and Myocardial Infarction. N Engl J Med ., 311:501–505.

IOP Conf. Series: Earth and Environmental Science **130** (2018) 012010 doi:10.1088/1755-1315/130/1/012010

15. Kesler C, Spitzer C, Stauske D, Mende S, Stadlmuller J, Walther R (1997) The Apolipoprotein E and β fibrinogen G/A -455 gene polymorphism are associated with ischemic stroke involving large vessel disease. *Arteriosclerosis, Thrombosis, and Vascular Biology.* 17: 2880-2884
16. Nishiuma S, Kario K, Yakushijin K, Maeda M, Murai R, Matsuo T, Ikeda U, Shimada K, Matsuo M (1998) Genetic variation in the promoter region of the beta-fibrinogen gene is associated with ischemic stroke in a Japanese population. Blood Coagul Fibrinolysis. 9:373-379.
17. Liu Y, Pan J, Wang S, Li X, Huang Y (2002) Beta Fibrinogen gene -455 A/G polymorphism and plasma fibrinogen level in Chinese Stroke patients. Chin. Med. J (Engl). 115: 214-216.
18. Blake, GJ, Schmidt C, Lindpaintner K, Ridker PM (2001) Mutation in the promoter region of the β-fibrinogen gene and the risk of future myocardial infarction, stroke and venous thrombosis. European Heart Journal.2001;22:2262-2266

Antimicrobial activity of lactic acid bacteria isolated from bekasam against *staphylococcus aureus* ATCC 25923, *escherichia coli* ATCC 25922, and *salmonella* sp.

Melia Sari[1], Dwi Suryanto[1*], Yurnaliza[1]

Department of Biology, Faculty of Mathematics and Natural Sciences, University of Sumatera Utara, Jl. Bioteknologi No. 1, Kampus USU, Medan 20155

*Email: dwisuryanto@usu.ac.id

Abstract. Bekasam is an Indonesian fermented food made of fish. As a fermented food, this food may contain some beneficial bacteria like lactic acid bacteria (LAB), which usually have antimicrobial properties such as organic acid, hydrogen peroxide, and a bacteriocin. A study on antimicrobial activity of LAB isolated from bekasam against some pathogenic bacteria has been conducted. The purpose of this study was to know the ability of crude bacteriocin produced LAB of bekasam against *Staphylococcus aureus* ATCC 25923, *Escherichia coli* ATCC 25922, and *Salmonella* sp. Bekasam sample was taken from South Sumatera. LAB isolation was done using de Man Rogosa and Sharpe agar. A bacterial colony with clear zone was selected and purified to get a single colony. The antagonistic assay of the LAB was conducted in Muller-Hinton agar Selected isolates with higher clearing zone were assayed for antibacterial effect of their crude bacteriocin of different culture incubation time of 6, 9, and 12 hours. The results showed that the crude extract bacteriocin of isolate MS2 of 9 hours culture incubation time inhibited more in *Staphylococcus aureus* ATCC 25923 with inhibition zone of 13.1 mm, whereas isolate MS9 of 9 hours culture incubation time inhibited more in *Escherichia coli* ATCC 25922 and *Salmonella* sp. with inhibition zone of 12.7 and 7.3 mm, respectively.

Keywords: crude extract bacteriocin, inhibition zone

1. Introduction

"Bekasam" (traditional food) is an Indonesian fermented made of fish. Bekasam is widely known in the areas of Southern Sumatera and Kalimantan, especially Central Kalimantan. Bekasam is produced by mixing freshwater fish mixed with salt, rice, or fermented tape as a source of carbohydrate. This usually takes 4-10 fermentation days [1]. It has been known that fermented food is often associated with special benefit fermenting bacteria such as a LAB. LAB produced antibacterial substances which is important in the utilization LAB as probiotics agent and food preservative. It is related to the production of metabolites such as organic acids, hydrogen peroxide, diacetyl, and bacteriocin [2]. Bacteriocin is one of the antimicrobial metabolites produced by the LAB.

Bacteriocin can be safely consumed and be applied as food preservative. It can be degraded by proteolytic enzyme in human digestion, so no harm to human health. It was known that bacteriocin has antagonistic properties against some pathogenic bacteria such as *Listeria*, *Clostridium*, *Staphylococcus*, *Bacillus* spp.,*Brochotrix*, *Aeromonas*, and *Vibrio* spp. [3]. Crude extract bacteriocin inhibited the *E.coli* and *S. aureus* [4].

Many bacteria produce bacteriocin. Gram-negative bacteria produce a relatively narrow inhibiting bacteriocin, while Gram-positive bacteria produce relatively broad inhibiting bacteriocin [5]. Many bacteriocins showed stable activity to the temperature of -20 to 100ºC, while it might be sensitive to pH. Bacteria may produce bacteriocin in synthetic media such as de Man Rogosa and Sharpe broth (MRS broth), Tryptone Glucose Extract Yeast (TGE), or other synthetic media. However, MRS broth is a common one [6].

Bacteriocin is potentially used in the control of contaminant bacteria in the food industry, but the availability of bacteriocin is very limited and the price is very high, therefore to find LAB bacteriocin producers from traditional fermented food like bekasam is still necessary.

2. Methods

2.1. Isolation and characterization of LAB from bekasam

Isolation of LAB of bekasam from South Sumatra was conducted in de Man Rogosa and Sharpe (MRS) agar added with 1% $CaCO_3$. The culture was incubated at 37°C for 24-48 hours. Different colonies growing on agar were purified. The colony was characterized by its shape and color. Selected isolates after assay antibacterial of crude extract bacteriocin were characterized for gram staining, motility test, catalase, and sugar fermentation on triple sugar iron agar.

2.2. Antagonistic assay of LAB to three pathogenic bacteria

Antagonistic assay was done on Muller Hinton agar. *Staphylococcus aureus* ATCC 25923, *Escherichia coli* ATCC 25922, and *Salmonella* sp. was sub-cultured in nutrient agar. All bacterial cultures were diluted with 0.9% NaCl to 0.5 MacFarland. A bacterial suspension of pathogenic bacteria was swab on Mueller-Hinton agar (MHA) using sterile cotton bud. A-10 μl suspension of the LAB was dropped on the blank disc. The blank disc was placed on the bacterial lawn on MHA. The culture was incubated at 37°C for 24 hours. The diameter of inhibition zone was measured using calipers.

2.3. Assay antibacterial of crude extract bacteriocin

Selected LAB isolates were sub-cultured in MRS broth at 37°C for 6, 9, and 12 hours. The culture was adjusted to pH 6.0 to inactivate antimicrobial activity of organic acids. The culture was spin at 10.000 rpm for 15 minutes and subjected to pH re-adjusted to 4.0 by using HCl 1N to release bacteriocin from the cell prior kept into the refrigerator at 4°C for 24 hours [7]. The cool suspension was filtered using millipore 0.20 μm [4]. Assay antibacterial of crude extract bacteriocin was conducted as previously described in the antagonistic assay of a LAB to some pathogenic bacteria but using filtered supernatant.

3. Results and Discussion

3.1. Isolation and characterization of LAB

Eleven isolates were obtained in MRS agar with different colony characteristics. Clear zone produced by the colony indicated that it was LAB colony [8,9,10]. LAB produces lactic acid and reacts with $CaCO_3$ to produce soluble lactate calcium, characterized by a clear zone around the growing bacterial colonies [11,12].Two selected isolates MS2 and MS9 were further characterized for their cell morphology and biochemistry (Table 1).

Table 1. Colony and cell characterization of LAB isolates

Isolates	Colony shape	Colony edge	Colony elevation	Colony color	Gram	Catalase	Motility	Oxidase	TSIA				Fermentation
									Slant	Butt	Gas	H_2S	
MS1	circular	Flat	convex	white									
MS2	circular	Wavy	convex	white	+	-	-	-	yellow	yellow	-	-	Hetero
MS3	circular	Irregular	flat	white									
MS4	circular	Flat	convex	white									
MS5	circular	Wavy	flat	white									
MS6	circular	Flat	convex	white									
MS7	circular	Flat	convex	white									
MS8	circular	Flat	flat	white									
MS9	circular	Flat	convex	creamy	+	-	-	-	yellow	yellow	-	-	Hetero
MS10	circular	Flat	convex	white									
MS11	circular	Wavy	convex	white									

3.2. Antagonistic assay of LAB to three pathogenic bacteria

Assay on antimicrobial activity of LAB against three pathogenic bacteria showed that the LAB isolates were capable to inhibiting the pathogen to some extent (Figure 1). This indicated that LABs produced antibacterial metabolites such as organic acids and bacteriocin during growth [13]. Isolates MS2, MS4, MS8, and MS9 showed to inhibit more (Figure 2). All four isolates showed relatively high inhibition zone [14].

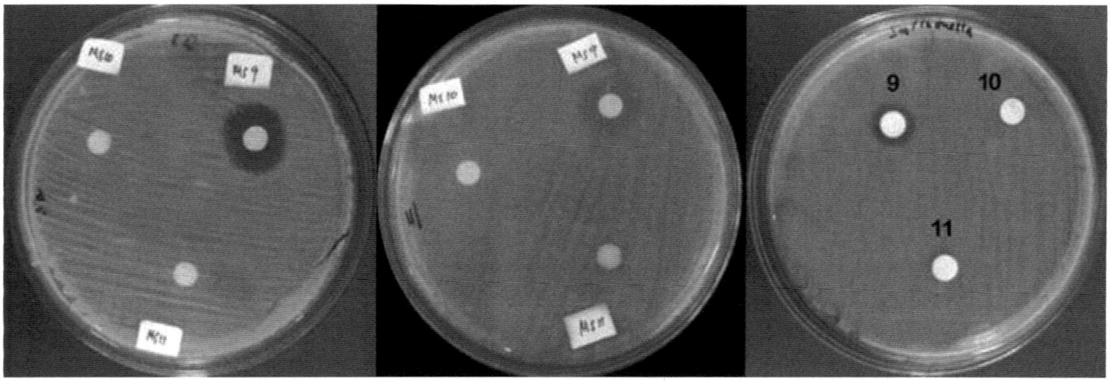

Figure 1. MS9 showed to inhibit (a). *S. aureus*, (b). *E. coli*, and (c). *Salmonella* sp.

MS9 showed to inhibit more to *S. aureus*, *E. coli*, and *Salmonella* sp. with an inhibition zone of 18.1, 9.6, and 7.5 mm, respectively which was higher compared to that of Desniar et al.(2012) and Indriati et al.(2006) studies [15][12]. However, it was lower in inhibiting *E. coli* compared to that of Nigam et al. [16] study.

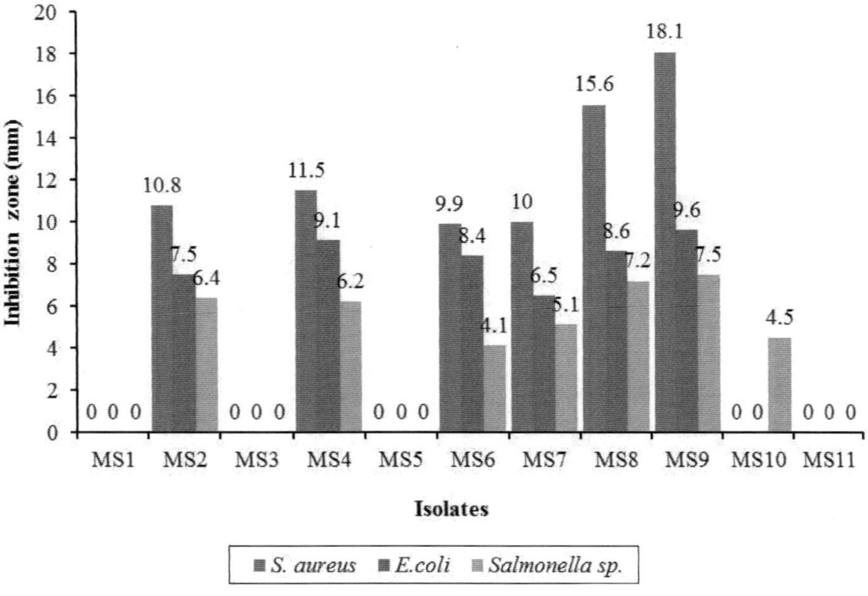

Figure 2. Diameter of inhibition zone of LAB to three pathogenic bacteria

3.3. Assay antibacterial of crude extract bacteriocin

In general, bacteriocin is produced optimally by a LAB in the logarithmic until the stationary phase [17]. In this study, there was a variation of incubation time of LAB culture. This was to determine the optimum producing time of bacteriocin of our LAB isolates. It was shown that supernatant of potential LAB culture inhibited all pathogenic bacteria of all incubated time of LAB, especially of 9 hours incubation time (Figure 3.). MS2 and MS9 supernatant inhibited more to three pathogenic bacteria compared to other two isolates of MS4 and MS8. Unlike others, MS9 inhibited more to *E. coli*. This indicated that bacteriocin produced by the isolates might different because bacteriocin has a specific binding site to the target bacteria, types of bacteriocin and bacteriocin concentrations used [18].

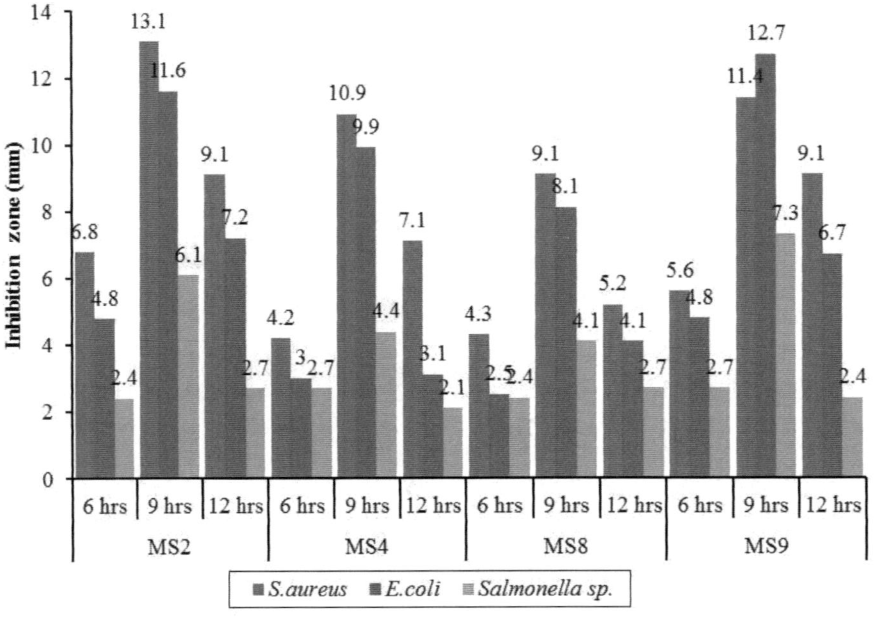

Figure 3. The diameter of inhibition zone of crude extract of bacteriocin of different incubation time to three pathogenic bacteria

Bacteriocin attacks cytoplasmic membrane of pathogenic bacteria through the formation of cytoplasmic membrane pores [19,20], penetrating cell membrane thus increasing the permeability of the cytoplasmic membrane or inhibiting septal formation. The formation of pores in the cell membrane stimulates membrane permeability and disrupts intracellular ADP/ATP balance due to inorganic phosphate leakage, in turn reduce the power of the proton [21] and the number of bivalent cation (Mg^{2+} or Ca^{2+}) causing neutralization of phospholipid negative charges [22], increases permeation of the ion (K^+ and Mg^{2+}), amino acid (glutamate and lysine acid), and ATP. The failure of proton motive force mechanism causes cell death through the cessation of all reactions that require energy [23].

4. Conclusions

Eleven LAB isolates were isolated from bekasam. Of that four isolates MS2, MS4, MS8, and MS9 showed to inhibit more to three pathogenic bacteria *S. aureus*, *E. coli*, and *Salmonella* sp. However, crude bacteriocin of isolates MS2 and MS9 inhibited more to the pathogens, especially in 9 hours incubation time. The bacteriocin produced by our LAB isolates might differ since it showed to inhibit differently to pathogenic bacteria species.

Acknowledgement

Special thanks to my parents for providing a scholarship and research fund, and Yenni Veronika for providing bekasam sample.

References

[1] Desniar R, Iman A, Suwanto A, Mubarik NR (2011) Aktivitas bakteriosin dari bakteri asam laktat asal bekasam. Jurnal Pengolahan Hasil Perikanan Indonesia. XIV(2): 124-133.
[2] Gálves A, H Abriouel R, L López, and NB Omar (2007) Bacteriocins-based strategies for food biopreservation. Int.J.Food Microbio. 120: 5
[3] Brillet A, Pilet MF, Prevost H, Cardinal M, Leroi F (2005) Effect of inoculation of Carnobacterium divergens V41, a biopreservative strain against Listeria monocytogenes risk, on the microbiological, and sensory quality of cold-smoked salmon. Int.J.Food Microbial. 104:309-324.
[4] Abubakar dan Arpah M (2015) Pengaruh suhu produksi terhadap aktivitas ekstrak kasar bakteriosin dari berbagai galur Lactobacillus sp. dalam menghambat Escherichia coli dan Staphylococcus aureus. Buletin Peternakan. 39(3): 189-198.
[5] De Vuyst L, and F Leroy (2007) Bacteriocins from lactic acid bacteria: production, purification, and food applications. J. Molecular Microbiol. Biotechnol. 13:194-199.
[6] Todorov SD, and LMT Dicks (2004) Influence of growth conditions on the production of a bacteriocin by *Lactobacilus lactis* subp. *Lactis* ST34BR, a strain isolated from barley beer. J.Basic Microbial. 44: 305-316.
[7] Usmiati S dan WP Rahayu (2011) Aktivitas hambat bubuk ekstrak bakteriosin dari Lactobacillus sp. galur SCG 1223. Prosiding Seminar Nasional Teknologi Peternakan dan Veteriner, Puslitbang.
[8] Chen YS, Wu HC, Yanagida F (2010) Isolation and Characteristics of Lactic Acid Bacteria Isolated from Ripe Mulberrien In Taiwan. Brazilian Journal of Microbiology. 41: 916-921.
[9] Djie MN, dan Sartini (2008) Isolasi, Identifikasi Bakteri Asam Laktat dari Kol Brassica oleracea L. dan Potensinya sebagai Antagonis Vibrio harveyi in vitro. Torani. 18(3): 211-216.
[10] Lawalata HJ, and Satiman U (2015) Identification of Lactic Acid Bacteria Proteolytic Isolated from An Indonesia Traditional Fermented Fish Sauce Bakasang by Amplified Ribosomal DNA Restriction Analysis (ARDRA). International Journal of ChemTech Research. 8(12) : 630-636.
[11] Indriati N, Setiawan, IP, dan Yulneriwarni (2006) Potensi antibacterial bakteri asam laktat dari peda, jambal roti, dan bekasam. Jurnal Perikanan VIII (2): 153-159.
[12] Romadhon, Subagiyo, Margino S (2012) Isolasi dan Karakterisasi Bakteri Asam Laktat dari Usus Udang Penghasil Bakteriosin sebagai Agen Antibakteria pada Produk-produk Hasil Perikanan. Jurnal Saintek Perikanan. 8 (1): 59-64.
[13] Theron MM, and Lues JFR (2011) Organic Acids and Food Preservation. United State: CRC Press. Hlm: 273.
[14] Pan J, Magoules F, and Biannic YL (2009) Executing multiple group by query in a mapreduce approach. In ICDB and ICCSNA. Hongkong.
[15] Desniar, Rusmana I, Suwanto A, dan Mubarik NR (2012) Senyawa antimikroba yang dihasilkan oleh bakteri asam laktat asal bekasam. Jurnal Akuatika. III No.2.
[16] Nigam A, Kumar A, HV Madhusudan, and Bhola N (2012) In-vitro Screening of antibacterial activity of lactic acid bacteria against common enteric pathogens. Journal of Biomedical Science. 1(42) : 1-6.
[17] Khoiriyah, H, Ardiningsih P, Jayuska A (2014) Penentuan Waktu Inkubasi Optimum Terhadap Aktivitas Bakteriosin Lactobacillus sp. RED4. JKK. 3(1): 7-11.
[18] Parada JS, Caron CR, Medeiros ABP, and Soccol CR (2007) Bacterocins from Lactic Acid Bacteria: Purification, Properties and use as Biopreservatives. Brazilian Archives of Biology and Technology. 50(3): 521-542.
[19] Chen H and DG Hoover (2003) Bacteriocins and their food application. Comprehensive reviews in Food Science and Food Safety. 2: 82-100.

IOP Conf. Series: Earth and Environmental Science **130** (2018) 012011 doi:10.1088/1755-1315/130/1/012011

[20] Sablon E, B Contreras, and E Vandamme (2000) Antimicrobial peptides of lactic acid bacteria: Mode of action, genetics and biosynthesis. Adv. Biochem Eng Biothenol. 68: 21-60.

[21] Eijsink VGH, L Axelsson, DB Diep, LS Havarstein, H Holo, and IF Nes (2002) Production of class II bacteriocins by lactic acid bacteria; an example of biological welfare and communication. Antonie van Leeuwenhoek. 81: 639-654.

[22] Cleveland J, TJ Montville, IF Nes, ML Chikindas (2001) Bacteriocins: safe, natural antimicrobials for food preservation. Intern. J. Food Microbial. 71: 1-20.

[23] Gajić O (2003) Relationships between MDR proteins, bacteriocin production and proteolysis in Lactococcus lactis. Dissertation University of Groningen. Netherlands.

The Potency of Local Bacterial Isolates Encapsulated Within Sodium Alginate in Carbofuran Degradation

Nunuk Priyani*, Dian Pratiwi, and Dwi Suryanto

Department of Biology, Faculty of Mathematics and Natural Science, University of Sumatera Utara. Jln. Bioteknologi No.1, Kampus USU, Padang Bulan, Medan 20155

*Email: nunuk@usu.ac.id

Abstract. Research on the viability of bacteria encapsulated within sodium alginate and their potential in carbofuran degradation has been done. A total of 8 bacterial isolates have been isolated from slaughter house waste. A 100 ml of *Bushnell-Hass Broth* (BHB) medium containing 146.982 ppm of carbofuran was used as a medium. As much as 2 gr of beads which equal to 10^8cells.ml^{-1} was inoculated into each medium culture and incubated for 15 days at ambient temperature and was shaken at 100 rpm. Analysis of carbofuran residues using *High Performance Liquid Chromatography* (HPLC) showed that the best 2 isolates, DN 1 and OR 2, were able to decrease carbofuran phenol concentration up to 30.37 % and 32.09% respectively compared to control. These results suggested that no significant different from the ability of free cell which decreased carbofuran phenol concentration up to 32.54% and 28.29%.

Keywords: biodegradation, carbofuran-phenol, immobilization, local isolate

1. Introduction

Indonesia is an agricultural country with high potency in agricultural and plantation industries. Many efforts such as the use of various types of pesticides, irrigation and fertilizer have been done to improve the quality and quantity of agricultural production. On the other hand, unwise application of pesticides results in accumulation of pesticides residue in the environment and causes a serious environmental problem. The main active compounds of pesticides such as demeton-S-methyl sulfonate, phthalimide, 1,2,3,6-tetrahydrophthalimide and others are carcinogens. Therefore their concentration must be strictly controlled [1].

Pesticides may undergo degradation naturally through photo-oxidation, as well as chemical and biological process. Biodegradation of pesticides by microorganism including bacteria, as well as fungi offers many advantages compared to chemical and physical degradation. Various species of bacteria such as *Pseudomonas, Flavobacterium, Arthobacter, Rhodococcus and Stenorophomonas* have been reported to degrade pesticides.

Lipolytic bacteria isolated from slaughter house waste such as *Bacillus* sp. *Pseudomonas* sp. [2] and *Halobacillus trueperi* and *Rhodobacteraceae bacterium* [3] breakdown lipid hydrocarbon compounds by producing surfactants. The effectivity of microorganism in degrading hydrocarbon compounds such as pesticides is influenced by various environmental factors, and therefore these microorganism has to have good endurance. In order to maintain bacterial capabilities from environmental influences, it is necessary to protect the bacteria, one of which is by encapsulation [4]. Research on the encapsulation of biosurfactant producing bacteria in degrading carbofuran has not been widely studied before. Bacterial encapsulation can sustain and even increase the potential and viability of bacteria in decomposing pesticides in the environment. Therefore, it is necessary to study the viability of local bacteria from slaughter house waste encapsulated in sodium alginate in carbofuran degradation in vitro.

2. Methods

2.1 Isolation and Characterisation of Biosurfactant- producing Bacteria from Slaughter House Waste

As much as 1 g of waste was diluted into 9 ml of distilled water, and homogenized using vortex. 0.1 ml of the suspension was inoculated on the surface of Nutrient Agar medium, incubated at ambient temperature for 24 hours. The grown colonies were observed, characterized and purified.

The purified bacteria were stored in Tryptic Soy Agar (TSA) for further use. Characterization of bacteria including macroscopic observation such as colony morphology (shape, color, edge, elevation) and microscopic observation such as cell shape, arrangement and Gram staining. The observed biochemical properties including starch hydrolysis test, citrate test, hydrogen sulfide test, motility test and catalase test. Bacterial isolates obtained were then grown and stored on TSA medium before used for further observation.

2.2 Measurement of Cell Growth

Two ml of bacterial suspension with the cell density of 10^9 cells.ml^{-1} was inoculated into 98 ml of BHB medium containing 146.982 ppm of carbofuran as the sole carbon source. The culture were incubated at orbital shaker at 100 rpm at room temperature. Cell growth were observed at day 5, 10 and 15 using Standard Plate Count (SPC) method.

2.3 Biosurfactant Activity Screening

Screening of biosurfactant activity was carried out using modified Drop Collapsing Test method [5]. Bacterial isolates were grown in BHB medium with 2% of dextrose. Two ml of bacterial suspansion with 10^9 cells.ml^{-1} was inoculated into 98 ml BHB medium containing 2% dextrose. The cultures were then incubated in an orbital shaker at 100 rpm at room temperature for 15 days. At the end of incubation, each culture in incubation medium was centrifuged at 6,000 rpm for 10 min, in which then 4 ml of the supernatant from each isolate was added with 4 ml of N-hexan and 2 ml of distilled water in test tube. The mixture was shaken vigorously using vortex for one minute. The stable emulsion formed in this process were then measured afterwards.

2.4 Production of Biosurfactants

To encourage bacteria to produce biosurfactant, isolates were grown in BHB medium containing 146.982 ppm of carbofuran. Two ml of bacterial suspansion with 10^9 cells.ml^{-1} was inoculated into 98 ml of BHB medium containing 146.982 ppm carbofuran [6]. The culture was then incubated at orbital shaker at 100 rpm at room temperature for 15 days and then centrifuged at 6,000 rpm for 10 minutes to separate the supernatant from free cells. Four ml of supernatant was extracted with 2 ml dietilether for 5 min. The ether layer was aspirated using a pipette, dried and then redissolved in 2 ml of sodium bicarbonate ($Na.HCO_3$) 0.05 M solution. The solution was then mixed well and carefully added with 3.6 ml of orsinol, heated to boil, cooled at room temperature for 15 minutes, and measured using spectrophotometer with a wavelength of 421 nm. Biosurfactant content was measured based on Rhamnose standart curve.

2.5 Bacterial Encapsulation

Two bacterial isolates were selected for further encapsulation based on cell growth, biosurfactant production, and biosurfactant activity. Bacterial encapsulation using sodium alginate was done using extruction method. Three gram of sodium alginate was dissolved in 100 ml of distilled water and boiled. The solution was sterilized at 121 °C with 2 atm pressure for 15 minutes and cooled to 30 °C. Bacterial cells suspension was added to the solution and mixed well afterwards. The mixture was loaded into 3 ml of syiringe and was carefully dropped the mixture into $CaCl_2$ solution to form alginate beads in which the bacterial cells have been trapped.

2.6 Effectiveness of Encapsulation

To evaluate the effectiveness of encapsulation, a modified viability test [4] was conducted. The viability test was performed after the beads have been stored for 5, 10, and 15 days at room temperature. One gram beads was dissolved in water and 0.1 ml of suspension was plated on the surface of NA medium to observe the colony growth.

2.7 Culture Condition For Carbofuran Degradation

To determine the potential of bacteria to degrade carbofuran, both isolate as free cells (2 ml bacterial suspension) as well as encapsulated cells (2 g of beads), was inoculated into 98 ml BHB medium containing 148.982 ppm of carbofuran as the sole carbon source. The culture were incubated for 15 days at room temperature on orbital shaker at 100 rpm [7]. The bacterial growth and the carbofuran residue were observed on day 5, 10, and 15 after inoculation. Bacteria released from beads were measured using TPC method, while the carbofuran residue was analyzed using High Performance Liquid Chromatography (HPLC). A BHB medium containing 148.982 ppm carbofuran without the addition of neither free cells nor beads was used as a control.

3. **Results and Discussion**

3.1 Bacterial Characteristic

Eight isolates were succesfully isolated from slaughter house waste. Each isolate showed different characteristic based on their colony and cell morphology as presented in Table 1 below.

Table 1. Morphology colony and cells and Gram staining

Bacterial Isolates	Colony morphology				Cell Morphology		Gram
	Form	Edge	Elevation	Color	Form	Setup	
DB 1	Irregular	Lobate	Flat	Beige	Basil	Diplobasil	+
DB 2	Rhizoid	Curled	Flat	White	Basil	Monobasil	+
DN 1	Irregular	Circular	Flat	Beige	Basil	Diplobasil	-
DN 2	Irregular	Lobate	Flat	White	Focus	Streptococcus	-
EM 1	Irregular	Filamentous	Raised	White	Basil	Monobasil	-
EM 2	Irregular	Filamentous	Flat	White	Focus	Monococcus	+
OR 1	Irregular	Entire	Flat	Beige	Basil	Streptobacil	+
OR 2	Rhizoid	Filamentous	Raised	White	Focus	Diplococcus	+

Biosurfactant-producing isolates from areas contaminated with petroleum showed a large amount of gram-negative bacterium that is as much as 155 isolates and a small portion is gram-positive bacteria as much as 17 isolates [8].

To ensure that they are different isolates, some simple and common biochemical tests have been done. The result are shown in the Table 2 below

Table 2. Biochemical Test of Eight Isolates

Bacterial Isolates	Biochemical Test						
	Starch Hydrolysis	Gelatin Hydrolysis	Citrate Test	Hydrogen Sulfide		Motility	Catalase
				Slant	*Butt*		
DB 1	+	-	+	Yellow	Yellow	+	-
DB 2	+	-	-	Yellow	Yellow	+	-
DN 1	+	-	+	Yellow	Yellow	+	-
DN 2	+	-	+	Red	Yellow	+	-
EM 1	-	-	+	Red	Yellow	+	-
EM 2	+	-	-	Yellow	Yellow	+	-
OR 1	-	-	+	Red	Yellow	+	-
OR 2	-	-	+	Red	Yellow	+	-

Note:(+) = Positive Test;(-) = Negative Test

Some isolates such as DB1, DB2, DN1, EM1, EM2, and OR1 shared common characteristics of biochemical test, but showed different results on colony and cell morphologies test.

The main objective of morphological characterization, biochemical test and gram staining is to distinguish isolates one from another. Each of the isolate should show different characteristics in each test performed. Bacterial identification can be based on morphology, staining properties, colony growth patterns and some growth reactions in carbohydrates and the use of amino acids [9].

3.2 Bacterial Growth

The bacterial growth on day 5, 10, and 15 was observed from the isolates, including control to ensure that there was no contamination during the treatment. All isolates shared the same pattern as shown in Figure 1 below.

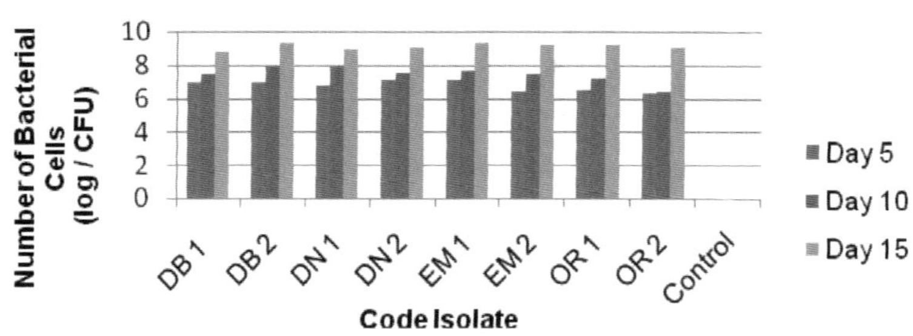

Figure 1. Bacterial Growth

According to Figure 1 above, all isolates grew from day 5 to day 15, but the growth of isolate OR2 on day 5 and 10 was not significant.The growth patterns of each isolate reached its peak on day 15. These results is similar with the results of research conducted by some isolates originating from agricultural land in Berastagi, North Sumatera, showing that bacterial cells continued to grow up to day 21. Increased cell growth that occured in each isolate showed that each isolate has a high adaptability in media containing propineb which is toxic [6].

IOP Conf. Series: Earth and Environmental Science **130** (2018) 012012 doi:10.1088/1755-1315/130/1/012012

Increased cell growth might be caused by the amount of nutrient content in the media was still abundant that isolates entered the logarithmic phase on day 14. It could also be due to the fact that each isolate is capable of producing high amounts of biosurfactant to utilize carbofuran as a source of carbon and nitrogen [10].

3.3 Biosurfactant Assay

The biosurfactant activity from each bacterial isolate can be observed by measuring the volume of emulsion produced upon the addition of N-Hexan to the supernatant.The result of emulsification test obtained from 8 isolates of bacteria are presented in Figure 2.

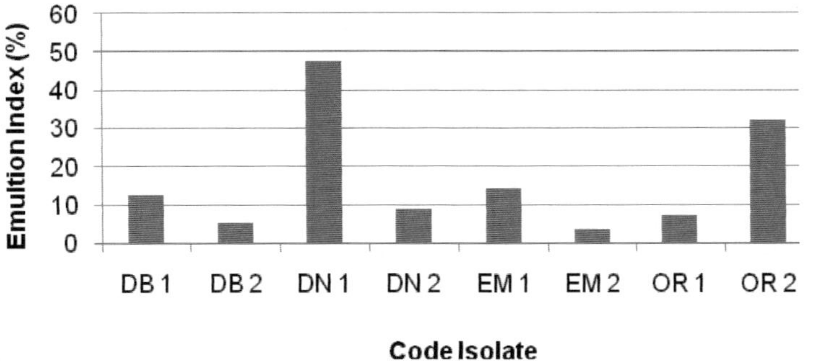

Figure 2. Bacterial Biosurfactant Activity

It can be seen from Figure 2 above that all isolates have different biosurfactant activities. Isolates with the high emulsion index is expected to represent high biosurfactant activity as well. The highest emulsion index was obtained from isolate DN1 which was 47.68%, followed by OR2 with 31,79%, while the lowest index was shown by EM2 with 3.53%. Differences in the emulsion index values produced by each isolate is due to the specific metabolic system from each isolate to survive under certain substrate conditions.The biosurfactants produced by each species may have different quantities and qualities. The higher the emulsion index produced by the isolates, the better ability it has to reduce the surface tension to helps the bacteria to utilize carbon compounds on the media for its growth.

The ability of bacteria to produce biosurfactant will determine the ability of these bacteria to use carbon from the growth substrate [10]. According to Kurniati [11], the type of biosurfactants produced, as well as its quality and quantity, are affected by the nature of the carbon substrate, the concentration of nitrogen, phosphorus, magnesium, iron and manganese ions in media, and culture conditions such as pH, temperature and agitation.

3.4 Biosurfactant Production Assay

Biosurfactant production assay was performed in BHB medium with 146.982 ppm of carbofuran as the sole carbon source. Results of biosurfactant production by each isolate is shown in Figure 3.

IOP Conf. Series: Earth and Environmental Science **130** (2018) 012012 doi:10.1088/1755-1315/130/1/012012

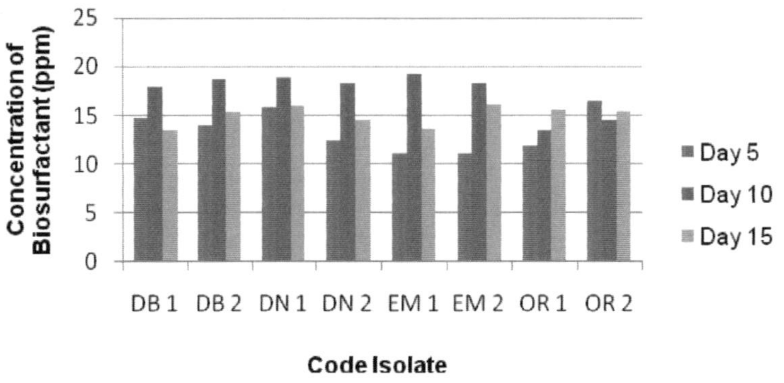

Figure 3. Production of Biosurfactant

The amount of biosurfactant produced by each isolate during the 15 days of incubation was varied one from another. Most isolates showed the same pattern, in which the maximum production occured 10 days after incubation. OR1 isolate was the only isolate that produced biosurfactant at highest volume on day 15.

Biosurfactant produced by microbes varied depending on the types of microbes and nutrients consumed. Similarly, for the same type of microbe, the amount of surfactant produced is different according to the nutrients it consumed [12]. Nutrient played an important role for the growth of microbes, including biosurfactant-producing bacteria. Several studies showed that macro element that played an important role in supporting the growth of biosurfactant-producing bacteria are carbon and nitrogen [13].

3.5 Encapsulation Effectiveness

Two out of eight isolates were chosen for further encapsulation. The isolates were selected based on bacterial growth, volume of biosurfactant produced, and biosurfactant activity. The effectiveness of encapsulation was examined by conducting bacterial viability test after the beads has been stored for four weeks at room temperature. The result is shown in Table 3 below.

Table 3. Effectiveness of Encapsulation After Four Weeks Storage

Code of Isolate	Initial Cell population	Cell Population at the end of storage
DN1	7.2×10^{10}	2.9×10^{7}
OR2	4.1×10^{10}	1.8×10^{7}

Result showed that bacterial cells could survive after being kept at room temperature for four weeks. The advantage of bacterial immobilization is to protect the cells from harsh environment.

The bacterial viability during application in medium BHB containing 146.982 ppm of carbofuran was also observed on day 5, 10, and 15 as the cells degrade carbofuran. The bacterial viability during 15 days of incubation is shown in Figure 4.

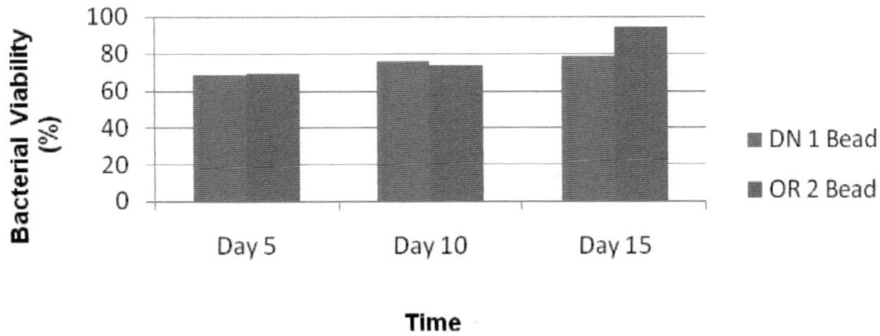

Figure 4. Viability of Bacteria in BHB medium containing146.982 ppm Carbofuran

The initial population of isolate DN1 in the bead was 7.2 × 10^{10} cell.g^{-1} while the initial population of isolate OR2 was 4.1 × 10^{10} cells. g^{-1}. Both isolates decreased in population after five days of incubation. This means that some bacteria were unable to adapt to utilize carbofuran as the carbon source. After 10 days of incubation, the bacterial population started to increase and at the end of the incubation, the bacterial population was almost the same as its initial population with the highest viability of 96.4% by isolate OR2. This information showed that alginate encapsulation was an excellent matrix that the bacteria showed high survivability and were able to utilize carbofuran as their carbon and energy sources.

Cell encapsulation is a technique used to maintain microorganisms in the matrix, to increase the efficiency of microorganisms to work as well as to maintain or protect microorganisms from environmental influences that are unfavorable to the cell [13]. Report on *Bacillus* encapsulated in alginate showed survivability in medium containing mercury and could reduce mercury up to 10 mg / L within 11 hours [14].

3.6 The Potency of Encapsulated Bacteria and Free Cell Bacteria in Degrading Carbofuran In-vitro

The ability of both isolate DN1 and OR2 as encapsulated and free cell to degrade carbofuran was examined in BHB medium containing carbofuran which was incubated for 15 days. The carbofuran residue was analyzed using HPLC on day 5, 10, and 15. The result is shown in Table 4 below.

Table 4. Carbofuran Residue

Bacterial Isolates	Carbofuran Residue (ppm)			
	Days After Incubation			
	0	5	10	15
DN 1 Bead	146.982	-	-	-
DN 1	146.982	-	-	-
OR 2 Bead	146.982	-	-	-
OR 2	146.982	-	-	-
Control	146.982	-	-	-

The result showed that no carbofuran can be detected on day 5 after application in all treatments of both isolates as encapsulated and free cells, as well as in control. This information showed that carbofuran can be degraded not only by microorganism but also spontaneously. The same result has been reported by Onunga et al. [15] in which there is a peak emerged right after carbofuran standard. It was thought that the peak was carbofuran phenol. Carbofuran phenol itself is a compound derived from carbofuran which has toxic propertiy in the environment. Bacteria will degrade carbofuran phenol further into 3-(2-hydroxy-2 methylpropyl) benzene-1,2-diol, in which afterwards it will be hydrolyzed into methylamine.

Pesticide degradation can occur naturally through several ways such as enzymatic degradation, photodegradation and spontaneous degradation. Based on the result of this study, the concentration of carbofuran phenol resulted from spontaneous degradation increased steadily during the incubation period. This means that no carbofuran phenol utilization was shown in the control treatment and therefore it can be assumed that this compound was not degraded further, but instead they are accumulated. On the other hand, different results was observed from the encapsulated and free cells treatment. The concentration of carbofuran phenol on isolate DN1 as encapsulated cells and as free cells showed to decrease during incubation period as shown in Figure 5 below.

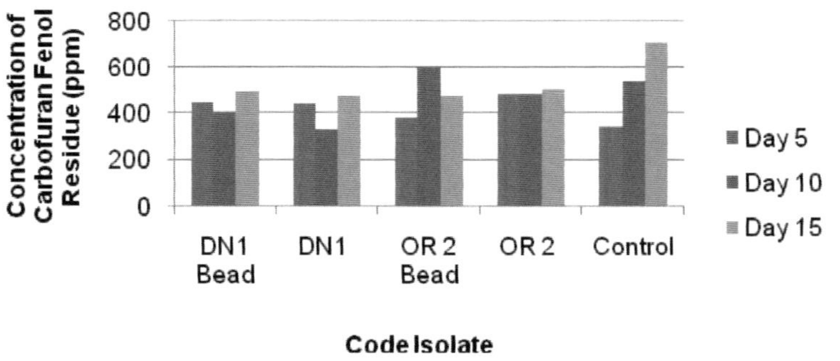

Figure 5. Carbofuran Fenol Residue

It can be assumed that both isolates, DN1 and OR2, were able to degrade carbofuran into carbofuran phenol, and degrade and utilize it further as a source of carbon and energy for their growth and other activities. This result is confirmed with the increase in population of both isolates as discussed earlier. Bacteria can use carbofuran as the sole source of carbon and nitrogen as a nutrient to grow and form carbofuran phenol [5]. The greater amount of carbofuran is degraded and consumed, the higher amount of carbofuran phenol produced [15]. The ability of isolate DN1 on degrading carbofuran phenol was slightly better than that of OR2. Furthermore, the ability to reduce carbofuran phenol of both isolates in encapsulated form was slightly higher compared to free cells. Encapsulated DN1 and OR2 isolates were able to decrease carbofuran phenol concentration up to 30.37% and 32.09% respectively compared to control. Meanwhile, the free cells of DN1 and OR2 decreased carbofuran phenol concentration to 32.54% and 28.29% respectively.

4. Conclusion

Based on the results of the research on Viability Of Bacteria Encapsulated Within Sodium Alginate In Carbofuran Degradation, it could be concluded that isolates DN1 and OR2 were able to degrade carbofuran into carbofuran phenol and decrease carbofuran phenol concentration up to 30.37% and 32.09% respectively. Bacterial encapsulation using alginate showed comparable result as free cells in carbofuran degradation.

Acknowledgments

The authors are grateful to acknowledge Indonesian Ministry of Research, Technology, and Higher Education for providing grant for this research on the fiscal year of 2016, as well as University of Sumatera Utara, Faculty of Mathematics and Natural Sciences, Department of Biology for providing facilities.

References

[1] Ignosa OE, Ajisebutu OS, Okoh IA (2007) Studies on Aerobic Biodegradation Activities of 2,4 dichlorophenoxy acetic acid by Bacteria Spesies Isolated From Petroleum Polluted Site.*African Journal of Biotechnology*. 6(12): 1426-1431

[2] Riupassa RM, Masdiana CP, Dyah KW (2013) *Isolasi dan Karakterisasi Bakteri Penghasil Biosurfaktan Asal Limbah Rumah Potong Ayam Tradisional di Kota Malang*. [Undergraduate Thesis]. Malang: Universitas Brawijaya, Animal Medicine Study Program.

[3] Najiyah D, Nuning VH, Cut NS (2013) *Manfaat Surfaktan dari Bakteri Laut Hidrokarbonoklastik untuk Akselerator Proses Hidrokarbon Minyak Bumi*. Lembar Publikasi Minyak dan Gas Bumi. 47(2):97-104.

[4] Cock LS, Vladimir VC (2013) Probiotic Encapsulation. *African Journal of Microbiology Research*. 7(40):4743-4753.

[5] Nisha P, Eldo P, George Prakash P (2016) Biodegradation of Carbofuran Using Bacteria From Pretreated Plantation Field. *European Journal of Experimental Biology*. 6(2): 30-35.

[6] Utami, DS, Nunuk P, Erman, M (2013) *Isolasi dan Uji Potensi Bakteri Tanah Pertanian Berastagi Sumatera Utara dalam Mendegradasi Fungisida Antracol Berbahan Aktif Propineb*. [Undergraduate Thesis]. Medan: Universitas Sumatera Utara, Department of Biology

[7] Fadhilah N (2015) *Isolasi dan Uji Potensi Bakteri Penghasil Biosurfaktan dari Laut Belawan Sumatera Utara dalam Mendegradasi Pestisida Berbahan Aktif Karbofuran pada Tanah*. [Undergraduate Thesis]. Medan: Universitas Sumatera Utara. Department of Biology.

[8] Batista SB, Mounteer AH, Amorim FR, Totola MR (2005) Isolation and Characterization of Biosurfactant/ Bioemulsifier-Producing Bacteria From Petroleum Contaminated Sites. Bioresource Technology. 97(2006):868-875.

[9] Lay B (1994) *Analisa Mikroba di Laboratorium*. PT Raja Grafindo Persada.Jakarta

[10] Zupliker H (2015) *Isolasi Bakteri Dari Akar Tomat (Solanum Lycopersicum L.) dan Uji Kemampuannya dalam Mendegradasi Insektisida Berbahan Aktif Karbofuran*. [Undergraduate Thesis]. Medan: Universitas Sumatera Utara, FMIPA, Departement of Biology.

[11] Kurniati TH (2016) *Bakteri Penghasil Biosurfaktan Dari Lingkungan Tercemar Limbah Minyak Dan Potensinya Dalam Mendegradasi Hidrokarbon Aromatik Polisiklik* (*HAP*). [Thesis]. Bogor: Sekolah Pascasarjana, Bogor Institute of Agriculture.

[12] Duvnjak Z, Cooper DG, Kosaric N (1983) Effect on N Sources on Surfactant Production by Arthrobacter Parafineus ATCC 19558 In :Microbial Enhanced Oil Recovery. Ed.

[13] Horowitz A, D Gutnick, and E Rosenberg (2005) Sequential Growth of Bacteria on Crude Oil: *Apllied Microbiology*. 30(1).10-19

[14] Sinha, Arvind, Pant, Kamal Kishore, and Khare, SK (2012) Studies on Mercury Bioremediation by Alginate Immobilized Mercury Tolerant Bacillus cereus Cells. International Biodeterioration & Biodegradation 71:1-8.

[15] Onunga DO, Kowino OI, Ngigi AN, Osogo A, Orata F, Getenga ZM, Were H (2015) Biodegradation of Carbofuran in Soil Within Nzoia River Basin, Kenya. *Journal of Environmental Science and Health*. 50: 387-397.

Isolation and analysis of a multifunctional triterpene synthase *KcMS* promoter region from mangrove plant *kandelia candel*

M Basyuni[1,3*], R Wati[1], N Sulistiyono[1], Sumardi[2], H Oku[3], S Baba[4], and H Sagami[5]

[1] Department of Forestry, Faculty of Forestry, Universitas Sumatera Utara, Jl. Tri Dharma Ujung No. 1 Medan, North Sumatera 20155, Indonesia
[2] Faculty of Pharmacy, Tjut Nyak Dhien University, Jl. Rasmi No. 28 Medan 20123, Indonesia
[3] Molecular Biotechnology Group, Tropical Biosphere Research Center, University of the Ryukyus, 1 Senbaru, Nishihara, Okinawa 903-0213, Japan
[4] International Society for Mangrove Ecosystems, Faculty of Agriculture, University of the Ryukyus, 1 Senbaru, Nishihara, Okinawa 903-0213, Japan
[5] Institute of Multidisciplinary Research for Advanced Material, Tohoku University, 2-1-1 Katahira, Aoba-ku, Sendai, 980-8577, Japan

*Email: m.basyuni@usu.ac.id

Abstract. Molecular cloning of *Kandelia candel KcMS* gene has previously been cloned and encoded a multifunctional triterpene synthase. In this study, the *KcMS* gene promoter was cloned through Genome walking, sequenced, and analyzed. A 1,358 bp genomic DNA fragment of *KcMS* promoter was obtained. PLACE and PlantCARE analysis of the *KcMS* promoter revealed that there was some regulatory elements in response to environmental signals and involved in the regulation of gene expression. Results showed that four kinds of elements are regulated by hormone binding, namely 2 MeJA-responsiveness elements (CGTCA-motif and TGACG-motif), the ABRE (TACGTG) involved in abscisic acid responsiveness, gibberellin-related GARE-motif (AAACAGA), and the TGA-element (AACGAC) as an auxin-responsive element. Several elements in the *KcMS* have been shown in other plants to be responsive to abiotic stress. These motifs were MBS (CAACTG), TC-rich repeats, and eight light responsive elements. The *KcMS* promoter was also involved in the activation of defense genes in plants such as HSE (AAAAAATTC) and four circadian control elements (CAANNNNATC). The presence of multipotential regulatory motifs suggested that *KcMS* may be involved in regulation of plant tolerance to several types of stresses.

Keywords: Abiotic stress, *cis*-acting elements, gene expression, leaf, terpenoid.

1. Introduction

Mangrove plants are distributed in intertidal tropical and subtropical regions and have long been well known a source of triterpenoids and phytosterols [1-2]. Molecular cloning of *Kandelia candel KcMS* gene has previously been cloned and accumulated a mixture of lupeol, β-amyrin, and α-amyrin. The *KcMS*, therefore, encoded a multifunctional triterpene synthase, a member of oxidoqualene cyclases (OSCs) gene. The open reading frame (ORF) of *KcMS* consists of 2286 bp that encodes a 761 amino acid polypeptide [3]. Molecular mechanism of salinity tolerance in mangrove plants *K. candel* has been reported: mRNA level of *KcMS* was enhanced with salt concentration in both roots and leaves of *K. candel* [4]. This elevated expression of salt-dependent change in mRNA level of *KcMS* was reversible after removal the salinity and transfer to freshwater [5].

Particular emphasis was on the triterpenoid because these compounds were used as biomarkers of organic matter from mangrove as well as lipid input into the estuarine sediments [2]. Furthermore, OSCs have been attracted our consideration because of their possibility to modify the chemical structures of triterpenoids, as well as, their significance as the primarily committed enzymes in the triterpenoid biosynthesis [3].

However, promoters region from mangrove plants and their expressions have not been studied yet. Apart from these metabolic shifts to overcome environmental stresses. Thus, the present study aimed to describe the isolation and analysis of a multifunctional triterpene synthase *KcMS* promoter region from mangrove plant *K. candel.*

2. Materials and Methods

2.1. Plant samples and DNA extraction

Fresh young leaves of *K. candel* were collected at Okukubi River, Okinawa, Japan and used for DNA extraction. Total genomic DNA was extracted from *K. candel* leaves using modified cetyl trimethyl ammonium bromide (CTAB) procedure [6]. The quality of total DNA (1.2 μg) was evaluated using 1% agarose gels and then quantified by UV-Spectrophotometer (Shimadzu, Kyoto, Japan). The material extraction was stored at -20 °C.

2.2. Isolation of KcMS promoter from mangrove plant K. candel by PCR-based genome walker

The Universal Genome Walker kit (Clontech, USA) was used to obtain promoter region of a multifunctional triterpene synthase *KcMS* gene as described in Figure 1. For isolation of *KcMS* promoter, the genomic DNA library was digested with *Dra*I (library 1), *EcoRV* (library 2), *Pvu*III (library 3), and *Stu*I (library 4) in separates tubes to make a blunt end. Genome Walker adaptors were ligated to the digests, and the ligated products were used as a template to amplify promoter regions of *KcMS* gene.

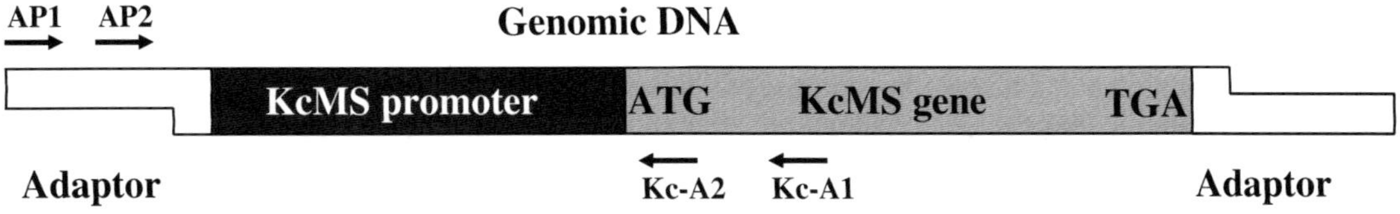

Figure1. Schematic isolation of *KcMS* promoter used Genome walking technique.

Based on the full-length sequence of *KcMS* gene, for isolation of *KcMS* promoter, two specific oligonucleotide primers: Kc-A1 (5'-GATTTCCTCACCATCCTGAACCTTCC-3') and Kc-A2 (5'-TCAAACTCCCATGTTTGCCTTCCCAGG-3'), were synthesized. These primers were used along with two outer adapter primers provided with the kit, AP1 (5'-GTAATACGACTCACTATAGGGC-3') and AP2 (5'-CCCCATCCCTTCGAAGGTCGCAATCGC-3'). The first PCR was performed using AP1 primer and Kc-A1 primer with following PCR conditions: 7 cycles at 94 °C for 25 s and 72 °C for 3 min; 35 cycles at 94 °C for 25 s, 67 °C for 3 min and a final extension 67 °C for 7 min. The nested PCR with AP2 and Kc-A2 primer was carried out with the first PCR product as template. PCR amplification scheme was 5 cycles for 25 s at 94 °C and 3 min at 72 °C, followed by 25 cycles of 25 s at 94 °C and 3 min at 67 °C, with a final extension of 7 min at 67 °C.

2.3. Promoter analysis

The PCR product of promoter fragment was separated using 1 % agarose gel GTG and purified by the Suprec-01 filter (Takara Bio Inc). The purified fragment was ligated into a plasmid vector of TOPO TA cloning vector (Invitrogen) and propagated in *Escherichia coli*, and sequenced by ABI PRISM™ 3100-Avant Genetic Analyzer (Applied Biosystems) with BigdyeR Terminator v1.1 Cycle Sequencing Kit (Applied Biosystems). The sequence identification for putative *cis*-regulatory elements in *KcMS* promoter was analyzed with PLACE (htpp://www.dna.affrc.go.jp/PLACE/signalup.html) [7] and PlantCARE database (http:// http://bioinformatics.psb.ugent.be/webtools/plantcare/html/) [8]. The transcription start site in *KcMS* was predicted online using Neural Network Promoter Prediction (htpp://fruitfly.org/seq_tools/promoter.html)

3. Results and Discussion

The results will be discussed in two subsections; they are *KcMS* promoter sequence analysis and the putative functions of important *cis*-regulatory elements

3.1. KcMS promoter sequence analysis

The promoter region of *KcMS* gene was cloned from *K. candel* genomic DNA. A 1,358 bp fragment upstream the start codon of *KcMS* was obtained (Figure 2). The sequence of the promoter of *KcMS* gene along with several identified core fragments is shown in Figure 2, which is homologous to the *cis*-regulatory elements of other plants and of importance for the promoter functions. Two predicted transcription start sites were found in *KcMS* promoter, at position + 493 to + 543 with a score of 0.93 and at position + 602 to + 652 with a score of 0.94 (Figure 2). The promoter sequence has been shown that transcription factors and RNA polymerase bind and regulate the genes [9]. The specific DNA was recognized by transcription factors as *cis*-acting elements and regulates the gene expression in environmental stimuli [10]. Environmental stresses, including salt stress, therefore, induce the expression of many genes including transcription factors [11] and *KcMS* gene [4-5]

3.2. The putative functions of important cis-regulatory elements

To elucidate the transcriptional regulation of *KcMS* promoter, the upstream was isolated and analyzed for various putative *cis*-acting elements using PLACE and PlantCARE databases. Table 1 shows *cis*-acting elements of *K. candel KcMS* promoter. PLACE and PlantCARE analysis of the KcMS promoter revealed that there was some regulatory elements in response to environmental signals and involved in the regulation of gene expression. As displayed in Table 1, four kinds of elements are regulated by hormone binding, namely 2 MeJA-responsiveness elements (CGTCA-motif at position -1341 and TGACG-motif at position +1341), the ABRE (TACGTG, position +753) involved in abscisic acid responsiveness. Furthermore, two gibberellin-related GARE-motifs (AAACAGA, positions -525 and +1061), and two TGA-elements (AACGAC, positions -337 and +1250) as auxin-responsive element MBS also related to ABA signaling (Figure 2).

The TGACG-motif had been previously reported as the binding site for transcription factors involved in the MeJA signal transduction pathway in barley [12], as two complementary sequences (CGTCA and TGACG) in *Catharanthus roseus* [13], also found in this study (Figure 2). ABRE *cis*-acting element (TACGTG) has been reported to be involved in the osmotic, cold, salt, drought, and ABA responsiveness [9-11], suggested that the importance understood regulatory gene in the stress response of *cis*-acting elements. Besides, GARE-motif was identified as gibberellin (GA) response element (GARE) play an important role in promoting flowering as a factor GA-regulated signaling pathways [15]. Several elements in the *KcMS* have been shown in other plants to be responsive to abiotic stress. These motifs were MBS (CAACTG), TC-rich repeats, and eight light responsive elements. The *KcMS* promoter was also involved in the activation of defense genes in plants such as a HSE (AAAAAATTC), involved in heat stress responsiveness and four circadian control elements (CAANNNNATC). The circadian element is known to be late elongated hypocotyls (LHYs) and circadian clock-associated 1 (CCA1) to increase morning expression of genes [14].

```
>KcMS Promoter
KcMS  +  TCTTCTGCATGAATTGGGGGACCTTCGATTAAAGTTTGATCTCGAATAATTTTGAGGTGT  60
      -  AGAAGACGTACTTAACCCCCTGGAAGCTAATTTCAAACTAGAGCTTATTAAAACTCCACA  60
                 TATA-box
KcMS  +  TTTATCTTTATAATTAAGCAGTTTAATCCACTAGAAGAGTAATGGACAACCGTTTAGAGT  120
      -  AAATAGAAATATTAATTCGTCAAATTAGGTGATCTTCTCATTACCTGTTGGCAAATCTCA  120
                                         CAAT-box TC-rich repeats TCT-motif
KcMS  +  AATGGACATTCGAGTCGGGCTCTCCTAGCAAATTCAATTCGTTTAGATTTCTTACAAGTA  180
      -  TTACCTGTAAGCTCAGCCCGAGAGGATCGTTTAAGTTAAGCAAATCTAAAGAATGTTCAT  180
                                                  CATT-motif     Spl
KcMS  +  TTGCAACTTTGGGTTTGGGCCCCCATGCCCACTAATATTGATGCATTCCCAAGTTTCCAC  240
      -  AACGTTGAAACCCAAACCCGGGGGTACGGGTGATTATAACTACGTAAGGGTTCAAAGGTG  240

KcMS  +  CCCGGCCAGTCGACCACGCCTGGCGTATCGTAAGGGCGAATTCGCGGCCGCTAAATTCAA  300
      -  GGGCCGGTCAGCTGGTGCGGACCGCATAGCATTCCNGCTTAAGCGCCGGCGATTTAAGTT  300
                               CAAT-box
KcMS  +  TTCGCCCTATAGTGAGTCGTATTACAATTCACTGGCCGTCGTTTTACAAGGTACTGACTG  360
      -  AAGCGGGATATCACTCAGCATAATGTTAAGTGACCGGCAGCAAAATGTTCCATGACTGAC  360
                          Circadian        TGA-element
KcMS  +  GGAAAACCCTGGCGTTACCCAACTTAATCGCCTTGCAGCACATCCCCCTTTCGCCAGCTG  420
      -  CCTTTTGGGACCGCAATGGGTTGAATTAGCGGAACGTCGTGTAGGGGGAAAGCGGTCGAC  420
                                                            TATA-box
KcMS  +  GCGTAATAGCGAAGAGGCCCGCACCGATCGCCCTTCCCAACAGTTGCGCAGCCTATACGT  480
      -  CGCATTATCGCTTCTCCGGGCGTGGCTAGCGGGAAGGGTTGTCAACGCGTCGGATATGCA
                                                  MBS
KcMS  +  ACGGCAGTTTAAGGTTTACACCTATAAAAGAGAGAGCCGTTATCGTCTGTTTGTGGATGT  540
      -  TGCCGTCAAATTCCAAATGTGGATATTTTCTCTCTCGGCAATAGCAGACAAACACCTACA  540
                 WBOXATNPR1                          Chs-CMA2a CAAT-box
KcMS  +  ACAGAGTGATATTATTGACACGCCGGGGCGACGGATGGTGATCCCCGCAATTCCGGTTCG  600
      -  TGTCTCACTATAATAACTGTGCGGCCCCGCTGCCTACCACTAGGGGCGTTAAGGCCAAGC  600
    Cicardian CAAT-box             CAAT-box
KcMS  +  CTTGCTGTCCAATAAACCGCCCAGTCTAGCAATCGCCATGTAAGCCCACTGCAAGCTACC  660
      -  CGAACGACAGGTTATTTGGCGGGTCAGATCGTTAGCGGTACATTCGGGTGACGTTCGATG  660
                           Spl
KcMS  +  TGCTTTCTCTTTGCGCTTGCGTTTTCCCTTGTCCAGATAGCCCAGTAGCTGACATTCATC  720
      -  ACGAAAGAGAAACGCGAACGCAAAAGGGAACAGGTCTATCGGGTCATCGACTGTAAGTAG  720
                                      ABRE-motif  GATA-box TATA-box
KcMS  +  CGGGGTCAGCACCGTTTTTGCGGACCGGCTTTTTACGTGAAAAGGATATAGGTGAAGATC  780
      -  GCCCCAGTCGTGGCAAAAACGCCTGGCCGAAAAATGCACTTTTCCTATATCCACTTCTAG  780
                                 TATA-box          GT1-motif
KcMS  +  CTTTTTGATAATCTCATGCCTGACATTTATATTCCCCAGAACATCAGGTTAATGGCGTTT  840
      -  GAAAAACTATTAGAGTACGGACTGTAAATATAAGGGGTCTTGTAGTCCAATTACCGCAAA  840
     Skn-1 motif                CAT-box
KcMS  +  TTGATGTCATTTTCGCGGTGGCTGAGATCAGCCACTTCTTCCCCGATAACGGAGACCGGC  900
      -  AACTACAGTAAAAGCGCCACCGACTCTAGTCGGTGAAGAAGGGGCTATTGCCTCTGGCCG  900
                    Skn-1 motif
KcMS  +  ACACTGGCCATATCGGTGGTCATCATGCGCCAGCTTTCATCCCCGATATGCACCACCGGG  960
      -  TGTGACCGGTATAGCCACCAGTAGTACGCGGTCGAAAGTAGGGGCTATACGTGGTGGCCC  960
            I-box
KcMS  +  TAAAGTTCACGGGAGACTTTATCTGACAGCAGACGTGCACTGGCCAGGGGGATCACCATC  1020
      -  ATTTCAAGTGCCCTCTGAAATAGACTGTCGTCTGCACGTGACCGGTCCCCCTAGTGGTAG  1020
       GC-motif CAAT-box TATA-box Cicardian GARE-box
KcMS  +  CGTCGCCCCGGCGTGTCAATAATATCACTCTGTACATCCACAAACAGACGATAACGGCTC  1080
      -  GCAGCGGGGCCGCACAGTTATTATAGTGAGACATGTAGGTGTTTGTCTGCTATTGCCGAG
           TATA-box                    TATA-box     MBS
KcMS  +  TCTCTTTTATAGGTGTAAACCTTAAACTGCCGTACGTATAGGCTGCGCAACTGTTGGGAA  1140
      -  AGAGAAAATATCCACATTTGGAATTTGACGGCATGCATATCCGACGCGTTGACAACCCTT  1140

KcMS  +  GGGCGATCGGTGCGGGCCTTTTCGCTATTACGCCAGCTGGCGAAAGGGGGATGTGCTGCA  1200
      -  CCCGCTAGCCACGCCCGGAAAAGCGATAATGCGGTCGACCGCTTTCCCCCTACACGACGT  1200
                                               G-box       TGA-element
KcMS  +  AGGCGATTAAGTTGGGTAACGCCAGGGTTTTCCCAGTCACGACGTTGTAAAACGACGGCC  1260
      -  TCCGCTAATTCAACCCATTGCGGTCCCAAAAGGGTCAGTGCTGCAACATTTTGCTGCCGG  1260
      Cicardian          TATA-box
KcMS  +  AGTGAATTGTAATACGACTCACTATAGGGCGAATTGAATTTAGCGGCCGCGAATTCGCCC  1320
      -  TCACTTAACATTATGCTGAGTGATATCCCGCTTAACTTAAATCGCCGGCGCTTAAGCGGG  1320
   TATA-box     WBOXATNPR1 TGACG-motif Start codon
KcMS  +  TTACTATAGGGCACGCGTGGTTGACGCCCGGGCTGGTATGTGGAGGCTTAAGATTGCAG  1380
      -  AATGATATCCCGTGCGCACCAACTGCCGGGCCCGACCATACACCTCCGAATTCTAACGTC  1380
     Spl  MNF1          CGTCA-motif HSE  CAAT-box
KcMS  +  AGGGCGGCGACAACCCTTACATATACAGCACAAACAATTTCCTGGGAAGGCAAAC    1440
      -  TCCCGCCGCTGTTGGGAATGTATATGTCGTGTTTGTTAAAGGACCCTTCCGTTTG    1440
```

Figure 2. The genomic sequence of the *KcMS* promoter showing various *cis*-acting elements for gene regulation predicted in the PLACE and PlantCARE databases are dashed and marked out.

Table 1. *Cis*-acting elements of *K. candel KcMS* promoter from Place and PlantCare databases

Cis-element	Sequence (5'-3')	Function
ABRE	TACGTG	Abscisic acid responsiveness element
CAAT-box	CAAAT	Promoter and enhancer regions
CAT-box	GCCATC	Meristem expression
CATT-motif	GCATTC	Light responsive
CGTCA-motif	CGTCA	MeJA-responsiveness element
G-box	CACGTC	Light response
GARE-motif	AAACAGA	Gibberellin-responsive element
GC-motif	GCCCCGG	Anoxic specific enhancer-like element
GT1-motif	GGTTAAT	Light responsive element
HSE	AAAAAATTC	Heat stress responsive element
I-box	GATATGG	Light responsive element
MBS	CAACTG	MYB binding site in drought-inducibility
MNF1	GTGCCC (A/T) (A/T)	Light responsive element
Skn-1-motif	GTCAT	Endosperm regulatory element
Sp1	GGGCGG	Light responsive element
TATA-box	TATA	core promoter element around -30 of a transcription start
TC-rich repeats	GTTTTCTTAC	Defense and stress-responsive
TCT-motif	TCTTAC	Light responsive element
TGA-element	AACGAC	Auxin-responsive element
TGACG-motif	TGACG	Salicylic acid (SA)- and MeJA-responsiveness element
chs-CMA2a	GCAATTCC	Light responsive element
Circadian	CAANNNNATC	Circadian control element
WBOXATNPR1	TTGAC	SA-responsive element

Furthermore, two TGA-elements are presents in *KcMS* promoter gene. The function of these putative auxin responses remains to be elaborated. Searching the regulatory elements in *KcMS* promoter such as TATA and CAAT boxes are indications of the functionality of sequence. Thirty-four putative TATA boxes are the type of core-sequence element essential for transcription initiation [9]. Twenty CAAT sequences were predicted upstream from the start site of transcription. Therefore, the presence of numerous putative *cis*-regulatory motifs is in this OSC promoter region as well as the *trans*-acting factors still need further investigation.

4. Conclusions

The presence of multipotential regulatory motifs suggested that *KcMS* promoter region may be involved in regulation of plant tolerance to several types of stresses. The new *KcMS* promoter region will further deepen research on the regulation and biosynthesis of triterpenoids in mangrove plants.

Acknowledgment

This study was supported by an International Research Collaboration and Scientific Publication Grant 2017 (No. 003/SP2H/LT/DRPM/IV/2017 to MB) from the Directorate for Research and Community Service, Ministry of Research, Technology and Higher Education, Republic of Indonesia.

References

[1] Koch BP, Rullkotter J, Lara RJ (2003) Evaluation of triterpenoids and sterols as organic matter biomarkers in a mangrove ecosystem in northern Brazil *Wetl. Ecol. Manag.* **11** 257–263.

[2] Basyuni M, Oku H, Baba S, Takara K, Iwasaki H (2007) Isoprenoids of Okinawan mangroves as lipid input into estuarine ecosystem *J. Oceanography* **63** 601–608.

[3] Basyuni M, Oku H, Inafuku M, Baba S, Iwasaki H, Oshiro K, Okabe T, Shibuya M, Ebizuka Y (2006) Molecular cloning and functional expression of a multifunctional triterpene synthase cDNA from a mangrove species *Kandelia candel* (L.) Druce *Phytochemistry* 67 2517–2524.

[4] Basyuni M, Baba S, Inafuku M, Iwasaki H, Kinjo K, Oku H (2009) Expression of terpenoid synthase mRNA and terpenoid content in salt-stressed mangrove *J. Plant Physiol.* 166 1786–1800.

[5] Basyuni M, Baba S, Kinjo Y, Putri LAP, Hakim L, Oku H (2012) Salt-dependent increase in triterpenoids is reversible upon transfer to fresh water in mangrove plants *Kandelia candel and Bruguiera gymnorrhiza J. Plant Physiol.* 169 1903–1908.

[6] Basyuni M, Baba S, Oku H (2017) Microsatellite analysis on genetic variation in two populations of red mangrove *Rhizophora mangle* L. (Rhizophoraceae) and its implication to conservation *IOP Conf. Ser.: Mater. Sci. Eng.* **180** 012243.

[7] Higo K, Ugawa Y, Iwamoto M, Korenaga T (1999) Plant cis-acting regulatory DNA elements (PLACE) database: 1999 *Nucleic Acids Res.* **27** 297–300.

[8] Lecot M, Dehais P, Thijs G, Marchal K, Moreau Y, Van de Peer Y, Rouze P, Rombauts S (2002) PlantCARE, a database of plant *cis*-acting regulatory elements and a portal to tools for *in silico* analysis of promoter sequences *Nucleic Acids Res.* **30** 325–327.

[9] Juven-Gershon T, Kadonaga JT (2010) Regulation of gene expression via the core promoter and the basal transcriptional machinery *Dev. Biol.* **339** 225–229.

[10] Yamaguchi-Shinozaki K, Shinozaki K (2005) Organization of *cis*-acting regulatory elements in osmotic- and cold-stress-responsive promoters *Trends Plant Sci.* **10** 88–94.

[11] Basyuni M, Kinjo Y, Baba S, Shinzato N, Iwasaki H, Siregar EBM, Oku H (2011) Isolation of salt stress tolerance genes from roots of mangrove plant, *Rhizophora stylosa* Griff., using PCR-based suppression subtractive hybridization *Plant Mol. Biol. Rep.* **29** 533–543.

[12] Rouster J, Leah R, Mundyt J, Cameron-Mills (1997) Identification of a methyl jasmonate-responsive region in the promoter of a lipoxygenase 1 gene expressed in barley grain *Plant J.* **11**: 513–523.

[13] Wang Q, Yuan F, Pan Q, Li M, Wang G, Zhao J, Tang K (2010) Isolation and functional analysis of the *Catharanthus roseus* deacetylvindoline-4-*O*-acetyltransferase gene promoter *Plant Cell Rep.* 29 185–192.

[14] Devlin PF (2002) Signs of the time: environmental input to the circadian clock *J. Exp. Bot.* **53** 1535–1550.

[15] Gocal GF, Sheldon CC, Gubler F, Moritz T, Bagnall DJ, MacMillan CP, Li SF, Parish RW, Dennis ES, Weigel D, King RW (2001) GAMYB-like genes, flowering, and gibberellin signaling in *Arabidopsis Plant Physiol.* **127** 1682–1693.

Viability of lactic acid bacteria coated as synbiotic during storage and gastro-intestinal simulation

It Jamilah*, Nunuk Priyani, and Santa Lusia Natalia

Department of Biology, Faculty of Mathematics and Natural Sciences, University of Sumatera Utara, Jalan Bioteknologi Nomor 1, Campus USU. Medan City, North Sumatera, Indonesia

*Email: it@usu.ac.id

Abstract. Lactic acid bacteria (LAB) has been added to various food products as a probiotic agent because it has been known to provide beneficial health effects in humans. In the application of LAB, cell viability often decreased as influenced by environment stresses. Encapsulation technique is one of the cell protection techniques using a coating material. Effective coating material is required to produce maximum protection of LAB cells. In this study, candidate of probiotic LAB (isolate US7) was encapsulated with alginate-mung bean flour and alginate-gram flour with inulin prebiotic by extrusion technique. Viability of encapsulated LAB cells were able to survive by up to 10^8CFU g^{-1} after 4 weeks of storage at 4 °C. Beads were incubated in simulated liquid gastric acid (pH=2) for 2 hrs and simulated intestinal fluid (pH=6) for 3 hrs at 37 °C. The results showed that encapsulated LAB cells maintained the survival rate of 97% with the number of cells at 9.07 Log CFU g^{-1}in the simulated liquid gastric acid and then followed by releasing cells in simulated intestinal fluid. In general, this study indicates that encapsulation with alginate-mung bean flour and alginate-gram flour with inulin successfullyprotect probiotic bacteria against simulated human gastro-intestinal conditions.

Keywords: Alginate-mung bean, encapsulation, lactic acid bacteria, probiotic

1. Introduction

Probiotics have been studied for its health promoting effects such as: lowering cholesterol, improving lactose tolerance, nutritional enhancement and preventing some cancers and antibiotic associated diarrhea [1]. The benefit of using probiotic is due to their nature on production of acid and bacteriocins, competition with pathogens rendering their adhesion to the intestine and enhancement of the immune system [2]. Probiotics are commonly lactic acid bacteria (LAB) with common examples like *Lactobacillus* and *Bifidobacterium* which are extensively investigated for their beneficial effects when incorporated in fermented foods [3,4]. Therefore, LAB have been considered as Generally Recognized as Safe or GRAS [5].

Due to the perceived health benefits of probiotics, there has been an increased use of probiotics in different health based products. Probiotics have been incorporated into a variety of food products. Probiotics have gained global popularity but several studies showed some generala weakness in terms of their cells viability in food products (especially fermented ones).When administered into physiological body, e.g. gastrointestinal conditions may become the barrier for their existence in human. These have encouraged researchers to innovate different methods on delivering probiotics. Encapsulation of the probiotic cells is one of the newest and highly efficient methods, which is now under the special attention and is being developed by various researchers. A variety of substrates have been employed as encapsulating materials for entrapping live microorganisms, including alginate, whey protein, pea protein, gelatin, methylcellulose, etc. [6,7]. Among these, alginate-gel beads have frequently been used for the immobilization of LAB [6]. In this study, Lactic acid bacteria (LAB) as a candidate of probiotic was isolated from North Sumatera freshwater fish farm and intestine of *Oreo chromisniloticus* L. Alginate, mung bean flour, gram flour and inulin were used to encapsulate probiotic for maintaining cell viability during processing, storage and in artificial gastric juice simulation. The aim of this study was to determine the effective material ingredients of encapsulation and to determine the effects of temperature, storage duration and tolerance within simulated gastric juice tocells viability of encapsulated probiotics.

2. Methods

2.1.Preparation of Lactic Acid Bacteria Culture

Pure cultures of isolated LAB namely AK1, AK3, EK2 and US7 were sub-cultured into MRS agar (Oxoid™, UK) medium and incubated at 37 °C for 24 hr.

2.2.Antimicrobial activity

Bacteria used as test organism in this study consisted of three pathogenic bacteria: *Escherichia coli* ATCC 25922, *Salmonella typhimurium* and *Staphylococcus aureus* ATCC 25922. The antimicrobial activity of LAB isolates was determined by disc diffusion method. Briefly, LABs were grown in MRS broth and incubated at 37 °C until cell density reached 10^8 CFU/ml then 20 µL of cell solution was pipette and dropped on the surface of the blank disc and waited for ±1hr until the bacterial cell diffused into the disc. Pathogenic bacteria were streaked on Muller Hinton Agar (Merck®, USA) then discs containing LAB cells regularly placed on the surface of the test medium incubated at 37 °C for 24 hr. The antimicrobial activity was recorded as growth-free inhibition zones (clear zone) diameter.

2.3.Preparation of probiotic LAB for encapsulation

Selected probiotic LAB were subcultured three times in MRS broth at 37 °C for 8 h. Cells were harvested by centrifugation at 3000 xg for 10 min at 4 °C.

2.4.Encapsulation procedure

Beads were produced by using extrusion method with modification [8]. Bacterial cells were dissolved in 100 ml of a mixture consisting of 5% glycerol (v/v), inulin 2% (w/v), and $CaCO_3$ 0.1% (w/v) for control treatment. For AM treatment bacterial cells were dissolved in 100 ml of a mixture consisting of mung bean flour 2% (w/v), 5% glycerol (v/v), inulin 2% (w/v), and CaCO3 0.1% (w/v), were trapped for 45 minutes in 100 ml of alginate solution 3% (w/v). Wall material–culture mixture was dropped through a 27G syringe needle into $CaCl_2$. After one hour the gel was transferred in 0.85 %saline solution to obtain a compact gel structure. Gel beads formed were transferred to distilled water and stirred slowly at 50 rpm for 1 hr to remove $CaCO_3$to obtain rigid beads. For AF treatment, treatment is then repeated with 2% gram flour (w/v), 5% glycerol (v/v), inulin2% (w/v), and $CaCO_3$ 0.1% (w/v). Beads were dried using hot air oven at 45 °C for 48 h.

2.5.Viability of encapsulated isolate US7 during storage

Viability test was done immediately after the encapsulation process was completed (week 0) and weekly for a month at cool 4 °C and mild temperature. Beads (1 g) were added to 9 ml of citrate buffers pH 6 homogenized by vortex for 10-15 minutes and performed serial dilutions. 0.1 ml of homogenized samples were removed, diluted to appropriate concentrations and spread into Plate Count Agar (Merck®, USA) then incubated at 37 °C for 24 h.

2.6.Viability of free cell and beads in gastrointestinal tract condition

The procedure for examination was modified from Woraharn et al. [9]. Free cell survival and cell encapsulated were added in simulated gastric juice (0.08 M HCl with 0.2% (w/v) NaCl; Raoet al.1989). Simulated gastric fluid (SGF) was prepared with pH 2 (adjusted with 0.5 N HCl). The choice of pH 2 for SGF took into account the activity of pepsin, which was maximal in a pH range of 1.7 to 3.0 [10]. Simulated intestinal fluid (SIF) was prepared with pH 6 (adjusted with 0.1 N NaOH). Beads were incubated in SGF for 120 min and subsequently transferred to SIF for 180 min at 37 °C. Control samples of free cells and beads were dissolved in phosphate buffered saline (PBS) of pH 7.4. Cell counts were determined by counting growth of colonies in Plate Count Agar. Treatment is done in duplicate.

3. Results and Discussion

The results will be discussed in 4 subsections which are Antimicrobial activity, Encapsulation of probiotic bacteria (US7), Viability of encapsulated LAB during storage, and Viability of free cells from encapsulated LAB during gastric acid simulation

3.1. Antimicrobial activity

The bacteria used as indicators in this study included Gram-positive bacteria (*S. aureus*) and Gram-negative bacteria (*E. coli* and *S. typhimurium*). Only two isolates, AK1 and US7 showed inhibition effect against all indicatorstrains (Table 1). Among them, US7 showed the higher inhibitory effect against all pathogenic bacteria. The results indicated that LAB isolates US7 could be determined as a candidate of probiotic and was later used in encapsulation experiment.

Table 1. The antimicrobial activity of LAB isolate against food-borne pathogenic bacteria

Isolate code	Diameter of inhibition zone (mm)		
	Escherichia coli	*Staphylococcus aureus*	*Salmonella typhimurium*
AK1	11,88	10,93	10,11
AK3	8,86	0	0
EK2	0	0	0
US7	10,8	16,32	13,54

From the results of antagonist test between LAB isolates and food pathogenic bacteria,onlytwo isolates namely, AK1 and US7 which showedantimicrobial activity against three indicatorstrains. There wasonly oneisolate(AK3) which inhibited *E. coli* with diameter of inhibition zone reaching 8.86 mm and one isolate(EK2) whichdid not exhibit any antimicrobial activity. Isolate US7 produced the largest inhibition zone measuring of 10.8 mm against*E. coli*, 16.32 mm to *S. aureus* and 13.54 mm to *S. typhimurium*. Diameter of inhibition zone produced by US7 isolate against pathogenic bacteria showed higher result than other LAB isolates reported in previous researches. Previous study tested antibacterial activity of some *Lactobacillus* strains against pathogenic bacteria such as *Listeria innocua* LMHAE-LI 107, *Escherichia coli* EC 108 LMHAESA, *Escherichia coli* ATCC25922, *Enterococcus faecalis* ATCC 25212, *Streptococcus* D, and *Klebsiella pneumoniae* CIP 53 153. Their results indicated that the average diameter of inhibition zone is greater than 6 mm on each pathogen using well diffusion method. Other studyalso reported that *L. acidophilus* showed the greatest inhibition zone against *E. coli* and *Cl. sordeli* with diameter of inhibition zone measuring 13 ± 0.8 mm and 14 ± 0.2 mm respectively using dots method.

3.2. Encapsulation of probiotic bacteria (US7)

In this study LAB isolate named US7 was selected for encapsulated by extrusion. The size and spherical shape of the beads is shown in Figure 1.

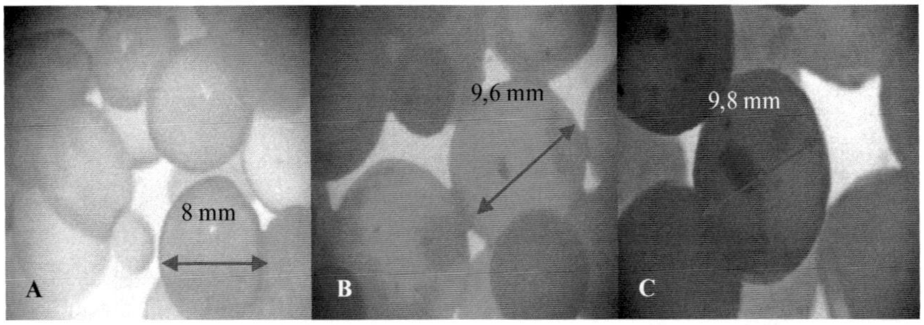

Figure 1. Beads of encapsulated US7 viewed under stereo microscope:
A. Control with alginate; B. Alginate + mung bean flour (AM); C. Alginate + gram flour (AF)

From the figure it can be seen that control beads produced smaller sizes compared to the treatments: AM (Alginate + Mung bean flour) and AF(Alginate + Gram flour) beads. This is due to the use of only one layer(alginate) as coating material whereas the other treatments used two layers, with mung bean flour and gram flour ascoating ingredients. All beads: A (Alginate), AM (Alginate + Mung bean flour) and AF (Alginate + Gram flour), have a considerably uniform spherical shape and compact structures, and a little difference in color. Beads A were transparent, beads AM were browny and beads AF were whitish in colour. Beads have a chewy texture and diameter size ranged between 9-10 mm. The size and shape of the resulting beads were affected by distance from injection, syringe needle size, concentration of alginate and concentration of $CaCl_2$ solution. Size of the injection, needle syringe and distance of syringe to $CaCl_2$ solution were closely related to the size of beads. Distance injection in the process of making the beads is 10-15 cm from the beads collecting solution and syringe needle with size of 23G x 1 1/4 mm was used. The concentration of 3% alginate could provide sufficient size and uniform spherical shape and chewy texture that were more stable against temperature treatment. Some studies reported also use of a varietyof coating materials such as whey proteins as capsule material [13,14,15], soybean oil as a capsule coating on Gum Arabic and gelatin mixture [16], and calcium chloride-to-alginate [17] capsule coating that also being used to coat probiotics in this study.

3.3. Viability of encapsulated LAB during storage

In this experiment, it was found that encapsulation technique could protect higher number of live cells. The viability of isolate US7 entrapped in alginate beads (A), Alginate + Mung bean flour (AM) beads and Alginate + Gram flour (AF) beads during alternating temperature of 4 °C and room temperature for 1-month storage is shown in Figure 2.

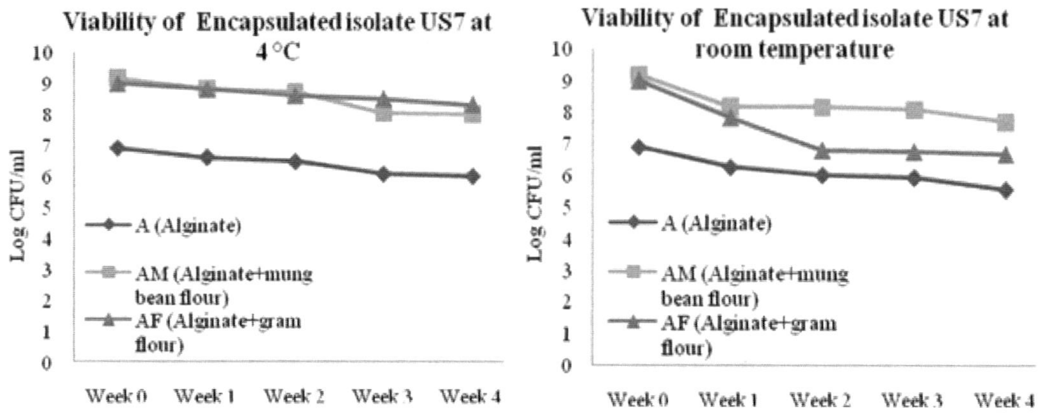

Figure 2. Viability of encapsulated probiotics

The survival of cells in A, AM and AFbeads gradually declined with increasing durationof storage for both temperatures. However, beads stored at room temperature have the highest decline reaching 19%, 16% and 25% for A, AM, and AF respectively. All beads were stored for 4 weeks at 4 °C and at room temperature/ambient (26-30 °C). All treatmentsshowed no considerable loss of viability with increasing storage period at 4 °C. Control treatment (A) decreased viability of 0.90 Log CFU.g^{-1}(13%), (AM) treatment of 1.17 Log CFU.g^{-1} (12.7%)and (AF) treatment of 0.68 Log CFU.g^{-1}(7.5%). All treatments decreased the viability of probiotic during prolongedstorage. The decrease of viability in room temperature was greater than in low temperature. The greatest loss on viability was found from AF treatment with 2.31 Log CFU.g^{-1} (25%). Control treatment (A) decreased viability of 1.36 Log CFU.g^{-1} (19%) while AM decreased until 1.47 Log CFU.g^{-1} (16%). From this results storage temperature at 4 °C is the optimumcondition for maintaining cell viability in beads. The effect of storage conditions on viability of Isolate US7 was similar to Cui et al. [18] who found that the stability of free flowing *Bifidobacteria*-loaded alginate poly-L-lysine microparticles was enhanced during storage at cool temperature (4 °C) when compared to *Bifidobacteria* cultures.

3.4. Viability of free cells from encapsulated LAB during gastric acid simulation

To evaluate the release and viability of isolate US7 beads in gastrointestinal condition, beads were tested in simulated gastric fluid (SGF) for 120 min and simulated intestinal fluid (SIF) for 180 min. The viability of encapsulated US7 under SGF and SIF is shown in Figure 3.

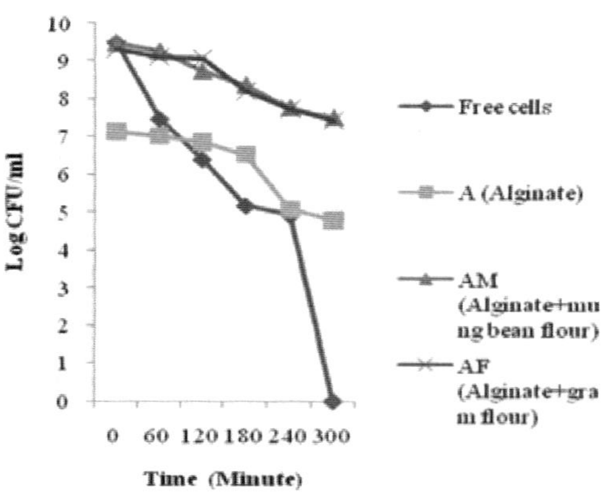

Figure 3. Viability of encapsulated probiotics during Simulated Gastric Fluid (SGF)

Three formulation of beads were disintegrated within 60 min in SGF.A viable count on free cells of US7decreased rapidly with increasing contact time in SGF (from ~ 9.49 to 6.39 log CFU.g^{-1}) and SIF (from ~ 5.17 - 9 to 0 log CFU. g^{-1}).However, in entrapped state or beads,probiotic survived well in gastrointestinal conditions compared to non-entrapped free probiotic cells. The survival of cells in gel beads gradually declined with increasing contact time in SGF and SIF for (A) (from ~ 7.13 log to 5-4 log CFU.g^{-1}), for (AM) (from ~ 9.48 log to 7 log CFU.g^{-1}) and for (AF) (from ~ 9.32 log to 7 log CFU.g^{-1}).For the viability of US7 beads in SGF and SIF, the results of our study corresponded well with other reports [17,19,20].Other study reported that *L. acidophilus* CSCC 2400, *Bifidobacterium longum* and *Bifidobacterium infantis* in calcium alginate survived better when exposed to SGF [20]. Furthermore, viability of encapsulated *L. acidophilus* CSCC 2400 in simulated gastric conditions increased asaddition to alginate gel concentration from 0.75% to 1.8% (w/v) [17]. Microscopy examination on microcapsules showed that bacteria are settled within the capsule material during SGF and were released in SIF. It seemed likely that in our experiment, gastric fluid entered the microparticles through the surface pinholes resulting in a loss of viability. Thus, the survival of US7 declined as increasing of incubation time due to the detrimental effects of low pH on their cells' environment. However, the dense membrane was expected to create diffusion resistance through the beads, which resulted in lower diffusion of SGF and SIF. Consequently, cell survival increased with increased of alginate gel concentrations.

4. Conclusions

Encapsulation of US7 in (AM) Alginate + Mung bean flour and (AF) Alginate + Gram flour with prebiotic inulin coating resulted in better survival of cells after gastro-intestinal simulation, as compared to free cells. Therefore the resultsfrom our study might prove beneficial to applytwo coating materials (mung bean and gram fluor) to improve delivery of probiotic cultures for further application. Beads formulated from Alginate + Gram fluor or AM provided the best protection to US7 cells and there was no survival of US7 free cell in the presence of simulated intestinal fluid due to its low acid resistance. Storage temperature at 4 °C is theoptimum conditiontomaintain cell viability in beads.

Acknowledgments

The authors would like to acknowledge to Diannita Harahap, Ulfa Yani Mayasari and Dewi Rulya Sitepu for using LAB isolates in this study.

References

[1] Fitton N and Thomas JS (2009) Gastrointestinal dysfunction. *J Surgery* **27**(11) 492-495.

[2] Chen KN, Chen MJ, Liu JR, Lin CW and Chiu HY (2007) Optimization of incorporated prebiotics as coating materials for probiotic microencapsulation. *J. Food Sci***70**: 260-266.

[3] Rokka S and Rantamaki P (2010) Protecting probiotic bacteria by microencapsulation: Challenges for industrial applications. *European Food Research and Technology* **231**: 1-12.

[4] Saranya S and Hemashenpagam (2011) Antagonistic activity and antibiotic sensitivity of Lactic acid bacteria from fermented dairy products. *Advances in Applied Science Research* **2**(4) 528- 534.

[5] Salminen S and von Wright A (1998) Current probiotics-safety assured? *Microbial Ecology in Health and Disease* **10**:68-77

[6] Sheu TY, Marshall RT, Heymann H (1993) Improving survival of culture bacteria in frozen dessert by microentrapment. *J Dairy Sci***76**: 1902-1907.

[7] Gombotz WR and Wee SF (1998) Protein release from alginate matrixes. *Adv. Drug Deliv.Rev.* **31**: 276–285.

[8] Reyed MR (2007) Novel hybrid entrapment approach for probiotic cultures and its application during lyophlization. *int. J. Biol. Anthr* **3**: 2.

[9] Woraharn S, Chaiyasut C, Sirithunyalug B and Sirithunyalug J (2010) Survival Enhancement of Probiotic *LactobacillusplantarumC*MU-FP002 by Granulation andEncapsulation Techniques. *African Journal of Microbiology Research***4**(20).

[10] Tobey NA, Hosseini SS, Caymaz-Bor C, Wyatt HR, Orlando GS, Orlando RC (2001) The role of pepsin in acid injury to esophageal epithelium. Am. *J. Gastroenterol***96**: 3062-3070.

[11] Jamaly N, BenjouadA and Bouksaim M (2011) Probiotic potential of *Lactobacillus* strains isolated from known popular traditional Moroccan dairy products. *British Microbiology Research Journal*, **1** (4): 79-94.

[12] Kotikalapudi BL (2009) Characterization and encapsulation of probiotic bacteriausing a pea-protein alginate matrix. Thesis Canada: Universitas of Saskatchewa

[13] Picot A and Lacroix C (2003a) Effect of Micronization on ViabilityandThermotolerance of Probiotic Freeze-Dried Cultures. *International Dairy Journal* **13**: 455-462.

[14] Picot A and Lacroix C (2004) Encapsulation of *Bifidobacteria* in Whey Protein Microcapsules and Survival in Simulated Gastrointestinal Conditions and in Yoghurt. *International Dairy Journal* **14**: 505-515

[15] Gildas GK, Thierry V, Saïd E and Eric M (2009) Microencapsulation of *Lactobacillus plantarum*spp. in an alginate matrix coated with whey proteins. *Int. J. Food Microbiol***129**: 103-105.

[16] Truelstrup-Hansen L, Allan-wojtas PM, Jin YL and Paulson AT (2002) Survival of Free and Calcium-Alginate Microencapsulated *Bifidobacterium*spp. in Simulated Gastro-Intestinal Conditions.*J. Food Microbiol***19**: 35-45.

[17] Chandramouli V, Kailasapathy K, Peiris P and Jonesb M (2004) An improved method of microencapsulation and its evaluation to protect *Lactobacillus* spp. in simulated gastric conditions. *J. Microbiol. Meth***56**: 27-35.

[18] Cui JH, Goh JS, Kim PH, Choi SH and Lee BJ (2000) Survival and stability of *bifidobacteria* loaded in alginate poly-*l*-lysine microparticles. *Int. J. Pharm***210**: 51-59.

[19] Lee KY, Heo TR (2000) Survival of *Bifidobacterium longum* immobilized in calcium alginate beads in simulated gastric juices and bile salt solution. *Appl. Environ. Microbiol***66**: 869 873.

[20] Ross C, Rangika W, Luz S, Maryann A (2006) Synbiotic microcapsules that enhance microbial viability during non-refrigerated storage and gastrointestinal transit. *Appl. Environ. Microbiol* **72**:2280-2282.

Growth characteristics of mangrove seedling in silvofishery pond – the allometric relationship of height, diameter and leaf abundance

Endah Dwi Hastuti[1,*], Rini Budi Hastuti[1]

[1]Biology Department, Faculty of Science and Mathematics Diponegoro University, Jl. Prof. Soedarto, SH Tembalang, Semarang, Central Java 50275

*Email: endah_pdil@yahoo.com; phone/fax: +62 24 70799494

Abstract. Dynamic environment condition of the silvofishery pond should provide an effect on the growth of mangrove seedling. This research aimed to observe the morphometric growth rate of mangrove seedling of *Avicennia marina* and *Rhizophora mucronata* planted in the silvofishery pond and to analyze the morphometric growth relationship of height, diameter and leaf number development of mangrove seedling. The research was conducted through field experiment involving mangrove species of *A. marina* and *R. mucronata* for 18 months during March 2015 to September 2016, both single structured and mixed structure. The observation was conducted every 13 weeks including seedling height, diameter and number of leaves. Data analysis was conducted by regression to provide the statistical relation between the growth of diameter – height, diameter – number of leaves and height – number of leaves. The result showed that the growth rate of *A. marina* in single structured pond was ranged from 0.38 – 3.00 cm.wk^{-1}, 0.0015 – 0.0969 cm.wk^{-1} and 0.1 – 13.7 leaves.wk^{-1} respectively for height, diameter and number of leaves, while in mixed structure was 0.23 – 1.69 cm.wk^{-1}, 0.0169 – 0.0731 cm.wk^{-1} and 0.5 – 14.0 leaves.wk-. The growth of *R. mucronata* respectively in single and mixed structure were 0.08 – 2.00 cm.wk^{-1} and 0.15 – 2.62 cm.wk^{-1}, 0.0031 – 0.1369 cm.wk^{-1} and 0.0008 – 0.0831 cm.wk^{-1} and 0.0 – 1.9 leaves.wk^{-1} and 0.0 – 1.6 leaves.wk^{-1} respectively for height, diameter and number of leaves. Data analysis showed that the growth of seedling height of Avicennia in the mixed structure was significantly affected by its diameter growth and the number of leaves of Avicennia in single structured was significantly affected by its diameter. While the height, diameter and number of leaves of *R. mucronata* both in mixed and single structured silvofishery ponds were independent to each other. This research concluded that mangrove seedling growth is varied among species and growth environment.

Keywords: growth, morphometry, seedling, silvofishery pond, variation

IOP Conf. Series: Earth and Environmental Science **130** (2018) 012015 doi:10.1088/1755-1315/130/1/012015

1. Introduction

Silvofishery is an artificial ecosystem developed by the human to integrate mangrove replantation into aquaculture activities. It aims to provide better environmental quality along with rehabilitation purposes to support sustainable aquaculture activities through the maintenance of suitable water quality [1]. Silvofishery pond ecosystem environmental quality varied by its age which indicates ecosystem maturity and stabilization [2]. Thus, selection of suitable fish species is required to achieve optimum cultivation activities. Mostly, the silvofishery ponds occupy Milkfish and Crab as the recommended cultivars. Even though, certain treatment could be applied to improve its suitability for a more economically valuable cultivars.

Within a silvofishery pond, mangrove vegetation has important role in providing nutrient supply for the cultivated organisms. According to [3], mangrove plants are capable of hindering pond water from pollution. Unfortunately, the capability of mangrove is limited by its age. Young mangrove could not absorb nutrient effectively which might risk to nutrient eutrophication. Thus, mangrove require certain age or size to provide optimum benefit to pond ecosystem and aquaculture activities.

The growth of mangrove plant within a silvofishery pond is an important factor to support silvofishery ecosystem establishment. The condition of mangrove plants such as its density, height, canopy coverage, and species define the pond nutrient balance and productivity [4]. Seedling growth is an important aspect of plant development. Seedling is the most vulnerable phase of plant development. Thus, the risk of environmental stress and mortality is higher than the established plants. Mangrove vegetation might experience physiological stress due to environmental changes [5]. Coastal area land conversion had been occurred for various purposes which effect to the loss of mangrove ecosystem [6].

The growth scales among plant parts are defined by allometric relation. The relations are not limited to plant morphometry but also to its grain mass which could be used to estimate the harvest productivity of the plant [7]. Allometric relation of plants usually relate plant height or diamet or to its above ground nor below ground biomass. This model is expected to provide estimation on potential crops [8]. As for mangrove, leaf production is the most valuable product regarding its role as a source of organic matter for the ecosystem.

The allometric relation is considered as an important aspect of plant growth since it indicates the possible physiological response to particular disturbances. The growth pattern of plants are varied according to their growing environment. The allometric relation of a plant is also affected by its population. More plant density would result a thinner plant stands compared to less dense population [9]. Altering competition over nutrient affects the plant metabolism which causes reduced growth. Thus, to achieve an optimal plant growth, plantation density should be considered to hinder plant competition. According to [10], several woody species seems to have a nearly isometric growth pattern among its leaf and stem biomass to root biomass, while aboveground biomass was also nearly isometric to below ground biomass.

Environment condition is the main factors affecting the allometric model of plant's growth [11]. Plants which grow in a dynamic environment might have inconsistent allometric model such as post burned shrubs [12]. This condition might also occur in mangrove ecosystem which dynamically changes due to tidal activities [13]. Ecosystem aging seems to provide a better suitability for plant growth. Thus, changing environment suitability lead to the development of ecosystem structure. In a restored mangrove such as in the silvofishery ponds, mangrove plants could not achieve stable growth without human assistance. Thus, the impact of particular environmental condition should be understood to provide best management practice of mangrove replantation, especially in silvofihsery ponds.

The structural composition of mangrove ecosystem mainly affects the environment condition of mangrove floor. Species composition defines the enrichment of organic carbon, salinities and grain size which further affect litter decomposition rate and nutrient availability [14]. In the other hand, mangrove species distribution is also defined by its environment setting. Thus, mangrove ecosystem might have various species composition or growth performance due to its growing environment [15].

Within silvofishery pond, this would affect the advantages of mangrove community on the fish culture optimation.

Silvofishery pond is a unique ecosystem which represents mangrove development in a limited closed aquatic system differing its original habitat. However, there are the least understanding on how mangrove response this ecosystem and its impact on the growth pattern. Thus, research on mangrove growth is required to understand the right mangrove plantation model within silvofishery pond to achieve optimum allometric growth. This research aimed to observe the growth of mangrove seedling in the silvofishery ponds and to analyze the relationships of seedling height, diameter and number of leaves in the silvofishery ponds with different mangrove structures.

2. Methods

This research was conducted from March 2015 – September 2016 in Mangunharjo Village, Tugu, Semarang. There were two mangrove species occupied including *Avicennia marina* and *Rhizophora mucronata*. Research treatments involved single species and combined species of mangrove plantation in the silvofishery pond canals. Morphometric measurements were conducted to 27 samples of mangrove seedlings respectively from each treatment.

Data collections were conducted through field observation with three months intervals for five times. Observations included the measurements of the seedling height, seedling diameter and the number of leaves. Data was processed to calculate the growth of mangrove seedling between observations.

Data analysis was conducted through regression analysis to estimate the relation of seedling diameter on seedling height, seedling diameter and leaf abundance and seedling height and leaf abundance. Data analysis utilized 95% of confidence interval.

3. Results and Discussion

Field observation resulted in various numbers of mangrove stand survival of particular species and treatments. The survival rate of *A. marina* was low, but in the single structured treatment, the survival was better than in the mixed structure, while *R. mucronata* achieved better survival in the mix structured treatment although the difference was not noticeable. However, the survival of *R. mucronata* was noticeably higher than *A. marina*. It indicates that the silvofishery pond environment was more suitable for *R. mucronata* than *A. marina*.

Observation on mangrove seedling in the silvofishery ponds showed there were variations of the height, diameter and leaf growth. Collected data from 5 observations were combined into one. The observed growth of mangrove seedling was varied due to seedling mortality. Table 1 showed the range and average of observed mangrove seedling growth.

Table 1. Growth of Mangrove Seedling Within Silvofishery Pond (Growth Period = 13 weeks)

No.	Mangrove Species	Observed Stands	Height (cm/period)	Diameter (cm/period)	Number of Leaves
A.	**Single Species**				
1.	*A. marina*	14	5 – 39 (19.86 + 11.38)	0.02 – 1.26 (0.37 + 0.35)	1 – 178 (56.08 + 53.43)
2.	*R. mucronata*	30	1 – 26 (11.77 + 6.90)	0.04 – 1.78 (0.43 + 0.35)	0 – 25 (4.78 + 7.15)
B.	**Mixed Species**				
1.	*A. marina*	8	3 – 22 (13.38 + 6.99)	0.22 – 0.95 (0.47 + 0.25)	7 – 182 (80.14 + 59.94)
2.	*R. mucronata*	35	2 – 34 (14.11 + 8.99)	0.01 – 1.08 (0.40 + 0.28)	0 – 21 (5.50 + 5.44)

According to Table 1, the growth of mangrove seedlings were not consistent among treatments. The height growth of *A. marina* was observed better in the single species treatment, while in the mixed species treatment the diameter growth was better. Inversely, the height growth of *R. mucronata* in the single species treatment was lower, but it has better diameter growth performance compared to the mixed species treatment. Observation on the leaf number showed that the pattern was quite similar with the height growth, seedling with better height growth performance showed better leaf development.

Collected data were then analyzed with regression analysis to understand the growth relation among parameters. Regression analysis was grouped into three groups, including diameter growth to height growth, diameter growth to leaf number development and height growth to leaf number development. The detailed analysis result of each group is presented in Table 2 – Table 4.

Table 2. Linkage of diameter growth and height growth of mangrove seedling in silvofishery ponds

No.	Mangrove Species	Relationship	R^2 (Sig.)
A.	**Single Species**		
1.	*A. marina*	$Y = 1.4506.X^{0.030}$	0.002 (0.866)
2.	*R. mucronata*	$Y = 1.689 + 0.212 \ln(X)$	0.118 (0.063)
B.	**Mixed Species**		
1.	*A. marina*	$Y = 0.0131.X^{-1.215}$	0.595 (0.025)*
2.	*R. mucronata*	$Y = 0.3956.E^{7.113(X)}$	0.044 (0.225)

Note: * significant at confidence interval of 95%

Analysis result of the linkage of seedling growth of diameter and height of mangrove planted in silvofishery pond showed that only *A. marina* planted in the mixed species treatment was significantly related. It indicated that most of the observed growth of mangrove has no specific relationship. The determination coefficient was 59,5% indicated a fair linkage between the growth of seedling diameter and height. Analysis of other treatments showed no significant effect of the diameter and height growth relationship.

Table 3. Linkage of diameter growth and leaf number development of mangrove seedling in silvofishery ponds

No.	Mangrove Species	Relationship	R^2 (Sig.)
A.	**Single Species**		
1.	*A. marina*	$Y = 51.9873.X^{0.797}$	0.351 (0.033)*
2.	*R. mucronata*	-	-
B.	**Mixed Species**		
1.	*A. marina*	$Y = 0.5582.X^{-0.593}$	0.080 (0.539)
2.	*R. mucronata*	$Y = 0.3746.E^{0.662(X)}$	0.000 (0.920)

Note: * significant at confidence interval of 95%

The linkage between seedling diameter growth and leaf number development was significant on *A. marina* in single species treatment. The analysis resulted determination coefficient of 35.1% indicated a low relationship, while in the single structured *R. mucronata* was not analyzed since the data distribution of leaf number development was not normal. Analysis on the relationship of diameter growth and leaf number development in mixed mangrove structure showed no significant effects.

Table 4. Linkage of height growth and leaf number development of mangrove seedling in silvofishery ponds

No.	Mangrove Species	Relationship	R^2 (Sig.)
A.	**Single Species**		
1.	*A. marina*	$Y = 1.8984.X^{0.833}$	0.134 (0.219)
2.	*R. mucronata*	-	-
B.	**Mixed Species**		
1.	*A. marina*	$Y = 5.1603.X^{0.765}$	0.301 (0.202)
2.	*R. mucronata*	$Y = 0.3856.X^{0.068}$	0.004 (0.743)

Analysis of the linkage between the growth of seedling height and leaf number development showed there were no any significant relationships among treatments. The height growth of mangrove was not related to the leaf development for all treatments. Analysis on single structured *R. mucronata* was not conducted since the data distribution of leaf number development did not have normal distribution.

The analysis resulted that the morphometric growth of particular plant parts was not always related to another plant parts. Only *A. marina* seedling showed the possibility of related plant developments. The absence of significant relation of seedling growth of *R. mucronata* might be caused by the variation of environmental factors which lead to various plant growth rate. Thus, the growth rate of diameter nor height were not consistent among stands.

The growth pattern of a plant is usually expressed in an allometric relationship. Allometric relations among plant parts such as height, diameter, and biomass indicated that plant has a proportional growth rate. According to the result, the various condition of growth linkage of mangrove seedling showed that the variation environment condition would lead to different growth patterns. Ecosystem and environment conditions were considered as the affecting factors for allometric measures and relations of mangrove trees [11].

The utilization of allometric relation of plant is considered to has a great potential for plant component estimation such as its measurements and biomass [11]. Estimation of the biomass of several plant parts such as the dry weight of leaf, branches, and trunks, nor total above ground dry weight had been conducted based on collar diameter, total height and canopy diameter [16]. High determination coefficients were achieved from the analysis indicating its capability on the estimation of tree productivity. Allometric relations is also utilized in the calculation of carbon stocks within mangrove forests [17].

The Plant growth is generaly proportional among parts such as height, diameter, above ground biomass and plant part biomasses [18]. Even though, the proportions might change as the plants grow. However, there is still limited information concerning the relation of growth rates among plant parts.

The relationship of mangrove growth showed that there was a significant linkage between diameter growth and height growth of *A. marina* in the single species treatment and between the diameter growth and leaf number development of *A. marina* in the mixed species treatment. Various factors might affect the growth performance of mangrove seedling which causes to inconsistent growth rate such as environment condition. Suitable environment condition should provide better growth performance of mangrove stands, inversely stressful environment conditions might decrease its growth performance [19].

The linkage of mangrove growth among parts should provide information concerning the development of the ecosystem. Faster mangrove ecosystem maturity is expected to provide better support for the environment [20]. The relation of growth rate among parts provided information concerning the optimum environment condition for best mangrove growth performance [21].

Various research has shown the significant role of environmental factors to the allometric growth of woody plants. According to [22], the consistent allometric pattern would only be achieved in an optimum growth environment. Even though, the change of plant size might also require different environment condition. The variation of environment condition also affected the biomass allocation of plants. The certain growing environment might affect to smaller or larger organ size [23], which might be caused by physiological stress leading to organ malfunction or adaptation activities causing unbalanced nutrient distribution [24,25]. The adaptation capabilities of plants also play an important role in maintaining the optimum allometric pattern of plant growth. Thus, an inappropriate allometric growth of plants might indicate that plants are experiencing environmental stress or unsuitable environment condition.

The inconsistent plant growth rate resulted in this research indicated that silvofishery ponds in Semarang coastal area had low suitability for mangrove growth. It was supported by low survival of mangrove, especially for *A. marina* during the research. The inconsistent growth rate of *R. mucronata* also showed the variation of environment condition among sites. Thus, mangrove seedling experienced various stress level.

4. Conclusion

This research concluded that mangrove growth rate was varied among species and treatments. The growth rate of *A. marina* height was higher in the single structured community, while *R. mucronata* was higher in mixed structure community. The diameter growth rate both for *A. marina* and *R. mucronata* was higher in the single sturctured community. The leaf number development of both *A. marina* and *R. mucronata* were faster in the mixed structure community. The allometric growth of mangrove seedling was only significant for *A. marina*, including diameter to height growth in the mixed structure and diameter growth to leaf number development in single structured.

Acknowledgement

The authors acknowledge Directorate of Research and Public Services, General Director of Higher Education, Ministry of Education and Culture for funding support through research grant **No. 022/SP2H/LT/DRPM/2016**.

References

[1] Hastuti ED, Budihastuti R (2016) Biotropia **23** (1): 58–63.
[2] Suwarto, Lahjie AM, Ruchaemi A, Simorangkir BDAS, Mulyadi F (2015) Global Journal of Agricultural Research **3** (1): 27–35.
[3] Purwiyanto AIS, Agustriani F (2014) Ilmu Kelautan **19** (2): 81–87.
[4] Gatune C, Vanreusel A, Troch M De (2017) Aquaculture Environment Interactions **9**: 73–85.
[5] Eddy S, Ridho MR, Iskandar I, Mulyana A (2016) *Jurnal Silvikultur Tropika* **7**(3): 42–47.
[6] Van TT, Wilson N, Thanh-Tung H, Quisthoudt K, Quang-Minh V, Xuan-Tuan L, Dahdouh-Guebas F, Koedam N (2015) Acta Oecologica **63:** 71–81.
[7] Qin X, Weiner J, Qi L, Xiong Y, Li F (2013) Field Crops Research **144:** 162–166.
[8] Deng J, Zuo W, Wang Z, Fan Z, Ji M, Wang G, Ran J, Zhao C, Liu J, Niklas KJ, Hammond ST, Brown JH (2012) PNAS **109**(22): 8600–8605.
[9] Li L, Weiner J, Zhou D, Huang Y, Sheng L (2013) Journal of Ecology **101:** 475–483.
[10] Cheng D, Ma Y, Zhong Q, Xu W (2014) Ecology and Evolution **20** (4): 3968–3977.
[11] Peters R, Vovides AG, Luna S, Gruters U, Berger U (2014) Ecological Modelling **283:** 53–61.
[12] Schafer JL, Mack MC (2014) Am. Midl. Nat. **172:** 266–284.
[13] Vovides AG, Vogt J, Kollert A, Berger U, Grueters U, Peters R, Lara-Dominguez AL, Lopez-Portillo J (2014) Trees **28:** 1413–1425.
[14] Kabir M, Abolfathi M, Hajimoradloo A, Zahedi S, Kathiresan K, Goli S (2014) AACL Bioflux **7**(4): 286–300.
[15] Satyanarayana B, Idris IF, Mohamad KA, Husain M-L, Shazili NAM, Dahdouh-guebas F (2010) Botanica Marina **53:** 79–88.
[16] Ghasemi A, Fallah A, Joibary SS (2016) İstanbul Üniversitesi Orman Fakültesi Dergisi **66**(2): 691–697.
[17] Sitoe AA, Júnior L, Mandlate C, Guedes BS (2014) 1967–1981.

[18] Medeiros TCC, Sampaio EVSB (2008) Wetlands Ecology and Management **16**(4): 323–330.
[19] Suwa R, Deshar R, Hagihara A (2009) Aquatic Botany **91**(2): 99–104.
[20] Salmo SG, Lovelock C, Duke NC (2013) Hydrobiologia **720**(1): 1–18.
[21] Su Y, Lin Y, Jing S, Hou PL (2011) Marine Pollution Bulletin **62**(7): 1455–1463.
[22] McCarthy MC, Enquist BJ (2007) Functional Ecology **21:** 713–720.
[23] Guo H, Weiner J, Mazer SJ, Zhao Z, Du G, Li B (2012) Journal of Ecology **100**: 452–458.
[24] Anfodillo T, Petit G, Sterck F, Lechthaler S, Olson ME (2016) Frontiers in Plant Science **7:** 1681
[25] Mao W, Zhang T, Li Y, Zhao X, Huang Y (2012) Journal of Arid Land **4**(2): 161–170

Exclusion of *pituitary homeobox 2* gene polymorphism in vertical mandibular asymmetry patients: a preliminary study

Ervina Sofyanti[1*], Trelia Boel[1], Benny Soegiharto[2], Syafruddin Ilyas[3], Lidya Irani Nainggolan[1], Elza Ibrahim Auerkari[2*]

[1]Lecturer, Faculty of Dentistry, Universitas Sumatera Utara, Jl. dr. Mansur Kampus USU Medan 20155
[2]Lecturer, Faculty of Dentistry, University Indonesia, Kampus Baru UI Depok Jawa Barat 16424
[3]Lecturer, Faculty of Mathematics and Natural Sciences, Universitas Sumatera Utara, Jl. dr. Mansur Kampus USU Medan 20155

*Corresponding author: Tel: +62-821-6131, Fax: +62-821-3421, e-mail: ervina.sofyanti@usu.ac.id/ Tel: +62-021-316-2821, Fax: +62-021-310-6705, e-mail: auerkarie@yahoo.com

Abstract. *Pituitary Homeobox 2 (PITX2)*, is an active gene as a paired-related homeobox gene that encodes multiple isoforms. Its Nodal pathway in determination of left-right patterning during embryogenesis has been reported in satellite cells and expressed in adult human skeletal muscle. *PITX2A* and *PITX2B* are produced by alternative splicing and used of different promoters. *PITX2C* uses an alternative promoter located upstream of exon 4. *PITX2D* is produced by *PITX2C* alternative promoter and differential splicing. The 5'-primers and 3'-antisense primer were unique for each isoforms. Variability measurement in vertical dimension showed stronger genetic component than sagittal. This study aims to obtain the genotype marker of vertical mandibular asymmetry related to *PITX2A and PITX2D* isoform by visualization of the amplified product on stained gel to allele specific oligonucleotide between the case and control with Restriction Fragment Length Polymorphism (RFLP). Determination of vertical mandibular asymmetry based on condylar height asymmetry index of pre-treatment panoramic radiograph using Kjellberg's technique whilst vertical mandibular growth pattern using lateral cephalogram. The differences of condylar height asymmetry in case-control based on vertical growth pattern was compared using *Pearson's chi-squared* test. DNA extraction of 129 out-coming orthodontic patients in Universitas Sumatera Utara Dental Hospital were obtained from Buccal swab. Then DNA samples were amplified by Polymerase chain reaction (PCR) and digested with *NciI* restriction enzyme prior to electrophoresis visualization. There was no significant statistical difference in vertical mandibular asymmetry compared to vertical mandibular growth pattern. The RFLP analysis did not show any polymorphism for *PITX2A* and *PITX2D* isoform. All of the samples showed wild type homozygote. Further analysis method, except RFLP, were required to understand the genetic factor in the variance of vertical mandibular asymmetry.

Keywords: *PITX2*, polymorphisms, vertical mandibular asymmetry

IOP Conf. Series: Earth and Environmental Science **130** (2018) 012016 doi:10.1088/1755-1315/130/1/012016

1. Introduction

Factors potentially contributing to timing and amount of mandibular vary considerably among individuals. One of the common craniofacial deformities is an asymmetric growth of the mandible or other diseases affecting the facial growth. It has a direct impact to aesthetic concern and functional problems of the stomatognathic system [1,2]. Factors potentially contributing to this individual masticatory action variation is related to the consistency of the diet and genetic predisposition that seems to be direct consequences of malfunction at the cellular level of condylar cartilage [3]. Some regulatory gene expressions have been reported in mandibular condyle growth and required further studies.

For the majority of malocclusion with mandibular asymmetry, we need to identify vertical, transversal, or sagittal discrepancy simultaneously. The study of malocclusion heritability about genetic contribution to asymmetry is quite rare nowadays. The different expression of masticatory muscle fibers that attached directly to mandibular condyle demonstrated stronger genetic component of variability measurement in the vertical dimension rather than sagittal [4-6]. Smooth and skeletal muscles in the temporomandibular joint attach directly to the mandibular condyle. The asymmetrical function and activity of jaw movement differ in both sides can contribute to the difference in condylar development. Any dental treatment during growth period might influence the downward and forward mandible growth potential. The displacement of the mandible towards the affected side such as asymmetric condylar growth could happen [2,3,7].

The emergence of Single Nucleotide Polymorphisms (SNPs) in genetic analysis can influence splicing in the intronic region and gene expression in the promoter. It can help to answer the association of DNA variants with either the presence or absence of different phenotypes among individuals from the different population, such as: multivariance of the vertical mandibular pattern [8,9]. *PITX2* gene is a bicoid-like homeobox transcription factor located in the 4q25 locus that is required for the early determination, development, and maintenance of extraocular muscles (EOM) compared to the limb skeletal muscle. This gene was also required for maintaining characteristic properties of the adult EOM phenotype [10]. Formerly, an inversion of this gene involved in phenotype of a child with translocation (t4;14)(q25;q13) who had mild craniofacial dysmorphism and agenesis of the corpus callosum without the limb or ocular abnormalities though not many studies have been done on *PITX2* gene and craniofacial expression [11]. Human *PITX2* gene mutations was genetic factor related to haploinsufficiency condition that affected the eyes, teeth and umbilicus with associated abnormalities of the craniofacial region (midfacial hypoplasia, prognathism) [12,13]. In syndromic patients related to Peters' anomaly, Axenfeld-Rieger Syndrome patients, CHARGE syndrome, iridogoniodysgenesis, and iris hypoplasia [11,13,14,15].

The *PITX2* locus produces five isoforms, designated *PITX2A-2E* that arises through differential splicing and alternative promoter usage. An alternative promoter in exon 3 and 4 of *PITX2C* gene resulted in isoform of *PITX2D* and *PITX2E* gene. Both alternative splicing promoter usage which overlapped and distinct expression patterns produced different isoforms [16]. Identification of gene expression in Nodal Signaling Pathway of *PITX2* expression gene was also reported to help promote development of asymmetry due to a central role in left-right asymmetry [6]. Suggestive associations ($p<0.05$) were also identified with *PITX2* as candidate gene by genotyping 222 single-nucleotide polymorphisms in 82 genes/loci asymmetric variations of 3-dimensional dentoalveolar phenotypes in subjects with malocclusion [17].

Various radiological modalities have been evaluated for imaging of the mandibular asymmetry from 2D to 3D measurements. The panoramic radiograph is relatively accessible and provides the bilateral view of the mandible compared other radiograph. Then, the panoramic radiograph can be used to evaluate condylar mandibular asymmetry with 6% difference by Habets' formula and 93% difference by Kjellberg's technique [2,7,18,19]. Several authors have used this method to assess mandibular asymmetry in patients with temporomandibular disorders (TMD), unilateral and bilateral posterior crossbites, various sagittal malocclusions, and cleft lip and palate [7,20,21]. The incidence of 13% in 605 primary school students in Genoa showed malocclusions, wrong posture, and ocular

IOP Conf. Series: Earth and Environmental Science **130** (2018) 012016 doi:10.1088/1755-1315/130/1/012016

divergences disorders and required multidisciplinary medical approach [22]. Therefore the similarity in asymmetry gene expression of any organ in the vertebrae might be considered as a genomic risk factor.

One type of polymorphisms method is restriction fragment length polymorphism (RFLP) that showed the presence of different fragment lengths after DNA samples digestion with specific restriction endonucleases. This method can detect the difference in homologous DNA sequences and can be used as molecular marker specific to a single clone/restriction enzyme combination. The RFLP probes are frequently used in genome mapping and analysis, such as: genotyping, forensics, paternity tests, and genetic disease diagnosis [9].

Recent studies reported that the *PITX2* gene expression in satellite cells of masseter muscle was related to asymmetry. Here, we test the hypothesis that *PITX2* expression is responsible for the functional difference in the condylar asymmetry that is related to the *PITX2* target gene in organogenesis, especially ramus height. The aim of this research is to identify the Single Nucleotide polymorphism (SNPs) of *PITX2* gene especially related to isoform *PITX2*A and *PITX2*D in condylar asymmetry by using Restriction Fragment Length Polymorphism (RFLP) method.

2. Method

This study was approved by Research Ethics Committee of Universitas Sumatera Utara Medical Faculty and conducted in Universitas Sumatera Utara Dental Hospital. This was a case-control study by analyzing pre-treatment panoramic and lateral cephalometric radiographs. Then DNA isolation was taken by buccal swab while patients came for orthodontics treatment

2.1. *Experimental inclusion*

The pre-treatment panoramic and lateral cephalometric radiographs of orthodontics patients between 11 to 25 year old were analyzed with Kjellberg's technique. There were no previous orthodontics, prosthodontics treatment, or occlusal adjustment history. There were also no traumatic facial injury or congenital disease. Then, 79 subject with asymmetry and 50 subjects without asymmetry regardless of temporomandibular joint status were recalled for genotyping analysis.

2.2. *Problem formulation*

Selection of samples for vertical mandibular asymmetry used Kjellberg's technique. The condylar height asymmetry was also estimated using 93% difference (Figure 1) [19]. The symmetry between the right and left condylar height was estimated using the following formula:

$$\text{Kjellberg Symmetry Index (SI)} = \frac{\left(\frac{CH}{RH_A}\right)}{\left(\frac{CH}{RH_B}\right)} \text{ X } 100$$

In figure 2, type of vertical mandibular growth pattern based on the SN-MP angle [23]. The validity and reliability of inter-rater and intra-rater digitized panoramic radiograph and cephalometry measurements using t-test. The differences of condylar height asymmetry in case-control based on vertical growth pattern was compared using *Pearson's chi-squared* test (SPSS software, version 22.0 for windows; SPSS, Chicago IL).

IOP Conf. Series: Earth and Environmental Science **130** (2018) 012016 doi:10.1088/1755-1315/130/1/012016

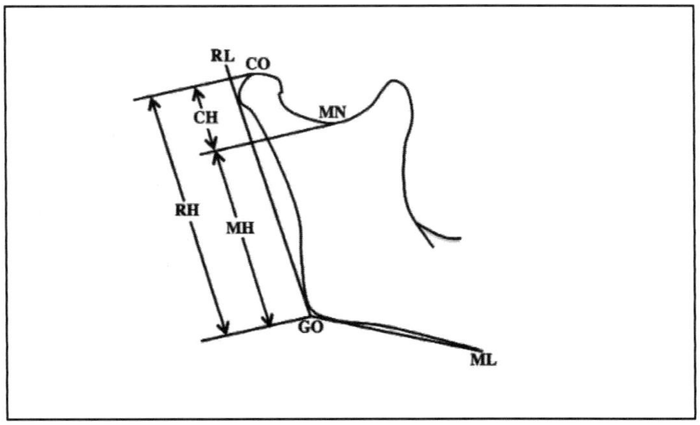

Figure 1. Ramus Height of Kjellberg's technique

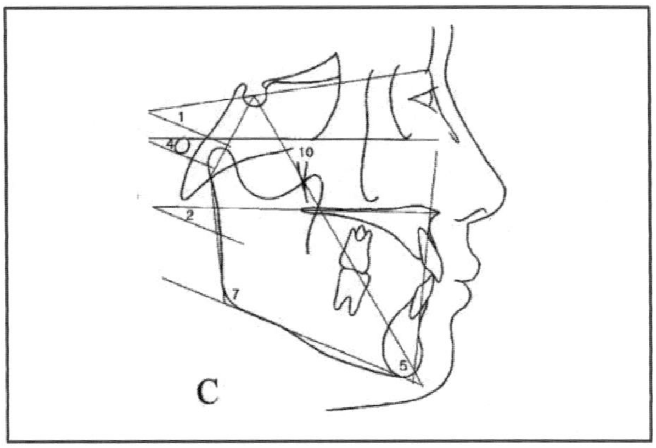

Figure 2. Vertical mandibular growth pattern

Patients' DNA was obtained using buccal cell swab procedure for genomic analysis using Presto TM (Geneaid, Taiwan). Primer pairs used for *PITX2*A were AAATCTCTGCTGACGTCACGT and CCAGACTCGCATTATCTCAC and PIT2XD, CAGCTCTTCCACGGCTTCT and TTCTCTCCTGGTCTACTTGG [13]. The annealing temperature for the PCR reaction was at 55°C. The target PCR results for *PITX2A* was 370 bp and was 374 bp for *PITX2D* [13]. Amplified products were then separated on agarose gel, reamplified, separated further on 2% agarose gels, purified with KAPA Taq™ EXtra HotStart ReadyMix with dye. The amplicons were subjected to restriction digestion using *NCiI* and then analyzed on a 3% agarose gel [15].

3. Results and Discussion

Table 1. Ramus height asymmetry based on different vertical mandibular growth pattern.

		Ramus Height		Total	P
		Symmetry	Asymmetry		
Vertical Mandibular Growth Pattern	Low Angle (<27°)	10 34.5%	19 65.5%	29	0.185
	Normal (27°-37°)	33 45.2%	40 54.8%	73	
	High Angle (37°)	7 25.9%	20 74.1%	27	
Total		50 38,8%	79 61,2%	129 100%	

* *Pearson's chi-square* test, $p<0.05$ significant.

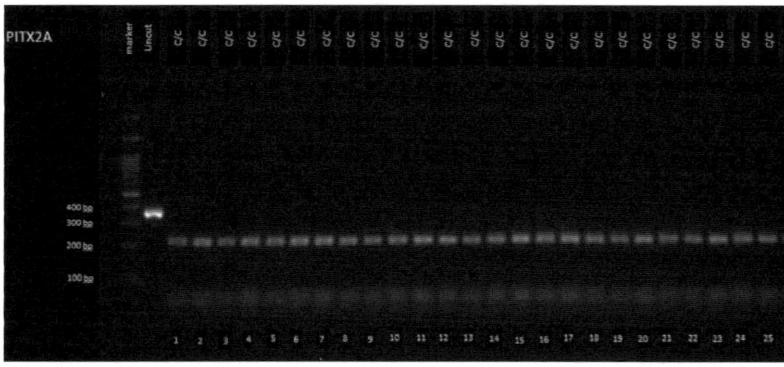

Figure 3. *PITX2A* gene of vertical mandibular patients on 3% agarose gel. 100 bp marker: wild type homozygote patterns of *PITX2A* gene

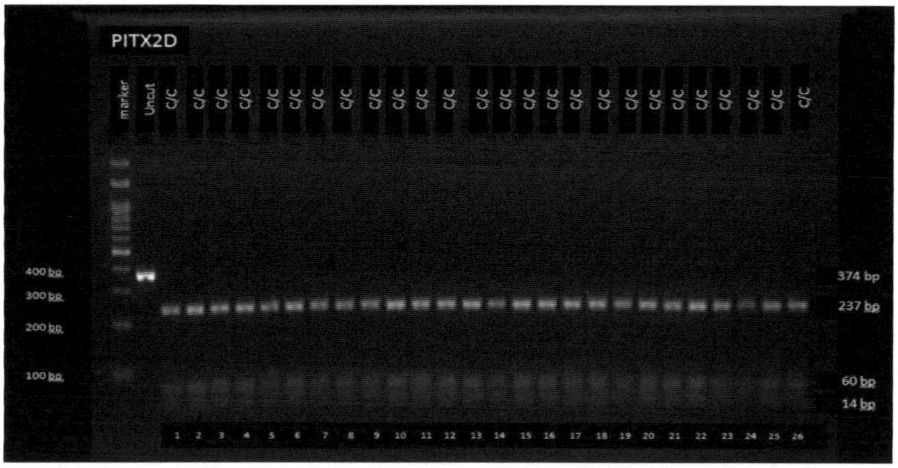

Figure 4. *PITX2D* gene of vertical mandibular patients on 3% agarose gel. 100 bp marker: wild type homozygote patterns of *PITX2D* gene

The mean age of 129 samples was 20.7±3.2 years old. There were no significant differences in condylar height asymmetry based on different vertical mandibular growth pattern (Table 1). We demonstrated the absence of *PITX2A* (Figure 3) *and PITX2D* (Figure 4) *gene* mutations in a series of 129 orthodontic patients with or without vertical mandibular asymmetry. The RFLP analysis did not show any polymorphism for *PITX2*. All of the samples showed wild type homozygote.

Since genetic factors were found to be responsible for only 40% of the skeletal and dental variations resulting in malocclusion, the genetic component was reported higher for the skeletal pattern than dental features. Individual differences in amount and direction of ramus growth and modelling can be explained by mandibular rotation and displacements. Multivariate assessments revealed that superior condylar growth and ramus modeling were mostly close to genetic factor.

The condylar height asymmetry using panoramic radiograph has been widely used to validate clinical tests of diagnostic categories in mandibular asymmetry. In our study there was no statistically significant differences for vertical mandibular asymmetry based on various mandibular growth pattern. The findings were different with previous CBCT imaging findings that reported a significant difference in hypodivergent patients, but not in normal and hyperdivergent patients [20]. The complexity of genetics in dentofacial variation explains why most treatments approaching malocclusion with mandibular asymmetry directed to the symptoms rather than its etiology. Despite the polygenic traits of mandible asymmetry, the study of genomic is fundamental for any clinicians in understanding the biology of underlying craniofacial growth and dental phenotype related to asymmetry [4,5,24].

The mechanism of bite force from the mandible to cranium suggest the magnitude of continuous joint loading is related to growth rotation and morphology of mandible condyle. The etiology of mandibular asymmetry is vast and often a combination of genetic and environmental factors. In this study, we analyze *PITX2A* and *PITX2D* isoform due to a difference in promoter and similar target base pair. The myogenic factor is one of the common causes in mandibular asymmetry related to Nodal signaling pathway (NSP) genes in masseter muscles. The *PITX2* is a key regulator in the development process and is expressed in the brain, heart, pituitary, mandibular and maxillary region, eye, and umbilicus. Alternative splicing and the use of different promoters produced these *PITX2* isoforms. *PITX2A* and *PITX2B* are generated by alternative splicing mechanisms. The *PITX2*C used an alternative promoter located upstream of exon 4. The *PITX2D* isoform acts to down-regulate the transcriptional activities of *PITX2A* and *PITX2C*. The identification of a new *PITX2* isoform (*PITX2D*) generated by the *PITX2C* alternative promoter and differential splicing found in human [16,25]. This data suggests that the differences in *PITX2* gene expression has emerged as a key element involved in the fine-tuning mechanism that regulates skeletal-muscle development as well as the differentiation and cell fate of satellite cells in adult muscle [6,11,25].

All isoforms can form homodimers, except for *PITX2B*, which can form heterodimers. The interactions between *PITX2* isoforms provide a mechanism to regulate gene expression. This *PITX2* isoform provided and controlled several mechanisms that interact in gene activation and repression [10]. The *PITX2* isoforms were expressed and differentiate to activate genes involved in the development of the brain, craniofacial region, and pituitary. These isoforms were expressed in humans and identified from a human craniofacial library [11]. These isoforms transcriptional activity is cell-dependent with the mechanism for the regulation of *PITX2* transcriptional activation through the action of a novel *PITX2* isoform, like in Axenfel-Rieger syndrome patient [14].

The screened probes for RFLPs using genomic DNA of different genotypes digested with restriction endonucleases. The previous study demonstrated the *NciI* enzyme used as RFLP marker could cut both alleles in homozygous wild type sample, thus revealing a unique blotting pattern characteristic to a specific genotype at a specific locus. Separation of the amplified fragment by PCR depends on enzyme restriction site. Our study focused on *PITX2A* (370bp) and *PITX2D* (374bp) by considering the sizes in base pairs of both isoforms almost equal. *PITX2D* is also produced by *PITX2C* alternative promoter and differential splicing. Modern orthodontists need to be aware of the basis of the genetic sciences due to rapid advances in genetic techniques and application in orthodontic practices. Genome-wide association studies, metabolic pathway analysis and candidate gene studies are necessary to further the evidence base for the practice of orthodontics to determine the diagnosis for each patient in the era of truly personalized orthodontics [5].

4. Conclusions

In the vertical mandibular asymmetry, we found the exclusion of *PITX2* polymorphism in the region of *PITX2A* and *PITX2D*. Further analysis, such as sequencing or SNPshot analysis will be required for polymorphism/mutations analysis to understand the genetic factor phenotype variance in order to obtain proper diagnostic in vertical mandibular asymmetry treatment.

Acknowledgments

The authors would like to thank Donny Nauphar and Tiar M. Pratamawati for their thoughtful comments in the manuscript. This work was partially supported by TALENTA grants from the Universitas Sumatera Utara (5338/UN5.1.R/PPM/ 2017). The authors declare no conflicts of interest in the authorship and/or publication of this article.

References

[1] Lin H, Zhu P, Lin Y, Wan S, Shu X, Xue Y, Zheng Y (2013) Mandibular asymmetry: a three dimensional quantification of bilateral condyles. Head and Face medicine. 9: 1-7. (doi.org/10.1186/1746-160X-9-42)

[2] Haraguchi S, Iguchi Y,Takada K (2008) Asymmetry of the face in orthodontic patients. Angle Orthod 78(3): 663-669. (doi: 10.2319/022107-85.1)

[3] Pirttiniemi P, Peltomaki T, Muller L, Luder HU (2009). Abnormal mandibular growth and the condylar cartilage. Eur J Orthod 31: 1–11. (doi: 10.1093/ejo/cjn117)

[4] Cakan DE, Ulkur F, Taner T (2012) The genetic basis of facial skeletal characteristics and its relation with orthodontics. Eur J Dent 6: 340-345.

[5] Hartsfield J, Morford LA, Otero LM (2012) Genetic Factors Affecting Facial Growth, Orthodontics - Basic Aspects and Clinical Considerations: InTech http://www.intochopen.com/books/orthodontics-basic-aspects-and-clinical considerations / genetic-factors-affecting-facial-growth, 125-150.

[6] Nicot R, Hottenstein M, Raoul G, Ferri J, Horton M, Tobias JW, Barton E, Gele P, Sciote JJ (2014) Nodal pathway genes are down regulated in facial asymmetry. J Craniofac Surg 25(6): 548-555. (doi: 10.1097/SCS.0000000000001076)

[7] Al-Taki AA, Ahmed MH, Ghani HA, Kaddah FA (2015) Impact of Different malocclusion types on the vertical mandibular asymmetry in young adult sample. Eur J Dent 9: 373-377. (doi: 10.4103/1305-7456.163233)

[8] Fareed M, Afzal M (2013) Single nucleotide polymophism in genome-wide association of human population: A tool for broad spectrum service. The Egyptian Journal of Medical Human Genetics 14: 123-134. (dx.doi.org/10.1016/j.ejmhg.2012.08.001)

[9] Schork NJ, Fallin D, Lanchbury JS (2000) Single nucleotide polymorphisms and the future of genetic epidemiology. Clin Genet 58(4): 250–264.

[10] Hebert SL, Daniel ML, McLoon LK (2013) The role of *PITX2* in maintaining the phenotype of myogenic precursor cells in the extraocular muscles. PLOS ONE 8(3): 1-13. (doi.org/10.1371/journal.pone.0058405)

[11] OMIM. 2017. PAIRED-LIKE HOMEODOMAIN TRANSCRIPTION FACTOR 2;*PITX2* (journal elektronik) accessed on 10th of March 2017; www.omim.org/entry/601542?search=*PITX2*%20gene&highlight=*PITX2*%20gene.

[12] Liu W, Selevar J, Lu MF, Martin JF (2003) Genetic dissection *PITX2* in craniofacial development uncovers new function in branchial arch morphogenesis, late aspects of tooth morphogenesis and cell migration. Development 130(25): 6375-6385. (10.1242/dev.00849)

[13] Martin DM, Probst FJ, Fox SE, Schimmenti LA, Semina EV, Hefner MA, Belmont JW, Camper SA (2002) Exclusion of *PITX2* mutations as a major cause of CHARGE association. Am J of Medical Genetics 111: 27-30. (10.1002/ajmg.10473)

[14] Tumer Z, Holm DB (2009) Axenfeld-Rieger syndrome and spectrum of *PITX2* and FOXC1 mutations. Eur J Hum Genet 17: 1527-1539. (10.1038/ejhg.2009.93)

[15] Weisschuh N, Dressler P, Schuettauf F, Wolf C, Wissinger B, Gramer E (2006) Novel Mutations of FOXC1 and *PITX2* in Patients with Axenfeld-Rieger Malformations.

[16] Cox CJ, Espinoza HM, McWilliams B, Chappell K, Morton L, Hjalt TA, Semina EV, Amendt BA (2002) Differential regulation of gene expression by *PITX2* isoforms. J Biol Chem: 25001-25010. (10.1074/jbc.M201737200)

[17] Weaver CA, Miller SF, Fontoura CSG, Wehby, Amendt BA, Holton NE, Allareddy V, Southard TE, Uribe LMM (2017). Candidate gene analyses of 3-dimensional dentoalveolar phenotypes in subjects with malocclusion. Am J Orthod Dentofacial Orthop 151(3): 539-558. (10.1016/j.ajodo.2016.08.027)

[18] Elslände DCV, Russett SJ, Major PW, Flores-Mir C (2007) Mandibular asymmetry diagnosis with panoramic imaging. Am J Orthod Dentofacial Orthop 134:183-192. (10.1016/j.ajodo.2007.07.021)

[19] Fuentes R, Engelke W, Bustos L, Oporto G, Borie E, Sandoval P, Garay I, Bizama M, Borquez P (2011) Reliability of Two Techniques for Measuring Condylar Asymmetry with X-Rays. Int. J. Morphol 29(3): 694-701.

[20] Celik S, Celikoglu M, Buyuk SK, Sekerci AE (2016) Mandibular vertical asymmetry in adult orthodontic patients with different vertical growth patterns : A cone beam computed tomography study. Angle orthod 86, 271-277. (doi.org/10.2319/030515-135.1)

[21] Uysal T, Sisman Y, Kurt G, Ramoglu SI (2009) Condylar and ramal vertical asymmetry ini unilateral and bilateral posterior crossbites patients and a normal occlusion sample. Am J Orthod Dentofacial Orthop 136: 37-43. (10.1016/j.ajodo.2007.06.019)

[22] Silvestrini-Biavati A, Migliorati M, Demarziani E, Tecco S, Silvestrin-Biavati P, Polimeni A, Saccucci M (2013) Clinical association between teeth malocclussions, wrong posture and ocular convergences disorders: an epidemiological investigations on primary school children. BMC Pediatrics, 13:12. http://www.biomedcentral.com/1471-2431/13/12.

[23] Kim JY, Lee SL, Kim TW, Nahm DS, Chang YI (2005) Classification of the Skeletal Variation in Normal Occlussion. Angle Orthod 75:301-11. (10.1043/0003-3219(2005)75[311:COTSVI]2.0.CO;2)

[24] Uribe IMM, Miller SF (2014) Genetics of the dentofacial variation in human malocclusion. Orthod Craniofac Res 18(1): 91-99. (10.1111/ocr.12083)

[25] Hernandez-Torrez F, Outerino LR, Franco D, Aranega AE (2017) *PITX2* in embrionic and adult myogenesis. Frontiers in Cell and Developmental Biology. www.frontiersin.org.

Plagiochila (*marchantiophyta*) of mount sibayak north sumatra

Etti Sartina Siregar[1*], Nunik S. Ariyanti[2], Sri S.Tjitrosoedirdjo[2,3]

[1]Department of Biology, Faculty of Mathematics and Natural Sciences University of Sumatera Utara, Jl. dr. Mansur Kampus USU Medan 20155
[2]Department of Biology, Faculty of Mathematics and Natural Sciences, Bogor Agricultural University, IPB-Campus Darmaga, Bogor
[3]SEAMEO BIOTROP, Jl. Raya Tajur km 6, Bogor

*Corresponding author: ettisartina@yahoo.com; (+62) 85313560038

Abstract Information on the liverworts (*Marchantiophyta*) flora of Sumatra is still less reported, including the knowledge on the genus of *Plagiochila* (*Plagiochilaceae*). Research is conducted to explore the diversity of *Plagiochila* in Mount Sibayak North Sumatra. Exploration was carried out in along the track of study site. Eighteen species of *Plagiochila* were found, of which two species were recognized as new records to Sumatra (*Plagiochila gracilis* and *Plagiochila laxissima*). Species description and the key to species of *Plagiochila* in Mount Sibayak North Sumatra are provided.

Keywords: Liverworts, *Marchantiophyta*, Mount Sibayak, North Sumatra, *Plagiochila*

1. Introduction

Plagiochila is the largest genus in *Marchantiophyta*, consisting of more than 1600 species ever assigned to the genus [1]. Asia and the Neotropics are centers of diversity for *Plagiochila*, estimated 480 species recognized from Asia [2]. The specific characters of this genus are: 1) plant medium to large sized, 2) dorsal leaf base decurrent, 3) leaf apex and margins toothed, 4) perianths compressed laterally or inflated at least on the basal half, 5) androecia with long spikes in cluster or solitary, terminal or intercalary [1,3].

The floristic work on *Plagiochila* in Indonesia is still very scarce. A monograph of *Plagiochila* from Southeast Asia was published by Inoue (1984), who listed 48 species from Southeast Asia, includes 31 species from Sumatra. After Inoue, no further study on *Plagiochila* in Sumatra. More comprehensive study on *Plagiochila* in Sumatra will find a greater number of species. Mount Sibayak is one of the suitable area for the liverworts including *Plagiochila*. However, knowledge on *Plagiochila* from this location is poorly understood. Exploration of the diversity of *Plagiochila* in Mount Sibayak North Sumatra is badly needed for comparing with the one by Inoue [1]. This paper is one of the series dealing with the *Marchantiophyta* in Mount Sibayak North Sumatra.

IOP Conf. Series: Earth and Environmental Science **130** (2018) 012017 doi:10.1088/1755-1315/130/1/012017

2. Methods

The study area is located in Mount Sibayak, North Sumatra, Indonesia, approximately 60 km from Medan city. The locality has an altitude of about 700–2050 m, annual rainfall of 2400–2800 mm/year, and relative humidity of at least 80–90 %. Exploration was carried out along the tracks of Mount Sibayak from lowland to the peak. The specimens collected were classified and identified based on morphological characters, using Inoue [1], Gradstein [3] and other publications of *Plagiochila.* The voucher collected specimens are deposited at the Herbarium Medanense (MEDA), Biology Department, University of Sumatra Utara. An identification key to the species of *Plagiochila* known from Mount Sibayak North Sumatra is provided.

3. Results and Discussion

There are eighteen species of *Plagiochila* found in Mount Sibayak North Sumatra; sixteen species are previously known in Sumatra, two of the species are new records, *Plagiochila gracilis* and *P. laxissima.*

3.1. Key to species of *Plagiochila* in Mount Sibayak North Sumatra:

1 Lateral leaves with a distinct tubular sac at the ventral base... 2
1 Lateral leaves without a distinct tubular sac at the ventral base...................................... 5
2 Free margin of sac entire..3
2 Free margin of sac toothed.. 4
3 Teeth on leaf-margin small, triangular, lacking on dorsal margin, free margin of sac strongly incurved.. 5. *P. clavato-saccata*
3 Teeth on leaf-margin strongly ciliate, present on dorsal margin, free margin of sac slightly incurved...4. *P. blepharophora*
4 Underleaves lacking.. 14. *P. sandei*
4 Underleaves present, ciliate ....3. *P. bantamensis*
5 Plants dendroid, flagelliform branches present..6. *P. dendroides*
5 Plant not dendroid, flagelliform branches absent...6
6 Stem with paraphyllia 7
6 Stem without paraphyllia...8
7 Paraphyllia dense on the ventral and dorsal side of the stem, leaves loosely imbricate, leaf-lobes broadly ovate...1. *P. abietina*
7 Paraphyllia dense only on the ventral side of the stem, leaves closely imbricate, leaf-lobes ovate-oblong ....12. *P. obtusa*
8 Branches frequent...9
8 Branches infrequent..10
9 Flabellately branched, ventral leaf base dilated, underleaves absent...2. *P. arbuscula*
9 Pseudodichotomously branched, ventral leaf base not dilated, underleaves vestigial...8. *P. javanica*
10 Leaves linearly oblong, usually 2 – 3 times as long as wide, teeth almost restricted to around apex...11
10 Leaves ovate or oblong-ovate, less than two times as long as wide, teeth on whole of leaf-margin (sometimes absent on dorsal margin)..12
11 The cuticle of leaf-cells is verrucose .... 16. *P. singularis*
11 The cuticle of leaf-cells is smooth .... 11. *P. laxissima*
12 Branches dominantly dichotomous..13
12 Branches dominantly lateral intercalary or rarely branched..15
13 Leaves always with two distinct, larger teeth at the apex.................... 9. *P. junghuhniana*
13 Leaves without two larger teeth at the apex..14

14 Teeth of leaves margin 7–12 in total, the apex of leaves often asymmetric bilobed.......................................13. *P. salacensis*
14 Teeth of leaves margin 18–35 in total, the apex of leaves symmetric, entire.......................................18. *P. teysmannii*
15 Teeth on leaf-margin are long ciliate .... 15. *P. sciophila*
15 Teeth on leaf-margin are spinose or short triangular....................................... 16
16 Leaves widest at or around the middle part, cells margin of leaves never elongated .... *P. gracilis*
16 Leaves widest at the 1/3 basal part, cells margin of leaves elongated.......................................17
17 Ventral leaves with a distinct keel on the shoot, leaf base dilated....................................... 17. *P. sumatrana*
17 Ventral leaves without a distinct keel on the shoot, leaf base not dilated ....10. *P. korthalsiana*

3.2. Species description

1. *Plagiochila abietina* (Nees) Lindenb.

Plant 25–50 mm long, about 1 mm wide, brownish in the the dry specimen, with a habit of *Mastigophora* species. Paraphyllia dense both on the dorsal and ventral side of stem and branch, 1–2 cells wide at the base and 2–10 cells long; branches very frequent, pinnate, terminal and *Frullania*-type, rarely lateral intercalary on lower; leaves loosely imbricate, lobes broadly ovate, 0.4–0.5 mm long and wide; dorsal margin strongly recurved, long decurrent along dorsal stem midline, ventral margin hardly or only shortly decurrent at the base; apex rounded; teeth present on the whole margin, stronger towards the ventral base, 5–10 x 2–4 cells. Leaf cells at the middle portion 20–32.5 µm x 15–20 µm, at leaf-base 20–35 µm x 12.5–15 µm, at the marginal 20–30 µm x 12.5–15 µm; trigones medium size, cuticle smooth. Underleaves are indistinguishable from paraphyllia.

Dioicous. Gynoecia on the terminal of branches, with one innovation, bracts similar to branch leaves in shape, 0.45 mm long, 0.35 mm wide, margin densely ciliate; perianth campanulate, 0.55 mm long, 0.7 mm wide, mouth slightly arched, the margin of mouth ciliate. Androecia not seen.

Ecology: found on the tree trunks and tree root in lowland forest.

Distribution: Australia, Borneo, Celebes, Fiji, Java, New Caledonia, Papua New Guinea, Sumatra, Solomon Islands, [1,4,5].

Specimens examined: Mount Sibayak, Etti Siregar 154, 304a, 1568

Note: *Plagiochila abietina* is easily recognized from 1) the very frequently and pinnately branched, 2) the densely spine-like paraphyllia on the stem, 3) the habit of *Mastigophora* species.

2. *Plagiochila arbuscula* (Brid. ex Lehm. & Lindenb.) Lindenb.

Plant robust, 30–70 mm long, about 5–6 mm wide, yellowish brown in the the dry specimen, with rhizomatous caulids; branches frequent, almost of the terminal and *Frullania*-type, sometimes with lateral intercalary on the lower portion of the shoot, forming a weakly flabellate habit. Leaves approximately imbricate, lobes ovate-oblong, 2.5–3 mm long, 1.75–2.0 mm wide; dorsal margin slightly recurved and moderately decurrent along dorsal stem midline; ventral margin shortly decurrent, base dilated; apex rounded; teeth present on whole margin, or sometimes absent on basal half of dorsal margin, 15–18 in total number, triangular, 3–7–(8–10) x 2–4–(6–7) cells. Leaf cells at the middle portion 20–45 µm x 12.5–17.5 µm, at leaf-base 25–50 µm x 12.5–17.5 µm, at the marginal 17.5–25 µm x 10–15 µm; trigones medium sized, triangular or nodulose, cuticle smooth; underleaves absent.

Dioicous. Androecia intercalary or terminal on branches, bracts small, 6–10 pairs, closely imbricate. Gynoecia not seen.

Ecology: found on the tree trunks in lowland to lower montane forest.

Distribution: Australia, Borneo, China, Java, Japan, Malaya, New Caledonia, Papua New Guinea, Philippines, Samoa, Solomon Islands, Sumatra (Mt. Singgalang, Mt. Sibayak), Taiwan, Thailand [1, 6, 7].

Specimens examined: Mount Sibayak, Etti Siregar 279, 560, 747a, 866.

Note: diagnostic characters of *P. arbuscula* are 1) plant robust with frequently branches, forming a weakly flabellate habit, 2) the broadly ovate-oblong leaves with a distinctly dilated ventral base, 3) the teeth present on the whole leaf margin. *Plagiochila arbuscula* is readily distinguished from other species of *Plagiochila* by the frequently terminal branches forming a weakly flabellate habit.

3. *Plagiochila bantamensis* (Reinw., Blume, and Nees) Mont.

Plant 20–40 mm long, 4–5 mm wide, greenish to pale brown in the dry specimen, with a rhizomatous caulids; branches very rare, if present almost of the lateral intercalary branches. Leaves closely imbricate; lobes oblong, 2.0–3.0 mm long, 1.25–1.5 mm wide; dorsal margin strongly incurved, moderately decurrent along dorsal stem midline; ventral margin not decurrent, the base with tubular sac; sac globose or broadly ovate, with ciliate margin, free margin not involute; apex of lobes rounded; teeth present on whole margin, 3–10 cells long, 2 cells wide at the base. Leaf cells at middle portion 17.5–32.5 µm x 17.5–25 µm, at leaf-base 25–37.5 µm x 17.5–25 µm, at the marginal 30–45 µm x 10–15 µm; trigones medium sized, triangular, cuticle smooth; underleaves large, margin strongly ciliate. Generative organ not seen.

Ecology: found on the tree trunks in lowland to lower montane forest.

Distribution: Borneo, Cambodia, Celebes, China, Indian mainland, Japan, Java, Malaya, Nicobar Islands, Melanesia, Philippines, Sumatra (Enggano, Padang, Jambi, Mt. Sibolga),Sri Lanka, Taiwan, Thailand [6,7,8,9].

Specimens examined: Mount Sibayak, Etti Siregar 147, 310, 333a, 351, 527, 1495, 1636.

Note: diagnostic characters of *P. bantamensis* are 1) the oblong leaves with ciliate margin, 2) the globose or broadly ovate sac at the ventral base of the leaf with ciliate free margin, 3) the rather large underleaves with ciliate margin.

Plagiochila bantamensis seems to be rather common in Mount Sibayak. This species can be readily distinguished from the rest of the species by the presence of ventral globular sac on mature leaves. *Plagiochila bantamensis* is variable in size of plants, leaves, underleaves, and sac on ventral leaf base. The sac on ventral leaf-bases sometimes are small or not well developed or only strongly incurved.

4. *Plagiochila blepharophora* (Nees) Lindenb.

Plant 25–40 mm long, 4–5 mm wide, pale greenish in the dry specimen, with a few rhizomatous caulids. Plants usually simple, or sometimes with lateral intercalary branches, shoot apex sometimes ceasing to grow further and produce 1–2 vegetative innovation, forming a dichotomous habit. Leaves moderately or loosely imbricate, lobes ovate–oblong, 2.5–3.25 mm long, 1.1–1.6 mm wide; dorsal margin revolute and decurrent along dorsal stem midline; ventral margin nearly straight, the base forming an oblong or ovate-oblong sac, free margin entire; apex rounded; margin spinose or ciliate on whole ventral and dorsal side of lobes, or sometimes absent from dorsal and basal half of ventral margin. Leaf cells at the middle portion 30–45 µm x 22.5–30 µm, at the eaf-base 30–50 µm x 25–35 µm, at the marginal 30– 0 µm x 12.5–15 µm; trigones small to medium size, acute, cuticle smooth; underleaves small, vestigial, base 6–10 cells wide, 2–4 lobed, lobes composed of uniseriate 1–3 cells.

Dioicous. Gynoecia on the terminal of stem, with one innovation, bracts similar to leaves in shape and size, wider at the base, two pairs, 3.5 mm long, 2.1 mm wide, with weakly incurved ventral base, margin ciliate, more dense and stronger than leaf-margin; perianth cylindrical, 1.87 mm long, 1.37 mm wide, mouth truncate, margin of mouth ciliate. Androecia on terminal or intercalary of the main stem, in long or short spicate, bracts (9) – 12 – 27 pairs, moderately to closely imbricate, distal half obliquely or nearly vertically spreading.

Ecology: found on the tree trunks in lowland to lower montane forest.

Distribution: Japan, Thailand, Philippines, Sumatra (Mt. Singgalang), Java, Borneo, Papua New Guinea [1, 5].

Specimens examined: Mount Sibayak, Etti Siregar 85, 272, 274, 307, 388, 491, 548, 718, 719, 724, 753, 1438, 1713, 1724, 1734.

Note: the diagnostic characters of *P. blepharophora* are 1) the ovate-oblong leaves, 2) the oblong or ovate-oblong sac at the ventral leaf-base, with entire free margin, 3) the vestigial underleaves.

Plagiochila blepharophora seems to be common in Mount Sibayak. This species is variable in the shape of tubular sac from ovate to ovate-oblong; wide variation in the size of toothed margin, spinose to ciliate. The sac at the ventral leaf-base sometimes is weakly developed or only incurved. The number of teeth on the leaf margin is also variable; toothed on the whole leaf margin or sometimes entire from the dorsal margin.

5. *Plagiochila clavato-saccata* Steph.

Plant robust, 30–60 mm long, 6–8 mm wide, yellowish in the dry specimen, arising from creeping rhizomatous caulids. Plants usually simple, branches very rare. Leaves densely imbricate, horizontally spreading, lobes oblong, 5–6 mm long, 2.2–2.5 mm wide; dorsal margin moderately revolute and recurved and long decurrent along the dorsal stem midline, ventral margin forming a distinct keel on the shoot, long decurrent along the ventral stem midline; the ventral base dilated, with tubular sac at the base; sac oblong, free margin strongly incurved and entire; apex obtuse or rounded nearly; margin of leaves toothed, but entire on dorsal margin. Leaf cells at middle portion 35–45 µm x 37–50 µm, at the marginal 10–12.5 µm x 65–75 µm; trigones small to medium sized, acute, cuticle smooth; underleaves present, bilobed to the middle, margin ciliate. Generative structures not seen.

Ecology: found on the tree trunk in lowland forest.

Distribution: Malaya, Sumatra [1, 10].

Specimen examined: Mount Sibayak, Etti Siregar 337.

Note: the diagnostic characters of *P. clavato-saccata* are 1) the large and robust plant, 2) the oblong leaves, 3) the small, triangular teeth on the leaf-margin (dorsal margin entire), 3) the oblong sac at the base of the ventral leaf-margin, with entire and strongly incurved free margin, 4) the small underleaves with ciliate margin.

Plagiochila clavato-saccata is closely to *P. sandei* but the later without underleaves, teeth on the whole of leaf-margin and the free margin of sac on the ventral leaf-base with irregular teeth. *Plagiochila clavato-saccata* was uncommon in Mount Sibayak, only one specimen was found.

6. *Plagiochila dendroides* (Nees) Lindenb.

Plant yellowish to pale brown in the dry specimen, 15–45 mm long, 1.7–2.2 mm wide, stem often with scale-like small leaves; branches very frequent of terminal and *Frullania*-type, forming a dendroid habit, apex of Frullania-type branches frequently elongated and becoming minute-leaved, flagelliform. Leaves remote, lobes ovate-oblong, 0.7–0.9 mm long, 0.4–0.5 mm wide; dorsal margin moderately revolute and shortly decurrent along dorsal stem midline; ventral margin shortly decurrent; apex of lobes with shallowly and asymmetrically bilobed, or with 2–(3) teeth; dorsal and ventral margin entire. Leaf cells at the middle portion 20–32 µm x 10–13 µm, at leaf-base 25–40 µm x 10–15 µm, at the marginal 17.5–25 µm x 10–12.5 µm; trigones indistinct, cuticle smooth. Underleaves vestigial, filiform.

Androecia terminal or intercalary of branches. Gynoecia terminal on branched or intercalary with 1 or 2 innovation; bracts ovate-oblong, 1.2–1.5 mm long, 0.5–0.7 mm wide; perianth 1.5 mm long, 0.5–0.8 mm wide, the margin of mouth rounded, with coarsely spinose.

Ecology: found on the tree trunks from lowland to lower montane forest.

Distribution: Borneo, Fiji, Java, Malaysia, Japan, New Caledonia, Papua New Guinea, Philippines, Sumatra (Mt. Singgalang, Mt. Talang), Taiwan, West Irian [1].

Specimens examined: Etti Siregar 151c**,** 215, 287, 296, 500, 1484, 1645, 1651, 1714, 1732.

Note: the diagnostic characters of *P. dendroides* are 1) the very frequent of branches, forming a dendroid habit, the flagelliform branches often present on the ventral of stem, 2) the small and remote leaves on the main stem, 3) the shallowly and asymmetrically bilobed leaves, or with 2–(3) teeth at the apex.

Plagiochila dendroides is very common in Mount Sibayak. This species is readily distinguished from other species of *Plagiochila* by the shallowly and asymmetrically bilobed leaves or with 2–(3) teeth at the apex.

7. *Plagiochila gracilis* Lindenb. et Gott.

Plant small 10–20 mm long, 2.5–3.5 mm wide, yellowish brown in the dry specimen; branches very rare, if present exclusively lateral intercalary. Leaves distant to contiguous, lobes oblong–ovate or obovate, widest at or around the middle part, 1.25–1.5 mm long, 0.6–0.8 mm wide; the dorsal margin slightly revolute, long decurrent along the dorsal margin; ventral margin not decurrent; the apex rounded, with 2–4 teeth (two theeth are often prominent when plants are young), the two prominent teeth show a tendency toward bilobed); margin entire on the dorsal side, ventral margin with 4–6 teeth on distal half ; teeth on leaf margin 6–10 in total number, triangular, 2–6 x 2–4 cells. Leaf cells at the middle portion 25–32.5 µm x 12.5–22.5 µm, basal cells 25–40 µm x 12.5–20 µm, marginal cells 22.5–30 µm x 12.5–17.5 µm; trigones medium sized, acute, cuticle smooth; underleaves very vestigial, filiform, 1–2 cells wide at the base, 2–3 cells long. Generative organ not seen.

Ecology: found on the tree trunks from lowland to lower montane forest.

Distribution: Buthan, China, India, Japan, Java, Nepal, Philippines, Sri Lanka, Sumatra (new record based on this study), Taiwan, Thailand, [6,8,10,12].

Specimens examined: Etti Siregar 782, 1739, 1949, 2006a.

Note: the diagnostic characters of *P. gracilis* are 1) the oblong–ovate or obovate leaves, widest at or around the middle part, 2) the teeth on the dorsal leaf margin absent, ventral margin with 4–6 teeth on distal half, 3) the rounded apex of leaf with 2–4 teeth (two are often prominent).

Plagiochila gracilis seems to be rare in Mount Sibayak, although we found this species in four collections, they were in a limited population. The species also seems to be quite rare in tropical Asia [1].

8. *Plagiochila javanica* (Swartz) Dum.

Plant green yellowish in the dry specimen, 30–80 mm long, 3–6 mm wide, with rhizomatous caulids; branches very frequent, predominantly of terminal and *Frullania*-type, often forming a pseudo dichotomous habit. Leaves contiguous, lobes ovate-oblong, 1.25–1.8 mm long, 1.0–1.2 mm wide; the dorsal margin revolute and long decurrent along dorsal stem midline; ventral margin straight, shortly decurrent at the base, slightly incurved along 1/3 basal; apex obtuse; teeth present on whole of leaf margin, except 1/3 basal of dorsal margin, marginal teeth small, triangular, 15–20 in number, 2–3 cells wide at the base, 2–5 cells long. Leaf cells at middle portion 17.5–32.5 µm x 12.5–20 µm, at leaf-base 20–37.5 µm x 20–25 µm, at the marginal 12.5–25 µm x 10–12.5 µm; trigones medium to large, nodulose, cuticle smooth; underleaves are small or vestigial. Generative organ not seen.

Ecology: found on the tree trunks from lowland to lower montane forest.

Distribution: Java, Philippines Sumatra (Mount Merapi and Mount Singgalang) [1].

Specimens examined: Etti Siregar 747b, 849

Note: the diagnostic characters of *P. javanica* are 1) the predominantly of a terminal and *Frullania*-type branches, often forming a pseudo dichotomous habit, 2) the ovate-oblong or oblong leaves, 3) the small, triangular teeth on whole of leaf margin with rather short terminal cells.

9. *Plagiochila junghuhniana* S. Lac.

Plant 25–50 mm long, about 2.5–3 mm wide, green to yellowish in the dry specimen; branches few to frequent, almost terminally and Frullania-type branched, often forming dichotomous habit. Leaves imbricate, lobes ovate-oblong to oblong, 1.3–1.6 mm long, 0.6–1 mm wide; dorsal margin long decurrent along stem midline; ventral margin shortly decurrent; apex rounded to truncate; two apical teeth are larger than others. The leaf cells at middle portion 17–28 µm x 14–21 µm, at leaf-base 24–44 µm x 12–19 µm, at the marginal 24–37 µm x 7–9.5 µm; trigones small, acute, cuticle smooth; underleaves are vestigial.

Dioicous. Androecia terminal or intercalary on branches, bracts 5–13 pairs. Gynoecia terminal or intercalary on branches with 1–2 innovation; bracts larger than lateral lobes and with strongly toothed, 27–30 teeth per bract, dorsal margin strongly revolute; perianth obovate, 1.5–2.4 mm long, 1.2–1.5 mm wide, mouth ciliate-dentate, 4–20 cells long.

Ecology: found on the tree trunks and rotten logs from lowland to lower montane forest.

Distribution: Java, Malaysia, Myanmar, Papua New Guinea, Philippines, Sumatra (Dolok Barus, Mount Singgalang), Thailand [1,6,8].

Specimens examined: Mount Sibayak, Etti Siregar 46b, 158, 159, 261, 275, 290, 302, 343, 385, 395, 438, 457, 508, 534, 632, 712, 740, 765, 779, 780b, 794, 815, 858, 884, 1418, 1429, 1560, 1696, 1723, 1738, 1744, 1771, 1785, 1829, 1881, 1934, 1958, 2028.

Note: the diagnostic characters of *P. junghuhniana* are 1) the frequently dichotomous, terminal and *Frullania*-type branches, 2) the ovate-oblong or oblong rectangular leaves, 3) the two prominently teeth on the leaf apex.

Plagiochila junghuhniana is variable in the size and shape of leaves from small to medium sized, from ovate-oblong to oblong rectangular or linear oblong. The ventral base of leaves variable from not dilated to dilated. *Plagiochila junghuhniana* seems to be the most common in Mount Sibayak.

10. *Plagiochila korthalsiana* Molkenb.

Plant 20–45 mm long, 2.3–3.0 mm wide, green yellowish in the dry specimen, with a few rhizomatous caulids; branches rare, the apex of main shoots with vegetative innovation. Leaf lobes ovate-oblong, widest at 1/3 of basal portion, 1.5–1.7 mm long, 0.7–0.8 mm wide; dorsal margin slightly recurved and decurrent along dorsal stem midline; ventral base shortly decurrent, not dilated; apex widely obtuse to rounded; margin with 9–12 teeth (dorsal entire or with 1–2 small teeth near apex, ventral margin toothed from base to apex or entire at the basal portion); teeth small, 1–2 cells wide at the base, and 1–3 cells long. The leaf cells at middle portion 21–38 µm x 20–24 µm, those of leaf-base 38–54 µm x 18–23 µm, at the marginal 18–29 µm x 13–18 µm; trigones small to medium sized, acute, cuticle smooth; without underleaves. Generative organ not seen.

Ecology: found on the tree trunk in lower montane forest.

Distribution: Java, Malaysia, Papua New Guinea, Philippines, Sumatra [1,2].

Specimen examined: Mount Sibayak, Etti Siregar 1768.

Note: the diagnostic characters of *P. korthalsiana* are 1) rarely with branches, the apex of main shoots sometimes with vegetative innovation, 2) leaves ovate-oblong, widest at 1/3 basal portion, 3) the small teeth were on the apical of a dorsal and ventral margin of leaves.

11. *Plagiochila laxissima* Schiffn.

Plant 15–20 mm long, about 3–3.5 mm wide, yellowish brown in the dry specimen, without rhizomatous caulids; branches very rare, if present almost of the lateral intercalary branched. Leaves distant, lobes linear-oblong, 1.5–1.8 mm long, 0.5–0.8 mm wide; dorsal margin weakly recurved and moderately decurrent along dorsal stem midline; ventral margin shortly decurrent; apex obtuse; teeth restricted to the apex of leaves, dorsal and ventral margin entire; toothed on distal portion, 6–9 in total number, triangular, 1–3 cells wide at the base and 1–4 cells long. Leaf cells at the middle portion 27–37 µm x 15–20 µm, at leaf-base 35–50 µm x 10–20 µm, at the marginal 25–35 µm x 7–12 µm; trigones small, acute, cuticle smooth. Underleaves vestigial, 1–4 cells long. Generative organ not seen.

Ecology: found on the tree trunk in lowland forest.

Distribution: Java, Malaysia, Philippines, Sumatra (new records) [1,10].

Specimen examined: Mount Sibayak, Etti Siregar 128.

Note: the diagnostic characters of *P. laxissima* are 1) the rarely branched plant 2) the distant and the linear-oblong leaves, 3) the teeth on leaves margin restricted to the apex, 4) underleaves are vestigial

Plagiochila laxissima is closely related to *P. singularis*, but the latter with verrucose cuticle cells. *Plagiochila laxissima* is uncommon in Mount Sibayak, only one sample was found.

12. *Plagiochila obtusa* Lindenb.

Plant yellowish to brownish in the dry specimen, 39–64 mm long 4–5 mm wide, with rhizomatous caulids; branches infrequent, somewhat dichotomous and often unbranched; paraphyllia are present along the ventral of stem, margin with long ciliate. Leaves imbricate, stem surface invisible; lobes ovate-oblong, 2.5–2.7 x 2.0–2.25 mm; dorsal margin slightly revolute and moderately decurrent along dorsal stem midline; ventral margin shortly decurrent, forming a distinct ventral keel on the shoot, ventral base dilated; margin of leaves wavy; apex widely obtuse or rounded; teeth present on whole leaves margin or sometimes absent from the basal half of dorsal margin, 25 – 28 in number; around the ventral base margin usually ciliate, 2–3 cells wide at the base and 7–15–(19) cells long; teeth on the other portion of leaf margin rather small, triangular, 2–3 cells wide at the base, 1–4 cells long. Leaf cells at the middle portion 24–38 µm x 14–23 µm, at leaf-base 28–43 µm x 14–22 µm, at the marginal 17–29 µm x 9–14 µm; trigones medium to large, triangular, acute or nodulose, often confluent, cuticle smooth. Underleaves absent. Reproduction organ not seen.

Ecology: found on the tree trunk in lower montane forest.

Distribution: Java, Sumatra (Padang: Mount Singgalang and Mount Talang), Taiwan [1,7].

Specimen examined: Mount Sibayak, Etti Siregar 781.

Note: the diagnostic characters of *P. obtusa* are 1) plant rarely branched, 2) the numerous paraphyllia on the stem, 3) the ovate-oblong leaves, teeth on the whole margin. *Plagiochila obtusa* is uncommon in Mount Sibayak, only one specimen was found.

13. *Plagiochila salacensis* Gott.

Plant 25 mm long, 2.5 mm wide, yellowish brown in the dry specimen. Branches moderate in number, terminal and *Frullania*-type, usually forming a dichotomous system and sometimes producing lateral intercalary. Leaves approximately imbricate, lobes ovate or oblong–ovate, 1.25–1.5 mm long, 0.75–1 mm wide; ventral base dilated, basal ventral margin closely appressed to the base of opposite leaves, forming a distinct keel on ventral side of shoot; the apex sub truncate or rounded, often with two larger (often showing a tendency toward asymmetric bilobed) teeth; teeth are absent on the dorsal margin of leaves, or with 1–2 small teeth near the apex, 2–4 cells wide at the base and 2–7 cells long. Leaf cells at the middle portion 19-31 µm x 12–2.5 µm, those of leaf-base 24.5-37 µm x 12–19 µm, at the marginal 22–34.5 µm x 10.5–13 (15) µm; trigones small, acute or sub nodulose, cuticle smooth; underleaves vestigial, often found on young branches.

Dioicous. Gynoecia terminal, without or with one innovation; bracts ovate, 1.5 mm long and 0.8 mm wide, margin ciliate; perianth about 2.25 mm long and 1.5 mm wide, mouth truncate, margin ciliate about 2–3 (4) cells wide and 5–13(14) cells long, mostly uniseriate on 2/3 distal. Male plants not found.

Ecology: found on a rotten log in lowland forest.

Distribution: China, India, Indonesia (Java, Bali, Sulawesi, Sumatra: Mount Singgalang Padang), Philippines,Thailand [6].

Specimen examined: Mount Sibayak, Etti Siregar 559.

Note: the diagnostic characters of *P. salacensis* are 1) forming a dichotomous branching system, 2) the leaves are imbricate, 3) the leaves are ovate with dilated ventral base, 3) the basal ventral margin closely appressed to the base of opposite leaves, 4) the dorsal margin entire and rather small number of teeth on the ventral and apical margin.

Plagiochila salacensis is variable in the size of lobes and the number of teeth on lobe margin. *Plagiochila salacensis* seems to be uncommon in Mount Sibayak, only one specimen was found.

14. *Plagiochila sandei* Dozy.

Plant robust, pendulous, 5–10 cm long, 8–12 mm wide, greenish in the dry specimen, with strong rhizomatous caulids. Plants usually simple (rarely with branches). Leaves closely imbricate, lobes oblong–ovate, 4–6 mm long, 2,5–4 mm wide; dorsal margin strongly recurved and long decurrent along dorsal stem midline; ventral margin with a tubular sac at the base; sac oblong in outline, free margin strongly inrolled, ciliate (variable in size and number); apex rounded; teeth present on whole of leaf margin, 1–2 (3–4) cells wide at the base, 4–6 cells long on ventral margin and 7–11 cells long on basal of dorsal margin. Leaf cells at the middle portion 35–45 μm x 20–25 μm, at leaf-base 37.5–67.5 μm x 20–25 μm, at the marginal 25–37.5 μm x 10–12.5 μm; trigones very small, acute, cuticle smooth. Underleaves absent.

Dioicous. Gynoecia on the terminal of stem or branches, with one innovation; bracts similar to leaves in shape and size, two pairs, with smaller tubular sac on ventral base; teeth on margin of bracts more dense and stronger than the leaf-margin; perianth long cylindrical, 6 mm long, 3.13 mm wide, mouth truncate, margin of mouth spinose. Androecia terminal on shoots, in long spicate, 7–10 mm long, 2–7 in number and gregarious at shoot apex; bracts 4–24 pairs, closely imbricate.

Ecology: found on the tree trunks and rotten log from lowland to lower montane forest.

Distribution: Borneo, Java, Philippines, Sulawesi, Sumatra [1,13].

Specimens examined: Mount Sibayak, Etti Siregar 325, 335, 336, 338, 734, 756, 1776, 2009.

Note: the diagnostic characters of *P. sandei* are 1) robust and pendulous plant, 2) the imbricate leaves with a ventral keel, 3) the tubular sac on the base of ventral leaves, with strongly inrolled free margin, 3) the long spicate and gregarious androecia at the shoot apex.

Plagiochila sandei is variable in the size of a tubular sac of ventral lobes base and the number of their cilia. This species is rather common in Mount Sibayak.

15. *Plagiochila sciophila* Nees ex Lindenb.

Plant 20–35 mm long, 2.5–3.5 mm wide, green to yellowish-brown in the dry specimen; rarely branches. Leaves distant or loosely to moderately imbricate, lobes short to long rectangular or oblong, 1.4–1.7 mm long, 0.6–0.8 mm wide; dorsal margin weakly recurved, short decurrent; ventral margin weakly recurved, hardly or only shortly decurrent; apex rounded, leaf apex seems bilobed; teeth are present on the whole of leaf margin, long ciliate, 8–18 in total number, 1–3 cells wide and 3–10 cells long, two teeth at the leaf apex are often larger than the other. Leaf cells at middle portion 25–37.5 μm x 17.5–25 μm, at leaf-base 27.5–37.5 μm x 17.5–25 μm, at the marginal 25–37.5 μm x 12.5–27.5 μm; trigones small, cell walls thin, cuticle smooth. Underleaves are usually vestigial, 2–3 lobes, 4–8 cells long of uniseriate cells. Generative organ not seen.

Ecology: found on the tree trunks and rotten logs from lowland to lower montane forest.

Distribution: Bhutan, China, India, Indonesia (Sumatra, Java, Borneo, Bali, Sulawesi), Japan, Korea, Nepal, Papua New Guinea, Philippines, Thailand, Vietnam [1,6,13,14].

Specimens examined: Mount Sibayak, Etti Siregar 86, 323, 334, 387, 564, 641, 852.

Note: the diagnostic characters of *P. sciophila* are 1) rarely branched, 2) the rectangular or oblong leaves, 3) teeth of leaf-margin long ciliate, present on the whole leaf margin. *Plagiochila sciophila* seems to be common in Mount Sibayak.

16. *Plagiochila singularis* Schiff.

Plant 25–35 mm long, 3 mm wide, pale yellowish in the dry specimen, branches very rare (most plants simple). Leaves distant, caducous, dropping off from sub basal portion, so most of the stem denuded, lobes oblong, 1.4–1.5 mm x 0.6–0.67 mm; dorsal margin plane, moderately decurrent at the base; ventral margin hardly decurrent at the base; apex obtuse, often with two large teeth; teeth on margin almost restricted to around the leaf-apex. Leaf cells at the middle portion 25–37.5 μm x 17.5–25 μm, at leaf-base 25–37.5 µm x 17.5–25 µm, at the marginal 25–37.5 µm x 12.5–20 µm; cell walls thin, trigones small, cuticle verrucose; underleaves are vestigial, 2–3 cells. Generative organ not seen.

Ecology: found on the tree trunk in lowland forest.

Distribution: Indonesia (Java, Sumatra: Mount Singgalang), Malaysia, Thailand [1,8].

Specimen examined: Mount Sibayak, Etti Siregar 250.

Note: the diagnostic characters of *P. singularis* are 1) very rarely branches, 2) leaves distant, oblong, 3) teeth almost restricted to around the leaf-apex, 4) cuticle cells verrucose.

Plagiochila singularis is closely related to *P. laxissima*, as mentioned before.

17. *Plagiochila sumatrana* Schiff.

Plant 30–60 mm long, 4.5–5.5 mm wide, with rhizomatous caulids; branches rather frequent, dominantly lateral intercalary. Leaves moderately imbricate, lobes broadly ovate, 1.75–2.25 mm long, 1.6–1.8 mm wide; dorsal margin strongly recurved with shortly decurrent along dorsal stem midline, ventral margin shortly decurrent at the base, dilated toward the base, closely appressed to that of opposite leaves, so the ventral surface of stem completely hidden, forming a ventral keel on shoot; apex rounded; dorsal margin entire, ventral margin toothed; teeth small, triangular, 15–23 in number, (1)–2–3 cells wide at the base, 2–4 cells long. Leaf cells at the middle portion 24.5–37.5 µm x 17.5–21.5 µm, at leaf-base 38–54.5 µm x 18.5–19.5 µm, at the marginal 24.5–37.0 µm x 9.5–12.0 µm; trigones medium sized, triangular or nodulose, cuticle smooth; underleaves absent.

Dioicous. Androecia terminal on shoots apex, in 3 gregarious long spicate; bracts 10–12 pairs, closely imbricate.

Ecology: found on the tree trunk in lowland forest.

Distribution: Borneo, Java, Papua New Guinea, Philippines, Sumatra (Mount Singgalang) [1,6].

Specimen examined: Etti Siregar 882a.

Note: the diagnostic characters of *P. sumatrana* are 1) the rather frequently branches, dominantly lateral intercalary, 2) ventral margin of leaves dilated toward the base, closely appressed to the opposite leaves, 3) forming a ventral keel on shoot, 4) the small teeth on the whole ventral margin and absent on the dorsal margin.

Plagiochila sumatrana seems to be uncommon at Mount Sibayak, only one specimen was found. The species has a wide variable in total number of marginal teeth of the leaf (15–23 marginal teeth per leaf).

18. *Plagiochila teysmannii* S. Lac.

Plant large, 30–120 cm long, 4–6.5 mm wide, yellowish to pale brown in the dry specimen, with a rhizomatous caulids; branches moderately in number, *Frullania*-type, dominantly terminal or sometimes with lateral intercalary branched, forming a distinctly dichotomous habit. Leaves loosely to closely imbricate, stem visible or invisible, lobes broadly ovate, 2.3–3.5 x 2.2–3.3 mm; dorsal margin strongly recurved and long decurrent along dorsal stem midline; ventral margin short decurrent, usually closely appressed to that of the opposite leaf and forming a ventral keel on shoot, ventral base dilated; apex broadly rounded; teeth present on the whole of leaf margin or entire on basal half of dorsal margin, coarse, 18–35 in number, 2– 4 cells wide at the base, 3–6–(10) cells long. Leaf cells at the middle portion 27.5–47.5 µm x 20–25 µm, at leaf base 42–55µm x 17.5–25 µm, at the marginal 15–37.5 µm x 12.5–20 µm; trigones large, nodulose, somewhat confluent, cuticle smooth; underleaves absent. Generative organ not seen.

Ecology: found on the tree trunks in lowland forest

Distribution: Borneo, Cambodia, Java, Sulawesi, Sumatra (Mount Singgalang, Padang), Philippines, Thailand [1; 13; 8].

Specimens examined: Mount Sibayak, Etti Siregar 241, 1986, 2008b; Etti Siregar and Nunik S Ariyanti 1444.

Note: the diagnostic characters of *P. teysmannii* are 1) the dominantly terminal and *Frullania*-type branches, forming a dichotomously habit, 2) the ventral keel on the shoot, 3) the broadly ovate leaves, 4) trigones large, nodulose, and somewhat confluent

4. Conclusions

In this study, 18 species of *Plagiochila* from Mount Sibayak were recognized. The species recorded in this study is 60 percent of the total number of *Plagiochila* in Sumatra reported by Inoue [1]. Two of the species were recognized as new records for Sumatra.

Plagiochila junghuhniana and *P. dendroides* were the most common species recorded at Mount Sibayak. The uncommon species of *Plagiochila* in Mount Sibayak, found only in a limited population are *P. gracilis*, *P. korthalsiana*, *P. laxissima*, *P. obtusa*, *P. salacensis*, *P. sumatrana* and *P. singularis*.

Acknowledgement

This research is part of the dissertation of the first author, funded by BPPS project of the Indonesian Ministry of Education (DIKTI). We would like to thank Prof. Mien A. Rifai as a supervisor of the first author, and to Narin Printarakul for some literature and discussion about *Plagiochila.* Thanks are also due to all in specimen collection in the field.

References

[1] Inoue H (1984) *The Genus Plagiochila (Dum.)Dum. in Southeast Asia*. Tokyo. Academia Scientific Boo Inc.

[2] So ML, Grolle R (2000) Checklist of *Plagiochila* (*Hepaticae*) in Asia. J Hatt Bot Lab. 88: 199–248.

[3] Gradstein SR (2011) Guide to the Liverworts and Hornworts of Java. Bogor (ID). SEAMEO-BIOTROP Regional Centre for Tropical Biology.

[4] Schiffner V (1900). *Die Hepaticae der Flora von Buitenzorg I. Erster Band. Ricciaceae, Marchantiaceae, Jungermaniaceae, Anakrogynae und Jungermaniaceae Akrogynae Unterfam. Epigoniantheae*. Leiden. E. J. Brill. 220 pp.

[5] Piippo S (1989). Bryophyte flora of the Huon Peninsula, Papua New Guinea. *Plagiochilaceae* (*Hepaticae*). Ann Bot Fennici 26: 183–236.

[6] So ML (2001). *Plagiochila* (*Hepaticae, Plagiochilaceae*) in China. Syst Bot Monogr 60: 1–214.

[7] Wang J, Lai MJ, Zu RL (2011). Liverworts and hornworts of Taiwan: an updated checklist and floristic accounts. Ann Bot Fennici 48: 369–395.

[8] Lai MJ, Zhu RL, Chantanaorrapint S (2008). Liverworts and hornworts of Thailand: an updated checklist and bryo floristic accounts. Ann Bot Fennici 45: 321–341.

[9] Verma PK, Rawat KK, Yadav A (2013). *Plagiochila bantamensis* (Reinw. *et al.*) Mont. of the Subgenus *Metaplagiochila* Inoue (*Marchantiophyta: Plagiochilaceae*) new to the liverwort flora of the Indian mainland. Taiwania 58 (2): 124–127. DOI: 10.6165/tai.2013.58.124

[10] Menzel M (1988). An annotated catalog of the Hepaticae and Anthocerotae of Borneo. J Hatt Bot Lab 65:145–206.

[11] Srivastava SC, Verma PK, Alam A (2006). *Plagiochila gracilis* Lindenb. & Gott. and *P. subtropica* St. in Western Ghats (Nilgiri Hills). Phytotaxonomy 6: 78–83.

[12] Alam A (2012) Liverworts flora of Parson's valley, Nilgiri hills (Western Ghats), South India. Arch Bryol 119: 1–19.

[13] Ariyanti NS, Gradstein SR (2007) Wallace's line and the distribution of the liverworts of Sulawesi. Crypt Bryol 28 (1): 3–14.

[14] Piippo S, He XL, Koponen T, Paul J, Redfearn JR, Li JX (1998). Hepaticae from Yunnan China, with a checklist of Yunnan Hepaticae and Anthocerotae. J Hatt Bot Lab 84: 135–158.

Survey on aedes mosquito density and pattern distribution of *aedes aegypti* and *aedes albopictus* in high and low incidence districts in north sumatera province

Fazidah A Siregar[1*], Tri Makmur[2]
[1]Faculty of Public Health, University of Sumatera Utara, Medan, 20155, North Sumatera, Indonesia
[2]Faculty of Medicine, Islamic University of Sumatera Utara, Medan, North Sumatera, Indonesia

*Email: fazidah@usu.ac.id

Abstract. Transmission and control of dengue hemorrhagic fever are related to its vectors. This study investigated vector density and distribution patterns of *Aedes aegypty* and *Aedes albopictus* in Medan and Langkat as high and low incidence district, respectively. An entomological survey was carried out in 304 households both in Medan and Langkat. The results showed that adult *Ae. aegypti* were predominantly in Medan, while adult *Ae. albopictus* was only in Langkat. Larvae indices (HI, CI, BI) for *Aedes* in Medan (35,13 and 43) were higher than langkat (22,8 and 30). Adult indices (AHI, AD, RR) for *Ae. aegypti* in Medan and for *Ae. albopictus* in Langkat were 20,38,24 and 3,5, and 5, respectively. Pattern distribution of *Aedes* larvae and adult mosquitoes in both district had similar pattern. *Aedes* larval indices and adult indices both in HIDs and LIDs were above the critical level, indicating potential high risk for DHF transmission. By multiple regression analysis, HI is predictor for DHF transmission in North Sumatera. Thus, in designing an effective control measures for dengue hemorrhagic fever, monitoring distribution and vector density is crucial.

Keywords: Aedes mosquito density, high and low incidence district, pattern distribution

1. Introduction

Dengue hemorrhagic fever (DHF) as Vector-borne diseases are global public health risks. DHF is caused by the dengue virus, which is transmitted by the bite of a female mosquito. The disease is found in the tropical and subtropical region with an approximately 50-100 million cases of dengue fever occur annually and 25,000 deaths [1]. North Sumatera Province is one of endemic area, and the Ministry of Health of Indonesia (2011) reported that North Sumatera ranked the third highest for DHF cases with 16 dengue endemic district [2]. The incidence of DHF annually increased from 24/100,000 population in 2007 to 66.2/100,000 in 2015 [3].

The factors are responsible for the incidence of DHF are complex. The dynamic of dengue infection are driven by complex interaction between host, vectors and viruses that are influenced by environmental. Furthermore, environment affects the *Aedes* mosquito breeding site as the main vector that plays a role in disease transmission. Many studies had been conducted and described the behavior of these vectors [4,5]. In addition, few indices were found to be useful as indicators of their abundance and linked to event relates to dengue fever and DHF. Among them are house index (HI), Breteau index (BI) and container index (CI). These indices are used to monitor *Aedes* population for dengue virus transmission [6,7,8]. In addition, adult mosquito density also calculated to measure *Aedes* density, including adult house index (AHI), adult density (AD) and the resting rate (RR) [9,10,11].

The lack of available vaccines and drugs, makes vector control has been the right choice for the prevention and control of dengue hemorrhagic fever [12]. Various efforts related to the eradication of vectors had been conducted in the prevention of dengue, such as mosquito nest eradication (PSN), health education and the use of insecticide fogging and Abatisasi, but the results have not been as expected.

In North Sumatera Province, DHF still endemic with incidence increased annually. Various efforts had been conducted however, have not optimum include the vector monitoring have not been continuously that might result in uneffective control strategies. In designing effective strategies for dengue hemorrhagic fever, one approach is the eradication of vectors. A better understanding regarding habitat, pattern distribution and vector density is crucial. Therefore, this study was undertaken to provide information regarding *Aedes* density to help health organizations develop effective control strategies for dengue hemorrhagic fever.

2. Methods

In this study, an entomology survey was carried out in Medan, a district with high DHF incidence and Langkat, a district with low DHF incidence, in North Sumatera province from 5th April, 2011 to 16th May, 2011. Entomology survey including larval survey and adult *Aedes* survey .

The larval survey was conducted to determine larval density. It was done based on WHO procedure by inspection of containers both indoor and outdoor in selected houses for the presence of *Aedes* larvae. For each positive, a larva was pipetted into a plastic cup or plastic bag as shown in Figure 1 and brought back to the laboratory for identification. The House Index (HI), Container Index (CI) and Breteau Index (BI) were calculated to determine the larval density.

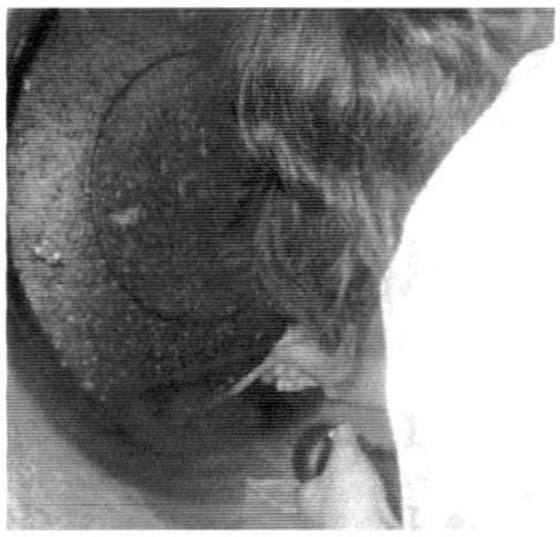

Figure 1. Activity in Larva survey

Adult *Aedes* survey was conducted to determine adult density and species of the mosquito. Aspirators were used to collect adult *Aedes* mosquitoes from the selected human dwellings, both inside and outside the house or premises where the mosquitoes found resting during the day time (9.00 am until 5.00 pm) as shown in Figure 2. The collected *Aede*s mosquitoes were kept in paper cups covered with gauze and fed with glucose while being transported to the laboratory for inspection and identification.

Adult house index (AHI), adult density (AD) and the resting rate (RR) were calculated to determine adult density.

Figure 2. Collection of *Aedes* mosquito using aspirator in adult survey

2.1 Sample

In this study, Multistage sampling was used to select the study population that was households in both districts with high DHF incidence (HIDs) and low DHF incidence (LIDs). In the first stage, one district with high DHF incidence (Medan) and one district with low DHF incidence (Langkat) were purposively selected based on DHF data in the previous year. In the second stage, two subdistricts, one with the highest and the other with the lowest DHF cases were again purposively selected for each district. In Medan, entomological survey was conducted in Marelan and Maimun, subdistricts with high and low DHF cases respectively. Like wise in Langkat, it was conducted in Stabat and Berandan. In the third stage, the same criteria as above were applied to select two villages from the subdistricts. For each village, two sub-villages were randomly selected. In the high subdistrict, households were systematically selected among the households with reported cases from the health office registry, whereas, in the low subdistrict, households were selected using systematic sampling based on the registry of the household from the sub-district administrative office.

2.2 Sample size

In this study sample size was calculated using the two-proportion formula[13]. An average proportion of DHF cases in the three districts with high DHF cases (P1) of 0.156 and an average proportion of DHF cases in the three districts with low DHF cases (P0) of 0.017. An allowable error of 5% and a power of study of 80%. By multistage sampling, the sample size was multiplied by the design effect of 2. Therefore, 304 households for districts with both high and low DHF cases were included in this study.

2.3 Study instruments

The tools used in the larval survey included a survey form, pipettes, plastic bottles, plastic bags, a specimen vial with stoppers, pens, a label and a flashlight. All of the indoor and outdoor containers at the selected houses had been inspected and from each positive container, a larva was pipetted into a plastic cup or plastic bag and brought back to the laboratory for identification. The findings were recorded on the survey form. Three larval indices, the House Index (HI), Container Index (CI) and Breteau Index (BI) were calculated. The HI is defined as the percentage of houses positive for *Aedes* larvae. The CI is defined as the percentage *of* containers positive for *Aedes* larvae. BI is defined as the number of containers positive for *Aedes* larvae per 100 houses examined.

The tools used in the adult survey include survey form, aspirator, flashlight, plastic cup/tube, paper cup covered with gauze and cotton and labels (masking tape). Aspirators were used to collect adult *Aedes* mosquitoes from the selected human dwellings, both inside and outside the house during the day time (9.00 am until 5.00 pm). The collected *Aedes* mosquitoes were kept in paper cups covered with gauze and fed with glucose and transported to the laboratory for identification. Adult house index (AHI), adult density (AD) and the resting rate (RR) were calculated to determine adult density. AHI is defined as the percentage of houses positive for *Aedes* mosquitoes. AD is defined as the number of *Aedes* mosquitoes per 100 houses examined. RR is defined as the number of female *Aedes* mosquitoes per 100 houses examined

2.4 Statistical analysis
The Statistical Package for Social Science (SPSS) program was used for data analysis (Release 22.0, SPSS Inc., Chicago, Illinois, USA). Larval *Aedes* indices and adult indices were calculated and tabulated for descriptive statistics. Multiple logistic regressions were used to analyze the association between larvae indices and adult indices with DHF transmission.

3. Results and Discussion
3.1. Results
The entomological survey had been conducted in Medan and Langkat district. The results revealed that larval indices (HI, CI and BI) for *Aedes* were higher in Medan than Langkat (35, 13 and 43 and 22, 8 and 30, respectively). Furthermore, *Aedes aegypti* larvae were predominantly found in Medan, whereas, *Aedes albopictus* larvae were predominantly found in Langkat. The larval indices (HI, CI and BI) for *Aedes aegypti* were higher in Medan than Langkat (33, 12 and 39 and 11,2 and 9 respectively), whereas, for *Aedes albopictus* were higher in Langkat than Medan (13, 5 and 18 and 3, 1 and 3, respectively) (Table 1).

Table 1. *Aedes* larvae indices in high and low incidence district in North Sumatera Province

	High incidence district			Low incidence district		
	Aedes spp	*Aedes aegypti*	*Aedes albopictus*	*Aedes spp*	*Aedes aegypti*	*Aedes albopictus*
Number of households	152	152	152	152	152	152
Number of positive households	55	50	5	33	16	19
Number of containers	519	519	519	594	594	594
Number of positive containers	65	60	5	45	14	28
Larval index						
HI	35.0	33.0	3.0	22.0	11.0	13.0
CI	13.0	12.0	1.0	8.0	2.0	5.0
BI	43.0	39.0	3.0	30.0	9.0	18.0

Then larval indices was compared to the WHO standard to determine density figure (DF) of larvae as shown in Table 2. The amount of HI, CI and BI scores divided by 3 and expressed on a scale of 1 -9 and categorized into DF = 1: low density, DF = 2-5: medium density and DF = 6 -9: high density. The result of this study showed that density figures for larval indices in Medan and Langkat were 4.7 and 2, respectively. It means that have a higher risk of DHF transmission. Furthermore, HI in both high and low DHF incidence districts were more than 10, indicates a high-risk of DHF transmission.

Table 2. Density figure of mosquito larvae

Density figure	House Index (HI)	Container index (CI)	Breteau index (BI)
1	1-3	1-2	1-4
2	4-7	3-5	5-9
3	8-17	6-9	10-19
4	18-28	10-14	20-34
5	29-37	15-20	35-59
6	38-49	21-27	50-74
7	50-59	28-31	75-99
8	60-76	32-40	100-199
9	77+	41+	200

Source: Queensland Government [14]

The presence of *Aedes* mosquitoes inside and outside the houses, a number of *Aedes* mosquitoes and females were detected and calculated as vector indices (AHI, AD, RR) were presented in Table 3. Adult indices (AHI, AD, RR) for *Aedes aegypti* in Medan were higher than Langkat district (20,38,24 and 9,13,10, respectively). It is supported that *Aedes aegypti* mosquito predominantly found in Medan. While adult *Aedes albopictus*, was only caught in Stabat and Berandan, two subdistrict in Langkat district. It might due to *Aedes albopictus* which preferred Berandan subdistrict with geographically more plants. The adult mosquito indices was 3, 5, 5 respectively.

Table 3. *Aedes* mosquito indices in high and low incidence district in North Sumatera Province

	High incidence district			Low incidence district		
	Aedes	*Aedes aegypti*	*Aedes albopictus*	*Aedes*	*Aedes aegypti*	*Aedes albopictus*
Number of households	152	152	152	152	152	152
Number of positive households	31	31	0	18	13	5
Aedes collected						
F	37	37	0	22	15	7
M	21	21	0	5	4	1
Adult index						
AHI		20	0		9	3
AD		38	0		13	5
RR		24	0		10	5

Furthermore, the larval indices (HI, CI, BI) and adult indices (AHI, AD, RR) were analyzed to determine if they were associated with DHF transmission. Of Simple Logistic Regression analysis revealed that larval indices (HI, CI, BI) and adult indices (AD, AHI, RR) were significantly associated with DHF transmission (Table 4). However, using the Multiple Logistic Regression analysis, house index

(HI) was the only significant factor for DHF transmission in North Sumatera Province as shown in Table 5.

Table 4. Association between vector indices and dengue hemorrhagic fever transmission in North Sumatera Province by Simple Logistic Regression

Variable	HIDs n(%)	LIDs n(%)	Crude OR (95%CI)	p value of Wald
House index				
House (+)	54(35.5)	29(19.1)	1.093(3.035,3.935)	0.001
House (-)	98(64.5)	123(80.9)		
Container index				
Container (+)	65(12.6)	40(7.0)	1.271(3.917,4.125)	0.009
Container(-)	452(87.4)	527(93.0)		
Breteau index				
Container(+)	65(39.9)	40 (24.5)	1.005(1.622,6.222)	0.027
Container (-)	98(60.1)	123(75.5)		
Adult house Index				
House (+)	31(13.8)	18(11.8)	4.189(9.765,17.97)	0.045
House (-)	121(86.2)	134(88.2)		
Adult density				
Aedes adult(+)	58(32.4)	25(15.7)	3.642(2.752,4.818)	0.012
Aedes adult(-)	121(67.6)	134(84.3)		
Resting rate				
Female aedes(+)	37(22.2)	21(13.3)	2.097(0.000,1.265)	0.115
Female aedes(-)	130(77.8)	137(86.7)		

Table 5. Association between vector indices and dengue hemorrhagic fever transmission in North Sumatera Province by Multiple Logistic Regression.

Variable	HIDs n(%)	LIDs n(%)	Adjusted OR (95 % CI)	p value
House indices	54(35.5)	29(19.1)	1.093(3.035,3.935)	0.001
House (+)	98(64.5)	123(80.9)		
House (-)				

Forward LR Multiple Logistic Regression was applied. There was no interaction and multicollinearity Classification table (overall correctly classified percentage=58.2) Area under the ROC curve (58.2%) were applied to check the model fitness.

The distribution pattern of *Aedes* larvae and adult mosquitos are presented in Figure 3- 6. Based on BI-AHI, BI-RR, BI-AD and HI-AHI, HI-RR, HI-AD, and CI-AHI, CI-RR, CI-AD; showed high subdistrict had the similar pattern with low subdistrict both in Medan and Langkat Districts (Marelan and Maimun, Stabat and Berandan) as presented in Figure 3 and Figure 4. Likewise, high subdistrict in district with high DHF incidence (Marelan) and high subdistrict in district with low DHF incidence (Stabat); low subdistrict in district with high DHF incidence (Maimun) and low subdistrict in district with low DHF incidence (Berandan) also had the similar pattern as presented in Figure 5 and Figure 6.

IOP Conf. Series: Earth and Environmental Science **130** (2018) 012018 doi:10.1088/1755-1315/130/1/012018

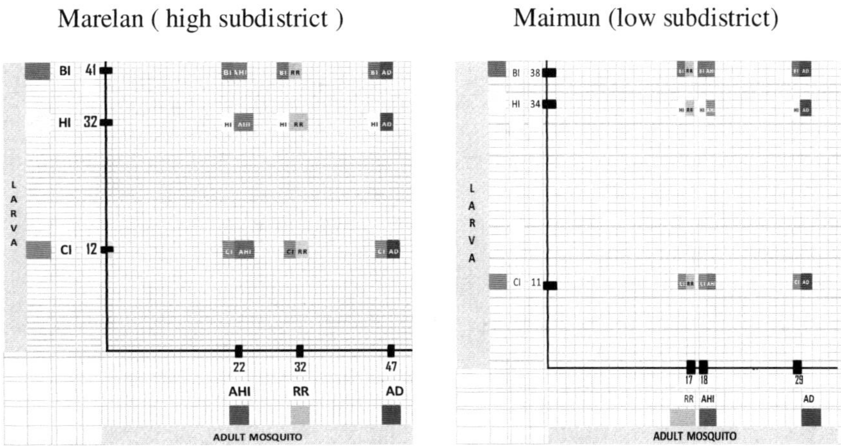

Figure 3. The distribution pattern of *Aedes* larvae and adult mosquito between high and low subdistrict in high incidence district.

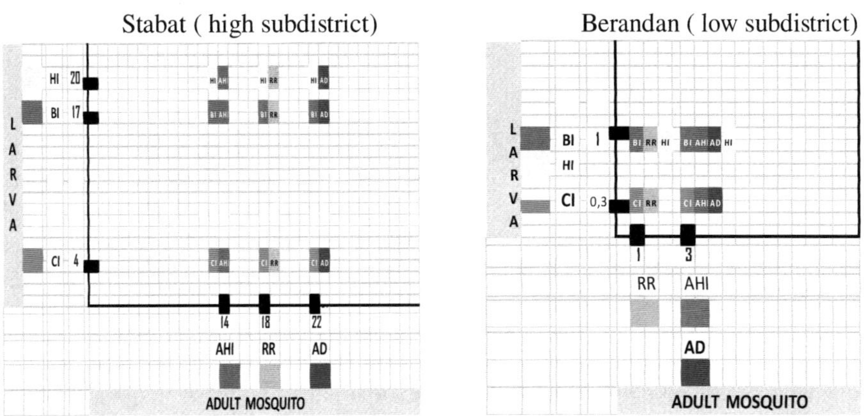

Figure 4. The distribution pattern of *Aedes* larvae and adult mosquito between high and low subdistrict in low incidence district

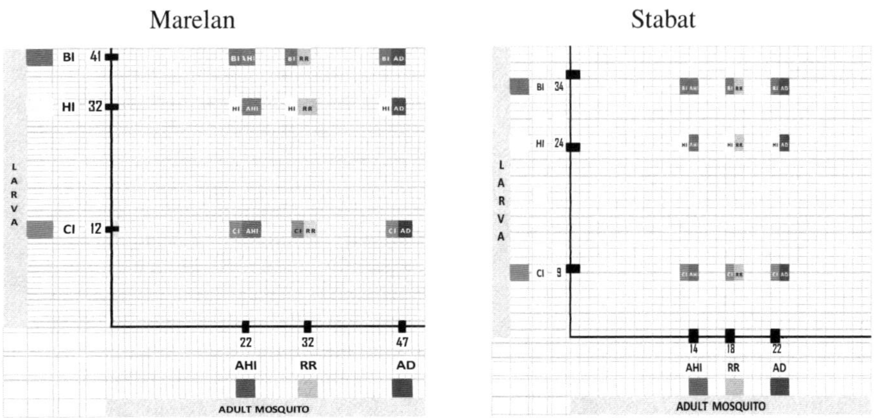

Figure 5. The distribution pattern of *Aedes* larvae and adult mosquito in high subdistrict between high and low incidence district

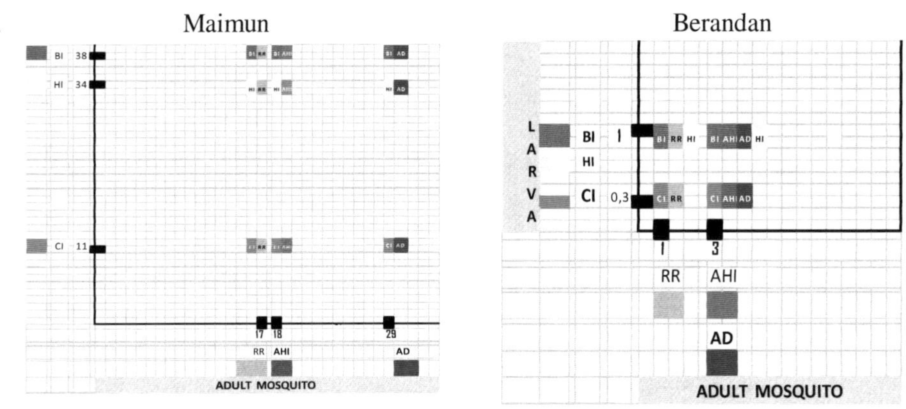

Figure 6. The distribution pattern of *Aedes* larvae and adult mosquito in low subdistrict between high and low incidence district.

The percentage of resting rate (RR) of *Aedes aegypti* in the high subdistrict more than low subdistrict both in Medan district and Langkat district (Figure 3 and 4). Likewise, the percentage RR of *Aedes aegypti* mosquitos in district high DHF incidence more than district low DHF incidence. It was indicated that the probability of *Aedes aegypti* biting and be a vector for DHF in both district high and low DHF incidence were differed. Therefore, DHF incidence in both districts were also differ. On the other hand, AHI and AD for *Aedes aegypti* in Berandan were lower than Maimun as shown in Figure 6. However, it could indicated that Berandan had a chance for DHF transmission.

3.2. Discussion

Transmission of DHF is directly related to its vectors. *Aedes aegypti* is the main vector for DHF, whereas *Aedes albopictus* is the potential vectors. Vector density, species identification and breeding places are essential for effective vector control [15]. In this study, *Aedes aegypti* larval indices (HI,CI,BI) in HIDs were higher than LIDs. On the contrary, *Aedes albopictus* larval indices in LIDs were higher than HIDs. It might be due to *Aedes aegypti* more common in urban area (Medan district), while *Aedes albopictus* preferred Brandan (Langkat district) that had geographical more plants. In the present study, *Aedes aegypti* and *Aedes albopictus* were found breeding at the household level in both HIDs and LIDs. A study by Rajendran *et al.* (2006) in Sulurpet India found HI,CI and BI were 9.45, 23.61 and 13.39 respectively. *Aedes aegypti* and *Aedes albopictus* were co-existed in the study area [16].

Furthermore, in this study larval indices were significantly associated with DHF transmission. A similar finding was reported by Pham *et al.* (2011) in Dak Lak Province in The central highlands region of Vietnam found that the incidence of DHF was significantly associated with House indices, container indices and Breteau indices [17]. Also, in this study, both districts have *Aedes* larval indices and adult indices were above the critical level, it means that have potential high-risk DHF transmission. Likewise, Balakrishnan *et al.* (2006) study on the 60 houses selected from three affected villages and 118 houses in the affected municipal ward in Tiruppur India found that *Aedes* larvae indices and adult density were above the critical level, indicating that potential outbreak would occur in the future [18].

The increased number of female *Aedes* mosquitoes would increase the probability of biting, thus increasing risk of DHF transmission [19]. *Aedes aegypti* adult indices (AHI, AD, RR) in Medan were higher than Langkat. Although RR of *Aedes aegypti* and *Aedes albopictus* in LIDs are low, however, it had a chance be a vector for DHF transmission. In the present study, adult house indices (AHI), adult density (AD) and resting rate (RR) were significantly associated with DHF transmission. A study by Sahani *et al.* in Senawang Negeri Sembilan, Malaysia from July 2009 to February 2010 found that the density of *Aedes albopictus* was high and could be a potential vector for DHF transmission [5].

According to the Health Ministry of Indonesia, HI was used as an indicator for DHF transmission due to these larval indices further illustrated the extent of spreading of DHF. Likewise, The Pan American Health Organization used HI as an indicator for dengue transmission [6]. In this study, in both districts, high and low incidence district have HI level was higher than 10, indicates a high-risk of DHF transmission. Further analysis with multiple logistic regression, HI was a predictor for DHF transmission in North Sumatera. Of this study, it could conclude that monitoring of vector density is essential not only monitor vector distribution, but could predict probability DHF transmission. Furthermore, provide accurate information for vector eradication and effective control strategies.

4. Conclusions

This study revealed that the *Aedes* larval indices (HI, CI and BI) were higher in HIDs than in LIDs. *Aedes aegypti* mosquito is predominantly in Medan district , meanwhile *Aedes albopictus* mosquito only found in langkat district. Adult indices (AHI, AD, RR) for *Aedes aegypti* in Medan and for *Aedes albopictus* in Langkat were 20,38,24 and 3,5, and 5, respectively. Pattern distribution of *Aedes* larvae and adult mosquito in both district had a similar pattern. Furthermore, both districts had larval indices and adult indices were above a critical level, it means that have potential high-risk dengue transmission. For an effective control strategy, should focused on vector control. Therefore, monitoring vector density is essential as a basis for health professional in designing more effective prevention strategies.

Acknowledgments

We are grateful to the head of the district health office and health facility for their assistance in this study. We are also thankful to all of the participants and people who were involved in this study. No conflict of interest involve in this manuscript.

References

[1]. Guzman MG, Halstead SB, Artsob H, Buchy P, Farrar J, Gubler DJ, Hunsperger E, Kroeger A, Margolis HS, Martinez E, Nathan MB, Pelegrino JL, Simmons C, Yoksan S and Peeling RW (2010) Dengue: a continuing global threat. *Nature Review Microbiology,* **8(12 Suppl),** S7-16.

[2]. Ministry of Health of Indonesia (2011) Report of dengue hemorrhagic cases in Indonesia, Jakarta, Indonesia.
[3]. MHI (2015) Report of dengue hemorrhagic cases in North Sumatera Province. Ministry of Health of Indonesia (MHI), Jakarta, Indonesia.
[4]. Rodriguez H, and de la Hoz RF (2005) Dengue and vector behaviour in Ca'queza, Colombia, 2004. *Revista de Salud Publica,* **7(1),** 1-15.
[5]. Sahani M, Othman H, Atiqah NMN, Rozita H, M Ali Z, Mohamad Rasid MN and CHOY EA (2012) *Kajian Ekologi Nyamuk Aedes di Senawang Negeri Sembilan*, Malaysia. *Sains Malaysiana* **41(2),** 261-269.
[6]. Sanchez L, Vanlerberghe V, Alfonso A, del Carmen Marquetti M, Guzman M, Bisset J, and van der Stuyft P (2006) Aedes aegypti Larval Indices and Risk for Dengue Epidemics. *Emerging Infectious Diseases,* **12(5),** 800-806.
[7]. Abdalmagid M and Alhusein H (2008) Entomological investigation of *Aedes aegypti* in Kassala and Elgadarief States, Sudan. *Sudanese Journal of Public Health,* **3(2),** 77-80.
[8]. Erlanger T, Keiser J and Utzinger J (2008) Effect of dengue vector control interventions on entomological parameters in developing countries: a systematic review and meta-analysis. *Medical and Veterinary Entomology,* **22(3),** 203-221.
[9]. Irlandez Cruz E, Salazar FV, Porras E, Mercado R, Oraisb V and Bunyic J (2008) Entomological survey of dengue vectors as basis for developing vector control measures in Barangay Poblacion, Muntinlupa City, Philippines.*Dengue Bulletin,* **32,** 167-170.
[10]. Singh R, Dhimana RC, Duaa VK and Joshib BC (2010) Entomological investigations during an outbreak of dengue fever in Lal Kuan town, Nainital district of Uttarakhand, India. *Journal Vector Borne Disease,***47,** 189-192.
[11]. Wongkoon S, Jaroensutasinee M, and Jaroensutasinee K (2005) Locations and Religious Factors Affecting Dengue Vectors in Nakhon Si Thammarat, Thailand. *Walailak Journal of Science & Technology,***2(1),** 47-58.
[12]. Kusriastuti R, and Sutomo S (2005) Evolution of dengue prevention and control programme in Indonesia. *Dengue Bulletin,* **29,** 1-7.
[13]. Naing NN (2010) A practical guide on determination of sample size in health science research.Malaysia: Pustaka Aman Press, p 36-57.
[14]. Queensland Government (2011) The Queensland Dengue Management Plan 2010-2015. Fortitude Valley : Queensland Health.
[15]. Zuhriyah L, Habibie I and Baskoro A (2012) The Key Container of Aedes aegypti in Rural and Urban Malang, East Java, Indonesia. *Health and The Environment Journal (HEJ): Mission Statement,***3(3)**, 51-58
[16]. Rajendran G, Amalraj D, Das L, Ravi R and Das P (2006) Epidemiological and Entomological Investigation of Dengue Fever in Sulurpet, Andhra Pradesh, India.*Dengue Bulletin*, **30**, 93-98.
[17]. Pham HV, Doan HT, Phan TT and Minh NNT (2011) Ecological factors associated with dengue fever in a Central highlands Province, Vietnam. *BioMed CentralInfectious Diseases,* **11(1),** 172,1-6.
[18]. Balakrishnan N, Venkatesh S and Lal S (2006) An entomological Study on the Dengue vectors during an outbreak of Dengue in Tiruppur town and its surroundings, Tamil Nadu, India. *Journal of Communicable Disease,* **38(2),** 164-168.
[19]. Halstead SB (2008) Dengue virus-mosquito interactions. *Annual Review Entomology,* **53,** 273-291.

Characterization of lactic acid bacteria from local cow's milk kefir

YS Ismail[1*], C Yulvizar[1], and B Mazhitov[1]

[1]Biology Department, Faculty of Mathematics and Sciences, Syiah Kuala University, Banda Aceh 23111, Indonesia

[*]E-mail: ysismail@unsyiah.ac.id

Abstract. One of products from milk fermentation is kefir. It is made by adding kefir grains which are composed of lactic acid bacteria and yeast into milk. The lactic acid bacteria are a group of bacteria that produce antimicrobial substances and able to inhibit the growth of pathogenic bacteria. In this research, the lactic acid bacteria were isolated from Aceh local cow`s milk kefir to determine the genus of the isolates. The methods used in the characterization of lactic acid bacteria are colony morphology, cell morphology, and biochemical tests which includes a catalase test; 5%, 6.5%, and 10% salt endurance tests; 37°C and 14°C temperature endurance tests, SIM test, TSIA test, MR-VP test, and O/F test. Of the four isolates found from the cow's milk kefir, two isolates were confirmed as lactic acid bacteria (isolates SK-1 and SK-4). Both isolates are Gram positive bacteria, and have negative catalase activity. From the observations of colony morphology, cell morphology, and biochemical tests, it was found that the genus of SK-1 is *Lactobacillus* and the genus of SK-4 is *Enterococcus*.

Keywords : characterization, Cow`s milk kefir, lactic acid bacteria, morphology,

1. Introduction

Fresh cow milk is one of many beverages with high nutrition, so it is not only benefits to humans but also benefit to decomposers. Bacteria contamination could spread rapidly and cause the quality of the milk goes below average and become no more consumable. To improve the functions, quality and economic factor of milk, managing technique is needed. One of the techniques which are very potential is milk fermentation. Some of the managed milk products which have been developed are fresh milk, yogurt and kefir pasteurization [1].

Kefir is fermented milk product. Kefir can be made by cows`, goats` or sheep` milk by adding kefir into one of those milk [2]. White particles are colonies of lactic acid bacteria, such as *Streptococcus* sp., *Lactobacilli* and some types of non pathogen yeast [3]; *Lactobacillus lactis, Lactobacillus delbrueckii* subsp. *bulgaricus*, and yeast [4]. Lactic acid bacteria are Gram positive bacteria, rod or coccus shaped cell, negative catalase, non-motile and produce lactic acid as the final result of carbohydrate fermentation [5].

Lactic acid bacteria in kefir are proved to have an antibacterial effect against pathogen bacteria that cause food poisoning such as *Salmonella, Helicobacter, Shigella, Staphylococcus* and *Escherichia coli* [6,7]. Food poisoning is a disease caused by consuming contaminated food. Food tends to be contaminated by pathogen bacteria, virus, parasite, natural chemical toxic like poisonous mushrooms

[8]. Food poisoning is a public health problem that happens a lot. Every year 2 million people, especially babies, and toddlers passed away because of food poisoning [9]. *Streptococcus sp.* and *Lactobacilli* which have been isolated from goat's milk kefir can prevent the growth of *Bacillus cereus* [10]. These bacteria cause food to become poisonous because of enterotoxin they produce. Kefir can prevent the growth of *E.coli* with 16.7 mm maximum area of prevention [11].

According to literature study, isolation of lactic acid bacteria from cow's milk kefir, especially in Aceh, have not done yet. Because of that, there is a need to isolate and characterize the lactic acid bacteria from Aceh local cow's milk kefir.

2. Methods

2.1. Instruments

The instruments were used in this study are as follows: petri dish, ose needle (inoculation needle), pipette, micropipette, Erlenmeyer flasks, measuring cylinder, beaker glass, analytical balance, stirring rod, spreader rod, autoclave, magnetic stirrer, test tubes, test tube rack, centrifugation, Bunsen lamps, glass object, glass cover, microscope, oven, refrigerator, incubator, laminar air flow cabinet, hotplate, digital camera, caliper and pencil.

2.2. Materials

The materials were used in this study are as follows : milk of Aceh's cow, kefir seeds from Kazakhstan, distilled water, alcohol 70% and 95%, KOH 3%, alpha-naphtol, Kovacs reagent, methyl red reagent, crystal violet, iodine, safranin, hydrogen peroxide (H_2O_2) 3%, physiological NaCl, 5% NaCl, 6.5% and 10%, de Mann Rogosa Sharpe agar (MRS agar) media, triple sugar iron agar (TSIA) media, sulfide indole motility (SIM) media, methyl red and Vogue Proskauer (MR-VP) media, oxidative/fermentative (O/F) media, and nutrient agar (NA) media.

2.3. Instruments and Materials Sterilization

All instruments washed until clean and dried. Then the instruments and materials were sterilized using an autoclave at a temperature of 121°C and a pressure of 15 psi for 20 minutes.

2.4. Milk Fermentation

The raw material is fresh milk kefir. Milk used in this study was the milk of Aceh's cow from the dairy farm. A total of 1 L of pasteurized cow's milk must be cooled at room temperature (28°C) and added by kefir seeds of Kazakhstan. Cow's milk was incubated at room temperature for 20-24 hours. Then, the milk became thick and formed cow's milk kefir.

2.5. Lactic Acid Bacteria Isolation

Isolation of lactic acid bacteria was done by suspending 1 ml kefir of fermented cow's milk into 9 ml of distilled water (dilution 10^{-1}). Then, 1 ml sample from dilution 10^{-1} was taken to dissolved in 9 ml of distilled water (dilution 10^{-2}), and so on, until the dilution 10^{-7}. Then 1 ml sample was taken from each dilution and grown in de Mann Rogosa Sharpe agar (MRS agar) media with the spread plate method. Furthermore, the media were incubated at 37°C for 24 hours. Isolates which were grown then retrieved and inoculated on MRS agar plate using ose needle and streak method to obtain pure cultures was incubated for 24-48 hours at 37°C in the incubator.

2.6. Lactic Acid Bacteria Characterisation

2.6.1. Morphological Observation of Lactic Acid Bacteria Colonies

Observations of colony morphology was done after getting a pure culture in a petri dish. These observations were made visually include shape, color, edges, and the elevation of bacterial colonies. The surface of colonies can be seen from the side and the edge of the colony can be seen from the above [12].

2.6.2. Morphological Observation of Lactic Acid Bacteria Cells (Gram Staining)
Object glass slides that will be used for Gram staining technique was sterilized by 70% alcohol. Furthermore, one of bacteria isolates was taken by ose needle aseptically and smeared on a slide and fixed several times on the Bunsen lamp fire. Then bacteria isolates were etched with two drops of crystal violet and allowed for 1 minute. Furthermore, isolates were washed with distilled water and dry-aired. Then, iodine was dropped on it and allowed for 1 minute, and isolates were washed with distilled water and dry-aired. Bacteria isolates were mixed with drops of 95% alcohol slowly for 30 seconds, then isolates were washed with distilled water and dry-aired. Isolates then spilled by safranin slowly for 30 seconds. Then, isolates were washed with distilled water and then dry-aired again. Bacteria isolates preparates were observed using a microscope. Observations included the color and shape of a bacterial cell. Gram-positive bacteria marked by the purple (or blue) color which indicates that the bacteria cell wall are capable of binding to crystal violet dye, whereas Gram-negative bacteria is characterized by the formation of pink color indicates that the bacteria cell wall are not able to bind the crystal violet dye, and only stained by safranin dye (as comparative dye).

2.6.3. Biochemical Tests of Lactic Acid Bacteria
2.6.3.1. Catalase Test
Catalase test was conducted by dripping two drops of hydrogen peroxide (H_2O_2) 3% at 24 hours aged cultures on an object glass slide. Catalase test positive reaction characterized by the formation of oxygen bubbles that indicate the bacteria produce the catalase enzyme which converts H_2O_2 to water (H_2O) and oxygen (O_2) [12].

2.6.3.2. Growth Test on Different Concentrations of Salt
One colony of bacteria was inoculated into the de Mann Rogosa Sharpe broth (MRS broth) media with a concentration of 5%, 6.5% and 10% NaCl. Then the cultures were incubated at 37°C for seven days. The bacteria growth showed by the sediment formation in the media.

2.6.3.3. Growth Test at Different Temperatures
One colony of bacteria was inoculated into the de Mann Rogosa Broth (MRS broth) media and incubated at temperatures 14°C and 37°C for seven days. The bacteria growth showed by the sediment formation in the media.

2.6.3.4. Motility Test
One colony of bacteria inserted into the sulfide indole and motility (SIM) media in a test tube. The test tube incubated for 48 hours at 37°C. The observation of motility test is the growth of bacteria on the media. The bacteria that only grow around inserted location show a negative result, while the bacteria that grow on the media surface or spread in the media show positive result [13].

3. Results and Discussions
Four isolates of milk kefir (SK-1, SK-2, SK-3, SK-4) were obtained from the dilution method. Dilutions of 10^{-1} until 10^{-7} of kefir were grown on selective media for lactic acid bacteria, de Mann Rogosa Sharpe agar (MRS agar).

3.1. Morphological Characterization of Lactic Acid Bacteria
Characterization is the basis for bacterial identification. Characterization can be done based on colony morphology, characteristic of cytology (cell type and Gram staining) and physiological characteristics [16]. Observed morphology in this study was colony morphology and cell morphology. Colony morphology includes color types, edges, and elevation of colonies that observed visually, while the cell morphology was observed by Gram stain using a microscope.

3.1.1. Colony Morphology
Colony morphology of 4 isolates (SK-1, SK-2, SK-3 and SK-4) bacteria are round (circle) shape (100%), has slippery (smooth) edge (100%), the form of convex elevation (100%), and milky white (75%) and cream (25%) of colony color (Table 1). Several isolates of lactic acid bacteria isolated from fresh goat's milk have circle type colonies, smooth edges, convex elevation and a milky white color [17].

Table 1. Characteristics of colony morphology and cell morphology of isolates

No.	Isolate Code	Colony Morphology				Cell Morphology	
		Type	Color	Edge	Elevation	Cell Shape	Gram Staining
1	SK-1	Circle	Milky White	Smooth	Convex	Bacilli (rod)	Positive
2	SK-2	Circle	Milky White	Smooth	Convex	Bacilli (rod)	Negative
3	SK-3	Circle	Milky White	Smooth	Convex	Bacilli (rod)	Negative
4	SK-4	Circle	Beige/Cream	Smooth	Convex	Cocci (spherical)	Positive

3.1.2. Cell Morphology
Cell morphology was done to determine the colored smear culture under a microscope and see how the shape of the cell, the characteristic of the Gram bacteria [18]. Gram staining results showed that isolates SK-1 and SK-4 are purple. These results showed that isolate SK-1 with rod shape of cell and SK-4 with cocci shape of cell belonging to the group of Gram-positive bacteria. While the results of isolates SK-2 and SK-3 are red. It shows these isolates belong to the group of Gram-negative bacteria with the rod shape of cell (Table 1). Isolates SK-1 and SK-4 with Gram-positive results included in the lactic acid bacteria group, whereas isolates SK-2 and SK-3 with Gram-negative results are not included in the group of lactic acid bacteria. These results are supported by researches [19, 20] which states that the lactic acid bacteria (LAB) are in the group of Gram-positive with the shape of cell cocci or bacilli. Research [21] also found five Gram-positive isolates that belong to the lactic acid bacteria, and five isolates of Gram negative excluding lactic acid bacteria.

3.2. Biochemical Tests
The biochemical test is a method to identify pure cultures of bacteria isolates through the characteristic of cell physiology. Biochemical tests performed in this study include catalase test, salt resistance test, test temperature resistance, SIM test, TSIA test, MR-VP test, and O/F test. The results of biochemical tests of bacteria isolated from milk kefir can be seen in Table 2.

IOP Conf. Series: Earth and Environmental Science **130** (2018) 012019 doi:10.1088/1755-1315/130/1/012019

Table 2. The biochemical tests' results of cow's milk kefir LAB isolates

No.	Biochemical Test		Isolate Code	
			SK-1	SK-4
1.	Catalase		-	-
2.	Salt endurance			
	- 5%		+	+
	- 6,5%		+	+
	- 10%		+	+
3.	Temperature endurance			
	- 14°C		+	+
	- 37°C		+	+
4.	Motility		-	-
5.	Indol		-	-
6.	TSIA	B/S	Y/Y	Y/Y
		Gas	-	-
		H_2S	-	-
7.	Sucrose		+	+
8.	Lactose		+	+
9.	Glucose		+	+
10.	MR		+	+
11.	VP		-	-
12.	O/F Test		F	F
	Identification result		*Lactobacillus*	*Enterococcus*

Information :
+ : Positive reaction - : Negative reaction
B : Butt S : Slant
Y : Yellow F : Characteristically fermentative

3.2.1. Catalase Test

Lactic acid bacteria do not produce catalase enzyme that converts hydrogen peroxide into water and oxygen. Some bacteria require oxygen to form hydrogen peroxide, which is a byproduct of aerobic metabolism that is toxic. While lactic acid bacteria do not require oxygen, because it includes in groups of anaerobic bacteria (bacteria that do not need oxygen to grow) [22, 23]. Catalase reaction is positive when the formation of air bubbles that indicate the formation of O2 gas, and negative catalase results if it does not indicate the presence of gas bubbles. These bacteria are not able to produce the catalase enzyme to convert hydrogen peroxide / H_2O_2 into water and oxygen, which is one of the characteristics of lactic acid bacteria [24]. Catalase is an enzyme that serves to decompose hydrogen peroxide into water and oxygen ($2H_2O_2 + O_2$) [25]. Research [26] found nine isolates were negative catalase lactic acid bacteria isolated from fermented goat milk.

3.2.2. Salt Endurance Test

Salt endurance test was done to determine the ability of bacterial growth at different salt concentrations. Salt endurance test results showed that isolates SK-1 and SK-4 may grow at the concentrations of 5%, 6.5%, 10% NaCl. Lactic acid bacteria that were grown included in the BAL group of moderately halophilic and extremely halophilic namely the need of salt for growth at a concentration of 5-30% [27]. Each genus of lactic acid bacteria has different tolerances to grow on media with different concentrations of NaCl salt [28]. Lactic acid bacteria of the genus *Lactobacillus, Pediococcus* and *Leuconostoc* are able to grow and thrive on fermented using high salinity [29].

Research [30] found four isolates of lactic acid bacteria that can live at NaCl concentration of 4% and one isolates at a concentration of 6.5% NaCl isolated from fermented milk.

3.2.3. Temperature Endurance Test
Temperature endurance test was done to determine the ability of bacterial growth at different temperatures. The test results showed the isolates SK-1 and SK-4 were able to survive at different temperatures. Both isolates can grow at temperatures of 14°C and 37°C. The bacterial growth showed by sediment formation in the media. Lactic acid bacteria are classified as mesophilic bacteria growth with a temperature range between 10-45°C [31]. According to [32, 33] the optimum temperature for growth of lactic acid bacteria is 30-45°C. This result is supported by research [34], getting three isolates of lactic acid bacteria from fermented goat's milk that can grow at a temperature of 14°C, but the better growth seen at 37°C. Each genus of lactic acid bacteria has different optimum growth temperature [28].

3.2.4. Motility Test
The aim of motility test is to determine the bacteria are motile or non motile, if there are flagella which function to use in movement. Motility testing of isolates SK-1 and SK-4 showed results are non motile. Isolates SK-1 and SK-4 are categorized as non motile because the growth was only along the line of inoculation. This result suggests that all the LAB isolates do not have flagella. Flagella is a filamentous structure composed of proteins attached to the cell surface and allows the movement of motile bacteria. Motility test is positive if the growth of the colony spreaded on agar [35]. Lactic acid bacteria isolates from buffalo milk curd is non motile [36]. Lactic acid bacteria of the genus *Lactobacillus* has a characteristic of non motile bacteria [35].

4. Conclusion
Based on analysis of the results, it can be concluded that the results of isolation and characterization of lactic acid bacteria from Aceh local cow's milk kefir got two isolated bacteria. Isolate SK-1 is bacteria from *Lactobacillus* genus and isolate SK-4 is *Enterecoccus* genus. Isolate SK-1 identified as lactic acid bacteria *Lactobacillus* genus with the characteristics are Gram positive, non-motile, rod shaped cell, negative catalase, and other physiological characters based on the biochemical tests. Isolate SK-4 identified as lactic acid bacteria *Enterococcus* genus with the characteristics are Gram positive, non-motile, coccus shaped cell, negative catalase, and other physiological characters based on the biochemical tests.

References
[1] Zakaria Y (2009) Pengaruh Jenis Susu dan Persentase Starter yang Berbeda terhadap Kualitas Kefir. *Agripet*. 9 (1): 26-30.
[2] Kosikowski (2011) Kajian Penggunaan Ekstrak Susu Kedelai Terhadap Kualitas Kefir Susu Kambing. *Jurnal Ternak Tropika*. 12 (1): 15-21.
[3] Usmiati S (2007) Kefir, Susu Fermentasi dengan Rasa Menyegarkan. *Jurnal Warta Penelitiandan Pengembangan Pascapanen Pertanian*. 29 (2): 12-15. Bogor.
[4] Albaarri AN and Murti TW (2003) Analisa pH, Keasaman dan Kadar Laktosa pada Yakult, Yoghurt, Kefir. *Proceeding Simposium Nasional Hasil-hasil Penelitian di Unika Soegijapranata*, Semarang 22 Maret 2003.
[5] Suryadjaja A (2005) Potensi Ubi Jalar Putih dan Merah (*Ipomoea batatas* L.) untuk Pertumbuhan Bakteri Asam Laktat dan Menekan Pertumbuhan Patogen. *Skripsi*. Fakultas Teknologi Pertanian, Institut Pertanian Bogor. Bogor.
[6] Ulusoy BS, Colak H, Hampikyan H and Erkan ME (2007) An *In Vitro* Study on the Antibacterial Effect of Kefir Against some Food-borne Pathogens. *Turk Mikrobioloji Cemiyeti Dergisi,* 37: 103-107.
[7] Raja A, Gajalakshmi, MMM, Raja and MM Imran (2009) Effect of *Lactobacillus lactis cremonis* Isolated from Kefir against Food Spoilage Bacteria. *American Food Technology,* 4: 201-209.
[8] Dorland (2011) *Dorland's Illustrated Medical Dictionary*, Deluxe Edition. London UK.

[9] Morgan GE, Mikhail MS and Murray MJ (2006) *Clinical Anesthesiology, 4th ed.* Lange Medical Books/McGraw-Hill. New York, USA.
[10] Suhartanti D, and Septian R (2014) Comparison of the Antibacterial Activity of Cow Kefir and Goat Kefir Against Bacteria *Bacillus cereus. KESMAS.* 8 (2): 71-76.
[11] Yesillik S, Yildirim N, Dikici A, Yildiz A, and Yesillik S (2011) Antibacterial Effects of Some Fermented Commercial and Homemade Dairy Products and 0,9% Lactic Acid against Selected Foodborne Pathogens. *Asian Journal of Animal and Veterinary Advances*. 6(2): 189-195.
[12] Hadioetomo RS (1993) *Mikrobiologi Dasar dalam Praktek*. Gramedia. Jakarta.
[13] Lay BW (1994) *Analisis Mikroba di Laboratorium.* Raja Grafindo Persada. Jakarta.
[14] Volk WA, and Wheeler, MF (1993) *Mikrobiologi Dasar Jilid I.* Diterjemahkan oleh Markham. PT.Gelora Aksara Pratama, Jakarta.
[15] Davis WW, and Stout TR (1971) Disc Plate Methods of Microbiological Antibiotic Assay. *Microbiology 22: 659-665.*
[16] Sudarsono A (2008) Isolasi dan Karakterisasi Bakteri pada Ikan Laut dalam Spesies Ikan Gindara (*Lepidocibium flavobronneum*). *Skripsi*. Institut Pertanian Bogor. Bogor.
[17] Ernawati (2010) Isolasi dan Identifikasi Bakteri Asam Laktat pada Susu Kambing Segar. *Skripsi*. Universitas Islam Negeri Maulana Malik Ibrahim, Malang.
[18] Pelczar MJ and Chan ECS (1986) *Dasar-dasar Mikrobiologi.* Diterjemahkan oleh Hadioetomo, R.S., Imas, T., Tjitrosomo, S.S. dan Angka, S.I. UI Press. Jakarta
[19] Syahrurachman (1994) *Buku Mikrobiologi Kedokteran*. Universitas Indonesia Press. Jakarta.
[20] Wibowo and Ristanto D (1988) *Petunjuk Khusus Deteksi Mikroba Pangan.*Pusat Antar Universitas Pangan dan Gizi Universitas Gadjah Mada. Yogyakarta.
[21] Velmurugan K, Sekar A, and Palingamoorthy G (2013) *Antagonistic Studies of Kefir Bacteria Against Food Pathogens.* LAMBERT Academic Publishing. Germany.
[22] Tadasse G, Ephraim E, and Ashenafi M (2005) Assessment of the Antimicrobial Activity of Lactic Acid Bacteria Isolated from Borde and Shamita, Traditional Ethiopian Fermented Beverages, on some Food-borne Pathogens and Effect of Growth Medium on the Inhibitory Activity. *Journal of Food Safety.* 8: 14-18.
[23] Battcoock M, and Ali SA (1998) *Fermented Fruits and Vegetables, A Global Perspective*. FAO Agricultural Services Bulletin No 134. Rome, Italy.
[24] Djide MN, and Wahyuddin, E (2008) Isolasi Bakteri Asam Laktat dan Air Susu Ibu dan Potensinya dalam Penurunan Kadar Kolestrol Secara *In Vitro. Majalah Farmasi dan Farmakologi*. 12 (3): 12-15.
[25] Fugelsang KC, and Edwards CG (2007) *Wine Microbiology, Practical Applications and Procedures*. Springer Science Bussiness Media LLC: New York, USA.
[26] Guessas B, and Kihal M (2004) Characterization of Lactic Acid Bacteria Isolated from Algerian and Zone Raw Goat's Milk. *African Journal of Biotechnology*. 3 (6): 339-342.
[27] Kuswanto KR and Sudarmadji S (1988) *Proses-proses Mikrobiologi Pangan.* Pusat Antar Universitas Pangan dan Gizi.Yogyakarta.
[28] Vos DP, Garrity GM, Jones D, Krieg NR, Ludwig W, Rainey FA, Schleifer KH, and Whitman, WB (2009) *Bergey`s Manual of Systematic Bacteriology*. Springer Dordrecht Heidelberg. New York, USA.
[29] Buckle KA, Edwards RA, Fleet GH, and Wootton, M (1987) *Ilmu Pangan*. Diterjemahkan oleh Purnomo dan Adiyono. UI Press. Jakarta.
[30] Yoshikazu I, Ikue S, and Seiichi S (1994) Identification and Characteristics of Lactic Acid Bacteria Isolated from Masai Fermented Milk in Northern Tanzania. *Journal Bioscience, Biotechnology, and Biochemistry*. 58 (4): 660-664.
[31] Fardiaz, S (1988) *Fisiologi Fermentasi*. Pusat Antar Universitas Lembaga Sumberdaya Informasi, Instititut Pertanian Bogor. Bogor.
[32] Stamer JR (1979) The Lactic Acid Bacteria. Microbes of Diversity. *J. Food Technology.* 1: 60 – 65.
[33] Ardhana MM, and Graham HF (2003) The Microbial Ecology of Cocoa Bean Fermentation in Indonesia. *International Journal of Food Microbiology*. 86: 87-99.

IOP Conf. Series: Earth and Environmental Science **130** (2018) 012019 doi:10.1088/1755-1315/130/1/012019

[34] Safriani SR (2015) Isolasi dan Karakterisasi Serta Uji Antimikroba Bakteri Asam Laktat dari Fermentasi Susu Kambing Peranakan Etawa. *Skripsi*. Universitas Syiah Kuala. Banda Aceh.
[35] Hasan ZH (2006) Isolasi *Lactobacillus*, Bakteri Asam Laktat dari Feses dan Organ Saluran Pencernaan Ayam. *Seminar Nasional Teknologi Peternakan dan Veteriner 2006*.
[36] Sunaryanto R and Bambang M (2012) Isolasi, Identifikasi dan Karakterisasi Bakteri Asam Laktat dari Dadih Susu Kerbau. *Jurnal Sains dan Teknologi Indonesia*. 14(3): 288-233.

Influence of exercise on plasma ammonia and urea after ingestion beverages of carbohydrate electrolyte

Gusbakti Rusip[1*], Sri Mukti Suhartini[2], *Ang Boon Suen[3]*
[1]Department of Physiology, Faculty of Medicine, Muhammadiyah University of Sumatera Utara
[2]Programme doctor, Faculty of Medicine, University of Indonesia
[3]Departement of Exercise Physiology, Faculty of Medicine, University of Sains Malaysia

* Email: gusbakti@nusa.net.id

Abstract. Ingestion of beverages with carbohydrate electrolyte during exercise can delay fatigue. Fatigue caused by the decreasing of glycogen deposit source and indefensible reproduced ATP result in the improvement of IMP and ammonia during fatigue. The aim of this research was to observe the alteration of plasma ammonia and urea before, during and after exercise, after ingestion beverages of carbohydrate - electrolyte. Ten male subjects (age 18-30 years) were subjected to there cycle ergometer at 60% of VO_{2max} with a pedal speed of 60 rpm until there is fatigued. The subject was given a drink of carbohydrate-electrolyte at a concentration of 6%, 12% and a flavored water placebo (P) to consume the volume of 3 ml/kg BW every 20 minutes. Blood samples were taken at rest and every 20 minutes until fatigue for analyzing plasma ammonia and urea. Mean exercise until fatigue show that no difference for three beverages. However, plasma ammonia and urea were significantly increase compared before and after exercise ($p<0.001$). Results of plasma ammonia before exercise for beverages CHO 12% (HC) (31.86±1.93μml/l vs 86.50±5.13μml/l), for CHO 6% (MC) (33.08±1.43μml/l vs 90.68±3.41μml/l), for no carbohydrate (P) (33.64±1.93μml/l vs 93.12 ± 2.91μml/l). Whereas plasma urea before exercise for beverages CHO 12% (4.75±0.12mmol/l vs 5.44±0.10mmol/l), for CHO 6% (4.88±0.20mmol/l vs 5.22± 0.10mmol/l), for Placebo (4.88±0.20mmol/l vs 5.54±0.24mmol/l). Conclusions that increase of plasma ammonia of during fatigue, can become the criteria for determining intensity exercise until fatigue results are better than plasma lactate.

Keywords: exercise cycling ergometer, plasma ammonia, urea

1. Introduction

Ingestion beverages carbohydrate-electrolyte during exercise can improve performance. Recently it has been reported that carbohydrate ingestion during prolonged strenuous exercise (i.e., 74% of max O2 uptake) can delay the development of fatigue [1,2]. Carbohydrate administration may result in increased utilization of blood glucose with a proportional slowing of muscle glycogen depletion [3]. Their fatigue during exercise is related to elevated levels of ammonia, which is one of the products of metabolism protein [4,5]. Fatigue usually occurs due to decreased levels of muscle glycogen to exercise at moderate intensity [2,6]. In the prolonged exercise, can a reduction in muscle glycogen resources and failure re-phosphorylation ADP. Source reduction of glycogen during fatigue cannot maintain re-synthesis ATP and causes increased inosine monophosphate (IMP) and ammonia. Several studies have manipulated the supply of carbohydrate to contracting the skeletal muscle in an attempt to alter ammonia production during prolonged exercises at moderate intensities [5,6,7,8]. The effect of low glycogen content on metabolism ammonia is currently unclear because one study reported that muscle ammonia, production was enhanced in this circumstance, whereas another found no change. Some studied trained cyclists the influence of carbohydrate ingestion combined with altered muscle glycogen content during 2 hours of cycling exercise [3,8,9]. However, during the first 20 minute of exercise when exercise intensities were identical, plasma ammonia concentration was elevated in the subjects depleted of carbohydrates, and higher plasma ammonia concentration was attributed an increased muscle ammonia production derived from amino acid catabolism.

Ammonia production intense exercise has often associated with the development of fatigue. Depletion of glycogen stores limits ATP re-synthesis, which impairs exercise capacity due to resulting increases in fatigue [10]. The main sources of exercise ammonia are adenosine monophosphate (AMP) deaminase and catabolism of branched chain amino acids. During high intensity exercise, ammonia mainly from the deamination of AMP.[11] The energy for muscular contraction is provided by the de-phosphorylation of ATP to ADP. When ATP use during exercise cannot be matched by re–synthesis, ATP depletion occurs due to irreversible deamination to form ammonia and inosine monophosphate (IMP) [12,13]. This is associated with a rising level in plasma ammonia, this terms " metabolic stress " because the impact on muscle contraction and accompanied by fatigue in healthy subjects [14,15]. The accumulation of plasma ammonia can be used as an indicator of ATP loss and the development of metabolic stress in exercising skeletal muscle [16,17]. Skeletal muscle depletion and plasma ammonia accumulation have been demonstrated during cycling ergometer in this study.

The purpose of this study was to investigate the changes in plasma ammonia and urea before and after exercise after ingestion beverages of carbohydrate electrolyte.

2. Methods

2.1 Subjects

Ten healthy recreational athletes were subjected to exercise on a cycle ergometer (Lode). Subjects were tested for maximal oxygen uptake (VO_2max) before any of the experimental trials initiated. Percentace age of body fat was estimated form skin fold. The solutions to be tested were two carbohydrate-electrolyte drinks and an artificially flavored water placebo. Each subject was tested in a random order under each of three conditions using a double-blind design has within two weeks the each trial. All subjects had given their consent before participation.

2.2 Materials

- Bicycle ergometer (Lode NVL-77)
- spectrophotometer
- Infusion catheter (butterfly needle)

2.3 Experimental design

Subjects were instructed to maintain a constant diet and training schedule throughout the experimental period, especially 2 days prior to each exercise bout. For all trials, subjects had reported to the laboratory in the morning after a period of 10–12 hours fasting. An 18-gauge catheter was insert into the superficial forearm vessel. Electrodes were secure to the skin for monitoring heart rate (HR) respectively. In the morning, subjects were allowed to consume water but no solid food. Five minutes prior to start cycling, subjects had consumed 3 ml/kg/body weight of MC, HC and water placebo and kept consuming the same schedule in during every 20 minutes exercise.

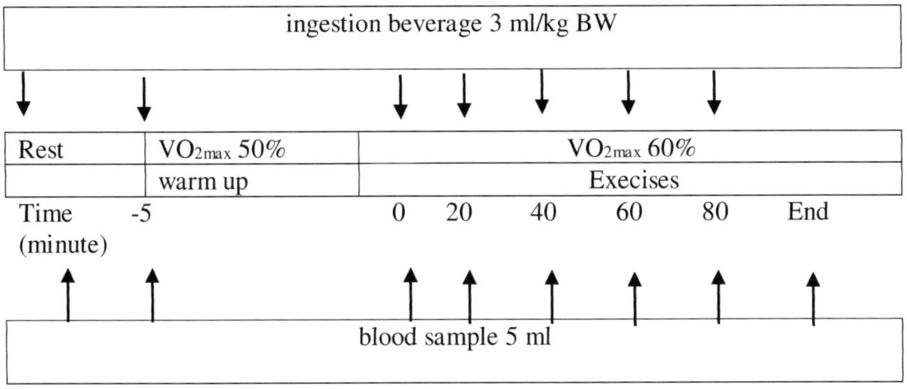

2.4 Biochemical analysis
Blood samples (5 ml) were obtain at rest and during exercise in every 20 minute until fatigued, centrifuged at 4°C, and plasma samples deposit temperature -70°C were used to analysis ammonia for measuring ammonia (Boehringer M.G,BH, ammonia) and Analysis urea (Boehringer M.G,BH, test combination urea S).

2.5 Statistical Analysis
All data were analyzed using student's t-test, was used for determining the level of significance ($P< 0,05$). All data reported as mean $\pm$ SE.

3. Results
The characteristic subjects age 24.6 ± 0.3 years, body weight 60.7± 2.3 kg, body height 166.3 ± 0.5 cm, capacity vital 4.0± 0.0 l, VO_{2max} 44.6 ± 0.5 $ml.kg^{-1}min^{-1}$, heart rate max 173.6 ± 1.1 $beat.min^{-1}$

3.1. Changes in plasma levels of ammonia
During exercises, plasma ammonia levels increase significantly on the three drinks were given, when compared before exercise, during exercise, and the end of the exercise showed a significant difference ($p <0.001$). Where the plasma levels of ammonia for drinks CHO 12% (HC) before 31.86 ± 1.93 μml /l and the end of exercise 86.50 ± 5.13 μml / l and CHO 6% (MC) before 33.08 ± 1.43 μml /l and the end of exercise 90.68 ± 3:41 μml /l, whereas the placebo drink without carbohydrate (P) before 33.64 ± 1.93 μml

/l and end of exercise 93.12 ± 2.91 μml /l. The three types of beverages no significant difference, but when compared to before exercise and the end of exercise seen a significant difference ($p < 0.001$).

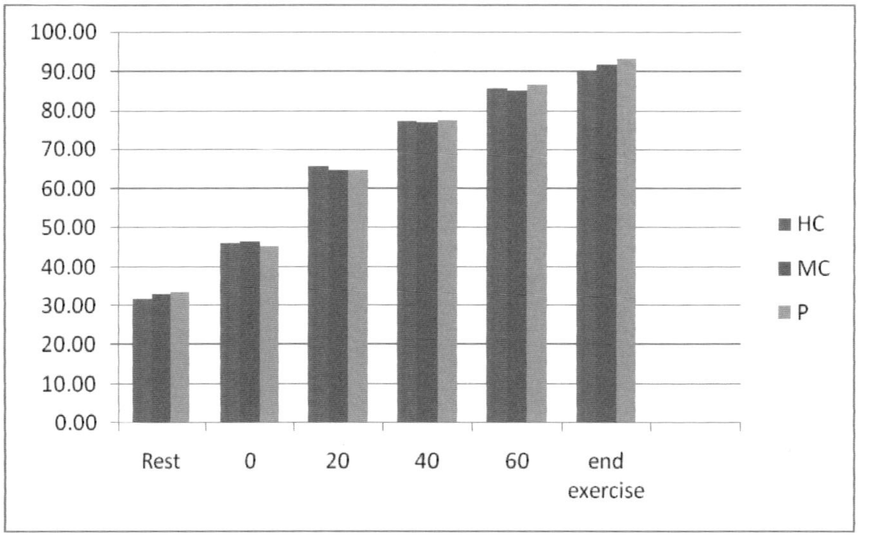

Figure 1. Changes plasma ammonia during exercise of cycling at VO_{2max} 60%, Rest, 0: warm up,end exercise, HC = high carbohydrate 12 %,MC: moderate carbohydrate 6%,P: no carbohydrate

3.2 Changes in plasma levels of urea

During exercises, plasma urea levels increase significantly on the three types of drinks given before exercise when compared the end of the exercise ($p < 0.001$). Urea levels before exercise for drinks CHO 12% (HC) 4.75 ± 0.12 mmol/l and the end of exercise 5.44 ± 0.10 mmol/l, and for drinks CHO 6% (MC) before 4.88 ± 0.20 mmol/l and the end of exercise 5.22 ± 0.10 mmol/l, whereas for placebo (P) before 4.88±0.20 mmol/l and the end of exercise 5.54 ± 0.24 mmol/l. From the following description does not a significant difference for all three types of drinks and significant of difference compare for before and the end of exercise (Figure 2).

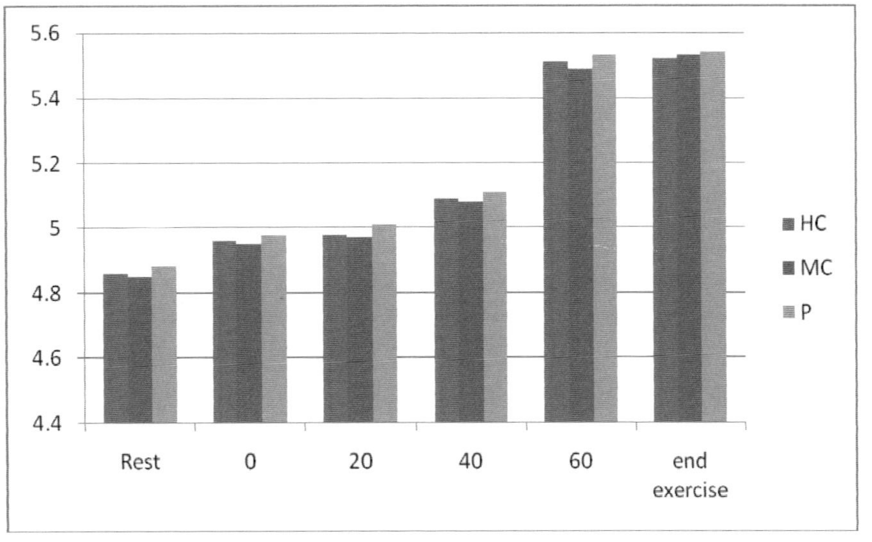

Figure 2. Changes plasma urea during exercise of cycling at VO_{2max} 60%, Rest, 0: warm up, end exercise, HC: high carbohydrate 12 %, MC: moderate carbohydrate 6%, P: no carbohydrate

4. Discussion

The predominant source of ammonia production during exercise, seems primary due to the deamination of AMP muscle by adenylate deaminase. Production of ammonia from other tissue has been observed experiment no specifically concerted with exercise e.q. chemical poisoning, oxygen toxicity, and also contribute to an elevated blood ammonia during exercise[18]. In this later study, blood samples were taken as long as 4 minute after the onset of exercise. The present study has investigated the kind course of blood ammonia concentration in man, in samples taken serially to a 20's interval during continuous ramp test to exhaustion on a bicycle ergometer [14,15,17].

In the study, it was clear that an increase in plasma ammonia was found during submaximal exercise intensity prolonged. One source of ammonia in the blood plasma is the activity of AMP deamination in muscle cells. Broberg and Sahlin, 1989 suggested that adenine nucleotide fragments may be caused low glycogen in muscle [5]. Brouns et al. (1990), also that the levels high of ammonia in the muscle is associated with muscle fatigue during prolonged exercise [15]. Although high levels of ammonia were secondary muscle glycogen levels decrease, but it is not clear whether the increased ammonia or low carbohydrate reserves is accelerating fatigue. In this study, an increase in plasma levels of ammonia on the three types of drinks are almost the same, although a little lower on the provision of carbohydrate drinks 12% (Figure 1). From the results of this study found an increase in plasma levels of ammonia caused by deamination of AMP may be more dependent on oxygen supply. Under normal circumstances 47-65 µml/l, during exercise can increase up to 20 folds. In a study undertaken by Babij et al. (1983) showed that in the resting state ammonia levels of arterial blood 50% taken up by the skeletal muscles, in addition Babij also estimates that only 28% of the total ammonia produced by deamination of adenine nucleotides that due to accumulate of ammonia in the blood and in the muscles of ammonia as a buffer to the synthesized of alanine, glutamic and glutamine [13,18]. Activities glucose-alanine cycle, in a derivative of glucose to form pyruvate alanine also found to increase during exercise. Alpha ketoglutarate and pyruvate in skeletal muscle is a precursor which showed an increase during exercise and tend to make ammonia as a buffer. Increased blood ammonia simultaneously with the addition of exercise intensity, may be a reciprocal relationship to changes in the synthesis of ammonia through glutamate, glutamine and alanine. The production of glutamine and alanine in skeletal muscle is affected by pH in the muscle during its formation. However, the expenditure of alanine and glutamine by muscle is greater than the expenditure of lactic acid, which contributes amino acids are controlled by pH. The time of recovery intensity of the exercise will result in a rapid increase, following a decline in blood ammonia and lactic acid blood. This situation has a significant connection between the workloads with increased blood ammonia and lactic acid. This increase is a reciprocal relationship between the production of substrates in the muscle, diffusion into the blood and changes the substrate through several metabolic pathways.

In this study, plasma levels of urea rise to the three types of drinks, increase significantly evident when compared before and end the exercise, the same result is also find in previous studies conducted by Fallowfied & William (1993). There is also found an increase in plasma urea during exercise and increased plasma urea concentration is a sign of metabolic protein [16]. Thus, increased plasma urea can estimate fatigue during exercise. Carbohydrate supplementation avoids the no use of amino acids. It shows that the role of the protein helps the final energy metabolism, and therefore known that changes in plasma urea have nothing correlation of carbohydrates during exercise.

5.Conclusion

The production of ammonia during exercise is an indicator of exercise intensity. The state exercise can increase the capacity of the muscles, changes in ammonia levels are associated with increased synthesis of glutamine. Ammonia and amino acid metabolism in the peripheral and central nervous system, an integration of physical activity, important that this increase may be affected by metabolic pathways during exercise. In this study found meaningless changes in plasma ammonia and urea after ingestion beverages of the three types, but significant changes were an apparent increase during exercise when compared before and the end exercise. Level blood lactic acid did not reach the summit due to the intensity of exercise until final few minutes of exercise and vary with blood ammonia levels that reach a maximum at the end of exercise and decreased slowly when exercise is stop. Therefore, increased levels of blood ammonia are equivalent to endpoint is fatigue. It appears that the criterion for the metabolism of proteins with the intensity of exercise until exhaustion occurs better than lactic acid.

References

[1] Coyle EF, Hagberg et al. (1983) Carbohydrate feeding during prolonged strenuous exercise can delay fatigue. J. App Physiol, 33: 230-35.
[2] Gusbakti R (2006) Ingestion of carbohydrate-electrolyte beverage improves exercise performance. Biomedical research. 17 (3): 183 – 87.
[3] Bergstrom JL, Hultman E (1967) A Study of the glycogen metabolism during exercise in man. Scand J Clin and Lab Invest. 19 (3): 218-23.
[4] Bromberg S, Sahlin B (1989) Adenine nucleotide degradation in human skeletal muscle during prolonged exercise. J Appl Physio. 67(1): 116-22.
[5] K Sahlin, S Bromberg, JM Ren (1989) Formation of inosine monophosphate (IMP) in human skeletal muscle during incremental dynamic exercise . Acta Physiol Scand. 136: 193-98.
[6] Adam Z, Malgorzata C, Adam M, Artur G, Jozef L (2015) Central and peripheral fatigue during resistance exercise. J Human Kinetics. 49: 159-69.
[7] Gorostiaga EM, Navarro-Amezqueta I, Calbet JAL. Sanchez-Medina I, Cusso M, Guerrero M et al. (2014) Blood ammonia and lactate as a marker of muscle metabolite during leg press exercise. J.Strength Cond Res., 28(10) : 2775-85.

IOP Conf. Series: Earth and Environmental Science **130** (2018) 012020 doi:10.1088/1755-1315/130/1/012020

[8] Rodney J, Snow MF et al. (2000) Effect of carbohydrate ingestion on ammonia metabolism during exercise in humans. J App Physiol. 88: 1576-80.

[9] Wagenmakers AJ, Beckers EJ et al. (1991) Carbohydrate supplementation, glycogen depletion, and amino acid metabolism during exercise. Am J Physiol Endocrinol metabolism. 260: E883- 890.

[10] Match BJC, Banister EW (1981)Ammonia metabolism in exercise and fatigue: A review. Med Sci Sport and Exerc. 15 (1): 41–50.

[11] Dudley GA, Terjungn RL (1985) Influence of aerobic metabolism on IMP accumulation in fast-twitch muscle. Am J Pysiol. 248(1 Pt 1): C37-42.

[12] Graham TE, MacLean DA (1992) Ammonia and amino acid metabolism in human skeletal muscle during exercise. Can J Physiol Pharmacol. 70(1): 132-41.

[13] Babij P, Matthews SM, Rennie MJ (1983)Changes in blood ammonia, lactate and amino acids about workload during bicycle ergometer exercise in man. Eur J Appl Physiol Occup Physiol. 50(3): 405-11.

[14] Nybo L, Dalsgaard MK, Steensberg A, Mielke M (2005) Cerebral ammonia uptake and accumulation during prolonged exercise in a human. J.Physiol.563(Pt 1): 285-90.

[15] Brouns F, Beckers E, Wagenmekers AJM and Saris WHM (1990) Ammonia accumulation during highly intensive long-lasting cycling: individual observations. Int J Sports Med. 11(Suppl 2), S78-S84.

[16] Fallowfield JL, William C (1993) Carbohydrate intake, and recovery from prolonged exercise. Int J Sport Med. 3(2): 150-64.

[17]Yuan Y, Chan KM (2004) A longitudinal study on the ammonia threshold in junior cyclists. Br.J.Sports Med. 38: 115-19.

[18] Demura S, Yamada T, Yamaji S, Komatsu M, and Morista KC (2010) The effect of L-ornithine hydrochloride ingestion on performance during incremental exhaustive ergometer bicycle exercise and ammonia metabolism during and after exercise. Eur.J.Clinical Nutr. 64: 1166-71.

In vivo thermoterapy: attempt to eliminate virus in potato tuber

Ida Ayu Astarini[1*], Deborah Margareth[1], I Gede Rai Maya Temaja[2]

[1]Biology Study Program, Faculty of Mathematics and Natural Sciences, Udayana University, Bali, Indonesia
[2]Plant and Diseases Study Program, Faculty of Agroecotechnology, Udayana University, Bali, Indonesia

[*]Email: iaastarini@unud.ac.id

Abstract. Potato is one of an important vegetable crop in Indonesia, including Bali. Main potato production areas in Bali are at Bedugul region, 1.200 m above sea level. Potato production in Bali continued to decrease due to diseases infection, such as early blight, late blight, black leg and virus diseases. Potato farmers in Bali usually set aside their harvest as seed potatoes, resulting in virus diseases being carried out on the next planting seasons and eventually would decrease potato production both in quantity and quality. Four types of virus were confirmed: PVY, PVX, PVS and PRLV. A number of studies have reported thermotherapy technique has been employed to eliminate potato virus *in vitro*. However, this technique is not readily available for farmers, since there is no established tissue culture laboratory to support. Therefore, there is an urgent need to develop a more practical method. The objective of this study was to eliminate virus on seed potatoes using thermotherapy on tuber. Seed potatoes with 1 cm sprout which were virus positive were placed on sterile charred rice paddy husk, and then put into a humidified incubator. Tubers were exposed to 37°C for four days followed by 34°C for three days alternately for two weeks and three weeks duration. Four tubers received heat exposure regime for each virus type. After thermotherapy, potato tubers were transferred to pots containing charred rice paddy husk and maintain for three weeks until new leaves emerge for virus analyses. Results show that seed tubers experienced delayed growth after thermotherapy. Control plants sprout one week after thermotherapy, while treated plants were not yet sprouting. Experiment is currently underway. It is expected that heat treatment on tuber will give a practical method for farmers to eliminate virus of seed potatoes.

Keywords: DAS-ELISA, heat treatment, *Solanum tuberosum*

1. Introduction

Potato (*Solanum tuberosum* L.) belongs to Solanaceae family. Potato is the fourth most important crop around the world including Indonesia. Potato is susceptible to various pathogens including fungi, bacteria, nematodes, and viruses. Virus is the most contributors for yield reduction in potato crop throughout the world. Among the viral diseases potato leaf-roll virus (PLRV), potato virus Y (PVY), potato mop top virus (PMTV), potato virus X (PVX), potato virus S (PVS) and potato virus A (PVA) is the major viral pathogens found throughout the world. As compare to the other pathogens, viruses damage plants and cause much more economic losses by a reduction in yield and quality of plant products. The severity of individual virus disease may vary with the locality, virus species, stage of infection and crop variety [1].

In Bali, potatoes grow at Bedugul highland region (1.200 m above sea level). Potato production in Bali continues to decrease, due to high production cost and diseases infections, resulting in declining number of farmers wanting to grow potatoes. Farmers could not afford to buy certified seed potatoes which can reach Rp. 50.000 per kg for third generation seed potatoes (G3). Farmers, therefore, would buy uncertified seed potatoes or set aside their harvest as seed potatoes. Using later generation (G3 and more) as seed potatoes has caused potato plants become very susceptible to pest and diseases, particularly virus [2].

Observation on some of potato fields in Pancasari Village, Bedugul region, revealed that approximately 70% of potato plants show virus symptoms on leaves and plant structure/habitus. Detection using DAS-ELISA technique shows four type of virus was confirmed, i.e., PVY, PVX, PVS and PLRV (data not shown). Tuber produces on these plants are mostly small, and many of them have a secondary infection from bacteria and fungi; causing tubers become easily rotten during storage. It is, therefore, an urgent need to find out practical methods to eliminate virus on potato tuber to be used for seed potato.

There are some successful methods have been employed to eliminate virus on potato, i.e. meristem culture, thermotherapy and cyrotherapy [3]. Encapsulation-dehydration followed by plunging in vitro shoot tips into liquid nitrogen (cryotherapy) was possible, but recovery rate was very low [4]. On in vitro thermotherapy technique, potato plantlets were incubated on 34°C/37°C night/day temperature for six weeks and then tested using ELISA for virus confirmation. If the virus was detected, repeat incubation until plantlets are virus free [5]. This repeat incubation slows down effort on providing virus free seed potatoes.

A more simple and practical method for virus elimination for local farmers need to be developed. An alternative method is to expose potato tubers with heat treatment. Evaluation of the effectiveness of thermotherapy to inactivate Potato leaf-roll virus (PLRV) from the potato tubers has been done [6]. The potato tubers were then treated with hot water at average 37°C for various intervals of time. It was concluded that thermotherapy at 37°C for 2 hours, 2½ hours and 3 hours in hot water treatment were in full or partially effective in elimination of PLRV from potato tubers. Exposing potato tubers to 35°C for 56 days or 36°C for 39 days resulting in significant effect on potato leaf roll virus elimination from small lots of seed of particular value in a breeding program; however, 50% of treated seed were dead [7].

The aim of this study was to find out response and survival rate of seed potato tubers after thermotherapy for different duration. This current work was conducted to explore the possibility of eliminating virus on potato tubers cultivar Granola using heat treatment regime. Using heat therapy close to death point of tuber might reduce the need for more complex technique to eliminate the virus.

2. Methods

The research was conducted at Biology Laboratory, Udayana University, Jimbaran campus, from October to August 2017. The equipment used in this experiment is a humidified chamber containing a thermostat, two light bulbs, and a thermometer.

Seed potato tubers cultivar 'Granola' was collected from farmer's field at Pancasari Village, Bedugul region. Collected seed were from plants with virus infection, and the virus type has been confirmed using DAS-ELISA method, i.e. PVY, PVX, PVS and PLRV. Seed potatoes were stored for ten weeks until sprouting. Seed potato tubers with 1 cm sprouts were placed in a humidified chamber and exposed alternately to 34°C temperature for three days, and 37°C for four days, for two and three weeks. Seed tubers were placed on charred rice paddy husk on the base to keep them moist. There are four tubers treated for each virus type. After thermotherapy treatment, tubers were transferred to 15 cm diameter pots containing charred rice paddy husk as the media and watered daily. Seed potato tubers with no virus and do not receive heat treatment were planted on the same media as the control.

3. Results and Discussion

Results show that there was no growth observed during thermotherapy on treatment 1 (37°C/34°C for 3 weeks) and 2 (37°C/34°C for 2 weeks). After one week exposure to heat, green sprouts started to change into brown color but tubers remain firm. On the second week, tubers on treatment 1 and 2 started to degrade due to heat. At the end of third week, all sprouts on treatment 1 were browning (Table 1). Only 50% of sprouts with leaf-roll and malformed symptom were survived on treatment 2 (Table 2).

Observation at one week after transferred to pots shows that only control tubers produced sprouts, started at five days after transferred. Heat treated tubers on treatment 1 was not survived; tubers were shrunk or dried out (Figure 1). Tubers receiving treatment 2 showed green sprouts but there was no further growth after 4 weeks in pots. Heat treatment delayed development of sprout and leaf emergence. This result is in agreement with [7] in which hot water treatment delayed shoot growth.

Exposing potato tubers rather than meristem tips has the advantage of less damaging and also allows use of larger meristem as the explant, and would be easier to develop into culture [9]. The mechanism of virus elimination using heat is still unclear, but induced changes in cell environment may inhibit virus multiplication. Heat-dependent alteration of cellular metabolism may cause the formation of o-quinones, activation of ribonucleases, and inhibition of replicases, and a reduction in ribosomes [3]. In this report, serological test DAS-ELISA has not been done.

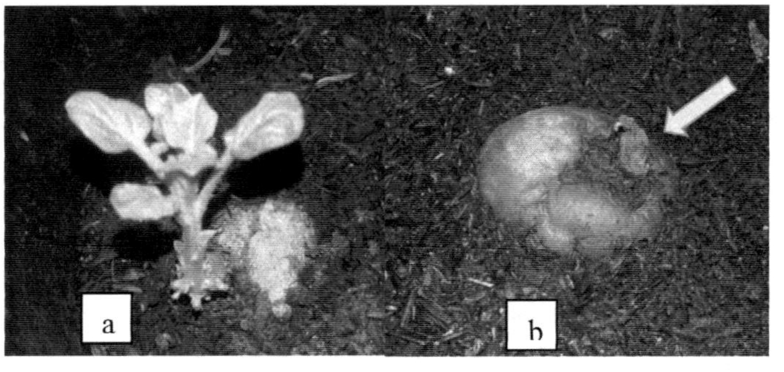

Figure 1. Post-thermotherapy tubers at two weeks after transplanting. a) Control tubers, b) Treated tubers. The arrow shows the browning sprouts.

Table 1. Seed tubers performance after exposure to heat treatment 37°C for 4 days followed by 34°C for 3 days alternately, for 3 weeks period.

No	Treatment 1	Potato tubers symptom, survival rate (%)		
		Mosaic	Leaf roll	Malformation
1	Green sprouts, firm tuber	-	-	-
2	Brown sprouts, firm tuber	-	25	50
3	Brown sprouts, squashy tuber	-	25	25
4	Brown sprouts, shrunk tubers	50	50	25
5	Brown sprouts, tuber rotten	50	-	-

Table 2. Seed tubers performance after exposure to heat treatment 37°C for 4 days followed by 34°C for 3 days alternately, for 2 weeks period.

No	Treatment 1	Potato tubers symptom, survival rate (%)		
		Mosaic	Leaf roll	Malformation
1	Green sprouts, firm tuber	-	50	50
2	Brown sprouts, firm tuber	50	25	50
3	Brown sprouts, squashy tuber	50	25	-
4	Brown sprouts, shrunk tubers	-	-	-
5	Brown sprouts, tuber rotten	-	-	-

4. Conclusion

Thermotherapy by exposing tubers to 37°C/4 days followed by 34°C/3 days for two weeks has the highest survival rate (50%). Exposing tubers for three weeks resulting in most tubers dead after therapy and survival tubers eventually died after two weeks transferred.

Acknowledgments

We would like to thank Ministry of Research, Technology and Higher Education, Republic of Indonesia, for the research grant under National Strategic Scheme 2017.

References

[1] Peter ET, Cliff EL, James CZ, Gary LR, Wojciech KK (2000) Extreme resistance to potato leaf roll virus in potato cv. Russet Burbank mediated by the viral replicase gene. Virus Res 71: 49-62.

[2] Damayanti TA, Kartika R (2015) Deteksi Virus-Virus pada Kentang di Jawa Barat dengan Menggunakan Teknik Molekuler. J Hort 25(2): 171-179 (DOI: http://dx.doi.org/10.21082/jhort.v25n2.2015.p171-179)

[3] Faccioli G (2001) Control of Potato Viruses using Meristem and Stem-cutting Cultures, Thermotherapy and Chemotherapy. *In* Loebenstein et al. (Eds.). Virus and Virus-like Diseases of Potatoes and Production of Seed-Potatoes, 365-390. Kluwer Academic Publishers.

[4] Wang B, Li JW, Zhang ZB, Wang RR, Ma YL, Blystad DR, Keller ERJ, Wang QC (2014) Three vitrification-based cryopreservation procedures cause different cryo-injuries to potato shoot tips while all maintain genetic integrity in regenerants. J Biotech 184:47-55

[5] Astarini IA, Chappell A, Scheuring D, Thompson SM, Julian Creighton Miller Jr, JC (2016) Optimasi Metode Cryotherapy untuk Mengeliminasi Virus pada Tunas Kentang In Vitro. J Hort 26(1): 97-102 (DOI: http://dx.doi.org/10.21082/jhort.v26n1.2016.p97-102)

[6] Zapata C, Miller Jr CJ, Smith RH (1995) An in vitro procedure to eradicate potato viruses X, Y, and S from Russet Norkotah and two of its strains. In vitro Cellular & Dev Biol. Plant 31(3): 153-159.

[7] Abbas A, Arif M, Ali A (2016) Use of Hot Water-Thermotherapy to Free Potato Tubers of Potato Leaf Roll Virus (PLRV). Int J Life Sci Scienti Res 2(2): 155-162.

[8] Fernow KH, Peterson LC, Plaisted RL (1962) Thermotherapy of potato leaf roll. Am J Pot Res 39 (12):445-451.

[9] Wang P and Hu C (1980) Regeneration of virus-free plants through in vitro-culture. In: Advances in Biochemical Engineering (ed.) A. Fletcher) Vol. 18, Plant Cell Culture II, pp 61-69. Springer-Verlag, Berlin, Germany.

Effect of incubation time of sago (*metroxylon sago)* waste by local microorganism "*ginta*" on ph, crude protein, and crude fiber content

Nurzainah Ginting[1*], E Pase[1]
[1]Animal Production Program Study, University of Sumatera Utara Jl. Dr. A. Sofian No.3, Padang Bulan, Kampus USU, Medan 20155, Indonesia

Email: nurzainahginting@gmail.com

Abstract. This study aims to examine the effect of incubation times of sago waste by local microorganism (MOL) "Ginta" to the crude protein and crude fiber content in relation to finding a cheap and good quality ruminants feed alternative. Incubation times were 0 hours to 144 hours. The data obtained were analyzed using Completely Randomize Design consisting of seven treatments and three replications. The result showed that the duration of incubation of sago waste by local microorganism (MOL) "Ginta" caused pH reduction, improved crude protein and crude fiber content. pH reduction was from 7.03 at 0 hour to 4.05 at 144 hours incubation. The highest increased in crude protein was H6U3 (5.58%) : 144 hours incubation and the lowest was H0U2 (3.22%) : 0 hour incubation while the highest crude fiber was H0U1 (19.99%) : 0 hour incubation and the lowest was H6U3 (18.23%) : 144 hours incubation. It can be concluded that incubation of sago waste triggered lower pH, higher crude protein and lower crude fiber than uninoculated. A recommendation could be given on using MOL 'Ginta" in order to produce a cheap and good quality ruminans feed alternative.

Keywords: sago waste, local microorganism (MOL) "Ginta", crude protein, crude fiber

1. Introduction

Sago waste which a by product of sago industry, has the opportunity and potential to be used as one of the alternative source of fibrous feed ingredients, because it has very high organic contents that potential as a source of energy. However, sago waste has low protein and high fiber content as for a feed [1].

According to Rumalatu [2] the processing of sago flour was obtained on flour and sago waste with a ratio of 1: 6. Sago waste was used to discharge in the location of sago flour processing which caused environmental pollution. Sago planting area in Indonesia has a land area of 1.6 million hectare BPS 2015 [3], where sago could be produced as much as 15 million tons because each sago stalk can produce 200 kg sago [4]. In Indonesia sago planting spread from Irian Jaya 1,406,469 Ha, Maluku 41,949 Ha, Sulawesi 45,540 Ha, Borneo 2,795 Ha, West Java 292 Ha to Sumatra 31,872 Ha [3].

Sago processing waste is classified as wet by-products because it still contains 75-80% moisture content, which can be degraded quickly if not processed immediately. Treatment by drying is commonly done before it is given to livestock. The nutrients contained in sago waste such as; states That the sago pulp contains 65.7 % starch, 14.8 % crude fiber, 1 % crude protein, and 4.1 % ashAbd-

Aziz[5]; The content of nutrients contained in sago waste are: crude protein of 3.36%, NDF 67.40%, ADF 42.11 and crude energy of 3,738 Kcal / kg, relatively comparable to grass [5]. With the nutrient content, sago waste is estimated to be only able to meet the basic needs of life, so for growth, pregnancy and lactation required additional feed to meet protein and energy needs of ruminants.

An alternative technique to improve nutrition quality of sago waste isby doing fermentation. In doing so, it is not only nutrition quality improved but also digestion and elimination of aflatoxin or toxic compounds. The fermentation process is done by adding for exmaplea starter of local microorganisms (MOL) "Ginta" which corresponds to the substrate and the purpose of the fermentation process. The use of mold of local microorganism (MOL) "Ginta" as a starter in the fermentation process is deemed most suitable and in accordance with the purpose of fermentation, namely to reduce the levels of crudefiber and at the same time to increase levels of crude protein. "Ginta" was created in the purpose that people in rural area could do fermentation without depend on commercial fermentor. This study aims to investigate effectiveness "Ginta" in fermenting sago waste on improving low protein content and high fiber.

2. Methods

This research had been conducted at Research and Technology Laboratory Faculty of Agriculture University of Sumatera Utara, Medan. Research started from December 2016 until February 2017.The materials used in this research were: sago waste, Local Microorganism (MOL) "Ginta" as source of isolate, aquadest as addition of water in MOL of "Ginta", alcohol as aseptic material, spiritus as bunsen fuel while working with media, media agar as new culture media for bacterial isolates. The tool were used autoclave, Erlenmeyer flask, petri dish, and ose loop.

The study was prepared based on Completely Randomized Design (RAL) pattern with 7 (seven) treatments with 3 (three) replications so that 21 units of experiments were obtained. The treatment arrangement based on the duration of incubation as follows [7]:

H0 : incubation time 0 hour, H1 : incubation time 24 hour, H2 : incubation time 48 hour, H3 : incubation time 72 hour, H4 : incubation time 96 hour, H5 : incubation time 120 hour, H6 : incubation time 144 hour. Fermentation of sago waste with local microorganism inoculum MOL "Ginta" was done with initial pH 6 and temperature 35 °C with long treatment incubation 0, 24, 48, 72, 96, 120 and 144 hours. The fermentation path can be seen in Figure 1.

The data obtained were analyzed using orthogonal polynomial test to see the influence trend of incubation duration on crude protein content and crude fiber of sago waste [7]. Furthermore, the data were analyzed statistically with the help of SPSS software Ver.13.0 to see the response curve. Parameters observed in this research were crude protein content [8] and crude fiber content [8].

Figure 1. Flow chart of Sago Waste Fermentation with Inoculum "Ginta"

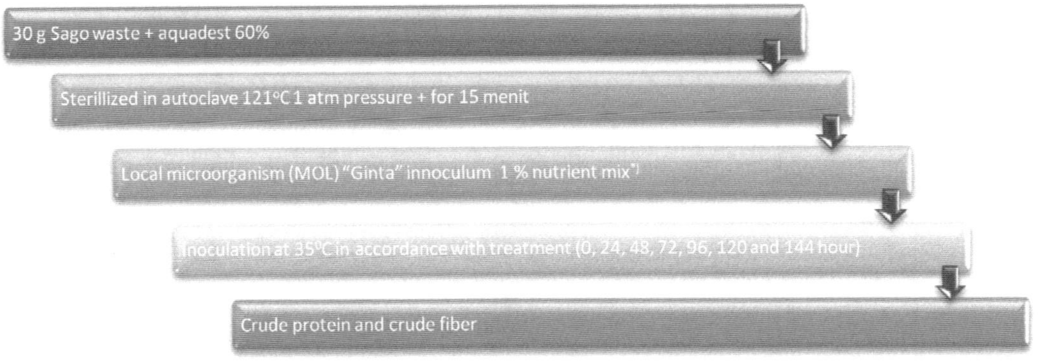

*)Nutrient mix consists of : Sucrose 25-50 g/l, Ammonium nitrate (NH_4NO_3) 2.25 g/l, Potassium hydrogen phosphate (KH_2PO4) 0.3 g/l, Magnesium ($MgSO_4$) 0.25 g/l.

3. Results and Discussion

The mean data of crude protein and crude fiber content of sago waste (*Metroxylon sagu*) inoculated with Local Microorganism (MOL) "Ginta" can be seen in Table 1. Inoculation with variation of time influenced pH, crude protein and crude fiber of sago waste.

3.1. pH

"Ginta" was created by adding kefir while it contained a lot lactic acid bacteria [9]. As the result bacteria metabolized carbohyrate became lactic acid which caused declined of pH. In this research, it was found that at 0 hour pH was 7.03 and at 144 hour pH was 4.05. Lower pH such as 4 was not a problem for sago to be ruminant feed as in the early stage of feed chewing, ruminant produced a lot of saliva which has an alkaline pH. As the result of chewing process was finely feed with normal pH Campbell et.al [10]. Research by Araujo [11] on youghurt while it also contained of BAL, there was a decrease in pH between 3.84 and 3.98. Sunarlim [12] found that there was a dominated bacteria amongs BAL which was Lactobacillusplantarum and that the benefits of BALs for a lot of health were included interfering with pathogenic bacteria, lowering cholesterol, anti-mutagenic, anti-carcinogenic and improving immune system.

3.2. Crude Protein

The influence of the incubation duration of sago waste (*Metroxylon sago*) by using Local Microorganism (MOL) "Ginta" on crude protein content based on response curve analysis can be seen in figure 2.Based on the results of the response curve analysis, it is known that crude protein gives polynomial response (quadratic) to the duration of incubation of sago pulp. The amount of correlation incubation duration of sago to crude protein content is 68.1%.The increment of crude protein content in the sago pulp of H6U3 is incubation for 144 hours due to the addition of microbial protein from local microorganism (MOL) "Ginta" which is called single cell protein. This is supported by the opinion of Tampoebolon [13], which states that the local microorganism (MOL) was a single cell protein (mold) containing high protein, so that like in this research the addition of starter with Local Microorganism (MOL) "Ginta" on sago waste added raw protein content of fermented sago waste. Najafpour [14] mentioned that microbial protein known as Single Cell Protein (SCP) or Single Cell Protein. Single Cell Protein is a term used for rough or pure protein derived from microorganisms, one of which is mold.

Table 1. Crude Protein (CP), Crude Fiber (CF) content of sago waste (*Metroxylon sagu*) inoculated with local microorganism *(MOL) "Ginta"* with variation of fermentation time

No.	Treatment	CP content (%)	CF content (%)
1.	H0U1 (0 h)	3.25	19.52
2.	H0U2 (0 h)	3.22	19.96
3.	H0U3 (0 h)	3.23	19.94
4.	H1U1 (24 h)	4.11	19.99
5.	H1U2 (24 h)	4.15	18.99
6.	H1U3 (24 h)	4.18	18.99
7.	H2U1 (48 h)	4.28	18.80
8.	H2U2 (48 h)	4.30	18.72
9.	H2U3 (48 h)	4.32	18.68
10.	H3U1 (72 h)	4.50	18.57
11.	H3U2 (72 h)	4.52	18.55
12.	H3U3 (72 h)	4.55	18.50
13.	H4U1 (96 h)	5.01	18.49
14.	H4U2 (96 h)	5.11	18.48
15.	H4U3 (96 h)	5.18	18.46

16.	H5U1 (120 h)	5.20	18.43
17.	H5U2 (120 h)	5.23	18.40
18.	H5U3 (120 h)	5.25	18.38
19.	H6U1 (144 h)	5.36	18.28
20.	H6U2 (144 h)	5.46	18.26
21.	H6U3 (144 h)	5.58	18.23

Figure 2. Respond Curve of The Influence of Incubation Duration of Sago (*Metroxylon sago)* Waste Using Local Microorganism (MOL) 'Ginta' against Crude Protein

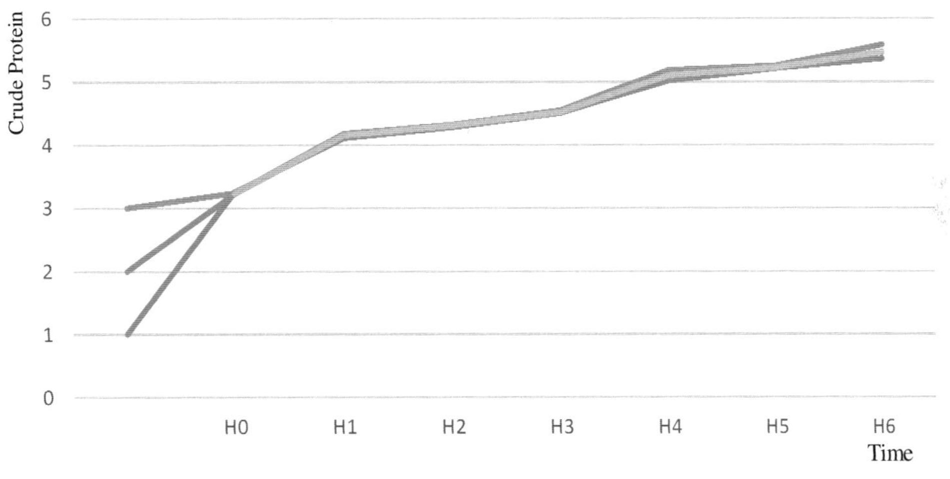

The content of crude protein in the sago waste begins to decrease in the treatment H0U2 (0 hours). This is because within 0 hours Local Microorganism (MOL) "Ginta" has not developed yet. The first and second day of the Local Microorganism (MOL) "Ginta" is in the lag phase, this phase is a microbial adjustment phase with the new environment. During this phase, cell division was slow. Day 2 (24 hours) to 3 (72 hours) growth of Local Microorganism (MOL) "Ginta" experiencing logarithmic phase, in this phase microbe is actively doing metabolism. Day 4 (96 hours) to 6th (144 hours) Local Microorganism (MOL) "Ginta" enters stationary phase, in this phase no longer develops. In this phase the cells become small because the cells keep splitting although the availability of nutrients in the medium has been greatly reduced. After undergoing the stationary phase the microbes begin to enter the phase of death. According to Vikineswary et.al [1], prior to inuculation, sago waste could be supplemented with source of nitrogen and that the result of microbial utilization was a good quality of animal feed.

3.3. Crude Fiber

The influence of the incubation duration of sago pulp (*Metroxylon sago*) by using Local Microorganism (MOL) "Ginta" on crude fiber content based on response curve analysis can be seen in Figure 3.

Figure 3. Respond Curve of The Influence of Incubation Duration of Sago (*Metroxylon sago)*Waste Using Local Microorganism (MOL) 'Ginta' against Crude Fiber Content

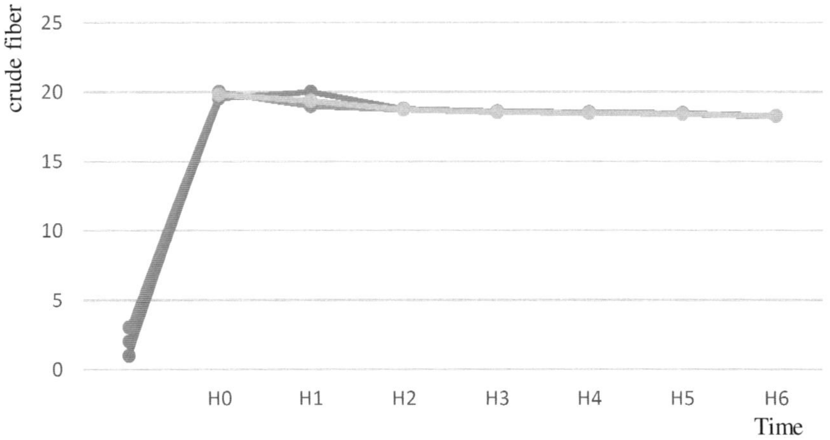

Based on the results of the response curve analysis, it was known that the crudefiber gives polynomial response (linear) to the duration of incubation of the sago pulp. The amount of correlation correlation duration of sago dregs incubation by using Local Microorganism (MOL) "Ginta" to crude fiber content is 92.7%.The decrease of crude fiber content of fermented sago waste with Local Microorganism (MOL) "Ginta" from day 0 to day 6 that is in treatment H0 until the treatment of H6 indicates that Local Microorganism (MOL) "Ginta" play an active role to produce enzyme Cellulase that serves to degrade the sago pulp. This is supported by the opinion of Tampoebolon[13] that the increase of incubation time causes the increase of local microorganism (MOL) "Ginta" to grow and fermentation, so that the incubation time will increase the chance of local microorganism (MOL) "Ginta" to degrade the higher. Cellulase enzyme is one of the enzymes produced by microorganisms that function to degrade cellulose into glucose. The decline in crude fiber in the sago fermentation results was due to the work of the cellulase enzyme produced by the local microorganism (MOL) "Ginta" which works to overhaul the crude fiber. This is supported by the opinion of Nurhayati [15], that the good growth of molds of local microorganisms (MOL) "Ginta" is expected to produce enzyme cellulase in large quantities so that it can be used to remodel and lower crude fiber.

4. Conclusion

The inoculation sago by local microorganisms (MOL) "Ginta" caused less pH, higher crude protein and lower crude fiber than non-inoculation. Increased duration of incubation from 0-144 hours in protein content gave a quadratic response where the highest peak of protein content was more in the duration of incubation of 120 hours.The increase in incubation duration from 0-144 hours on crude fiber content gave a linear response wherein the lowest crude fiber content was seen at the time of incubation of 144 hours.

References

[1] Vikineswary S, Shim JJ. Thambirajah, N Blakebrough (1994) Possible microbial utilization of sago processing waste. J. Resources, Conservation and Recycling. 11, Isssues 1-4, 30 June 1994, Pages 289-296.
[2] Rumalatu (1981) *Potensi dan Pemanfaatan Sagu***.** Yogyakarta: Kanisius.
[3] BPS-Statistic Indonesia (2015) Production Of Sago by Province (ton), 1993-2015.
[4] Flach M (1997) Yield Potential of The Realisation. Proc. Sago Conference in Serawak. Malaysia.

IOP Conf. Series: Earth and Environmental Science **130** (2018) 012022 doi:10.1088/1755-1315/130/1/012022

[5] Aziz, abdul (2002) Sago starch and its utilization. *Journal of Bioscience and Bioengineering*. 94(6):526 529. http://doi.org/cjwwkx.
[6] Trisnowati and Kurniawati (1991) *Kecernaan in vitro Ampas Sagu Metroxylon yang Diperlakukan Secara Biologis.* [Undergraduate Thesis]. Bogor Institute of Agriculture.
[7] Gazpersz V (1991) *Metode Rancangan Percobaan.* CV Armico, Bandung.
[8] AOAC (2005) Official Methods of Analysis of AOAC International. AOAC International.
[9] Dertli, Enes and Ahmet Hilmi (2017) Microbial diversity of traditional kefir grains and their role on kefir aroma. LWT- Food Science and Technology, Vol 89, Part A, November 2017, Pages 151-157.
[10] Campbell JR, MD Kenealy & KL. Campbell (2006) Animal Science. 4th Edition. McGraw-Hill, New York.
[11] Araújo DD, Alessandro B Amorim, Mayra AD Saleh, Felipe C, Pedro L. Perdig_on, Silvio J Bicudo, Dirlei A Berto (2016) *J of Animal Nutrition* 2: 149-153.
[12] Sunarlim R, Triyantini, Abubakar, M Poeloengan, and H Setiyanto (1999) *Peningkatan Teknologi Pembuatan Inokulum Mikroba Pengolahan Dadih untuk Menunjang Agroindustri Pedesaan.* Laporan Penelitian Balai Penelitian Ternak Bogor TA 1998/1999.
[13] Tampuebolon (2009) *Seminar Nasional Kebangkitan Peternakan*. Semarang.
[14] Najafpour D Ghasem (2007) CHAPTER 14-Single-Cell Protein. *J Biochemical Engineering and Biotechnology*. 332-341.
[15] Nurhayati (2010) *Bungkil Sawit dan Onggok Pakan Ternak Berkualitas.* http://www.polteklampung.ac.id/home/index.php?option=com_content&view=aticle&id=88%3A penelitian&catid=27%3Apenelitian&Itemid=6. Accessed on [Sunday, 21 March 2010].

Community structure, phytoplankton density and physical-chemical factor of batang palangki waters of sijunjung regency, west sumatera

Gusmaweti[1*], L. Deswati[1]
[1]Biology Education Study Program, Faculty of Teacher Training and Education University of Bung Hatta, Padang, West Sumatera, Indonesia

*Email: gusma_weti@yahoo.co.id

Abstract.The long-term goal of this study is to provide an overview of the presence of phytoplankton in support of its functions in the waters of Batang Palangki as a conservation area of information on river water management, especially for Batang Palangki stakeholders. Specific targets to be achieved in achieving these objectives are (1) to know the density of phytoplankton, index of diversity of species, equitabilty index, domination index, and in Batang Palangki waters, and (2) to analyze the chemical and physical factors of the waters. The sampling method of phytoplankton is *purposive sampling*. The phytoplankton sampling is done By filtering 100 liters of water into the net plankton no 25 and filtered into the 25 cc, and then identified. The determination of water quality such as water temperature, water pH and watercolour. dissolved oxygen (DO) and BOD, and Hg content (mercury). The results showed that phytoplankton found from each of station was 370 individualis per liter with the highest density found in the station I of 155. The number of genus was 7, namely *Neidium, Gyrogsima, Synedra, Frustulia, Fragillaria, Nitzschia* and *Peridinium*. The diversity index averaged at 0.45, equabilty index averaged at 0.54, while the dominance index averaged at 0.28. Physical and chemical factor measurement results found that water temperature averaged at 26 ^{o}C, transparency ranged from 12 - 30 cm, velocity speed ranged from 8 - 15 m/s, while chemical factors such as DO, BOD, and COD ranged from 5.25 to 5.96 mg/L, 3.28 - 3.49 mg/L, and 47.05 - 76.25 mg/L respectively. Likewise, TOM measured in this research was 9.61 - 2.10 mg/L while Hg content ranged from 0.098 - 0.208 mg/L.

Keywords: Community Structure, Phytoplankton, Waters, Physical-Chemical Factors

1. Introduction

Phytoplankton is water plants that have very small size and live floating in water Phytoplankton can act as one of the ecological parameters that can describe the condition of water. One characteristic of phytoplankton organisms is the basis of the feed chain in the waters [1]. Therefore, Its presence in a water can describe the characteristics of waters whether the water are in a fertile state or not.

The composition and abundance of phytoplankton will change at various levels in response to changes in environmental conditions both physical, chemical, and biological [2]. Diversity and abundance of phytoplankton in the waters affected by several environmental parameters and characteristics of a physiologically between factors physicochemical waters such as light intensity, dissolved oxygen, temperature stratification, and the availability of nutrients nitrogen and phosphorus, while the biological aspect is the activity of predation by animals, mortality Natural, and decomposition.

Classes are living as Phytoplankton is Chlorophyceae, Cyanophyceae, Euglenophyceae, Chrysophyceae, and Bacillariophyceae [3]. Chlorophyceae is a green alga. Has chloroplasts containing chlorophyll a, chlorophyll b and carotenoids. As an example of the genera of each order are: 1) an example of the genera Chlorococcales, *Pediastrum* and *Clorococcum,* 2) Ulotrichales: *Ulothrix* 3) Cladoporales: *Cladopora*, 4)Volvocales: *Poliblepharides*, *Clamidomonas* and *Volvox* and 5) *Conjugates, Clostridium, Desmidium, Spyrogira, Mougeotla* and *Zygnema* [4].

Batang Palangki is one of the river located in Ampek Nagari Sub-district of Sijunjung Regency of West Sumatera, with a length of a river of approximately 30 km and width of 20 meters. People activity conducted in a long the river is the mining and processing of gold, sand quarrying, household activities such as bathing, to wash and a fish pond and farms of the population where the water comes from the river. The activity of the population causes the turbid river water conditions and the addition of matrial into the river, And at the same time there is a change in river ecosystem. Changes in river ecosystems lead to changes in the type and density of organisms

Surveys and interviews conducted by the Wali Nagari and some people Batang Palangki District of Ampek Nagari Sijunjung on December 1, 2015 that generally livelihoods of the people tend to mine gold along the Batang Palangki, but it is also the activities of mining the sand, although officially permit mining by government No local, household activities such as bathing, washing and etc. As a result the activities carried out by the community caused the addition of material into the river. The addition of materials into the waters certainly will affect the river water conditions either biological, physical or chemical means. Changes in water content determined by the content of chemical compounds and materials that go into a body of water and is an important factor in studying the development of aquatic communities, especially phytoplankton

Research on the abundance of plankton and phytoplankton in the waters of the lake from India showed that the phylum Crysophyta of types Meloisira classified highest abundance was found, but the species of *Diatoma* sp, *Cymbela* sp and *Synedra* sp little find because of the type tend to live waters flow [5,6]. The Shannon-Wiener's diversity index shows the level of complexity of a community structure. Diversity also shows the pattern of distribution of a species in a community. The abundance of phytoplankton has a positive relationship with water fertility. If the abundance of phytoplankton in a water is high, then it could mean that they tend to have high productivity.

Phytoplankton can act as one of the ecological parameters that can describe the condition of water quality. Phytoplankton is the primary producer of food chain in the waters (4). Its presence in the water can describe the status of a water, whether the waters are polluted or not. Classify the fertility of the waters based on the abundance of plankton as follows. Waters with oligotrophic fertility are in abundance <2000 ind/L, Mesotrophic with abundance 2000-15000 ind/L and eutrophic> 15,000 ind /L [7].

Based on the description, we conducted a research entitled "Phytoplankton Community Structure in Batang Palangki Waters of Sijunjung Regency, West Sumatera"

This study aims to analyze: (1) phytoplankton abundance, (2), diversity index, uniformity index and phytoplankton dominance index, (3) physics and chemical factor waters of Batang Palangki, West Sumatera during research

2. Methods

The research activity was conducted on March to Mai 2016 in the waters of Batang Palangki Ampek Nagari Subdistrict, Sijunjung Regency, West Sumatra.

The location of sampling is determined by purposive sampling that is determining the sampling point, it determine from sampling point. They are the station I part of the mining and gold processing center, the upstream II station and the station III downstream part river. Phytoplankton samples were collected by taking 100 L of water and into the net plankton no. 25 The filter was put into a 25 cc sample bottle and fed 3 doses of formalin and labeled. Subsequently brought to the laboratory for identification.

IOP Conf. Series: Earth and Environmental Science **130** (2018) 012023 doi:10.1088/1755-1315/130/1/012023

Determination of water quality was measured at each station as water temperature was measured using a water thermometer, the pH of water using pH meters and watercolour was determined visually. Measurements of dissolved oxygen content (DO) and BOD were measured by DO meters, while free CO_2 content was measured with Acidimetry. The content of Hg (mercury) is measured using the AAF method.

2.1 The density of phytoplankton

Density phytoplankton in the water can be calculated using the following formula [8]:

$$N = \frac{(ax\ 1000)Xc}{L} \tag{1}$$

With:

N = abundance of phytoplankton (number of individuals per liter)
A = number of similar individuals per ml,
C = volume of water taken,
L = volume of water taken by bottle (L).

2.2 Diversity Index (H)

Diversity is the unregulated inherence of individual genera derived from a population. The variety of phytoplankton types determined using the Shannon-Wiener index:

$$\mathrm{H}' = \mathrm{pi\ Log\ pi} \tag{2}$$

With:

H' = Index of Shannon-Wiener's
Pi = The proportion of type i within community (i = 1,2,3, ... s)
S = Number of types

2.3. Equitability Index (E)

$$E = \frac{\mathrm{H}'}{\mathrm{Hmax}} \tag{3}$$

With:

E = Equitability Index
Hmax = Maximum diversity index
H' = Shannon-Wiener's diversity index
S = Number of taxa

2.4. Dominance Index (D)

$$D = \sum \frac{(ni)^2}{N} \quad (4)$$

With:
C = Indeks dominasi Simpson
Ni = Proportion of i in community (i=1,2,3,....,s)
N = Total number of individuals

3. Results and Discussion

3.1 Phytoplankton Density

Table 1. Density of phytoplankton (individual/liter) Batang Palangki the water of Sijunjung Regency of West Sumatera

No	Genea	Density			Number
		Station I	Station II	Station II	
1	*Neidium*	20	0	0	20
2	*Gyrosigma*	20	10	15	45
3	*Synedra*	40	0	0	40
4	*Frustulia*	75	0	0	75
5	*Fragillaria*	0	120	30	150
6	*Nitzschia*	0	20	0	20
7	*Peridinium*	0	0	20	20
	Number of indvidual.	155	150	65	370
	Number of types	4	3	3	

Table 1 shows the density (individual/liter) of phytoplankton in Batang Palangki waters, with 7 genus of phytoplankton found in total, i.e *Neidium, Gyrogsima, Synedra, Frustulia, Fragillaria, Nitzschia,* and *Peridinium* with density 370. The highest density was found at station I of 155, and the lowest density was found at station III which was 65. The abundance of phytoplankton in Batang Palangki waters was found to be between 65 - 155, with an average of 370 stations. Classifition water fertility based on plankton abundance as follows. Waters with oligotrophic fertility are in abundance <2000, Mesotrophic with abundance 2000-15000 and eutrophic > 15,000 [7].

Table 2. Relative Density (KR) and Relative Frequency (FR) Phytoplankton in Batang Palangki Waters Sijunjung Regency West Sumatra

No	Genera	Relative Density (KR) (%)			FR (%)
		Station I	Station II	Station III	
1	*Neidium*	12.9	0	0	33.3
2	*Gyrosigma*	12.9	6.72	23.1	100
3	*Synedra*	25.8	0	0	33.3
4	*Frustulia*	48.4	0	0	33.3
5	*Fragillaria*	0	80.0	50.0	66.6
6	*Nitzschia*	0	13.3	0	33.3
7	*Peridinium*	0	0	20.0	33.3

Table 2 shows the relative density (KR) and the relative frequency (FR) of phytoplankton in Batang Palangki waters of Sijunjung District. The highest relative density found in the genus Fragillaria is 80%. While the relative frequency of 100% found in Genus *Gyrogsima*.

3.2 Phytoplankton Type Diversity

Table 3. Average Diversity Index (H), Equitabiliy Index (E) and Domination Index (D)

Parameter	Station I	Station II	Station III	Average
H	0.59	0.35	0.42	0.45
E	0.66	0.44	0.52	0.54
C	0.42	0.41	0.001	0.28

Table 3 shows the average value of Diversity Index (H), Equitability Index (E), and Dominance Index (D) of the phytoplankton in Batang Palangki water of Sijunjung Regency of West Sumatra at station I, II and III respectively of 0.45, 0.54 and 0.28. Overall, it can be seen that diversity index of phytoplankton in Batang Palangki water ranged from 0.35 to 0.59 with an average of 0.54 (Table 3). States that waters with diversity indexes 0 - 1 include unstable the water, I - 3 including almost stable and large the water of 3 including stable the water [9]. The results of this research diversity index is in the range 0-1, which means including unstable the water. The level of river pollution can be seen from the diversity or diversity and growth rate of the structure of the plankton. If the plankton diversity in the high ecosystem signifies the quality of the water is good, but if the plankton diversity is small then indicates the water is polluted [7].

The equitabiliy index (E) of the phytoplankton in Batang Palangki ranged from 0.44 to 0.66 with an average of 0.54 (Table 3). The equitability of phytoplankton is said to spread evenly when the value reaches 1, meaning that all samples in the station have the same number of organism types based on existing criteria. The results of this study showed an average equitability index of 0.54, which means that the water of Batang Palangki have different types of phytoplankton organisms at each station.

The dominance index (D) of phytoplankton on the three research stations averaged at 0.28 (Table 3). Based on the criteria of dominance the value of index approaching the value of 1 indicated a higher dominance by certain species. The high value of existing dominance shows a low diversity with uneven distribution on each station. Also, the high value of dominance at each station indicates a change in the living environment of phytoplankton. The high dominance indicates that the place has a low type of wealth with uneven distribution, meaning within the observed community structure found species that dominate other species. Thus, these conditions can reflect the diversity of phytoplankton.

3.3 Physical Factor of Batang Palangki

Table 4. The result of Physics factor measurement of Batang Palangki the water

Station	Water temperature (°C)	Transparency (Cm)	Humidity (Atm)	Water colour	Water velocity (m/sec)
I	25	12	66	Murky	12
II	25	28	65	Murky	12
III	26	12	65	Murky	8

Based on table 4 shows the results of the measurement of the physics parameters of Batang Palangki waters of Sijunjung Regency, West Sumatra. The measurement result of the three stations for water temperature ranges from 25 – 26 0C, transparency ranges from 12 - 28 cm, the highest transparency is at station II (upstream), with watercolour of each cloudy station. Strong current parameters at station I of 12 m/sec, station II of 12 m/sec and station to III 8 m/sec. The movement of phytoplankton is strongly influenced by currents. Current velocities in streams are grouped into three categories, ie fast, when current velocities range from 0.5-1 m/sec, while current velocity ranges from

0.25-0.49 m/sec and slow when current velocity ranges from 0.1-24 m/sec [10].

3.4 Chemical Factor of Batang Palangki

Table 5. The result of the measurement of the water chemical factor of Batang Palangki water

Parameter Station	DO (mg/L)	BOD (mg/L)	COD (mg/L)	TOM (mg/L)	(Hg) (mg/L)	The water pH
I	5.41	3.40	60.33	11.07	0.174	6.00
II	5.25	3.49	76.25	12.10	0.098	7.00
III	5.96	2.80	47.05	9.61	0.208	7.60
Average	5.54	3.23	61.21	10.93	0.16	6.86

Table 5 shows the result of measurement of chemical factors (milligram per liter) of Batang Palangki the water of Sijunjung Regency, West Sumatera that the results of measurement of each parameter for all three stations are as follows: dissolved oxygen content (DO) averaged at 5.54, BOD at 3.23, and COD at 61.2. The total organic content (TOM) averaged 10.93. Furthermore mercury (Hg) averaged 0.16, the highest Hg levels were found in station III (downstream) that is equal to 0.208

Dissolved carbon dioxide (COD) ranges from 47.05 - 76.25. Carbon dioxide is produced from the respiratory processes of both animal and plant organisms. Waters undergoing rapid decomposition process will produce high CO2 content and lack of oxygen. High carbon dioxide content in the waters causes death for animals [11].

The degree of acidity shows the concentration of H^+ ions in the solution. The pH of water qualified for organism life ranges from 6.0-7.6. Changes in pH of water bodies can greatly disturb the life of plants, animals and decomposing organisms that live in these water bodies [2]. Furthermore, the optimum pH for plankton life ranged from 5.5 to 8.5 [10].

Dissolved oxygen content of Batang Palangki waters averages 5.4 mg/l According to [12], that the minimum dissolved oxygen required to support the aquatic organism's life normally is 2 ppm with no record in the water Toxic.

4. Conclusion

From the results and discussions that have been described, it can be taken some conclusions as follows:

1. The abundance of phytoplankton (individual/liter) ranges from 65 to 155 consisting of 7 Genus phytoplankton: Genus Neidium, Gyrogsima, Synedra, Frustulia, Fragillaria, Nitzschia, and Peridinium
2. Diversity Index (H), Uniformity Index (E) and Domination Index (C) were 0.45, 0.54 and 0.28, respectively
3. Chemical factors including dissolved oxygen (DO) averaged 5.54 mg/L, BOD 3.23 mg/L. COD averaged 61.21 mg/L. The total organic content (TOM) averaged 10.93 mg/L. Mercury content (Hg)
   averaged 0.16, the highest Hg levels were found in station III (downstream) that is equal to 0.208 mg /L
4. The above results: Batang Palangki waters are at oligotrophic and unstable levels of fertility

References

[1] Dawes CJ (1981) Marine Botany. A Willey Interscience Publ:628 p
[2] Moersidik S and Hardjojo (1998) Water Analysis and Quality. Karunika. Jakarta
[3] Sachlan M (1974) *Planktonologi*. Cierspondensi Cours Center, Jakarta
[4] Tjitrosoepomo G (1981) Taxonomy of Special Plants. Bhatara Aksara, Jakarta.

IOP Conf. Series: Earth and Environmental Science **130** (2018) 012023 doi:10.1088/1755-1315/130/1/012023

[5] Gusmaweti (2006) The impact of the floating net on the distribution and abundance of plankton in the waters of Maninjau lake. Journal of Academics *APTISI.* 10(2) October 2006 ISS N0.854 - 4336 Kopertis Wilayah X Padang.

[6] Deswati L (2000) The composition and structure of the phytoplankton community as well as the primary productivity of Mraninjau lake waters. Thesis. Postgraduate . Andalas University. Padang

[7] Basmi J (1999) Planktonology: Plankton as Bioindicator of Water Quality. Bogor: Faculty of Fisheries and Marine Sciences, Bogor Agricultural University.

[8] Michael P (1994) Ecologycal Methods for Field Investigation and Laboratory. Jakarta: University of Indonesia Press

[9] Barus A (2000) Introduction to Limnology. USU. Medan

[10] Welch E (1980) Limnology. Mc. Graw-Hill. Book Company. New York

[11] Afrizal (1992) Diatom Perifiton on atificial substates in westrn Cimahi river Unpublished Thesis results, post graduate program, Bndung Institute of Technology

[12] Lisanti (2000) Distribution of Plankton Disewakan Jujuhan Village of Batu Kangkung Kerinci Seblat National Park Sawahlunto Sijunjung Regency of West Sumatra Province. Thesis Faculty of Fisheries University of Bung Hatta Padang.

Selection of forages by timor deer (*cervus timorensis* blainville) in menjangan island, bali

I Ketut Ginantra[1*], Ida Bagus Made Suaskara[1], I Ketut Muksin[1]

Department of Biology, Faculty of Mathematics and Natural Sciences, Udayana University, Bali-Indonesia

Email: ketutgi@yahoo.com

Abstract. This study was conducted to determine the selection of forages plants by Timor deer (*Cervus timorensis*) on Menjangan Island and its relation to the availability, chemical and physical properties of feed plants. The study was conducted in July-September 2016 in savanna and monsoon forest habitats. The availability of habitat feed plants in the habitat was determined by the quadrat method, and the species of plant eaten by Timor deer was determined through the microhistological analysis of the fecal sample. The food selection index is determine by the Ivlev index. Energy contents of forages plants by bomb calorimeter apparatus, crude protein analyzed by Semi-Micro Kjeldahl technique, NDF, ADF and lignin levels refer to the method of Goering and Van Soest. Mineral content of calcium (Ca) and phosphorus (P) by using atomic absorption spectrophotometer. Determination of tannin content with Folin Denish reaction. Physical properties determine are water regain capacity and water solubility. The relationship between availability with the utilization of plants by Timor deer was analyzed with the similarity index. Multiple regression statistic to test the relationship between index selection with nutritional value factor and physical characteristic of plant species. The result showed that Timor deer selected 32 plants species of graminoids, forbs and woody plants. Feeding selection of Timor deer is strongly influenced by the availability of forage plants in habitat. The feeding selection was significantly influenced by three predictor variables i.e. positive nutritional value is crude protein and negative nutritional value were lignin and tannins. Selection of forage plant Timor deer is positively correlated with the physical properties of feed plants.

Keywords: Chemical and physical characteristic of plants, feeding selection, plant availability, Timor deer

1. Introduction

Timor deer (*Cervus timorensis*) is one of the wildlife found in the area of Menjangan Island Bali. This area is part of Bali Barat National Park (BBNP), located on the north western tip of Bali Island. An area of 175 Ha island consists of savanna habitat, monsoon forest and mangrove forest [1]. Different habitat types have consequences for different of vegetation constituents. These phenomena affect the abundance of forages species, productivity and quality. These differences will have implications for the feeding selection of Timor deer, especially in the effort to meet the need for nutrients.

Selection of feed plants in deer in wildlife is a complex behavior and is a consequence of interactions among several factors. Selection of feed species on deer involves several factors, such as nutritional value (positive nutrient, negative nutrient), allelochemical compound (negative value), physical characteristics of the plant itself, availability in habitat and also the presence of competitors or predators [2], [3], [4].

Several studies of forage selection on Timor deer in the BBNP have been conducted. Ginantra *et al.* [5], found more than 30 species of plants selected by Timor deer in two habitat units (in the monsoon forest and savanna) in the BBNP, Masy'ud *et al.* [6] found 12 species of plants eaten by Timor deer in Tanjung Pasir habitat unit of BBNP. These studies were conducted in the park area of Bali, but not in the area of Menjangan Island. Given the area of Menjangan island is a separate habitat for Timor deer, then the selection study, the availability of feed plants the need to be done.

The focus of this study was to determine the feeding selection of Timor deer (*C. timorensis*) with factors: Availability of botanical composition in the habitat; Nutritional value of feed plants including protein content, energy, minerals of Calcium (Ca) and Phospor (P), acid detergent fiber (ADF), neutral detergent fiber (NDF), lignin, allelochemistry (Tannin) and physical characteristics of feed plants that include water regain capacity and water solubility.

2. Methods

2.1 Study Area

Field research was conducted for the study of the composition of forage species and the collection of Timor deer feces. Based on the preliminary study, four habitat units were used by Timor deer. Sampling was conducted in four habitat units of Menjangan Island, i.e. S1, S2 and S3 locations for savanna habitat types and location of MF for vegetation type of monsoon forest (Figure 1). Field research was conducted in July-September 2016.

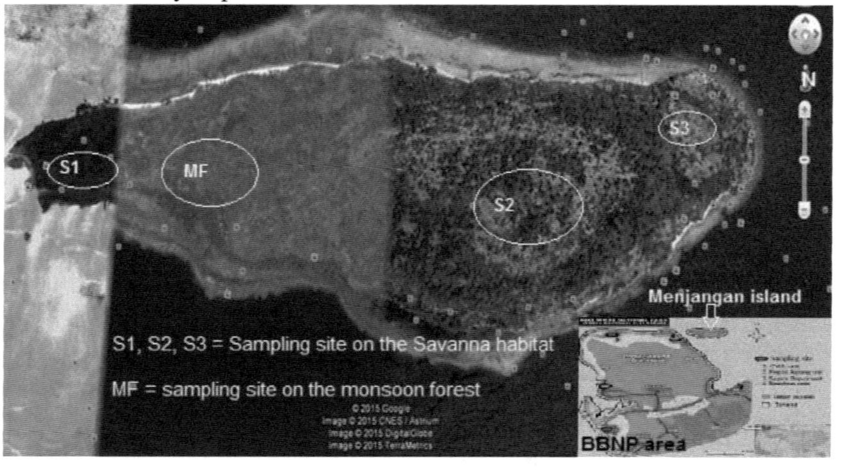

Figure 1. Locations of study area at Menjangan Island

2.2 Analysis of Botanical Composition in the habitat

The botanical composition availability in the habitat was determined by the quadrat method. Size of 0.25 m^2 plot for grass, herb and 1 m^2 for shrubs and 25 m^2 for trees. The number of plots for each habitat unit is ten times. The parameters measured were the ground cover of each plant species. For vegetation of shrubs and tree parts of plants that measured were percent cover of the shoot that count up to 1.2 m height (height level can be accessed by Timor deer) [7]. Botanical composition/forage species in habitat by percent cover [8].

2.3 Analysis of fecal microhistology

The composition of plants species eaten by Timor determined by techniques micro-histological fecal sample (Holechek *et al.*, 1990). Feces collected from five groups of pellets in four habitat units, each done as much as one time in each month. Immediately collected fecal sample was dried (oven 70 ^{0}C) and put into an envelope, then in refrigerant before micro-histological analysis done in the laboratory to avoid decomposition [9].

Epidermal fragment preparations of each plant species available in the habitat are made as reference slides to identify the species of edible plants. Five individual pellets were randomly selected from 5 fecal groups collected in each habitat unit. The fecal sample was destroyed, then immersed in NaClO (6%) for ± 7 minutes. The washed stool part taken for preparing preparations. The number of epidermal fragments of each plant species is calculated from 20 microscope slides at 400x magnification. The epidermal fragment was computed only by indentifiable fragment to reduce the bias between graminoid and nongraminoid. Observations were made at the plant taxonomy laboratory of Udayana University Bali. The botany composition of deer feed was determined based on the percentage of epidermal fragments of each plant on 20 slides.

2.4 Analysis of chemical and physical properties of forage plants

Each plant species in each plot in the field was cut about 5 cm representative of the animal eaten portion for the sample of feed plants [10]. Dry weight (DW) samples was finely milled (a particle size of about 1-2 mm screen) used in proximate analysis: dry matter (DM), gross energy (GE), crude protein (CP), calcium minerals, phosphorus, and tannin content.

Determination of DM is DW sample oven until temperature 105 ^{0}C for 2 x 24 hours. GE contents by bomb calorimeter apparatus, CP analyzed by Semi-Micro Kjeldahl technique, NDF, ADF and lignin levels refer to the method of Goering and Van Soest [11]. Mineral content of calcium (Ca) and phosphorus (P) by using Atomic Absorption Spectrophotometer (AAS) [12]. Determination of Tannin (total tannin) with Folin Denish reaction [13]. The analysis was conducted at the Laboratory of Livestock Research Center, Bogor.

Dry weight samples of each finely ground feed species up to 1-2 mm screen size to measure physical properties. Physical properties are water regain capacity (WRC) and water solubility (WS), was determined according to the methods of Suhartati *et al.* [14]. The analysis were done at laboratories of plant taxonomi, Biology of Udayana University.

2.5 Data Analysis

Selection of forage plants is determined by selection index (Si) "Ivlev" [15].

$$Si = \frac{(Ui - pi)}{(Ui + pi)} \quad (1)$$

pi = composition (%) of species-i in habitat
Ui = composition (%) species-i in deer diet

The value of Si ranges from -1 to 1, with the category of preferred degree (palatability) is as follows: preferably (0.1 to 1), proportional (0.09 to -0.09), avoidance (- 0.1 to -1).

The relationship between availability with the utilization of plants by Timor deer was analyzed with the similarity index (S), with the formula S = Yi (Yi = minimum value of plant species composition in the habitat (pi) and the composition of plant species eaten (ui)). If S > 50% showed a high similarity score and S≤50% showed a low similarity score [15]. The relationship between index selection with nutritional value factor and physical characteristic of plant species was tested by multiple regression statistics with the stepwise method. Test using the "IBM SPSS 20" software.

3. Results and discussion

3.1 Availability and selection of plant species in habitat

The species of plants available in four habitat units are as many as 39 species, 32 of which are selected by Timor deer. The species of plants selected are quite diverse both from graminoids, forbs and woody plants (Table 1).

In monsoon forest habitats, some plant species are not available, but in Timor deer diet (in fecal analysis). These include *Dactyloctenium aegeptium*, *Digitaria* sp., *Vernonia patula*. This suggests that the species of forages is eaten in other habitats (on savannas) and when in monsoon forest habitat deer remove feces. The distance between one habitat unit and another near enough (savanna S1 unit with monsoon forest, or between monsoon forest with savanna S2 only 1 to 2 km) to allow the deer to move or still be home range of timor deer. Allison [16], states that the home range of deer is generally 2.62 km^2.

Table 1. Composition (%) of plant species selected by timor deer in Menjangan Island, Bali.

No	Plants Species	Botanical composition in deer diet on the habitat unit				Ui	Pi	IS	Selection category
		S1	MF	S2	S3	%	%		
	Graminoids								
1	*Dactyloctenium aegeptium*	9.62	2.29	9.71		5.41	1.77	0.51	F
2	*Eragrostis amabilis*	6.51	8.62	4.5	8.31	6.99	2.38	0.49	F
3	*Themeda arguens*			1.49		0.37	0.59	-0.23	Av
4	*Heteropogon* sp.	1.21	1.31	2.56	1.6	1.67	2.64	-0.22	Av
5	*Digitaria* sp.	7.11	3.23	4.39	6.41	5.29	1.78	0.50	F
6	*Cyperus* sp.	2.13		3.26	5.13	2.63	1.61	0.24	F
7	*Imperata cylindrica*	1.26				0.32	0.86	-0.46	Av
8	*Andopogon aciculatus*				4.44	1.11	1.05	0.03	P
9	*Phragmites* sp.	8.11	7.89	9.12	9.41	8.63	13.25	-0.21	Av
	Forbs								
1	*Desmodium heterophylum*			2.75		0.69	0.42	0.24	F
2	*Ammania bascifera*	3.96				0.99	0.65	0.21	F
3	*Vernonia patula*	11.51	1.2		8.92	5.41	5.37	0.00	P
4	*Euphorbia hirta*		1.41			0.35	0.51	-0.19	Av
5	*Cleome viscosa*	4.21	4.01	5.29		3.38	3.45	-0.01	P
6	*Tephrosia pumila*	3.41				0.85	0.68	0.11	F
7	*Justicia* sp.	2.71		3.69	4.25	2.66	1.69	0.22	F
8	*Ocimum* sp.								
9	*Tridax procumbens*								
	Woodys								
1	*Leucaena leococephala*			5.62		1.41	0.81	0.27	F
2	*Schoutenia ovate*		10.5	8.82	5.78	6.28	4.67	0.15	F
3	*Grewia koordersiana*		6.3		4.11	2.60	1.38	0.31	F
4	*Scleichera oleosa*			3.31		0.83	0.85	-0.01	P
5	*Datura metel*	5.12			6.21	2.83	1.57	0.29	F
6	*Phyllanthus emblica*			3.22	2.49	1.43	2.70	-0.31	Av
7	*Eupatorium odoratum*	13.71	14.82	12.92	13.91	13.84	10.91	0.12	F
8	*Lantana camara*	2.5	2.14	1.29	1.32	1.81	5.21	-0.48	Av
9	*Spondias* sp.	2.78	7.22		3.11	3.28	3.28	0.00	P
10	*Cleistanthus laevis*	4.65	6.96	4.65	4.67	5.23	6.97	-0.14	Av

11	*Acacia leucophlea*	2.49			1.73	1.06	1.35	-0.12	Av
12	*Albezia lebbeckioides*	1.46	6.52	3.46	1.21	3.16	4.31	-0.15	Av
13	*Pasiflora foetida*		4.11			1.03	0.79	0.13	F
14	*Cleistanthus* sp.		5.52			1.38	1.57	-0.06	P
15	*Zysiphus mauritiana*			1.42		0.36	0.59	-0.24	Av
16	*Flacourtia indica*								
17	*Breynia oblongifolia*								
18	*Lannea gramdis*								
19	*Phempis acidula*								
20	*Azadirachta indica*								
21	*Pongamia pinnata*								

Note : Blank column : that's available plant species but not eaten by timor deer
F= preferably, Av= avoidance, P= proportional

The feed botanical composition of Timor deer in Menjangan Island consists of three groups of forbs, graminoids and woody plants. Woody plants show the highest botanical composition, and the lowest is forbs category (Figure 2).

This result is different from the composition of deer Timor feed plants in the park's main land (BBNP), the simplest botanical composition of woody plants. Meanwhile, on the Menjangan island composition of the highest woody plants compared with graminoids and forbs. The high composition of deer feed on plant groups of woody plants is associated with the low availability of forbs and grasses in Menjangan Island. Lack of botanical composition in forbs group and grasses are compensated by increasing woody plants. This suggests that the availability of habitat feed plants in the habitat determines the botanical composition in Timor deer feed.

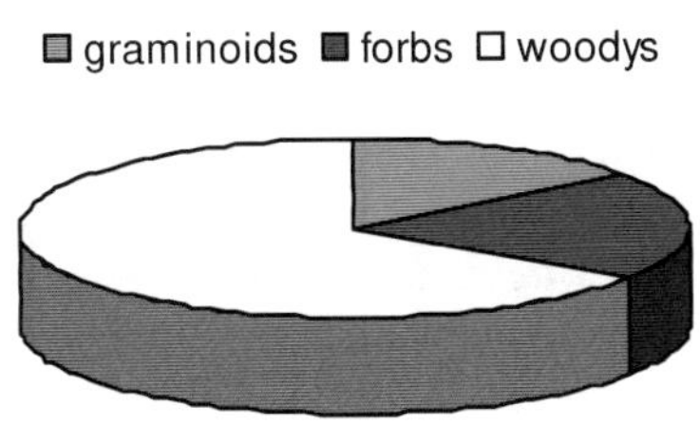

Figure 2. The Botanical Composition of forages by Timor Deer on the Menjangan Island.

Timor Deer can adapt to selective grazing and browsing, depending on the availability of feed in the habitat. Patisellano and Arobaya [17] reported that the composition of Timor deer feed in upland kebang Manokwari more towards grazer because grass is available higher in the habitat. DeGarine-Wichatitsky *et al.* [18], also found a similarity to the feed botanical composition of Timor deer in New Caledonia, in vegetation savanna higher botanical composition in deer feed is graminoids, whereas in forest vegetation the botanical composition in deer diet is dominated by broad leaf plants.

3.2 Relation of availability to plant utilization by Timor deer

In four habitat units observed, Timor deer showed plant species relationships that were eaten with the availability of vegetation in the habitat, both graminoids, forbs and woody plants, with index values above 50%. The category of forbs showed the highest similarity index value of 87.64% compared with woody and grass plants (Figure 3). This shows that the availability of feed plants of forbs election category high. This is because the availability of grass in the menjangan island habitat is very limited to meet the needs.

In contrast to the results of the study [5] that the Timor deer in the mainland BBNP area showed similar low value (<50%) in woody plants. In the area of the park's mainland, the availability of grasses and forbs more abundant than in Menjangan Island. As the availability of woody plants increases, while the herbaceous plants and the grasses are low, the Timor deer shows the composition of feed on high woody plants.

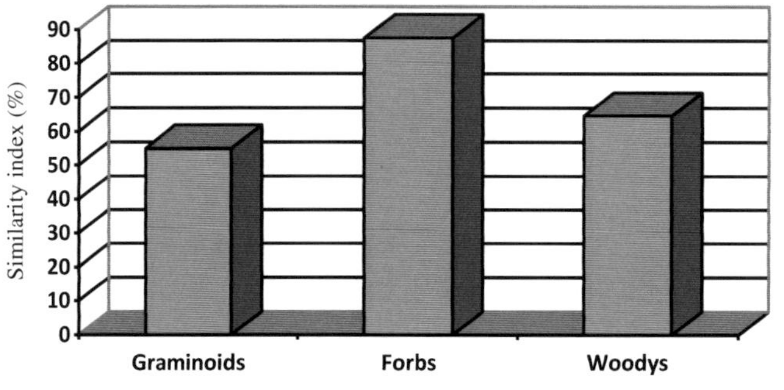

Figure 3. Similarity of the available plant composition in habitat with plants utilized by timor deer in Menjangan Island

The availability of plants species is one of the factors that influence feeding selection on herbivor in wild habitat, besides other factors like acceptability, plant digestibility and chemical composition of feed type. Nugent [19] suggest that feeding selection reflects the relationship between animals and vegetation in the habitat. Homolka [7] also reported a high similarity index between feed composition and availability food supply which is dominated by grass in red deer (*Cervus elaphus*).

3.3 Content of Nutrition and Tannins of feed plants

The nutrient content of feed plants, i.e. dry matter (DM), crude protein (CP), gross energy (GE), minerals (Ca and P) are presented in Table 2. The content of CP species of feed plants in forbs groups and woody plants shows the range of content which is higher than the graminoids group. GE content of plant species of grasses and woody plants show a higher range of energy content than forbs. Mineral content of Ca and P plant species group of forbs and woody plants showed the range of mineral content (Ca and P) is higher than the graminoids group. However, tannin content of woody plants and forbs relatively higher than the grass group.

In general, the tannin content (mainly condensed tannin) of feed plants is mostly at moderate level, i.e. less than 4% [20]. Grass plants of all species of feeds exhibit moderate tannin content. The content of tannins of plant species in forbs is also mostly a moderate level, but there is one species of plant, namely *Ammania bascifera* tannin content is high enough that is 5.46%. several woody plants show high tannin content, including *Schoutenia ovate* (5.86%), *Datura metel* (4.85%), *Eupatorium odoratum* (5.54%), *Spondias* sp. (5.14%), *Cleistanthus laevis* (5.61%), *Albezia lebbeckioides* (8.74%), *Cleistanthus* sp. (14.35%) and *Zyziphus mauritiana* (15.95%). Holechek [21], also reported that woody plants tend to have secondary metabolite content including tannins that are higher than forbs and grasses plant.

The results of nutrient content analysis of feed plants in Menjangan island habitat, showed that the quality of available plant feed varies. Herbaceous plants content of CP and minerals (Ca and P) is higher, grass content is lower, but its GE content is higher than forbs plants. Woody plant content of CP and GE is high and its availability tends to be relatively longer than forbs. The same is also reported by Amiri and Shariff [22], that the content of CP, GE and minerals varies by plant species (dicoty herbs, grasses and woody plants).

Table 2. Chemical content of feed plants

NO	Plants species	DM %	water %	CP %	GE Kcal/kg	Ca %	P %	Tanin %
	Graminoids							
1	*Dactyloctenium aegeptium*	18.02	81.98	13.44	3735.16	1.13	0.27	0.35
2	*Eragrostis amabilis*	22.00	78.00	12.50	4243.35	0.42	0.31	0.59
3	*Temeda arguens*	33.15	66.85	7.40	4142.21	0.21	0.20	0.34
4	*Heteropogon* sp.	27.62	72.38	7.05	3515.32	0.03	0.14	2.06
5	*Digitaria adscendens*	21.33	78.67	11.19	4183.34	0.56	0.27	0.32
6	*Cyperus haspan*	20.84	79.16	9.81	4140.84	0.79	0.21	0.43
7	*Imperata cylindrica*	28.83	71.17	9.43	4322.52	0.26	0.22	0.25
8	*Andopogon aciculatus*	22.17	77.83	9.98	4072.42	0.75	0.30	0.41
9	*Phragmites* sp.	21.88	78.12	8.47	4470.38	0.36	0.26	0.29
	Forbs							
1	*Desmodium heterophylum*	19.63	80.37	21.91	3936.58	1.48	0.45	0.52
2	*Ammania bascifera*	22.34	77.66	19.62	4235.00	0.42	0.30	5.46
3	*Vernonia patula*	17.08	82.92	13.97	4235.21	1.25	0.26	3.20
4	*Euphorbia hirta*	18.89	81.11	22.07	3585.22	1.93	0.52	*
5	*Cleome viscosa*	18.96	81.04	10.93	3278.99	9.64	0.15	2.33
6	*Tephrosia pumila*	24.34	75.66	21.77	4509.12	1.53	0.30	0.73
7	*Justicia* sp.	16.86	83.14	15.32	3917.81	2.48	0.09	2.77
	Woodys							
1	*Leucaena leococephala*	24.80	75.20	27.65	4832.55	1.21	0.27	2.16
2	*Schoutenia ovate*	15.90	84.10	16.44	4835.47	0.59	0.23	5.86
3	*Grewia koordersiana*	31.70	68.30	17.34	4546.49	1.33	0.27	0.41
4	*Schleichera oleosa*	31.41	68.59	14.78	4803.21	1.11	0.20	1.11
5	*Datura metel*	14.73	85.27	15.96	5284.13	1.54	0.18	4.85
6	*Phyllanthus emblica*	31.25	68.75	11.97	4715.90	0.60	0.07	2.75
7	*Eupatorium odoratum*	17.71	82.29	24.63	4459.80	2.24	0.14	5.54
8	*Lantana camara*	33.26	66.74	16.76	3779.49	1.43	0.22	*
9	*Spondias* sp.	27.69	72.31	16.91	4338.24	2.32	0.37	5.14
10	*Cleistanthus laevis*	36.62	63.38	9.96	5285.70	0.98	0.12	5.61
11	*Acacia leucophlea*	27.69	72.31	22.79	5057.89	0.75	0.11	*
12	*Albezia lebbeckioides*	27.72	72.28	11.87	4539.50	1.14	0.15	8.74
13	*Pasiflora foetida*	14.77	85.23	12.17	4286.03	2.41	0.09	1.79
14	*Cleistanthus* sp.	27.81	72.19	13.88	4660.03	2.72	0.12	14.35
15	*Zysipus mauritiana*	34.032	65.968	15.95	3558.69	2.47	0.12	15.95

** tannin data is not analyzed*

3.4 Content of ADF, NDF, lignin and physical properties of plant species of feed

The ADF content of the forbs and the woody plants is relatively lower than that of the grasses, but the lignin content of the grass plants is relatively lower (Table 3). The results of this study are a logical consequence of lower ADF in forbs compared with other plants as a result of low coarse fiber. Other researchers also found almost the same thing that grass (graminoids) is a group of plants whose cell wall content (ADF and NDF) tend to be higher than dicotyl herbs [22].

Water absorption and water solubility are affected by plant cell wall content (ADF and NDF fibers), this statement can be provened by referring to a negative correlation between ADF feed content with water regain capacity and water solubility (r = -0.094 and r = - 0.413) and negative correlation also between NDF content with water regain capacity and water solubility (r = -0.076 and r = -0.294). That is, the higher the cell wall content (ADF and NDF) plants, the water absorption and solubility of the lower. Behgar *et al.* [23] also stated that the physical properties of feed plants, especially their water holding capacity, are negatively correlated with ADF and NDF plant contents.

Table 3. Cell wall content and physical properties of feed plants

NO	Plants species	NDF %	ADF %	Lignin %	WRC	WS
	Graminoids					
1	*Dactyloctenium aegeptium*	60.80	50.93	8.80	5.76±1.16	0.37±0.12
2	*Eragrostis amabilis*	75.04	54.66		4.69±0.72	0.40±0.10
3	*Temeda arguens*		59.30		3.66±0.16	0.16±0.08
4	*Heteropogon* sp.	75.46	60.56	5.44	3.41±0.33	0.29±0.05
5	*Digitaria adscendens*	58.98	54.00	9.25	4.76±0.10	0.32±0.05
6	*Cyperus haspan*		56.83		4.51±0.20	0.31±0.03
7	*Imperata cylindrica*		50.38		3.37±0.25	0.25±0.05
8	*Andopogon aciculatus*		53.50		3.19±0.17	0.21±0.18
9	*Phragmites* sp.		57.88		3.30±0.22	0.21±0.09
	Forbs					
1	*Desmodium heterophylum*	44.29	40.85	15.58	6.42±1.16	0.46±0.09
2	*Ammania bascifera*	50.63	48.14	15.72	4.20±0.41	0.29±0.04
3	*Vernonia patula*	52.18	46.45	30.16	3.20±0.08	0.27±0.08
4	*Euphorbia hirta*		41.78		3.91±1.19	0.44±0.06
5	*Cleome viscosa*	43.79	43.39	17.73	5.20±0.48	0.40±0.02
6	*Tephrosia pumila*		43.11		3.41±0.79	0.34±0.18
7	*Justicia* sp.	40.81	34.93	8.93	4.22±0.56	0.35±0.08
	Woodys					
1	*Leucaena leococephala*	39.63	32.57	20.55	4.20±0.87	0.48±0.06
2	*Schoutenia ovate*	49.50	47.15	20.24	4.86±0.72	0.55±0.06
3	*Grewia koordersiana*	61.08	58.76	26.71	4.24±0.98	0.37±0.07
4	*Schleichera oleosa*	63.67	57.80	30.83	3.77±0.81	0.31±0.04
5	*Datura metel*	35.22	33.03	12.39	3.38±0.81	0.29±0.09
6	*Phylanthus emblica*	49.20	45.33		3.40±0.61	0.35±0.06
7	*Eupatorium odoratum*	61.72	59.55	28.13	3.22±1.12	0.33±0.06

8	*Lantana camara*				3.08±0.81	0.25±0.19
9	*Spondias* sp.	53.87	43.90	16.62	3.78±0.85	0.32±0.05
10	*Cleistanthus laevis*	64.33	51.87	21.10	3.48±0.37	0.25±0.04
11	*Acacia leucophlea*				2.72±0.76	0.36±0.05
12	*Albezia lebbeckioides*	49.11	47.89	19.60	3.80±0.86	0.33±0.08
13	*Pasiflora foetida*	43.50	42.99	18.89	3.56±0.07	0.38±0.07
14	*Cleistanthus* sp.	44.11	30.38	3.95	3.62±0.81	0.45±0.05
15	*Zysipus mauritiana*				2.71±0.76	0.23±0.07

3.5 Selection of forages plants in relation with chemical and physical properties of plants

Timor deer in Menjangan island habitat selects a variety of plants. Not only the species of high-quality nutrients are selected, but also selecting the species with the quality of the intermediates. This indicates that in the selection of feed in addition to nutrient fulfillment in quality, quantity is also important to maximize nutrient intake for their needs. Lopes-Coba *et al.* [24] state that in herbivore feeding selection it can make decisions to quickly determine the types of feed to optimize its consumption rate (intake) for a trade-off between the quality and quantity of feed availability.

The result of "stepwise" regression correlation test between feeding selection with CP, GE, Ca, P, ADF, NDF, lignin, tannin, WRC, WS species of feed plants, it was found that the feeding selection showed significant regression correlation with three predictor variables. The three variables are crude protein (X_1), lignin content (X_2) and tannin content (X_3). The model of regression of feeding selection with three variables of predictors are as follows:
$Y = 0.93 + 0.18X_1 - 0.052X_2 - 0.092X_3$

The value coefficient of R determinant (R^2) is 0.82, meaning that together the CP content factor, lignin, and tannin have an effect of 82% on the selection of plant species in the habitat, and the rest is due to other factors.

Furthermore, when examined the physiological aspect that selection of feed plants on deer in nature is a complex behavior and is a consequence of interactions among several factors [25]. Selection of feed species on deer involves several factors, such as nutritional value, allelochemical compounds (negative value), physical characteristics of the plant itself, availability in habitat and also the presence of competitors or predators. Similarly, olfactory or animal flavoring organs can be used to determine the chemical or nutrient compounds that are important to their life [26].

4. Conclusions

Feeding selection of Timor deer is strongly influenced by the availability of habitat feed plants (with similarity index> 50%). The feeding selection was significantly influenced by three predictor variables, i.e. positive nutritional value (X_1 = crude protein) and negative nutritional value (lignin = X_2 and X_3 = tannins). The model of regression of feeding selection with three variables of the predictors is $Y = 0.93 + 0.18X_1 - 0.052X_2 - 0.092X_4$.

Acknowledgements

We would like to thank the Research Center and Community Service of Udayana University who has funded this research through the udayana university flagship grant. Thank you to the hall manager BBNP over all facilities for the conduct of research in the field and to Mr. Ahmad (BBNP staff) for assisting in the field research.

References

[1]. Hall BBNP (2010) Inventory Timor deer (*Cervus timorensis*) in the Bali Barat National Park Region (Report on Activities). Cekik

[2]. Hanley TA (1997) A Nutritional View of Understanding and Complexity in the Problem of Diet Selection by Deer (Cervidae). Oikos 79:209-218, Copenhagen.

[3]. Littlefield KA, Mueller, JP, Muir JP and Lambert BD (2011) Correlation of Plant Condensed Tannin and Nitrogen Concentrations to White-Tailed Deer Browse Preferences in the Cross Timbers. *The Texas Journal of Agriculture and Natural Resource* 24:1-7.

[4]. Basha NAD (2012) Feeding behaviour, diet selection of goats and nutritive value of browse species in sub-humid subtropical savannah]. (Dissertation). School of Agricultural, Earth and Environmental Sciences, University of KwaZulu-Natal, Pietermaritzburg. South Africa.

[5]. Ginantra IK, S Putra, W Suarna and IW Kasa (2014) Botanical Composition of Forages by Timor Deer (Cervus timorensis Blainville) in A Monsoon Forest and Savanna of West Bali National Park. *Int. J. Pure App. Biosci.* 2 (5): 205-213 (2014)

[6]. Masy'ud B, Kusuma IH, and Rahcmadani Y (2008) Potential of Forage Vegetation and Effectiveness of Timor Deer Habitat Improvementr (*Cervus timorensis* de Blainville 1822) in Tanjung Pasir Bali Barat National Park. Media Konservasi Vol. 13. No. 2 Agustus 2008: 59 – 64.

[7]. Homolka M (1996) Foraging Strategy of large Herbivore in Forest Habitat. Folia Zoologica-45(2): 127-236, Czech Republic.

[8]. Morrison JI (2008) Using Microhistological Techniques to Predict Botanical Composition of Horse Diets on Cool-Season Grass pasture. (A thesis submitted) College of Agriculture at the University of Kentucky.

[9]. Maia MJ, Rego F, and Machado FS (2003) Determining Optimal Sampling Schemes to Study Red Deer Diets by Fecal Analysis. Silva Lusitana 11 (1):91-99

[10]. Grimaud P, Sauzier J, R Bheekhee, and Thomas P (2008). Nutritive value of tropical pastures in Mauritius Trop. Anim Health Prod (2006) 38: 159–167.

[11]. Ranjhan SK and G Krishna (1980) *Laboratory* Vikas Publishing House Pvt. Ltd., New Delhi. *Manual for Nutrition Research*.

[12]. Sinaga Y (1997). Elephant grass as Substitute Control Macro Mineral Analysis on Forage. Livestock research centers Ciawi, Bogor.

[13]. Suhardi (1997) Polyphenol Compounds Analysis Products Fruits and Vegetables. Vol.3. Processing Biochemistry Laboratory, Faculty of Agricultural Technology. Gadjah Mada University, Yogyakarta.

[14]. Suhartati FM, Suryapratama, W, and Rahayu R (2004) Analysis of the Physical Properties of a local Grass s. Animal Production, Vol. 6 No: 1: 37-42

[15]. Krebs JC (1989) EcologIcal Metodology. Harper Collins Publishers, New York.

[16].Allison, K (2011) Feeding Behavior of Deer. Available from: http://www.ehow.com/search.html.

[17]. Pattiselanno F, and Arobaya AYS (2009) Grazing Habitat of the Rusa Deer (Cervus timorensis) in the Upland Kebar, Manokwari. Jurnal Biodiversitas ISSN: 1412-033X (printed edition), Volume 10, Number 3, July 2009 ISSN: 2085-4722 (electronic), Pages: 134-138.

[18]. DeGarine-Wichatitsky M, Soubeyran Y, Maillard D, and Duncan P (2005) The Diets of introduced Rusa Deer (*Cervus timorensis russa*) in a Native Sclerophyll Forest and a Native Rainforest of New Caledonia. New Zealand Jurnal of Zoology, Vol. 32:117-126.

[19]. Nugent G (1990) Forage availability and the diet of fallow deer (*Dama dama*) in the blue mountains, Otago. *New Zealand J. of Ecol.* 13: *83-95*

[20]. Cannas A (2008) Tannins: fascinating but sometimes dangerous molecules. Cornel University. Department of Animal Science. Available from:http://www.ansci.cornell.edu/plants/toxicagents/tannin.html

[21]. Holechek JL, Vavra M and Pieper RD (1990) Methods for Determining the Botanical Composisition, Similarity, and Overlap of Range Herbivore Diets. National Reasearch Council/National Academy of Sciences, London
[22]. Amiri F, and Shariff ARBSM (2012) Comparison of nutritive values of grasses and legume species using forage quality index. Songklanakarin J. Sci. Technol.34 (5), 577-586.
[23]. Behgar M, Valizadeh R, Mirzaee M, Naserian AA, and Nasiri MR (2009) Correlation between the Physical and Chemical Properties of Some Forages and Non-Forage Fiber Sources. Journal of Animal and Veterinary Advances, Volume: 8 Issue: 11 Page No.: 2280-2285.
[24]. Lopez-Coba E, Sandoval-Castro CA, and Montes-Perez RC (2007) Intake and Digestibility of Tree Fodders by White Tailed Deer (*Odocoileus virginianus* yucatanensis). Journal of Animal and Veterinary Advances 6(1): 39-41.
[25]. Moser B, Schu¨tz M, and Hindenlang KE (2006). Importance of alternative food resources for browsing by roe deer on deciduous trees: The role of food availability and species quality. Forest Ecology and Management (226): 248–255. Available from: www.sceincedirect.com.
[26]. Ceacero F, Castillejos TL, García AJ, Estévez JA, and Laureano G (2010) Physiological variables explain mineral intake in Iberian red deer. Physiology & Behavior 100 (2010) 122–127.

Development of extract library from indonesian biodiversity: exploration of antibacterial activity of mangrove *bruguiera cylindrica* leaf extracts

K A Audah[1*], J Amsyir[1], F Almasyhur[1], A M Hapsari[1], and H Sutanto[2]

[1]Department of Biomedical Engineering, Swiss German University, Prominence Tower, West Sutera Road No. 15. Alam Sutera, Tangerang, Banten 15143, Indonesia
[2]Department of Chemical Engineering, Swiss German University, Prominence Tower, West Sutera Road No. 15. Alam Sutera, Tangerang, Banten 15143, Indonesia

*E-mail: kholis.audah@sgu.ac.id, Tel/ Fax: +622129779596/ 9598

Abstract. Antibacterial drugs derived from natural sources play significant roles in the prevention and treatment of bacterial infections since antibiotics have become less effective against many infectious diseases. Mangroves are very potential natural antibacterial sources among great numbers of wild medicinal plants. *Bruguiera cylindrica* is one of the many mangroves species which spread along Indonesian coastline. The aim of this study was to explore the antibacterial activity of *B. cylindrica* wet and dried leaf extracts. The wet extracts study was conducted with three different solvents system (water, ethanol, and n-Hexane) against *Escherichia coli* and *Staphylococcus aureus*. While, the dried extracts study was conducted with four different solvents system (water, ethanol, chloroform and n-Hexane) against three types of bacteria, *Escherichia coli*, *Staphylococcus epidermidis* and *Staphylococcus aureus*. The study showed that ethanol was the best solvent for extraction of phenolic and flavonoid. Antibacterial actitivity was measured by zone of inhibition which obtained from agar-disk diffusion method. The widest area of zone of inhibition was showed by wet extracts with ethanol against *S. aureus* and E. coli are 14.30 and 13.30 mm, respectively. While, the zone of inhibition dried extracts with ethanol against *S. aureus*, *S. epidermidis* and *E. coli* are 9.32, 6.59 and 6.20 mm, respectively. In conclusion, both type of extracts showed significant antibacterial activity against gram-positive bacteria as crude extracts.

Keywords: antibiotics, antimicrobial, infectious diseases, medicinal plants, natural products

1. Introduction

Antibiotics are the most important drugs for fighting bacterial infections [1,2]. However, antibiotics have become less effective against many infectious diseases not only because many of them produce toxic reactions to the patient but also due to the emergence of drugs-resistant bacteria [3,4,5,6,7]. The need for new, effective and affordable drugs to treat infectious diseases in the developing world is one of the major issues facing global health today.

Antibacterial drugs derived from natural source play significant roles in the prevention and treatment of human infection diseases [1,8,9]. Natural bioactive compounds would be best solution to overcome the effects of antibiotics resistance bacteria with no or less side effects [10,11]. In recent decades, many studies have been carried out on different plant species to discover compounds of possible interest for medicinal application against bacterial infections [12,13,14]. Among these studies, several have focused on the biological and phytochemical properties of different species of mangrove [15,16,17].

There are about 3.6 million hectares of mangrove forests which grew along 95,000 km of Indonesian coastline and it represent approximately 23% of the total mangrove forests in the world [18,19]. *Bruguiera cylindrica* is one of the many mangroves species which spread along Indonesian coastline. However, the antibacterial potentials of this Indonesian mangroves are yet to be discovered.

Another factor that renewed the exploration of mangrove is the rapid rate of plant species extinction in the past 20 years and the most threatened plant species are found in the tropics [20]. Mangrove forests suffered a shrinkage of about 1-2% every year [21]. Before any medicinal use or other important features can be assessed, it is likely that some species will extinct before they are even discovered.

The aim of this study was to explore the antibacterial activity potency of *B. cylindrica* leaf extracts against *Escherichia coli*, *Staphylococcus aureus* and *Staphylococcus epidermidis*.

2. Methods

2.1. Plant Materials

The leaves of *B. cylindrica* harvested from Kawasan Ekowisata Mangrove, Pantai Indah Kapuk, North Jakarta, Indonesia with the permission from Department of Marine, Agriculture, and Food Security Special Capital Region of Jakarta. *B. cylindrica* was identified by Indonesian Institute of Sciences (LIPI) taxonomist at Cibinong Science Center. Before used, the leaves were firstly cleaned with tap water to remove dirt and then rinsed with distilled water. The leaves were cut into transverse of approximately 2 cm width and dried by using oven (Memmert, Germany) at 45 °C for 12 hours. The dried leaves were ground into powder using blender (Cucina Philips, Indonesia) and screened by using mesh number 35 (CISA Cedaceria Industrial, Spain). The powder was then transferred into glass bottles with rubber stopper, wrapped in almunium foil and stored in the freezer at -20 °C prior to extraction.

2.2. Extraction

The method was based on method described by Dhayanithi et al. [22] with slightly modification. The leaf powder transferred into an erlenmeyer flask and immersed in solvent (1:10) for 24 hours in a shaker set at 125 rpm. The erlenmeyer flask mouth was sealed with aluminium foil and Parafilm and the whole flask was covered with aluminium foil to protect the content from light. After 24 hours maceration, the extracts were filtrated through cotton fine meshed cloth. The filtrates were then centrifuged (Hettich Rotina 35R centrifuge, Germany) for 10 minutes at 10,000 rpm, 25 °C. The supernatants were then collected by using graduated pipette and prepared for further phytochemical analysis.

2.3. Determination of total phenolic content

The determination of total phenolic content (TPC) of extracts obtained was adapted from Banerjee et al., 2008 [23] with slightly modification and calibrated against galic acid as the reference standard. A 0.3 ml sample was mixed with 1.5 ml of Folin-Ciocalteu reagent and 1.2 ml of sodium carbonate (7.5%), consecutively. The mixture of each step should be mixed well by using vortex (Genie 2 mixer, Scientific Industries, USA) and allowed to stand for one hour in a dark chamber. Absorption was measured by using UV-Vis spectrophotometer (Genesys 10 UV-Vis spectrophotometer, Thermo Electron Corporation, USA) at 765 nm. The standard curve galic acid was prepared by diluting the stock standard with the extraction solvents to yield 50 to 200 ppm TPC. The results were calculated according to the calibration curve for galic acid and crude extract of TPC derives from quadruplicate analyses and expressed as galic acid equivalents (GAE mg/ g) of dry material (DM).

2.4. Determination of flavonoid content

The determination of flavonoid content (FC) was adapted from Do et al. [24] with slightly modification and calibrated against quercetin as the reference standard. The Leaf powder was first diluted with its solvent to reach dilution factor of 25 for test the FC in the samples. 1.5 ml of methanol in test tube was prepared. A 0.5 ml sample was mixed with 0.1 ml of 10% aluminium chloride, 0.1 ml of 1 M potassium acetate and 2.8 ml distilled water, consecutively. The mixture of each step should be mixed well by using vortex. The mixtures were incubated in a dark chamber for 30 minutes and absorbance was measured by using a UV-Vis spectrophotometer at 415 nm. The standard curve quercetin was prepared by diluting the stock standard with the extraction solvents to yield 20 - 200 ppm of FC. The results were calculated according to the calibration curve for quercetin and crude extract of FC derives from quadruplicate analyses and expressed as quercetin equivalents (QE mg/ g) of dry material (DM).

2.5. Antibacterial activity

The study of antibacterial activity was conducted in two conditions, the wet extracts study which held at Swiss German University and conducted by using three different solvents (water, ethanol, and n-Hexane) against two types of bacteria, *E. coli* and *S. aureus*. While the dried extracts study which held at Biopharmaca Tropica Research Center, LPPM IPB and conducted by using four different solvents based on the level of polarity (water, ethanol, chloroform and n-Hexane), against three types of bacteria, *E. coli*, *S. epidermidis* and *S. aureus*.

The Agar disk diffusion assay has been widely used to assay plant extract for antibacterial activity [25]. *E. coli*, *S. epidermidis* and *S. aureus* were adjusted to certain concentration, 1 x 10^8 CFU/ml or 0.5 mcFarland's standard [26], were inoculated onto the entire surface of a Mueller-Hinton agar (MHA) plate with a sterile cotton-tipped swab to form an even lawn. In this method, 6 mm sterilized filter papers disks (Whatmann No. 1) were soaked into *B. cylindrica* extract and air dried for one hour at bio safety cabinet. Each type of extracts replicated by three. The paper disks which was impregnated with diluted extracts were placed on the surface of each MHA plate by using a sterile pair of forceps/ pincers. Then the plates were incubated at 37°C for 24 hours and diameter of the zone of inhibition (ZOI) obtained was measured by a ruler or caliper. The bigger diameter of the ZOI, the more susceptible of the microorganism to the extracts.

2.6. Statistical analysis

Data result was analyzed by Analysis of Variance (ANOVA) single factor. The P-two tail or P-Value below 0.05 ($P < 0.05$) indicated that the data were significantly different, while if P-Value above 0.05 ($P > 0.05$), the data were insignificantly different. The data graphic was built using Microsoft Excel 2010.

3. Results and Discussion

Extraction of plant materials depends on various factors such as solvents, methods, and extraction time to separate different quality and quantity of bioactive components in the crude extracts [27]. Polarity of solvent plays key role in the outcome of bioactive compound extracted from *B. cylindrica*. The wet extracts with ethanol showed the highest total phenolic content compared with water and n-Hexane, consecutively (Figure 1). It was because ethanol has the ability to bind with hydrophilic compounds such as flavonoid and phenolic compound [28]. Ethanol is chosen when water-insoluble constituents need to be extracted. However, higher ethanol percentages do not necessarily mean higher extraction activity [29].

The wet extracts with ethanol also showed the highest flavonoid content compared with water and n-hexane, consecutively (Figure 2). Less polar flavonoids (e.g., isoflavones, flavanones, methylated flavones, and flavonols) are best extracted with chloroform, dichloromethane, diethyl ether, or ethyl acetate, while flavonoid glycosides and more polar aglycones are best extracted with alcohols or

IOP Conf. Series: Earth and Environmental Science **130** (2018) 012025 doi:10.1088/1755-1315/130/1/012025

alcohol–water mixtures [30,31]. The combination of water and organic solvent facilitate the extraction of chemicals that are soluble in water and/or organic solvent [24].

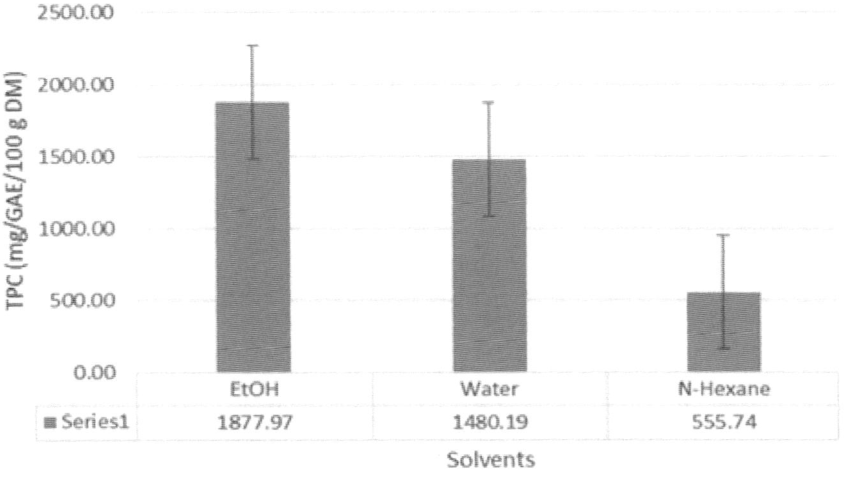

Figure 1. Total phenolic content of *B. cylindrica* leaf extracts

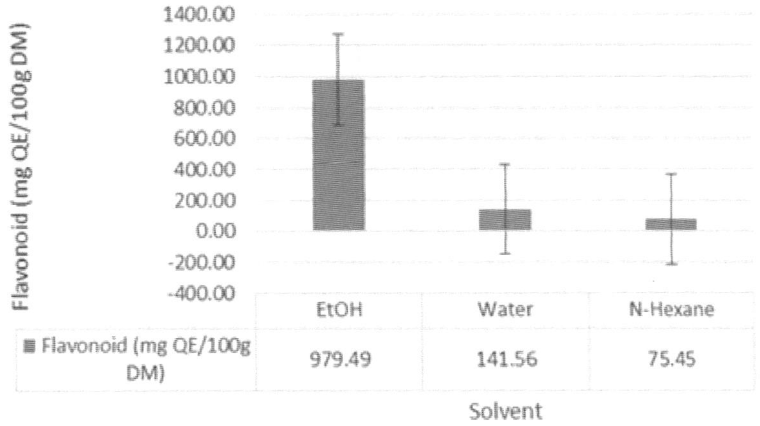

Figure 2. Flavonoid content of *B. cylindrica* leaf extracts

The mechanisms for phenolic toxicity to microorganisms include adsorption and disruption of microbial membranes, interaction with enzymes, and metal ion deprivation [32,33,34]. The higher FC was also expected to show the higher antibacterial activity since flavonoids are synthesized by plants in response to microbial infection [35,36,37,38]. Flavonoid involved in inhibition nucleic acid biosynthesis, inactivate microbial adhesins enzymes and cell envelope transport proteins [39,40]. Lipophilic flavonoids may also disrupt microbial membranes [13,36,41].

The ZOI obtained of the wet extracts with ethanol against *S. aureus* and *E. coli* were 14.30 and 13.30 mm, respectively (Figure 3). The dried extracts study also showed a consistent results with the wet extracts study. The ZOI of the dried extracts with ethanol against *S. aureus*, *E. coli*, and *S. epidermidis* were 9.32, 6.20 and 6.59 mm, as showed in Table 1. Taken altogether, the leaf extracts can be categorized as bacteriostatic due to the capability of bacterial growth inhibition. Quantitatively, the leaf extract can be classified as weak-to-moderate antibacterial since the ZOI obtained around 10 to 15 mm. While, the ZOI obtained by Ciprofloxacin, tetracycline and chloramphenicol were 41.2, 22.10 and 20.80 mm.

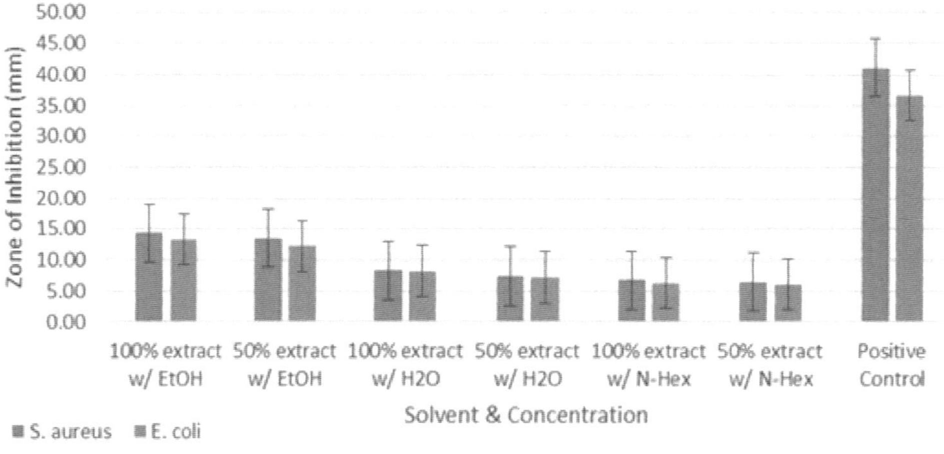

Figure 3. Zone of inhibition of wet extracts

Table 1. Antibacterial test of *B. cylindrica* leaf extracts againts *E. coli* and *S. aureus*

Conc	(+) mm		Extraction						(-) mm
(%)	Tetra	Chloram	b. Cyl ethx (mm)			b. Cyl chlx (mm)			DMSO
	22.10	20.80	SA	EC	SE	SA	EC	SE	
0.4			9.32	6.20	6.59	7.32	6.88	7.42	
0.2			8.32	6.00	6.32	7.00	6.52	6.26	
0.1			8.08	6.00	6.16	6.72	6.00	6.00	
0.05			6.40	6.00	6.20	6.56	6.02	6.52	
									6.00

Conc : concentration, tetra : tetracycline, chlor : chloramphenycol, b. cyl ethx : *B. cylindrica* ethanol extract, b. cyl chlx : *B. cylindrica* chloroform extract, r. muc ethx : *R. mucronata* ethanol extract, r. muc chlx : *R. mucronata* chloroform extract, DMSO = DMSO 20%, SA: *S. aureus*, EC: *E. coli*, and SE: *S. epidermidis*.

Since all the extracts consist of many different compounds, they presumably contain single compounds with higher activity than the value obtained as crude extract. In addition, isolation and characterization of the antibacterial compounds presented in these extracts seemed to be worthwhile [42]. The potential for developing antimicrobial agents from mangrove species appears rewarding, as it may lead to development of phytomedicine against pathogenic microbes [43].

Study with different sampling time, age of plant, plant part, solvent, method of extraction and time required for extraction would be necessary [44,45]. Because, many of the phytochemicals in plants can be detected at different concentrations because the amount and composition of secondary metabolites are not constant and their concentration depends on the tissue type and the age of the plant [46,47].

In general, both type of extracts showed significant antibacterial activity against gram-positive bacteria. This matter may be correlated with the cell envelopes composition. The cell envelopes of most bacteria fall into one of two major groups. Gram-negative bacteria are surrounded by a thin peptidoglycan cell wall, which itself is surrounded by an outer membrane containing lipopolysaccharide. Gram-positive bacteria lack an outer membrane but are surrounded by layers of peptidoglycan many times thicker than is found in the Gram-negatives [48]. Therefore, gram-negative bacteria is more resistant than gram-positive due to the lipopolyproteins and peptidoglycan composition in their cell envelope [49]. Gram-negative cell walls is strong to withstand three atm of turgor pressure, tough to endure extreme temperatures, pH and elastic to be capable of expanding several times their normal surface area [50].

Conclusions

The study of *B. cylindrica* leaf extract suggested that ethanol was the best solvent. The widest area of ZOI was showed by wet extracts with ethanol against *S. aureus* and *E. coli* were 14.30 and 13.30 mm, respectively. While, the ZOI dried extracts with ethanol against *S. aureus*, *S. epidermidis* and *E. coli* were 9.32, 6.59 and 6.20 mm, respectively. In conclusion, both type of extracts showed considerably high antibacterial activity against gram-positive bacteria as crude extracts.

Acknowledgments

The authors gratefully acknowledge that the present research is supported by the Central Research Fund of the Swiss German University with Contract Number A/REC/0073/IX/2015 and by Ministry of Research and Technology and Higher Education Republic of Indonesia with Contract Number SP DIPA-042.06.1.401516/2017. We also thank Directorate of Marine, Agriculture and Food Security, DKI Jakarta Province for giving us permission to collect mangrove leaves at the Mangroves Sanctuary Region, Pantai Indah Kapuk, North Jakarta, Indonesia. We thank our research collaborators Dr. Irmanida Batubara and staffs at Biopharmaca Tropica Research Center, Bogor Agricultural University and Dr. Doddy Kustaryono and staffs at STKIP Surya, Tangerang. We also thank the Director and staffs of Academic Research and Community Services Swiss German University for their support throughout the project.

References

[1] WHO (2014). Antimicrobial resistance: global report on surveillance. WHO Press. Geneva. Switzerland

[2] CDC (2013). Antibiotic resistamce threats in the United States. US Department of Health and Human Services. Atlanta: Georgia

[3] Kart A, Bilgili A (2008). Ionophore antibiotics: toxicity, mode of action and neurotoxic aspect of carboxilic ionophores. J Animal Vet Adv. 7(6):748-751.

[4] Bassetti M, Ginocchio F, Mikulska M. New treatment options against gram-negative organisms (2011). Critical Care. 15: 215

[5] Grill MF, Maganti RK (2011). Neurotoxic effects associated with antibiotic use: management considerations. British journal of clinical pharmacology. 72(3):381-93.

[6] Drouet M, Chai F, Barthelemy C, Lebuffe G, Debaene B, Decaudin B, et al., (2015). Endothelial Cell Toxicity of Vancomycin Infusion Combined with Other Antibiotics. Antimicrobial agents and chemotherapy. 59(8):4901-6.

[7] Ventola C (2015). The Antibiotic Resistance Crisis Part 1: Causes and Threats. P & T. 40(4):277-83.

[8] Bhalodia NR, Nariya PB, Shukla VJ. (2011). Antibacterial and Antifungal activity from Flower Extracts of Cassia fistula L.: An Ethnomedicinal Plant. International Journal of PharmTech Research. Vol. 3, No.1, pp 160-168,

[9] Prihanto AA, Firdaus M, Nurdiani R (2012). Anti-Methicillin Resistant *Staphylococcus aureus* (MRSA) of Methanol Extract of Mangrove Plants Leaf: Preliminary Report. Drug discovery today. 4(8):439-440

[10] Bansal D, Bhasin P, Punia A, Sehrawat AR (2011). Evaluation of antimicrobial activity and phytochemical screening of extracts of Tinospora cordifolia against some pathogenic microbes. J. Pharm. Res. 5(1):127-129.

[11] Priyatharsini S, Sivagurunathan P, Uma C, Bhuvaneswari M, Aruljothi S (2015). Assessment of antibacterial activity of halophytic plants against uropathogens. Asian J Pharm Sci Tech. 5(2): 102-105.

[12] Chan BCL, Lau CBS, Jolivalt C, Lui SL, GanemElbaz C, Paris JM, Litaudon M, Fung KP, Leung PC, Ip M. 2011. Chinese medicinal herbs against antibiotic-resistant bacterial pathogens. in Science against microbial pathogens: communicating current research and technological advances. Formatex

[13] Djouossi MG, Tamokou JD, Ngnokam D, Kuiate JR, Tapondjou LA, Harakat D, et al., (2015). Antimicrobial and antioxidant flavonoids from the leaves of *Oncoba spinosa* Forssk. (Salicaceae). BMC complementary and alternative medicine. 15:134.

[14] Voukeng IK, Beng VP, Kuete V (2017). Multidrug resistant bacteria are sensitive to *Euphorbia prostrata* and six others Cameroonian medicinal plants extracts. BMC research notes.10(1):321.

[15]. Ravikumar S, Gnanadesigan M, Suganthi P, Ramalakshmi A (2010). Antibacterial potential of chosen mangrove plants against isolated urinary tract infectious bacterial. Int J Med Medical Sci. 2(3): 94-99.

[16] Saad S, Taher M, Susanti D, Qaralleh H, Awang AFIB (2012). In vitro antimicrobial activity of mangrove plant *Sonneratia alba*. Asian Pacific Journal of Tropical Biomedicine. 2(6):427-9.

[17] Ramasubburayan R, Prakash S, iyapparaj P, Sumathi S, Thaddaeus BJ, palavesam A, immanuel G (2015). Investigation on antibacterial, antifungal and cytotoxic properties of chosen mangroves. Indian J Geo-Marine Sci. 44(11): 1769-1777.

[18] Giesen W., Wulffraat S., Zieren M., Scholten L (2006). Mangrove guide book for Southeast Asia. FAO and Wetlands International, Thailand: Dharmasarn Co., Ltd.

[19] Giri C (2016). Observation and Monitoring of Mangrove Forests Using Remote Sensing: Opportunities and Challenges. Remote Sensing. 8(9):783.

[20] IUCN 2010. New study shows over one fifth of the world's plants are under threat of extinction. http://www.iucnredlist.org/news/srli-plants-press-release

[21] Locatelli T, Binet T, Kairo JG, King L, Madden S, Patenaude G, et al., (2014). Turning the tide: how blue carbon and payments for ecosystem services (PES) might help save mangrove forests. Ambio. 43(8):981-95.

[22] Dhayanithi N, Kumar T, Murthy R, Kathiresan K (2012). Isolation of antibacterials from the mangrove, *Avicennia marina* and their activity against multi drug resistant *Staphylococcus aureus*. Asian Pac. J. Trop. Biomed. 2(3):1892-1895.

[23] Banerjee D, Chakrabarti S, Hazra AK, Banerjee S, Ray J, Mukherjee B (2008). Antioxidant activity and total phenolics of some mangroves in Sundarbans. Afr. J Biotech. 7(6):805-810.

[24] Do QD, Angkawijaya AE, Tran-Nguyen PL, Huynh LH, Soetaredjo FE, Ismadji S, et al., (2014). Effect of extraction solvent on total phenol content, total flavonoid content, and antioxidant activity of *Limnophila aromatica*. Journal of food and drug analysis. 22(3):296-302.

[25] Ergene A, Guler P, Tan S, Mirici S, Hamzaoglu E, Duran (2006). Antimicrobial and antifungal activity of *Heracleum sphondylium* subsp. artivinense. Afr J Biotechnol. 5(11):1087-1089.

[26] Baris O, Gulluce M, Sahin F, Ozer H, Kilic H, Ozkan H, Sokmen M, Ozbek T (2006). Biological activities of the essential oil and methanol extract of *Achillea biebersteinii* Afan. (Asteraceae). Turk. J. Biol. 30:65-73.

[27] Azmir J, Zaidul ISM, Rahman MM, Sharif KM, Mohamed A, Sahena F, et al., (2013). Techniques for extraction of bioactive compounds from plant materials: A review. Journal of Food Engineering. 117(4):426-36.

[28] Sasidharan S, Nilawatyi R, Xavier R, Latha LY, Amala R (2010). Wound Healing Potential of *Elaeis guineensis* Jacq Leaves in an Infected Albino Rat Model. Molecules. 15(5):3186-99.

[29] Morgan M (2009). Ethanol in Herbal Medicine. Mediherb 129

[30] Brglez Mojzer E, Knez Hrncic M, Skerget M, Knez Z, Bren U (2016). Polyphenols: Extraction Methods, Antioxidative Action, Bioavailability and Anticarcinogenic Effects. Molecules. 21(7).

[31] Bouhoreira A, Dadamoussa B, Gherraf N (2016). Phytochemical study and evaluation antibacterial activity of flavonoid excerpts: Male spathe of date palm. Der Pharmacia Lettre. 8 (13):171-176

[32] Fattouch S, Caboni P, Coroneo V, Tuberoso C, Angioni A, Dessi S, Marzouki N, Cabras P (2007). Antimicrobial Activity of Tunisian Quince (*Cydonia oblonga* Miller) Pulp and Peel Polyphenolic Extracts. J. Agric. Food Chem. 55(3):963- 969.
[33] Maddox CE, Laur LM, Tian L (2009). Antibacterial Activity of Phenolic Compounds Against the Phytopathogen *Xylella fastidiosa*. Current Microbiology. 60(1):53-8.
[34] Fu L, Lu W, Zhou X (2016). Phenolic Compounds and In Vitro Antibacterial and Antioxidant Activities of Three Tropic Fruits: Persimmon, Guava, and Sweetsop. BioMed research international. 2016:4287461.
[35] Baba SA, Malik SA (2015). Determination of total phenolic and flavonoid content, antimicrobial and antioxidant activity of a root extract of *Arisaema jacquemontii* Blume. Journal of Taibah University for Science. 9(4):449-54.
[36] Kumar S, Pandey AK (2013). Chemistry and biological activities of flavonoids: an overview. TheScientificWorldJournal. 2013:162750.
[37] Mierziak J, Kostyn K, Kulma A. (2014). Flavonoids as Important Molecules of Plant Interactions with the Environment. Molecules. 19(10):16240-65.
[38] Sofiane G, Wafa N, Abbas K, Amar O (2015). Antioxidant and antimicrobial activities of flavonoids extracted from *Thymus ciliatus* (Desf.) Benth. Der Pharmacia Lettre. 7(7):358-363.
[39] Tkachenko H, Buyun L, Terech-Majewska E, Osadowski Z (2016). Antibacterial activity of ethanolic leaf extracts obtained from various ficus species (Moraceae) against the fish pathogen, *Citrobacter freundii*. J Ecol Protection of the Coastline. 20: 117-136.
[40] Qwarse M, Sempombe J, Mihale MJ, Henry L, Mugoyela V, Sung'hwa F (2017). Cytotoxicity, Antibacterial, and Antifungal Activities of Five Plant Species Used by Agro-pastoral Communities in Mbulu District, Tanzania. Int J Res Pharm Chem, 7(1): 1-14.
[41] Mishra MP, Rath S, Swain SS, Ghosh G, Das D, Padhy RN (2017)In vitro antibacterial activity of crude extracts of 9 selected medicinal plants against UTI causing MDR bacteria. Journal of King Saud University - Science. 29(1):84-95.
[42] Eloff JN, Famakin JO, Katerere DRP (2005). *Combretum woodii* (Combretaceae) leaf extracts have high activity against Gram-negative and Gram-positive bacteria. Afr. J. Biotechnol. 4(10):1161-1166.
[43] Patra JK, Thatoi H, Panigrahi TK, Rath SK, Dhal NK (2009). Phytochemical Screening and Antimicrobial Assessment of Leaf Extracts of *Excoecaria Agallocha* L.: A Mangal Species of Bhitarkanika, Orissa, India. Adv Natur Appl Sci, 3(2): 241-246
[44] Sasidharan S, Chen Y, Saravanan D, Sundram KM, Latha LY (2011). Extraction, isolation and characterization of bioactive compounds from plants extracts. Afr J Tradit Complement Altern Med. 8(1):1-10.
[45] Okoduwa SIR, Umar IA, James DB, Inuwa HM, Habila JD (2016). Evaluation of extraction protocols for anti-diabetic phytochemical substances from medicinal plants. World Journal of Diabetes. 7(20):605.
[46] Silva-Beltrán NP, Ruiz-Cruz S, Cira-Chávez LA, Estrada-Alvarado MI, Ornelas-Paz JdJ, López-Mata MA, et al., (2015). Total Phenolic, Flavonoid, Tomatine, and Tomatidine Contents and Antioxidant and Antimicrobial Activities of Extracts of Tomato Plant. International Journal of Analytical Chemistry. 2015:1-10.
[47] Wouters FC, Blanchette B, Gershenzon J, Vassão DG (2016). Plant defense and herbivore counter-defense: benzoxazinoids and insect herbivores. Phytochemistry Reviews. 15(6):1127-51.
[48] Silhavy TJ, Kahne D, Walker S. The bacterial cell envelope. (2010). Cold Spring Harbor perspectives in biology. 2(5):a000414.
[49] Lopez D, Kolter R. Functional microdomains in bacterial membranes (2010). Genes & development. 24(17):1893-902.
[50] Beveridge TJ. Structures of Gram-Negative Cell Walls and Their Derived Membrane Vesicles (1999). J Bacteriology. 18(16): 4725±4733

Detection of *meca* gene from methicillin resistant *staphylococcus aureus* isolates of north sumatera

Gabriella Septiani Nasution[1,2], Dwi Suryanto[1*],R Lia Kusumawati[3,4]

[1]Department of Biology, University of Sumatera Utara, Jln. Bioteknologi No. 1, Kampus USU, Medan, 20155
[2]Medan Health Polytechnique, Ministry of Health, Jln. Jamin Ginting Km. 13.5, Lau Cih, Medan
[3]Department of Microbiology, Faculty of Medicine, University of Sumatera Utara, Jln. Universitas No. 1, Kampus USU, Medan 20155
[4]Clinical Research Unit, University of Sumatera Utara Hospital, Jln. Dr. T. Mansyur No. 66, Kampus USU, Medan 20154

*Email: dwisuryanto@usu.ac.id

Abstract. *Methicillin Resistant Staphylococcus aureus* (MRSA) is a major pathogen associated with hospital-acquired infections (nosocomial infections). MRSA is a type of *S. aureus* resistant to the sub-group of beta-lactam antibiotics such as penicillin, cephalosporin, monobactam, and carbapenem. MRSA is resistant because of genetic changes caused by exposure to irrational antibiotic therapy. This study aimed to detect *mec*A gene in North Sumatra isolates of MRSA and to determine the pattern of antibiotic resistance in *S.aureus* isolates classified as MRSA by Vitek 2 Compact in the Central Public Hospital Haji Adam Malik, Medan. Samples were 40 isolates of *S. aureus* classified as MRSA obtained from clinical microbiology specimens. DNA isolation of the isolates was conducted by a method of freeze-thaw cycling. Amplification of *mec*A gene was done by PCR technique using specific primer for the gene. PCR products were visualized using mini-gel electrophoresis. The results showed that all MRSA isolates showed to have 533 bp band of *mec*A. Antibiotics test of Vitek 2 Compact showed that despite all isolates were resistant to beta-lactam antibiotics groups; the isolates showed multidrug resistant to other common antibiotics, such as aminoglycosides, macrolides, and fluoroquinolones. However, they were still sensitive to vancomycin (82.5% isolates), linezolid (97.5% isolates), and tigecycline (100% isolates).

Keywords: antibiotic resistant, beta-lactam, detection, nosocomial infections, PCR

1. Introduction

Methicillin Resistant Staphylococcus aureus (MRSA) has become one of the worldwide health problems and is a major pathogen associated infections acquired in the hospital (nosocomial infections) [1]. MRSA is a type of *S. aureus* that is resistant to the sub-group of beta-lactam antibiotics such as penicillin, cephalosporin, monobactam and carbapenem [2]. This resistant caused by genetic changes because of irrational antibiotic therapy exposure.

MRSA strains were first reported in 1961 in the United Kingdom (UK) [3]. Prevalence of MRSA in hospitals in different parts of the world diverse, ranging between 2-70%, 20% in average [4,5,6]. The percentage of strains of MRSA is relatively high in Asia, such as 60% in Taiwan reached, 20% in China, 70% in Hong Kong, 5% in Philippines, and 60%in Singapore. In Indonesia in 2006, MRSA prevalence is 23.5% [7]. A study in Dr. Soetomo Hospital, Surabaya showed that of 643 patients there were 52 MRSA (8%) [8]. In contrary, a study in Central Public Hospital Haji Adam Malik Medan showed quite higher prevalence of MRSA, in which in January-June 2015 there were 56 isolates (67%), in July- December 2015 there were 48 isolates (57%), and January-June 2016, 58 isolates (45%) were MRSA.

Genotypes examination for MRSA resistance has been conducted to know antibiotics resistance gene such as *mecA* [9]. The gold standard to determine MRSA genotypes is to detect conserved genes (fixed/ preserved) constantly found in *mec*A gene, which is within range of a particular chromosome in Staphylococcal Cassette Chromosome (SCC*mec*) [10]. MRSA resistance is due to the mutant protein of penicillin-binding protein 2a (PBP2a or PBP2 ') encoded by *mec*A gene. PBP is a group of enzymes in the cell membrane of *S. aureus* that catalyzes the trans-peptidation for peptidoglycan chain (cross-linkage) formation. The affinity of PBP2a is so low that MRSA stays alive in high concentration of antimicrobial exposure [11]. Amplification of *mec*A can be done by using polymerase chain reaction (PCR), which is the gold standard for the detection of *mec*A [12]. No information on the distribution of *mecA* on MRSA in North Sumatra is available. Therefore, a study on this gene is urgent.

2. Methods

2.1. Bacterial isolates and phenotypic characterization

This research was conducted in the central public hospital of Haji Adam Malik Medan. *S. aureus* isolates were obtained from clinical microbiology specimens sent to Hospital Diagnostic Laboratory Installation H. Adam Malik, Medan from March to May 2017. A total of 40 Vitek 2 Compact classified MRSA were used in this study.

2.2. DNA isolation

MRSA was sub-cultured on blood agar and incubated at 37°C for 18-24 hours. The cell was broken by a freeze-thaw method as previously described [13]. The freeze-thawed solution was spin at 13.000 rpm for 5 minutes. The supernatant was separated from cell debris, and subjected to DNA purification check.

2.3. Amplification and detection of mec*A gene using PCR technique*

*mec*A gene amplification was performed by PCR technique using specific primer *mec*A-F: AAA AAA GGT GGT ATC GAT TGG C and *mec*A-R: AGT TCT GCA GTA CCG GAT TTG C [14] [15]. Amplification was done in PCR solution of 12.5µl GoTaq® Green Master Mix, 1µl of forward and reverse primer each, 8.5µl nuclease-free water, and 2µl bacterial DNA. Thermocycling reaction was conducted for denaturation at 95°C for 3 minutes, annealing for 55°C for 30 seconds, extension at 72°C at 60 seconds, and extended extension at 72°C for 6 minutes. The reaction was done for 30 cycles. PCR product was visualized in mini gel electrophoresis and documented in UV Reader/Gel Documentation System. MRSA ATCC43300 and MSSA ATCC25923 were used for positive and negative control, respectively.

3. Results and Discussion

3.1. Bacterial isolates and phenotypic characterization

A total of 40 isolates MRSA were collected from March-May 2017 isolated from various clinical specimens of Diagnostic Laboratory Installation. *Staphylococcus aureus* is a common type of bacteria found in the skin, nasal cavity healthy person or patients, and in pus. The prevalence of MRSA in the sample was between 2.5-40%. Most MRSA was isolated from pus (40%), sputum (22.5%), and blood (17.5%)(Table 1). A study carried out by Islam *et al.* (2011) isolated MRSA mostly from pus (7.5%) [16]. Pournajat *et al.* (2014) found that most MRSA was isolated from pus (29%) [15].

Table 1. Prevalence of MRSA from various clinical specimens

No	Clinical specimens	Total sample (n)	Prevalence (%)
1	Brain fluid	1	2.5
2	Tissue	1	2.5
3	Urine	1	2.5
4	Pleural fluid	2	5
5	Swab	3	7.5
6	Blood	7	17.5
7	Sputum	9	22.5
8	Pus	16	40

Phenotype test of antibiotic resistant using Vitek 2 Compact showed that all isolates were resistant to cefoxitin. Data from Clinical and Laboratory Standards Institute (CLSI) 2015 showed that isolates resisted to cefoxitin were also resistant to 52 other types of antibiotics. Cefoxitin including second-generation Cephalosporin is a potent inducer of *mec*A regulatory system that is used extensively as a marker for the detection of the *mec*A gene [17]. The percentage for antimicrobial sensitivity can be seen in Table 2 below.

Table 2. Antimicrobial sensitivity of MRSA isolates

No.	Antimicrobial Agents	Antibiotic sensitivity of 40 isolates (%)		
		Sensitive	Intermediates	Resistant
1.	Amoxicillin	0	0	100
2.	Ampicillin/Sulbactam	0	0	100
3.	Cefalexin	0	0	100
4.	Cefazolin	0	0	100
5.	Cefuroxime	0	0	100
6.	Ciprofloxacin	2.5	0	97.5
7.	Clindamycin	52.5	2.5	45
8.	Erythromycin	47.5	7.5	45
9.	Gentamycin	7.5	5	87.5
10.	Levofloxacin	0	2.5	97.5
11.	Tetracycline	37.5	0	62.5
12.	Vancomycin	82.5	2.5	15
13.	Trimethoprim/Sulfamethoxazole	72.5	0	27.5
14.	Moxifloxacin	2.5	2.5	95
15.	Linezolid	97.5	0	2.5
16.	Tigecycline	100	0	0

It was shown that all isolates were resistant to beta-lactam antibiotics of amoxicillin, ampicillin/sulbactam, and cephalosporins group such as cefalexin, cefazolin, and cefuroxime. High resistant was also showed to ciprofloxacin (97.5%), gentamycin (87.5%), and levofloxacin (97.5%). In contrary, the isolates were still sensitive to vancomycin (82.5%), trimethoprim/sulfamethoxazole (72.5%), linezolid (97%), and tigecycline (100%). For antibiotics treatment, these antibiotics may still be used.

Another study in Indonesia showed similar pattern of antibiotic resistant in which *S. aureus* of clinical samples was resistant not only to penicillin and methicillin, but also resistant to tetracycline

(24.5%), oxacillin (2%), gentamicin (1%), erythromycin (5.1%), chloramphenicol (9.2%), and trimethophrim/sulfamethoxazole (7.1%) [18]. However, the percentage of isolates resistant to other antibiotics was relatively lower compared to that of this study. Al Ruaily and Khalil (2011) study in Saudi Arabia showed *S. aureus* isolates were resistance to cephalosporins (95%), gentamycin (95%), ciprofloxacin (87%), vancomycin (100%), and penicillin (100%) [19]. A similar study of Elhassan *et al.* (2015) isolated from different clinical specimens in Shendi City, Sudan showed that all isolates of *S. aureus* were resistance to methicillin, penicillin, and ampicillin [20]. Resistant to gentamycin, ciprofloxacin, and clindamycin were showed in 80% of the isolates, while resistant to trimethoprim/sulfamethoxazole was showed in 74% of the isolates. However, the isolates resistant to linezolid were relatively low (13%). Sudigdoadi (2010) showed that 89% of his isolates were resistant to gentamycin [21].

Beta-lactam antibiotics kill bacteria by inhibiting cell wall synthesis. MRSA resistance to beta-lactam group is due to a protein mutant of penicillin-binding protein 2a (PBP2a or PBP 2') encoded in *mec*A gene. PBP is a group of enzymes in the cell membrane of *S. aureus* that catalyzes the transpeptidation for the formation of peptidoglycan chain webbing (cross-linkage). Affinity PBP2a antimicrobial beta-lactam group is so low that MRSA remains alive in a high concentration of antimicrobial exposure in [11].

It was interesting that the isolates showed resistant to vancomycin although only a few. The resistant to vancomycin (VRSA) is associated with changing and resetting bacterial cell wall. In addition, overproduction of Penicillin Binding Protein-2 (PBP-2) is also considered as an important factor for the expression of resistance to vancomycin. It is known that resistance to vancomycin is mediated by specific gene *van*A to glycopeptides [22]. Vancomycin resistant isolates is likely due to spontaneous mutations, occur acquisition of resistant factors from elsewhere, or from the surrounding enteric bacterial population [21].

3.2. Detection of mecA gene

MRSA detection is essential for proper patient care and infection control. Examination of MRSA can be done either phenotype or genotype. The gold standard to determine information about the genotype distribution of MRSA is to detect genes conserved constantly found. *mec*A gene is located within chromosome in a structure called Staphylococcal Cassette Chromosome (SCC*mec*) encodes mutant PBP2a or PBP2' of 76 kDa [10]. The PCR results of all isolates were shown in Figure 1.

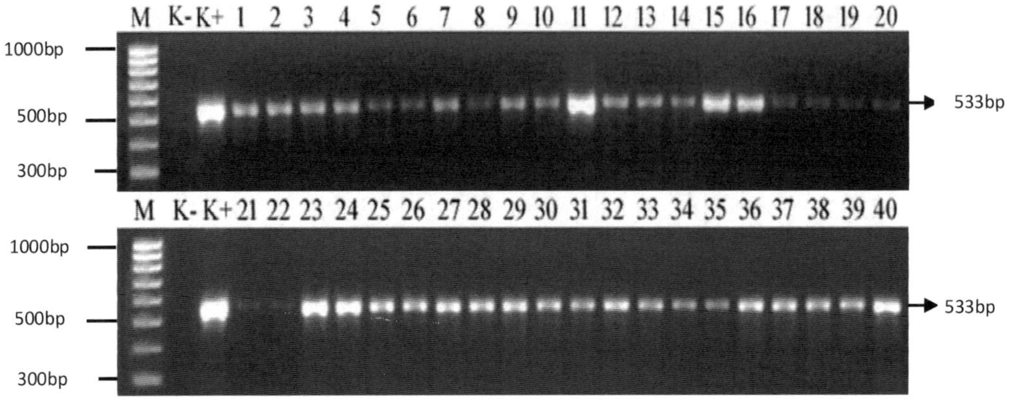

Figure 1. Distribution of *mec*A gene in isolates of MRSA (533bp product) (M=100 bp, K- = negative control (MSSA ATCC25923), K+ = positive control (MRSA ATCC43300), lane 1-40 = MRSA isolates of clinical samples)

The presence of *mec*A gene is generally to indicate the potential resistance to beta-lactam group and used as a marker to identify MRSA. In this study, PCR product was shown as 533 bp amplicon in all resistant isolates using a primer designed by Pournajat *et al.* (2015) [15]. The similar result showed by Sudigdoadi (2010) in which all 45 isolates of MRSA were investigated, certainly has a *mec*A gene which is found in a 20-100 kb called staphylococcal cassette chromosome (SCC*mec*) [23] [24]. MRSA resistance to methicillin and all beta-lactam group antimicrobial is due to changes in normal penicillin binding protein PBP 2 to PBP 2a. Mutation to PBP 2a showed that the change in the binding site resulted in lower affinity to beta-lactam group [25], therefore if the bacteria are cultured in medium containing a high concentration of beta-lactams, they still survive and grow.

4. Conclusions

Test of antibiotic sensitivity using Vitek 2 Compact of 40 isolates showed that the isolates were not only resistant to beta-lactam antibiotics groups, but also to other antibiotics such as aminoglycosides, macrolides, and fluoroquinolones. Gene detection of *mec*A showed that all isolates carried the gene.

Acknowledgments

Special acknowledgment is given to Ministry of Health, The Republic of Indonesia for providing a scholarship and research fund.

References

[1] Kaur H, Saini A, Purwar S, Kaur H, Kardesai SG, Kholkute SD and Roy S (2012) Susceptibility testing and resistance phenotypic detection in *Staphylococcus aureus* by conventional and molecular methods: importance of automated (Vitek 2) system. *International Journal of Engineering Research and Development*. 3(10) : 68-74.

[2] Dellit T, Duchin J, Hofmann J, and Olson EG (2004) *Interim Guidelines for Evaluation & Management of Community Associated Methicillin-resistant Staphylococcus aureus Skin and Soft Tissue Infection in Outpatient Settings*.

[3] Ajmal AN, Mir F, Aslam M, Hafeez R, and Attique R (2009) Nosocomial methicillin-resistant *Staphylococcus aureus* frequency in a tertiary care hospital, Lahore, Pakistan. *Biomedica*. 25:97-100.

[4] Bell JM and Turnidge JD (2002) High prevalence of oxacillin-resistant *Staphylococcus aureus* isolates oxacillin-resistant *Staphylococcus aureus* isolates from hospitalized patients in Asia-Pacific and South Africa: results from SENTRY Antimicrobial Surveillance Program, 1998–1999. *Antimicrob Agents Chemother*. 46(3):879-81.

[5] Denis O, Deplano A, Nonhoff C, De Ryck R, de Mendonca R, Rottiers S, Vanhoof R. and Struelens MJ (2004) National Surveillance of methicillin-resistant *Staphylococcus aureus* in Belgian Hospitals indicates rapid diversification of epidemic clones. Antimicrob Agents Chemother. 48(9):3625-3629.

[6] Vos MC, Ott A, and Verbrugh HA (2005) Successful search and-destroy policy for Methicillin-Resistant *Staphylococcus aureus* in the Netherlands. *J Clin Microbiol.* 43(4):2034-5.

[7] Sulistyaningsih (2010) Uji kepekaan beberapa sediaan antiseptic Terhadap bakteri *Staphylococcus aureus* dan *Staphylococcus aureus* resisten metisilin *(MRSA)* [Tesis]. Bandung : Universitas Padjajaran.

[8] Kuntaman K, Hadi U, Setiawan F, Koendori EB, Rusli M, Santosaningsih D, Severin J, and Verbrugh HA (2016) Prevalence of Methicillin-Resistant *Staphylococcus aureus* from nose and throat of patients on admission to medical wards of Dr. Soetomo Hospital, Surabaya, Indonesia. *Southeast Asian J Trop Med Public Health.* 47(1) : 1 - 5.

[9] Sjahrurachman A (2011) Cara Genetis untuk Menentukan Kepekaan Bakteri terhadap Antibiotik. *CDK.* Vol. 38(7) : 498 – 502.

[10] Jain A, Agarwal A, and Verma RK (2008) Cefoxitin disc diffusion test for detection of meticillin-resistant staphylococci. *Journal of Medical Microbiology.* 57: 957–961.

[11] Felten A, Grandry B, Lagrange PH, and Casin I (2002) Evaluation of three techniques for detection of low- level Methicillin-Resistant *Staphylococcus aureus* (MRSA): a disk diffusion method with cefoxitin and moxalactam, the vitek 2 system, and the MRSA – screen latex agglutination test. *Journal of Clinical Microbiology.* 40 (8) : 2766 – 2771.

[12] Jonas D, Speck M, Daschner FD and Grundmann H (2002) Rapid PCRBased Identification of Methicillin-Resistant *Staphylococcus aureus* from Screening Swabs. *J Clin Microbiol.* 40: 1821-1823.

[13] Bennimath VD, Gavimath CC, Kalburgi PB, and Kelmani C (2011) Amplification and sequencing of *mec*A gene from Methicillin Resistant *Staphylococcus aureus. International Journal of Advanced Biotechnology and Research.* 2(3): 310-314.

[14] Kumurya AS, Gwarzo MY, and Uba A (2015) One Step PCR for Detection of *Staphylococcus aureus* specific Sequence gene and *mec*A gene. *International Journal of Advanced Materials Research.* 1(3):73-79.

[15] Pournajat A, Ardebili A, Goudarzi L, Khodabandeh M, Narimani, and T Abbaszadeh H (2014) PCR based identification of Methicillin-resistant *Staphylococcus aureus* strains and their antibiotic resistance profiles. *Asian Pacific Journal of Tropical Biomedicine.* 4: S293-S297.

[16] Islam MA, Alam MM, Uddin MS, Kobayashi N, and Ahmed MU (2011) Detection of Methicillin Resistant *Staphylococcus aureus* (MRSA) from animal and human origin in Bangladesh by Polymerase Chain Reaction. *Bangl. J. Vet. Med.* (2011). 9(2):161-166.

[17] Anand KB, Agrawal P, Kumar S, and Kapila K (2009) Comparison of cefoxitin disc diffusion test, oxacillin screen agar, and PCR for *mec*A gene detection of MRSA. *Indian Journal of Medical Microbiology.* 27(1) : 27-29.

[18] Lestari ES, dan Severin JA (2009) *Antimicrobial Resistance in Indonesia (Prevalence, determinants, and genetic basis).* Rotterdam: Erasmus University.

[19] Al - Ruaily MA, and Khalil OM (2011) Detection of (*mec*A) gene in Methicillin Resistant Staphylococcus aureus (MRSA) at Prince A/ Rhmansidery hospital, Al-Jouf, Saudi Arabia. *Journal of Medical Genetics and Genomics.* 3(3) : 41-45.

[20] Elhassan MM, Ozbak, HA, Hemeg HA, Elmekki MA and Ahmed LM (2015) Absence of the *mec*A Gene in Methicillin Resistant *Staphylococcus aureus* Isolated from Different Clinical Specimens in Shendi City, Sudan. *BioMed Research International.* 1-5

[21] Sudigdoadi S (2010) Analisis Tipe Staphylococcal Cassette Chromosome *mec* (SCC*mec*) Isolat Methicillin Resistant *Staphylococcus aureus* (MRSA). *MKB.* 42 (4): 149 - 154.

[22] Hiramatsu K (2001) Vancomycin-resistant *Staphylococcus aureus*: a new model of antibiotic resistance. *Lancet Infect Dis*. 1(3):147-55.

[23] Arai KK, Kondo N, Hori S, Suzuki ET and Hiramtsu K (1996) Suppression of methicillin resistance in a *mec*A containing pre methicillin resistant *Staphylococcus aureus* strain is caused by *mec*I mediated repression of pbp 2' production. *Antimicrob Agents Chemother*. 40:2680-2685.

[24] Parvez MA, Shibata H, Nakano T, Niimi S, Fujii N, Arakaki N, and Higuti T (2008) No relationship exists between PBP 2a amounts expressed in different MRSA strains obtained clinically and their beta-lactam MIC values. *J Med Invest*. 55(3-4):246-53.

[25] Memmi G, Filipe SR, Pinho MG, Fu Z and Cheung A (2008) *Staphylococcus aureus* PBP4 is essential for beta-lactam resistance in community-acquired methicillin-resistant strains. *Antimicrob Agents Chemother*. 52(11): 3955-66.

Growth and Root Development of Four Mangrove Seedlings Under Varying Salinity

M Basyuni*, D A Keliat, M U Lubis, N B Manalu, A Syuhada, R Wati and Yunasfi

Department of Forestry, Faculty of Forestry, University of Sumatera Utara, Jl. Tri Dharma Ujung No. 1 Medan, North Sumatera 20155, Indonesia

*Email: m.basyuni@usu.ac.id

Abstract. This present study describes four mangrove seedlings namely *Bruguiera cylindrica*, *B. sexangula*, *Ceriops tagal*, and *Rhizophora apiculata* in response to salinity with particular emphasis to root development. The seedlings of four mangroves were grown for 5 months in 0%, 0.5%, 1.5%, 2.0% and 3.0% salt concentration. Salinity significantly decreased the growth (diameter and plant height) of all mangrove seedlings. Root developments were observed from the tap and lateral root. The number, length and diameter of both roots-typed of *B. cylindrica, B. sexangula* and *C. tagal* seedlings significantly decreased with increasing salt concentration with optimum development at 0.5% salinity. By contrast, the number, length, and diameter of tap root of *R. apiculata* seedlings were significantly enhanced by salt with maximal stimulation at 0.5%, and this increase was attenuated by increasing salinity. On the other hand, lateral root development of *R. apiculata* significantly thrived up to 1.5% salinity then decreasing with the increasing salinity. The different response of root development suggested valuable information for mangrove rehabilitation in North Sumatra and their adaption to withstand salt stress.

Keywords: Main root, rehabilitation, Rhizophoraceae, salt stress, water stress.

1. Introduction

Mangroves are halophytes that thrive in the inter-tidal zone of tropical and subtropical climates. One striking feature of mangrove plants is their ability to grow in various levels of salinity, ranging from freshwater to hypersaline environments [1]. Thus, the salinity environment of mangrove habitat may vary day to day or month to month. The mangrove root hence faces transient increases and decreases in soil salinity under natural field conditions. It is important therefore to get more insight into the physiological aspects of mangrove plants to cope with salinity.

Indonesia is one of the world's great mangrove nations, which is 22.6 percent of the global total. However, the Indonesian mangrove area has been lost from 4.2 million in 1980 to 3.1 million in 2011 [2]. Mangrove forests in North Sumatera distributed in the east coast of Sumatera Island mainly in the Karang Gading and Langkat Timur Laut Wildlife Reserve (KGLTLWR) and are rapidly threatened due to anthropogenic activities such as replacement of mangroves to aquaculture and oil palm plantation, filling and urban development [3]. Mangroves are one of the most threatened ecosystems all over the world today due to the deforestation and degradation [4]. Conservation efforts of mangrove implicate not only to protect the coastal areas and human communities from seawater intrusion and potential changes in the sea-level rise but also to make sure the availability of mangrove resources for future use through adaptation and mitigation to shifting environments.

IOP Conf. Series: Earth and Environmental Science **130** (2018) 012027 doi:10.1088/1755-1315/130/1/012027

The response of mangrove to salinity has mainly been studied in leaves [5-7], and limited information is available on salt stress-dependent changes of root development from mangroves in North Sumatran. It is important to note that the success of rehabilitation and conservation effort in the region. Our previous studies shed light on the optimal growth that 0.5% salinity was the optimal growth for *Kandelia candel* and *Bruguiera gymnorrhiza* seedlings grown in a glass house [7]. It is important to get more insight into the morphological aspects of mangrove seedlings to cope with salinity, which may experience transient periods of exposure to high salt concentrations. The present study, therefore, aims to investigate the response four mangrove seedlings namely *Bruguiera cylindrica, B. sexangula, Ceriops tagal*, and *Rhizophora apiculata* in response to salinity with a particular reference to root development.

2. Materials and Methods

2.1. Experimental design

Mature and healthy propagules of four mangrove species namely *Bruguiera cylindrica* Blume, *B. sexangula* (Lour.) Poir., *Ceriops tagal* C. B. Rob., and *Rhizophora apiculata* BL. (Rhizophoraceae), were collected from Pulau Sembilan, Langkat, North Sumatra, Indonesia. These propagules were planted in bottle pots with sand under varying salinity with exposure to natural temperature and sunlight in a greenhouse of Faculty of Agriculture, University of Sumatera Utara, Medan, Indonesia. Mature propagules of *B. cylindrica* were green to purplish green in color, 0.5-1.0 cm diameter and 10-15 cm long; mature propagules of *B. sexangula* were green to purple-tinted brown, 1.5-2 cm diameter, 6-12 cm long. On the other hand, mature propagules of *C. tagal* were green to brown hypocotyl, yellow cotyledonary collar, 0.8-1.2 cm diameter, up to 25 cm or more long. While mature propagules of *R. apiculata* were green to brown hypocotyls and red cotyledonary collar, 20-25 cm long and 1.3-1.7 cm diameter.

An artificial seawater solution was prepared by dissolving a commercial salt powder (Tropical Marine Sea Salt, Thailand) to adjust the salinity concentrations to 0.0%, 0.5%, 1.5%, 2.0% and 3.0% (equal to sea water level) according to the manufacturer's protocol. Each pot was irrigated with 1000 mL artificial seawater solution. The salinity in this study was defined as the mass of salt powder/weight of solution [8]. In each pot, the treatment of salt concentration was checked every week during the experiments using an S/Mill-E salinity refractometer (Atago Co. Ltd., Tokyo, Japan) and adjusted accordingly. Ten plants per treatment were grown for five months. After five months cultivation, the four mangrove seedlings were harvested, washed and the leaves and roots were immediately stored at -4 °C for further analysis.

2.2. Growth measurement

The growth of *B. cylindrica, B. sexangula, C. tagal* and *R. apiculata* seedlings under varying salt concentration was determined by the stem height and diameter of the plants. Thus, the stem heights and diameters of four mangrove species (10 seedlings each treatment) after five months of cultivation were the indices of growth in this experiment.

2.3. Root development

Root development of *B. cylindrica, B. sexangula, C. tagal* and *R. apiculata* seedlings under varying salt concentration were determined by a number of total roots, roots length and roots diameter as previously described [9]. A number roots were measured manually using counter after five months seedlings harvested. Measurement of root length was performed manually after harvesting. The diameter of root was measured after five months collected using the digital caliper. Root anatomy parameters were separated into tap root, and lateral roots as previously reported [9].

2.4. Statistical analysis

Data were analyzed by one-way analysis of variance (ANOVA) followed by Dunnett's test for comparisons of all treatments against the control. The value of $P < 0.05$ was selected as a limit of statistical significance. All statistical analyses were performed using the SAS 9.1 statistical software program (SAS Institute Inc. Cary, NC, USA).

3. Results and Discussion

3.1. Effect of salinity on the seedlings growth

Mangrove seedling growth was measured by the height and diameter of the plants as described in Materials and method. The growth of *B. cylindrica* decreased in the presence of salt (Fig. 1 A and B). Similarly, salt stress also significantly reduced the height and diameter of *B. sexangula* seedlings as well as *R. apiculata.* In contrast to this observation, the height of *C. tagal* seedlings thrived up to 0.5% salt concentration of both height and diameter of seedlings, and this increase then was attenuated by salinity when enhanced more than 1.5 % salt concentration (Figure 1).

Mangrove species often show growth stimulation at low salinity (25% seawater/0.5% salt concentration) and intermediate salinity (50% seawater/1.5% salinity) and then a decline in growth with further elevating in salt concentration [10-12]. However, the range of salinity in which the mangrove plant can survive varies according to the species [11]. This study suggested that mangrove seedlings survived to adapt either in saline condition or fresh water.

The present study also well agreed with the previous reports on the optimal growth at 0.5% salinity in *Kandelia candel* and *Bruguiera gymnorrhiza*, respectively. The growth of both species slightly increased after removal to salinity [7]. Furthermore, the optimum growth at 1.5% salt concentration found in *Avicennia marina* and *Rhizophora stylosa* seedlings grown in a greenhouse [8].

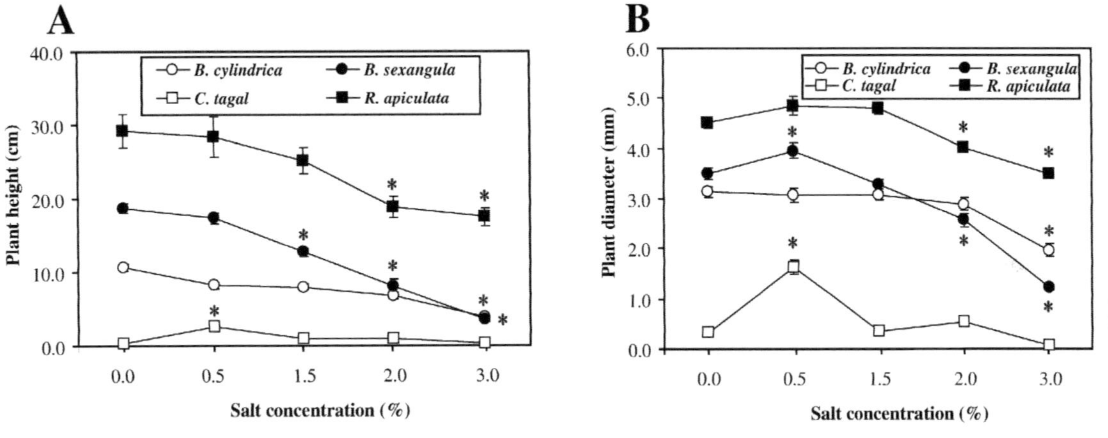

Figure 1. Effect of salinity on plant height (A) and diameter (B) of *B. cylindrica* (○), *B. sexangula* (●), *C. tagal* (□), and *R. apiculata* (■) seedlings. Data are expressed as the means ± SE (n=10) for growth measurements. The asterisk indicates a statistically significant difference from 0% at $P < 0.05$ using Dunnett's test.

The tolerance of mangrove species to salinity, therefore, followed the order of *C. tagal* < *B. cylindrica* <*B. sexangula* < *R.apiculata,* which is by their habitat zonation in North Sumatran mangroves. Our current result, therefore, suggested important information mangrove adaptation and to the success of rehabilitation and conservation efforts in the region.

3.2. Effect of salinity on the root morphology

The root characteristic of the five-month-old seedlings of four mangrove species under varying salt concentration is summarized in Fig. 2. *B. cylindrica, B. sexangula,* and *R. apiculata* seedlings consist of the precise type of root, namely tap (main) and lateral root. On the other hand, only *C. tagal* seedlings were distinguishable of uncompleted root anatomy, even though they are originated the same family Rhizophoraceae (Figure 2).

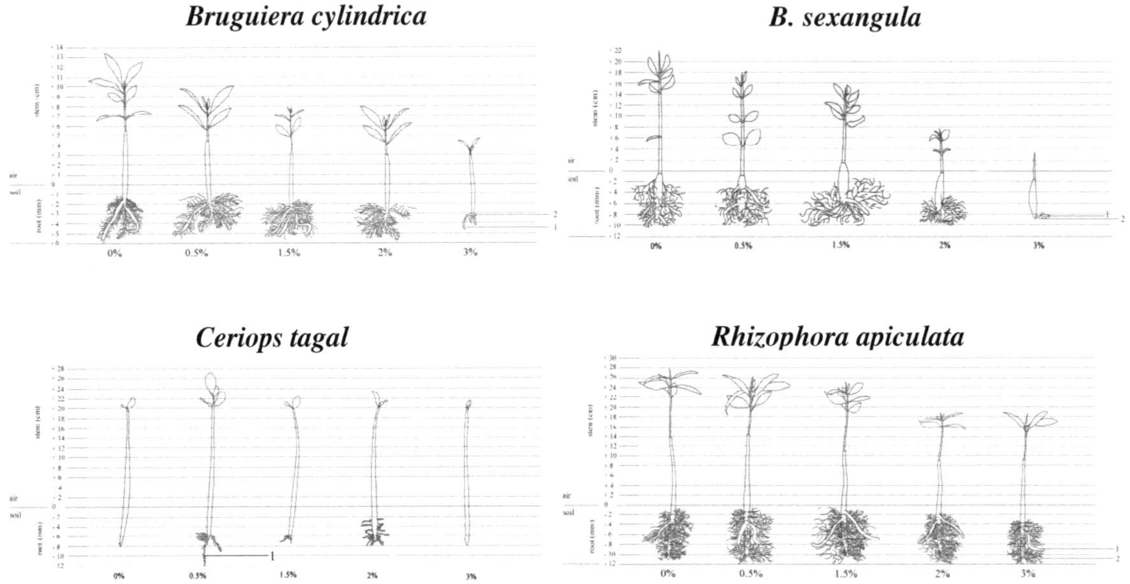

Figure 2. Effect of salinity on root morphology and depth (cm) of six-month seedlings of four mangroves grown with exposure to natural temperature and sunlight in a glass house. Root type: 1. tap root; 2. lateral root.

In case of *C. tagal* roots, the only seedling with 0.5% salinity has been showed distinct tap root. However, it is noteworthy that leaves are present in *C. tagal* under varying salinities. It has been reported that *C. tagal* from Pakistan showed an optimal growth at 50% seawater salinity [12].

3.3. Effect of salinity on the root development

Knowledge of root seedling in response to salinity is essential to an understanding of regeneration pattern of mangrove. Tables 1-4 summarizes the effect of salinity on the root development of *B. cylindrica, B. sexangula, C. tagal* and *R. apiculata*. For *B. cylindrica* seedlings, the significant decrease was found only 3% salinity in parameters of some taps and lateral root, both tap, and lateral root length as well as tap and lateral root diameter (Table 1).

Table 1. Effect of salinity on the root development of *B. cylindrica* seedlings

Salinity	Number of roots		Root length		Root diameter	
	Tap	Lateral	Tap	Lateral	Tap	Lateral
0.0%	10.0±0.6	92.6±4.3	6.8±0.2	2.7±0.3	2.4±0.3	0.2±0.0
0.5%	9.4±0.5	91.6±7.6	5.4±0.6	2.5±0.2	1.7±0.1	0.2±0.0
1.5%	8.7±0.4	87.4±6.3	5.2±0.6	2.5±0.3	1.7±0.1	0.2±0.0
2.0%	8.0±0.9	62.0±16.9	4.9±0.6	2.3±0.4	1.7±0.2	0.2±0.0
3.0%	2.4±0.3[a]	1.8±5.3[a]	2.4±0.2[a]	0.2±0.1[a]	1.5±0.1[a]	0.1±0.0[a]

Data are represented as the means ± SE (n=10)

[a]Significantly different from 0% at P <0.05 using Dunnett's test

Similarly, number, length, and diameter of both roots-typed of *B. sexangula* and *C. tagal* seedlings were significantly decreased with increasing salt concentration with optimum development at 0.5% salinity (Table 2-3). By contrast, the number, length, and diameter of tap root of *R. apiculata* seedlings were significantly enhanced by salt with maximal stimulation at 0.5%, and this increase was attenuated by increasing salinity. On the other hand, lateral root development of *R. apiculata* significantly thrived up to 1.5% salinity then decreasing with the increasing salinity (Table 4).

Table 2. Effect of salinity on the root development of *B. sexangula* seedlings

Salinity	Number of roots		Root length		Root diameter	
	Tap	Lateral	Tap	Lateral	Tap	Lateral
0.0%	3.5±0.4	61.5±5.7	12.7±0.8	4.2±0.2	3.8±0.3	0.5±0.1
0.5%	4.3±0.5	68.0±6.4	14.1±1.0	4.2±0.2	3.7±0.3	0.6±0.1
1.5%	3.9±0.4	50.8±4.6	12.4±6.3	3.5±0.2	2.8±0.3[a]	0.5±0.1
2.0%	3.1±0.6	19.8±2.2	4.5±0.4	2.8±0.4	2.2±0.2[a]	0.6±0.1
3.0%	1.4±0.4[a]	1.9±0.9[a]	0.9±0.2[a]	0.4±0.2[a]	1.2±0.2[a]	0.2±0.1[a]

Data are represented as the means ± SE (n=10)

[a]Significantly different from 0% at P <0.05 using Dunnett's test

Table 3. Effect of salinity on the root development of *C. tagal* seedlings

Salinity	Number of roots	Root length	Root diameter
0.0%	0.0±0.0	0.0±0.0	0.0±0.0
0.5%	3.2±0.5	1.9±0.2[a]	1.2±0.1[a]
1.5%	3.6±0.9	0.8±0.3	0.2±0.1
2.0%	5.2±0.6	1.2±0.1	1.0±0.1
3.0%	2.8±1.0	0.5±0.2	0.2±0.1

Data are represented as the means ± SE (n=10)

[a]Significantly different from 0% at P <0.05 using Dunnett's test

Table 4. Effect of salinity on the root development of R. *apiculata* seedlings

Salinity	Number of roots		Root length		Root diameter	
	Tap	Lateral	Tap	Lateral	Tap	Lateral
0.0%	1.1±0.1	39.9±1.6	3.7±0.4	2.1±0.1	3.3±0.3	1.4±0.1
0.5%	1.8±0.2	46.6±2.9[a]	5.3±0.6[a]	2.4±0.1	4.4.±0.4	1.3±0.1
1.5%	2.3±0.4	56.1±3.2[a]	3.3±0.3	2.7±0.1[a]	3.6±0.4	1.3±0.1
2.0%	2.3±0.3	37.8±2.0	5.2±0.4[a]	3.3±0.1[a]	3.0±0.1	1.4±0.1
3.0%	1.2±0.1	18.9±1.7[a]	4.2±0.3	2.7±0.2[a]	2.8±0.2[a]	1.5±0.1[a]

Data are represented as the means ± SE (n=10)

[a]Significantly different from 0% at P <0.05 using Dunnett's test

Root tissues are in contact with salt and are the primary site involved in the perception of salinity stress; they are assumed to have remarkable ability to control their Na^+ dan Cl^- content [7]. When different species compared, the root development declined under varying salinity in the order of *C. tagal* > *R. apiculata* > *B. sexangula* > *B. cylindrica.* The different response of root growth among mangrove species suggested the critical study of salt tolerance mechanism of individual genera or species [13].

4. Conclusion

The different response of seedling growth, as well as root development from four mangrove seedlings, suggested valuable information for mangrove rehabilitation in North Sumatra and their adaption to withstand salt stress. Our research indicated the critical study of salt tolerance mechanism of individual genera or species.

Acknowledgements

This work was supported in part by a BPPTN Research Grant (No. 6049/UN5.1.R/PPM/2016 to MB) from University of Sumatera Utara.

References

[1] Tomlinson PB (1986) The Botany of Mangroves. 1st ed. Cambridge: Cambridge University Press.

[2] Giri C, Ochieng E, Tieszen LL, Zhu Z, Singh A, Loveland T, Masek J, Duke N (2011) Status and distribution of mangrove forests of the world using earth observation satellite data. *Global Ecol. Biogeogr.* **20** 154–159

[3] Basyuni M, Putri LAP, Murni MB (2015) Implication of land use and land-cover change into carbon dioxide emissions in Karang Gading and Langkat Timur Wild Reserve, North Sumatra *J. Manaj. Hut. Trop.* **21** 25–35.

[4] Duke NC, Meynecke JO, Dittmann S, Ellison AM, Anger K, Berger U, Cannicci S, Diele K, Ewel KC, Field CD, Koedam N, Lee SY, Marchand C, Nordhaus I, Dahdouh-Guebas F (2007) A world without mangroves? *Science* **317** 41–42.

[5] Takemura T, Hanagataa N, Sugihara K, Baba S, Karube I, Dubinsky Z (2000) Physiological and biochemical responses to salt stress in the mangrove, *Bruguiera gymnorrhiza*. *Aquat. Bot.* **68** 15–28.

[6] Parida A, Dam AB, Dam P (2002) NaCI stress causes changes in photosynthetic pigments, proteins, and other metabolic components in the leaves of a true mangrove, *Bruguiera parviflora*, in hydroponic cultures *J. Plant Biol.* **45** 28–36

[7] Basyuni M, Baba S, Kinjo Y, Putri LAP, Hakim L, Oku H (2012) Salt-dependent increase in triterpenoids is reversible upon transfer to fresh water in mangrove plants *Kandelia candel* and *Bruguiera gymnorrhiza* *J. Plant Physiol.* **169** 1903–1908.

[8] Basyuni M, Baba S, Kinjo Y, Oku H (2012) Salinity increases the terpenoid content of a salt-secretor and a non-salt secretor mangrove *Aquat. Bot.* **97** 17–23.

[9] Pi N, Tam NFY, Wu Y, Wong MH (2009) Root anatomy and spatial pattern of radial oxygen loss of eight true mangrove species *Aquat. Bot.* **90** 222–230.

[10] Burchett MD, Clarke CJ, Field CD, Pulkownik A (1989) Growth and respiration in two mangrove species at a range of salinities. *Physiol. Plant* 75: 299–303,.

[11] Ball, MC (1988) Salinity tolerance in the mangroves, *Aegiceras corniculatum* and *Avicennia marina*.Water use in relation to growth, carbon partitioning and salt balance. *Aust. J. Plant Physiol.* 15:447–464.

[12] Khan MA, Aziz I, (2001) Salinity tolerance in some mangrove species from Pakistan. *Wetlands Ecol. Manag.* 9: 219–223.

[13] Basyuni M, Kinjo Y, Baba S, Shinzato N, Iwasaki H, Siregar EBM, Oku H (2011) Isolation of salt stress tolerance genes from roots of mangrove plant, Rhizophora stylosa Griff., using PCR-based suppression subtractive hybridization *Plant Mol. Biol. Rep.* **29** 533–543.

The influence of altitude of the place on the vegetative growth of several varieties of sugarcane (*saccharum officinarum* l.)

Lince R. Panataria [1*], Ramli Lubis [1]
[1]Department of Agronomy, Faculty of Agriculture, University of Methodist Indonesia, Jl. Harmonika Baru, Pasar 2, Tj.Sari, Medan 20155

*Email: meddy.siregar@yahoo.com

Abstract. This study is aimed at finding out vegetative growth of some varieties of sugarcane on different altitudes. The study applied Factorial Randomized Block Design comprising two factors. The first factor was varieties of consisting of three levels: BM 13579, Ps. 92-1, and Ps. 89-1. The second factor was altitudes of the land with three different levels of altitudes: Kebun Saentis (15 MASL), district of Pancurbatu (±300 MASL), and district of Sibolangit (±700 MASL). The study resulted in that variety of sugarcane, altitudes, and the interaction of both variables affected these: the leaf area on observations of 21, 42, 63, and 84 daps, number of tillers, root volume, nutrient uptake on observation 84 daps. The best quality of the sugarcane varieties was variety Ps. 92-1 on altitude 15 MASL (Kebun Saentis), followed by an altitude ±300 MASL (district of Pancurbatu), and then on ±700 MASL (district of Sibolangit); while the worst quality of the variety was variety BM 13579 of altitude 15 MASL (Kebun Saentis).

Keywords: height, location, plant

1. Introduction

Sugarcane (*Saccharum officinarum* L.) is the main source for production of commercial sugar. Sugar is one of the strategic commodities; which is needed by people in Indonesia as basic daily consumption [1]. The need for sugar today is increasing in line with the growing number of people in Indonesia. Meanwhile, some recent years it was reported that the production of sugar has been going down on its productivity [2]. The decrease of this productivity has been caused by not achieving the raw factors of cultivation techniques of the product. One of the determinant factors in sugarcane productivity is the application of superior varieties which is implemented in the arrangement program of varieties based on the suitable typology of land, maturity characteristics, planting time, and harvesting time. The selecting process of the sugarcane varieties should be on attention, because of the limitation of productive time of the superior varieties which lasts only five years, because of the availability of location-specific varieties of sugarcane, and of the difficulty to obtain the introduction of foreign sugarcane [3].

The growing of sugarcane plant is affected by weather condition in a certain place. The height of it very much affects the growing of all plants. When we talk about the height of its/ altitude, it means that it involves the air temperature, sun rays, and air humidity. These aspects affect the growth of all plants. What it means by the height of the land is the height of the land above the level (elevation). The height of the land affects the change of the weather temperature. The higher the place, like

IOP Conf. Series: Earth and Environmental Science **130** (2018) 012028 doi:10.1088/1755-1315/130/1/012028

Mountains, the lower the temperature/ or, the colder the place. In reverse, the lower is the place; the higher is the temperature/ or the hotter is the place.

The regional distinction of topography, geography, and weather cause the existence of differences in plants, planting phase, and methods of cultivating. Constant planting of some plants and the suitable weather will increase the growth of the plant. The height of the place above the sea level determines the temperature and the light intensity which is received by the plant. The higher is the place; the lower is the temperature. Thus the sun ray intensity decreases. Temperature and illumination will be used to categorize which plants are suitable to grow on lowland and highland.

Every variety of plants responds differently toward different temperature and the height of the place and thus resulted in vary phases of growth. Based on this fact, this study was conducted to derive the effect of altitude on vegetative growth of some varieties of sugarcane (*Saccharum officinarum* L.).

2. Methods

This research was conducted in three locations namely in Kebun Saentis on 15 MASL, region of Pancurbatu on ±300 MASL, and district of Sibolangit on ±700 MASL.

The study applied the factorial Randomized Block Design, which consists of two factors. The first factor is the varieties of sugarcane: V_1= BM 13579, V_2= Ps. 92-1, V_3= Ps. 89-1. The second factor is the altitude: H_1= Kebun Saentis (15 MASL), H_2= district of Pancurbatu (±300 MASL), and H_3= district of Sibolangit (±700 MASL). In total, there were 9 treatments combined, and each treatment was repeated every third time, and thus accumulated 27 combined treatments. If the effect of the treatment is differently real on the variety print (ANOVA), then the further test is continued by using the next distance Duncan test on the level of 5%. Variables observed were the leaf area at observation on 21, 42, 63, and 84 daps, the number of tillers, root volume, and nutrient uptake at observation on 84 daps.

3. Results and Discussion

3.1. Leaf Area

The leaf area affects the photosynthetic process in producing assimilates. This assimilation affects the vegetative growth of the plant. The transmission of the result of assimilation process to the leaf will grow the leaf bigger and broader. The bigger leaf will allow the plant to have the capability to do photosynthesis. The differences between the three varieties of sugarcane are affected by the genetic factor, different planting of the plants, and also the environment condition [4]. The difference in genetic characteristics also causes the difference in reactive response toward the variety of surrounding conditions, so that growing activities appear different. Three different varieties of sugarcane if planted in the same condition will even grow differently [5].

The effect of the height of the place on the leaf area of some varieties of the sugarcane can be seen in Figure 1.

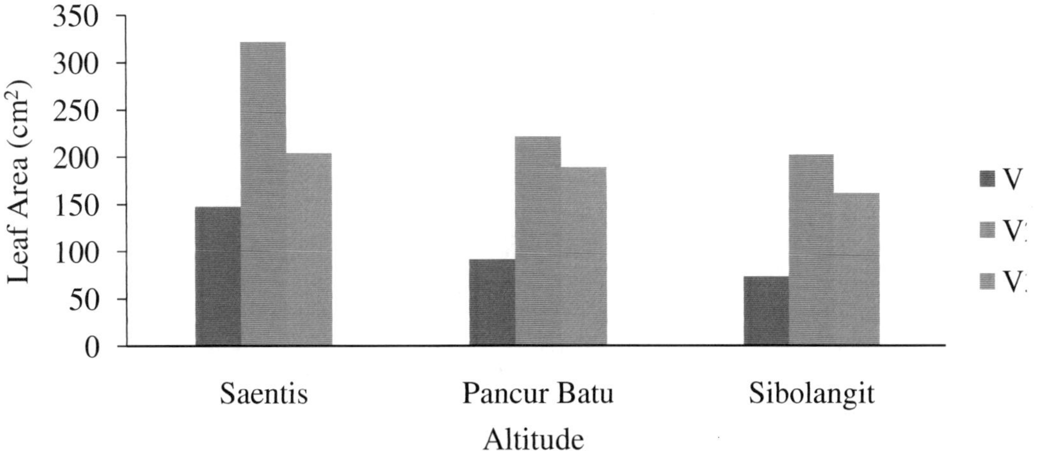

Figure 1. The Leaf Area of Some Varieties of Sugarcane on Different Altitudes of Land, At Age 84 daps

Figure 1 shows that the Ps. 92-1 (V_2) has the largest leaves on those three different heights, followed by a variety Ps. 89-1 (V_3); meanwhile, the smallest leaf of sugarcane is variety BM 13579 (V_1). Those three varieties grow better in lower places.

3.2. The Number of Tillers

Sugarcane plants have the capability of growth to produce tillers in a hill. Budding of tillers is regarded an important cycle in the growth of the plant because at this phase the plant is said to produce good plants or not. The process of budding of tillers is often called *tillering* and lasts when the sugarcane is between 3 or 4 months. In Figure 2 it can be seen that the number of tillers mostly found in a variety Ps. 92-1 (V_2) followed by Ps. 89-1 (V_3) and then BM 13579 (V_1). Variety Ps. 92-1 (V_2) adapts better with every different height of land. It is shown from the number of tillers growing. Tiller of the sugarcane is one of the variables which is important to increase the productivity of the sugarcane. The productivity of the sugarcane per unit of land is determined by the abundant number of tillers formed by the sugarcane [6]. The more is the tillers formed; the more is the harvest of the sugarcane obtained. Tillers of the sugarcane then are the main factor to get the high productivity of the plant [7].

The effect of the height of the place on the number of tillers of the sugarcane in different varieties of the plant is seen in Figure 2.

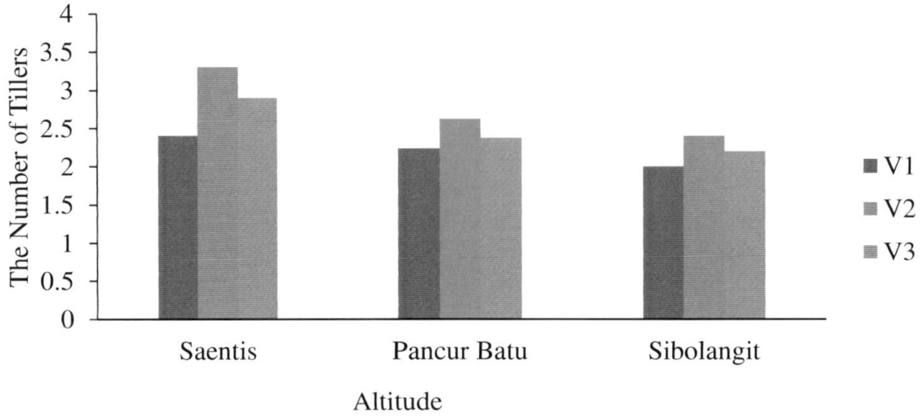

Figure 2. The Number of Tillers in Some Varieties of Sugarcane on Some Different Altitudes of Land, at Age 84

Figure 2 shows that variety of Ps. 92-1 (V_2) has the most number of tillers, which are planted on the three different altitudes, followed by the variety of Ps. 89-1 (V_3), while the fewest number of tillers exist in the variety of BM 13579 (V_1). The three varieties grow better on the lower altitude (Kebun Saentis).

3.3. Root Volume

Roots are the main absorptive element to uptake nutrient, mineral salts, and water from the soil. The growth of roots will positively correlate with the growth of other organs of the plant. The growth of roots can be indicated in their volume. The better is the growth of the roots, the better their capability to reach food nutrient from the soil to suffice the plants' Nutrient needs and the more optimal they will grow. In this study variety 92-1 (V_2) were more adaptive in the lowland, that means that conditions fluctuate with the rainfall, temperature and nutrient availability. Here variety 92-1 (V_2) stays capable of growing better. If the root volume is bigger, so the power to reach nutrient is more regarding absorbing nutrient and water from the soil so that transference through transportation pathway all over the plant tissues will increase [4].

The impact of the height of the land on the volume of sugarcane root in several varieties of the plant can check in Figure 3.

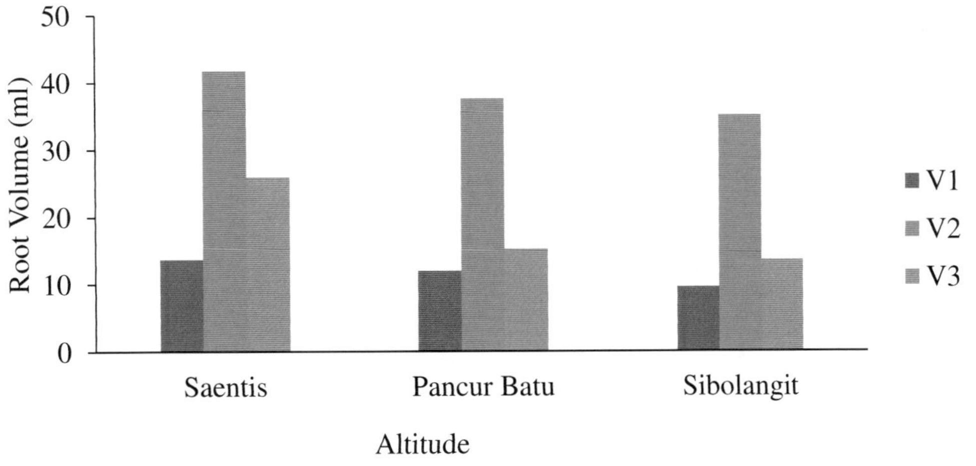

Figure 3. Root Volume of Some Varieties of Sugarcane on Some Different Altitudes at Age 84

Figure 3 shows that sugarcane variety Ps. 92-1 (V_2) has bigger root volume which is planted in the three altitudes, followed by sugarcane variety Ps. 89-1 (V_3), while the least volume of the sugarcane root was variety BM 13579 (V_1). Those three varieties grow better and have bigger root volume on lands which are low (Kebun Saentis).

3.4. Nutrient Uptake N (mg N/tan)
During vegetative phase of the growth, plants need a nutrient element to do metabolism process. The nutrient elements absorbed from the surrounding can make the plants grow leaves, tillers and better root volume, so that photosynthetic process can last more optimally [8]. At germination phase, nutrient uptake is slow. The uptake lasts faster when the roots and leaves grow which is the same time happens in blossoming process. In Figure 4 it can be seen that each highest nutrient uptake N exists in plants planted in the location of Kebun Saentis of variety Ps. 92-1 (V_2) and the slowest nutrient uptake exist in plants planted in a Sibolangit location that is variety Ps. 89-1 (V_3) at observation on 84 daps. Nitrogen is one of the main nutrient elements for plants. Nutrient N is absorbed by the roots of plants and can force vegetative growth like lengthening the branch and enlarging the leaves, making bigger diameter, producing more roots, tillering, and reducing dry leaves. In sugarcanes, nitrogen determines result and quality of the plant because it affects leaf area, leaf area indexes, and leaf area period, closing earlier canopy and photosynthesis speed which all of those can increase biomes production [9].

The impact of the land heights on nutrient Nitrogen uptake of plants like sugarcanes of varieties of sugarcane is seen in Figure 4.

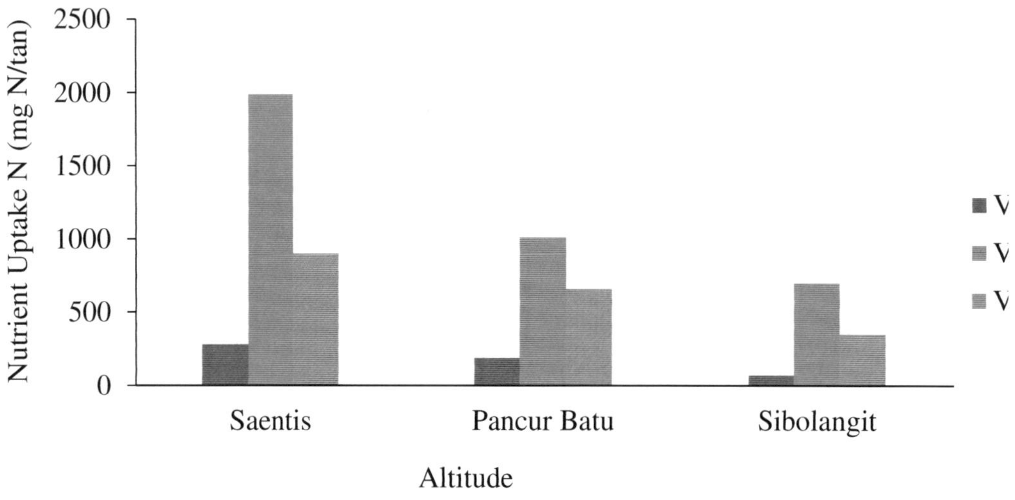

Figure 4. Nitrogen Nutrient Uptake of Some Varieties of Sugarcane on Some Different Altitude of Land, at Age 84

Figure 4 indicates that sugarcane variety Ps. 92-1 (V_2) has Nitrogen nutrient uptake which is higher on those three altitudes of land, followed by sugarcane variety Ps. 89-1 (V_3), while the lowest Nitrogen nutrient uptake exists in sugarcane variety BM 13579 (V_1). Those three varieties of sugarcane have better nutrient uptake when planted on lower lands (Kebun Saentis).

4. Conclusions

1. The heights of the land have a real effect on the vegetative growth of sugarcanes. Sugarcanes have vegetative growth in Kebun Saentis (15 MASL), followed by in district of Pancurbatu (±300 MASL), and in district Sibolangit (±700 MASL).
2. The varieties of sugarcane have a real effect on the vegetative growth of the plant. The sugarcane variety Ps. 92-1 (V_2) has better vegetative growth than variety BM 13579 (V_1) and variety Ps. 89-1 (V_3) do.
3. There happened interactions of altitudes of land with sugarcane varieties and with a vegetative growth of the plant. Sugarcane variety Ps. 92-1 (V_2) has better vegetative growth than sugarcane variety BM 13579 (V_1) and variety Ps. 89-1 (V_3) do, even when they are planted on different altitudes.

Acknowledgements

The writer extends his/her gratitude to Indonesia Methodist University their moral and material support.

IOP Conf. Series: Earth and Environmental Science **130** (2018) 012028 doi:10.1088/1755-1315/130/1/012028

References

[1] Priyono (2008) Analysis of Policies of National Sugar Industry by Econometric Model. Journal of Science and Technology BPPT Planning. No.6 (2): 1-12.

[2] Prihartono A, A Sudirman and A Aziz (2016) Vegetative Growth Responses of Some Varieties of Sugarcane (*Saccharum officinarum* L) to Mikoriza Arbuskular. AIP Journal Volume 4 (1): 12-20.

[3] Department of Plantation (2004) Technology to Improve Productivity of Sugarcane and Introduction of Superior Varieties in North Sumatra. A project of Food Staple Improvement in Plantations located in North Sumatra. Medan.

[4] Rusprasita Y, L Agustina, Syekhfani, and ME Premono (2001) Status and Efficiency of Nitrogen during Sugarcane Growth (*Saccharum officinarum* L.) Fertilized by imipramine. Journal of BIOSAIN. Volume 1 No. 1. Pasuruan.

[5] Mariah A, T Hidayat, N Husna (2012) The Effect of Some Varieties and Spacing on the Growth of Soybean (*Glycine Max* L.) Merril Agrista. Volume 16 (1): 22- 28.

[6] Rothman H, Taryono, Supriyanta (2014) Tillers Number and Cane Sucrose Content of Six Sugarcane (*Saccharum officinarum* L.) Clones from Stem Cutting, Bud Set, and Bud Chip Seedlings. Vegetalika, Vol.3, No.3: 89-96.

[7] Natarajan US (2011) Tillering in SSI-Emergence, Factors Affecting, Constraints, and Solutions. First National Seminar on Sugarcane Sustainable Initiative. p.21-23.

[8] Rizqiani FN, E Ambarwati, NW Yuwono (2007) The Effect of Dosage and Frequency of fertilizing Liquid Organic Fertilizer on the Growth and Product of Bean (*Phaseolus vulgaris* L.) in Lowlands. Soil and Environment Science. Volume 7 (1): 43- 53.

[9] Hunsinger G (1993) Production of Sugarcane, Theory, and Practice. Springer. Verlag. New York.

Dadih bamboo ampel (*bambusa vulgaris*) and bamboo gombong (*gigantochloa verticilata*) 2 and 3 days fermented : effect on salad dressing hedonic quality

Nurzainah Ginting *

*Animal Production Program Study, University of Sumatera Utara Jl. Dr. A. Sofian No.3, Padang Bulan, Medan Baru, Kota Medan, Sumatera Utara 20155

E-mail: nurzainah@usu.ac.id

Abstract. The study aims to find time of fermentation of *dadih* and hedonic quality of *dadih* salad dressing. Goat milk was fermented in two kinds of bamboo: bamboo Ampel (*Bambusa vulgaris*) and bamboo Gombong (*Gigantochloa verticilata*) with different days; i.e. 2 and 3 days which will then became *dadih* while the *dadih* then were used as a raw material for making salad dressing. In Indonesia today there is an increasing on vegetable salad demand due to understanding of the benefits of consuming vegetables. One form of vegetable preparation is vegetable salad that is generally used as non local dressings. This research was conducted from April to May 2017 using Factorial Completely Randomized Design with 2 factors; i.e factor 1 (2 and 3 days fermented *dadih*) and factor 2 (bamboo types : bamboo Ampel and bamboo Gombong) with 4 replications. The parameters were flavor, color, aroma and texture (hedonic evaluation) where there were 25 panelists in doing evaluation. The results showed that 2 days fermented in bamboo ampel significantly (P <0.05) were preferred.

Keywords: bamboo ampel, bamboo gombong, *dadih*, salad dressing

1. Introduction

Dadih is a traditional food from West Sumatera, especially in MinangKabau, Indonesia. *Dadih* is a processed product of traditionally fermented milk buffalo, the form of *dadih* is like tofu, soft textured, white color and *dadih* served as a side dish. *Dadih* processing is done traditionally with the basic ingredients of pure milk buffalo that is inserted into the bamboo tube. In West Sumatera, the use of buffalo milk into *dadih* is caused people have buffalo as traditional livetosck and instead of consume fresh milk, people also processed it into *dadih*. Buffalo milk contains good nutrition so that eventually people make *dadih* as part of side dishes.

Table 1. Nutritional composition of various types of milk

Composition	Buffalo Milk	Goat Milk
Fat	7,4	3,3
Protein	3,8	3,3
Water	83,1	83-87,5

Source Warner [1]

Nowadays in North Sumatera people began to process goat milk into *dadih* due to the increasing number of people who keep *Peranakan Etawa* goat. Goat milk processing is also occurred due to the nutrient content in goat's milk which include the presence of antibody so that people assume will be healthier when consuming product related with goat milk. *Etawa* goat milk contains no beta-lactoglobulin, an allergen compound that often triggers reactions such as asthma, ear infections, redness of the skin, and food digestive disorders. Etawa goat milk also does not contain agglutinin, which is a compound that makes fat molecules clot like in cow's milk, so that etawa goat milk is easily absorbed by the small intestine [2].

Dadih is made in bamboo tube and type of bamboo which is used is bamboo Ampel (*Bambusa vulgaris*). The selection of this type of bamboo by the community is due to the bitter taste of this bamboo that the processed milk will not be contaminated by ants. This bamboo is filled with milk, spontaneously initiates the development of microorganisms while according to Ammermen [3] microorganisms that play a role in the fermentation process is thought to originate from the surface of the inner bamboo tube, the surface of the cover, and from the buffalo milk that is in use. These microorganisms consist of bacteria and yeasts of bacteria around 10^6- 10^7 and yeasts about 10^5. In making *dadih*, the bamboo was covered with banana leaves that have been loosened with fire and fastened by rubberband [4]. After 2 or 3 days due to the process of fermentation, milk turns its texture into a soft coag like a young cheese. Making *dadih* by using freshly cut bamboo can produce a unique blend of bamboo aroma and has a thick white color [5]. Bamboo that was used has a diameter of ± 4-5 cm and cut into ± 15 cm. Milk filled into bamboo as much as ± 250 ml per bamboo [6].

At this time various types of cheese processed into a component in making salad dressing. Increasing public awareness of health leads to increased salad dressing demand in which the salad dressing is added to the selected vegetable mixtures. This type of food is always provided in a banquet at the hotel or at the buffet or at home. Various kinds of salad dressings such as blue cheese, thousand island, barbecue and others are imported products. As for the non-imported salad dressing, they are a kind of franchise and therefore independence in making the salad dressing becomes a challenge. This research was designed to study salad dressing by using *dadih* from goat milk with different days of *dadih* ripening, i.e. 2 dan 3 days ripening.

2. Materials and Methods

This research was conducted in Animal Production Laboratory and in Food Technology Technology Laboratory Department of Food Science and Technology Faculty of Agriculture University of North Sumatera. This research was conducted from April to June 2017. The materials used in this research were fresh goat milk of two types of bamboo namely, bamboo ampel (*Bambusa vulgaris*) and bamboo gombong (*Gigantochloa verticilata*) as a tool for making *dadih*; Banana leaf as end cover on bamboo during fermentation. Panelists as many as 25 people as a tester that were a lecturer of Animal Production Program that were used to consume salad. Reseach design was Factorial Completely Randomized Design with 2 factors; i.e factor 1 : 2 and 3 days fermented *dadih* and factor 2 : bamboo types : bamboo Ampel (*Gigantochloa verticilata*) and bamboo Gombong (*Gigantochloaverticilata*) with 4 replications. The research parameters included the quality of *dadih* 2 and 3 days fermented and organoleptic test [7] on salad dressing. The parameters observed were flavor, color, aroma and texture. The determination of hedonic quality was tested based on the priciple evaluation of the panelists involved which in this research there were 25 panelists. The range of score used in the sensory evaluation of hedonic quality was 1 to 5. Skala 1 = do not like, 2 = somewhat dislike, 3 = neutral, 4 = like, 5 = very like. Panelist were provided by explanation and guidance for hedonic quality tests for flavor, color, aroma and texture.

The research was first conducted by making *dadih* of fermented goat milk on Ampel bamboo (*Bambusa vulg*aris) and bamboo Gombong (*Gigantochloa verticilata*). Each bamboo was taken the day before there fore, bamboo was in fresh condition and filled with 300 ml pasteurized goat milk. Then the bamboo waslid with banana leaves that had been softened by fire first and tied using a rubber band. Further fermented 2 and 3 days. After 2 days, *dadih* was opened, weighed. The volume of whey also measured and pH data were taken. Similarly, the fermented *dadih* 3 days, carried out the same treatment.

Figure 1.Scheme in making *Dadih* Goat milk

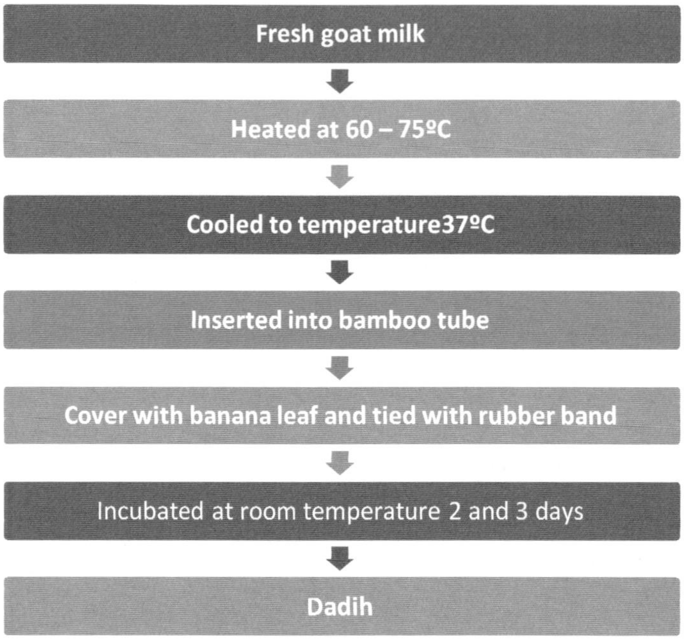

Figure 2. Scheme in Making *Dadih* Salad Dressing

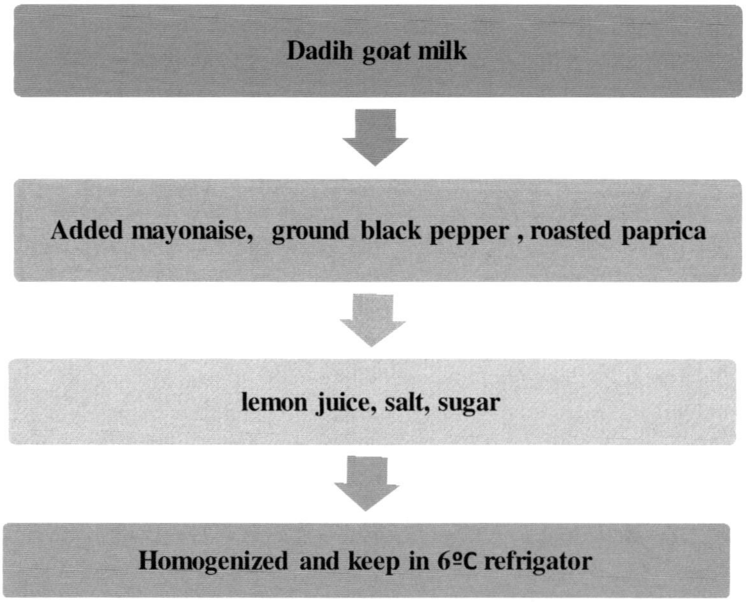

IOP Conf. Series: Earth and Environmental Science **130** (2018) 012029 doi:10.1088/1755-1315/130/1/012029

3. Results and Discussion

Dadih that was fermented within 2 days and 3 days produced *dadih* with the approximate weight of 50 g and whey volume about 230 ml. The initial volume of milk used to make yogurt was 300 ml. The occurrence of the difference in volume between the initial milk and *dadih* was due to the water absorbed by the bamboo stems through the pores [8] as well as the metabolism of microorganisms also required water [9]. Taufik [6] stated that lactose and casein were a component of milk that plays a role in milk coagulation. Lactose as a source of carbon and energy was metabolized by bacteria into lactic acid so that the increased acidity of the *dadih* is shown by a decrease in pH. The pH of the *dadih* that was fermented at 2 days on average around 4.592 to 4.649 while Usmiati and Setiyanto [8] found 4.68 to 4.77 with fermentation for 40 hours. Results by Pato [9] showed that there were 36 strains of lactic acid bacteria (BAL) dominant in *dadih* dominated by *Lactobacillus plantarum* Sunarlim et al. [10] reported that benefits of BALs for a lot of health include interfering with pathogenic bacteria, lowering cholesterol, anti-mutagenic, anti-carcinogenic, improving immune system.

Table 2. *Dadih* goat milk fermented within 2 and 3 days

	Dadih					
	2 Days fermentation			3 Days fermentation		
	Weigth of *Dadih* (g)	Volume of Whey (ml)	pH	Weigth of *Dadih* (g)	Volume of Whey (ml)	pH
Bamboo Ampel	50	230	4.592	50	230	4.926
Bamboo Gombong	50	230	4.649	50	230	4.952

The finished *dadih* was then processed into salad dressing with reference to a conventional formula of dressings with ingredients such as raw cheese. Some ingredients added include mayonnaise, lemon, sugar, salt, roasted paprica powder as well as black pepper powder. Then organoleptic test was done to 25 panelists with 3 kinds of salad dressing :commercial dressing containing cheese namely blue cheese salad dressing, dressing bamboo ampel, dressing bamboo gombong.

Table 3.Dressing bluechees, dressing bamboo ampel and bamboo gombong.

Variabel	2 day	Mean±SD	3 day	Mean±SD
Flavour	Dressing Bluecheese	3.26±0.82	Dressing Bluecheese	3.31±0.94
	Dressing Bamboo ampel	4.65±0.42[a]	Dressing Bamboo ampel	2.25±0.52
	Dressing Bamboo Gombong	4.45±0.47[a]	Dressing Bamboo Gombong	2.23±0.50
Color	Dressing Bluecheese	3.73±0.68	Dressing Bluecheese	3.83±0.67
	Dressing Bamboo ampel	3.59±0.96	Dressing Bamboo ampel	3.56±0.54
	Dressing Bamboo Gombong	3.59±0.96	Dressing Bamboo Gombong	3.56±054
Aroma	Dressing Bluecheese	3.20±0.83	Dressing Bluecheese	3.22±0.82
	Dressing Bamboo ampel	3.35±0.95	Dressing Bamboo ampel	3.23±0.81
	Dressing Bamboo Gombong	3.21±0.80	Dressing Bamboo Gombong	3.28±085
Texture	Dressing Bluecheese	4.32±0.50[a]	Dressing Bluecheese	4.32±0.50[a]
	Dressing Bamboo ampel	4.20±0.45[a]	Dressing Bamboo ampel	4.23±0.53[a]
	Dressing Bamboo Gombong	4.21±0.46[a]	Dressing Bamboo Gombong	4.22±0.52[a]

Note: Means in the same column and row with different superscripts differ significantly (P<0.05)
Flavour: 1 = very sour; 2 = sour; 3 = sourish; 4 = tasteless; 5 = very tasteless.
Color: 1 = reddish; 2 = slight reddish; 3 = whitish; 4 = white; 5 = very white.
Aroma: 1 = very rancid; 2 = putrid; 3 = less putrid; 4 = special milk; 5 = not putrid
Texture: 1 = very hard not padded; 2 = hard; 3 = paddish; 4 = soft, padded; 5 = very padded

IOP Conf. Series: Earth and Environmental Science **130** (2018) 012029 doi:10.1088/1755-1315/130/1/012029

The results of organoleptic test in Table 3 showed the score from panelist on the flavour of salad dressing. The *dadih* fermented within 2 days was 3.26-4.65 which was perceived as neutral until like where bluecheese dressing was perceived as neutral and *dadih* bamboo ampel dressing and *dadih* bamboo gombong was perceived as like. However, dressing with *dadih* bamboo ampel was preferable as its score was higher than dressing with bamboo gombong. Salad dressing where *dadih* was fermented within 3 days, the score was 2.23-3.31 for salad dressing which was perceived as less on both *dadih* dressing due to sour and bitter flavour. According to Usmiyati and Setiyanto [8] sour flavour occurred due to metabolism intensity of microorganism as the enzim kept metabolized lactoce became lactic acid while bitter flavour cause by contamination from the bamboo.
Traditional *dadih* production involves several kinds of microorganisms such as lactic acid bacteria (BAL), mold and yeast. The presence of yeasts in traditional fermented products can contribute positively during the fermentation process and to the final product such as providing growth factors for lactic acid bacteria and enhancing flavor. In addition secondary metabolites produced by yeasts such as acetate, succinate, propionate, fumarate and pyruvate have a good influence on the flavor of *dadih* [12]. In conclusion, *dadih* which fermented within 2 days caused better salad dressing than fermented within 3 days.
The score from panelist on the color of salad dressing was 3.59-3.73 on 2 days fermentation and 3.56-3.83 for 3 days fermentation which was perceived as neutral. However, the score on bluecheese dressing was slightly better as panelists mentioned that they like it more because its whitish meanwhile *dadih* dressing slightly red due to roasted paprica addition as dressing ingredient.
The score from panelist on the aroma of salad dressing was 3.20-3.35 on 2 days fermentation and 3.22-3.28 for 3 days fermentation which was perceived as neutral. According to panelists, aroma was important to be observed however, flavour was the most important.
The score from panelist on the texture of salad dressing was 4.21-4.32 on 2 days fermentation and 4.22-4.32 for 3 days fermentation which was perceived as neutral. Texture is a palpation or touch-associated sense in this research texture was associated with sense of human tongue papillae. The organoleptic test on salad dressing texture was soft. The soft texture of salad dressing resulted from a high water content. According to Yerlikaya and Karagozlu [12] texture was one of consumer ratings for determining the food product quality and as a direct observable parameter. In terms of salad dressing, panelists prefered more on soft product.

4. Conclusion
In conclusion, panelists prefer salad dressing where ingredients were *dadih* bamboo ampel followed by salad dressing where ingredients were bamboo gombong than blue cheese salad dressing. One of panelist mentioned that salad dressing from *dadih* reminded him of salad dressing in Al Baik restaurant in Arab possibly this related with aroma of *dadih* which made with goat milk. This research could initiated a local salad dressing which ingredient was *dadih* and was perceived by panelists as good as commercial salad dressing.

References
[1] Warner JN (1976) Principles of Dairy Processing.Wiley Eastern Limited. New Delhi
[2] Darmajati (2008) *Informasi Susu Kambing Etawa. Buletin Pikiran Rakyat.* Himpunan Studi Ternak Produktif. Jawa Tengah
[3] Ammermen, GR (1987) Effect of equal lethal heat threatments at varios times and temperature uponslected food counstituent . Prude University Lafayette.
[4] Risfaheri, Usmiati Sri (2012) *Pengembangan Dadih Sebagai Bahan Pakan Fungsional Probiotik Asli Sumatera Barat. J. Litbang Pert.* 32**(1)**:20-29
[5] Usmiati et al. (2011) *Sifat Fisiokimia Dadih susu Sapi: Pengaruh Suhu Penyimpanan dan Bahan Pengemas.* Jurnal Penelitian dan Pengembangan Pascapanen Pertanian.
[6] Taufik E (2004) *Dadih Susu Sapi Hasil Fermentasi Berbagai Starter Bakteri Probiotik yang Disimpan pada Suhu Rendah :Karakteristik Kimiawi.* Media Peternakan. 27**(3)** : (88-100).

[7] Usmiatidkk S, W Broto and H Setiyanto (2011) *Karakteristik Dadih Susu Sapi yang Menggunakan Starter Bakteri Probiotik.* Bogor: Balai Besar Penelitian dan Pengembangan Pascapanen Pertanian Jitv 16(**2**): 141-153.

[8] Buckle KA, RA Edwards, GH. Fleet, and M Wooton (1985) *Ilmu Pangan.* Terjemahan: Purnomo, H. danAdiono. UI Press, Jakarta.

[9] Pato U (2003) *Potensi Bakteri Asam Laktat yang Diisolasi dari Dadih untuk Menurunkan Resiko Penyakit Kanker. J. Natur Indonesia* 5(**2**):162-166

[10] Sunarlim R, Triyantini, Abubakar M Poeloengan, and H Setiyanto (1999) *Peningkatan Teknologi Pembuatan Inokulum Mikroba Pengolahan Dadih untuk Menunjang Agro Industri Pedesaan.* Laporan Penelitian Balai Penelitian Ternak Bogor TA. 1998/1999

[11] Coghe SK, Benoot F, Delvaux B, and Vanderhaegen (2004) Ferulic acid release and 4-viny lguaiacol formation during brewing and fermentation. J Agric Food Chem 52 (3): 602-608.

[12] Yerlikaya O and C Karagozlu (2011) Effects of ripening period on textural and sensory properties of capper cheeses. *J. Anim. Vet.* Adv. 10: 1171-1176. https:doi.org/10.3923/javaa.2011.1171.1176

IOP Conf. Series: Earth and Environmental Science **130** (2018) 012030 doi:10.1088/1755-1315/130/1/012030

Phytoplankton community as bioindicator of fertility in belawan river

Mayang Sari Yeanny*
1.Department of Biology, Faculty of Mathematics and Natural Sciences, University of Sumatera Utara, Jl. Bioteknologi No.1 Kampus USU Padang Bulan Medan – 20155

*E-mail: mayang_sy@yahoo.com/mayang@usu.ac.id Phone. (+62) 81349750016

Abstract Belawan River is an important river for the Medan residents and its surroundings. It serves as the main raw material for the local drinking water company, as well as domestic, industrial, hotel and tourism. Many human activities had led to the declining condition of water in the river throughout the year. One way to approach the concept of bioindicator is by knowing Abundance, Relative Abundance, Frequency of Attendance, equitability, dominance, and diversity of the phytoplankton itself. Results indicated that the phytoplankton community was from 3 different classes: Chlorophyceae, Bacillariophyceae, and Cyanophyceae. Phytoplankton individual abundance was around 2612 to 17755 ind / L. The diversity index was around 2.15 to 2.58, which is considered to have low to moderate diversity with high pollution level. Equitability Index was approaching 0, with relatively high domination from *Sphaeroplea* and *Asterionella*. The water quality that influences the diversity of phytoplankton as bioindicator was dissolved oxygen.

Keywords: Aquatic biota, Sumatera Utara, water quality

1. Introduction

Belawan River is the main river that crosses through Medan. Its flow passes through human residences, industries, electric steam power plant, local water company and aquaculture. With those activities around, waste water is often directly discharged into the water body, and thus can cause a negative impact to the aquatic environment. The utilization of the river itself as a waste water disposal site can cause changes in environmental factors that will be bad for the life of its biota. The quality change in water can greatly affect the life of the biota living in it.

Belawan River is an important river for the people of Medan and its surroundings. The local water drinking company even uses it as a main source of the raw material. Water pollution increases the cost for water treatment, as dissolving pollutants will undergo a longer process with higher cost, or else the quality of drinking water will decrease.

Biota who spends their lives in Belawan River, such as phytoplankton, is a producer in the water and their productivity, can be used as environmental bioindicator. By this nature, environmental changes can heavily affect the abundance and diversity of aquatic biota. This abundance and diversity itself are highly dependent on their tolerance and sensitivity to environmental change. They require a certain minimum amount of oxygen in order to support their lives [1].

Changes in the quality of water are closely related to the potential water viewed from the abundance of phytoplankton. The presence of phytoplankton in waters can provide information about

IOP Conf. Series: Earth and Environmental Science **130** (2018) 012030 doi:10.1088/1755-1315/130/1/012030

aquatic conditions. Phytoplankton can be used as an indicator to evaluate the quality and fertility level of a water body. Phytoplankton gives the highest contribution in term of oxygen in the water. The ability of phytoplankton as the initial binding of solar energy makes phytoplankton play an important role in the life of a water body.

The existence of phytoplankton can be used as an indicator of the quality of a water body, describing the number of phytoplankton species living in the water as well as the dominating species, the presence of phytoplankton that can live from blooming substances, can give a picture of the actual state of the water.

The use of aquatic organisms such as phytoplankton as a water bioindicator has several advantages, such as (a) providing relevant information of existing water quality conditions and can be done in an easy and relatively short time; (b) provides an overview of self purification in anaerobic or aerobic circumstances and may recognize toxic effects on the structure of the existing organisms; (c) provide important information, not only on pollution caused by waste water in aquatic environment, but also to complete the special factor, that is, the change of organism's life structure as a result of the existence of various organisms in the water; And (d) provide an overview of the state of water quality over a relatively long period, despite periodic changes. Thus, it is necessary to conduct a research entitled Phytoplankton Community as Bioindicator of Fertility In Belawan River.

This study aims to see the phytoplankton community as a bioindicator of the quality of Belawan River waters with various approaches, namely:
(1) to know the structure of the phytoplankton community including abundance, diversity, equitability and dominance of phytoplankton.
(2) the relationship between phytoplankton and water quality.
(3) to know the water pollution by using phytoplankton as bioindicator.

2. Methods

Sampling was conducted in August-September 2015. Five sample locations were picked based on the difference of community activities with 3 repetitions, from upstream to estuary site of Belawan river

Station I	Salam Tani Village, Pancur Batu Sub-district, Deli Serdang District	No activity/ Upstream Control
Station II	Sunggal Kanan Village, Sunggal Sub-district, Deli serdang District	Habitation, domestic
Station III	Kampung Lalang Village, Sunggal Sub-district, Deli Serdang District	Market, Hotel
Station IV	Kelambir Village, Hamparan Perak Sub-district, Deli Serdang District	Soy industries, Paper industries
Station V	Sicanang Village, Medan Belawan Sub-district, Medan Belawan City	Downstream (estuary)

2.1. Water Quality Sampling

From the measured water quality can be seen in the following table:

No	Water quality	Instrument	Place of measurement
1.	Temperature (^{o} C)	Thermometer	In-situ
2.	Light Penetration (m)	Secchi disk	In-situ
3.	Light intensity (cd)	Luxmeter	In-situ
4.	Flow Rate (sec / m)	Manual	In-situ
5.	pH	pH meter	In-situ
6.	DO (Dissolved Oxygen)	Winkler Method	In-situ
7.	BOD	Winkler & incubation Method	Laboratory
8.	COD	Reflux Method	Laboratory

IOP Conf. Series: Earth and Environmental Science **130** (2018) 012030 doi:10.1088/1755-1315/130/1/012030

2. 2. Phytoplankton sampling

Samples were taken using 5-liter bucket for 5 times using plankton net. Samples trapped in the plankton net was poured into the film bottle, then two drops of Lugol were added. Plankton was observed under a microscope and identified using the reference books [2,3,4,5]. Abundance (K), Relative Abundance (KR), Presence Frequency (PF), Equitability (E), Dominance (D), and Diversity (H) of the phytoplankton were then measured and analyzed in order to see the current pollution level of the Belawan river.

2.3. Data Analysis

Abundance (A)

$$A = \frac{T}{L} x \frac{P}{p} x \frac{V}{v} x \frac{1}{W} \quad (1)$$

With:
- A = Abundance of plankton per liter
- T = Surface Area of Haemocytometer
- L = Area of view
- P = Number of enumerated Plankers
- P = Number of field observed
- V = The volume of plankton concentration in the bucket
- V = Volume of concentration under the cover glass
- W = The volume of the medium water filtered with plankton net

Relative Abundance (RA)

$$RA = \frac{ni}{\sum N} x100\% \quad (2)$$

With:
- ni = the number of individuals of a kind
- ΣN = total of all individuals

Presence Frequency (PF)

$$PF = \frac{\text{The number of plots occupied by a genera}}{\text{Total number of plots}} \text{ x } 100 \quad (3)$$

With: PF =
- 0-25 (Very rarely)
- 25-50 (Rarely)
- 50-75 (Many)
- > 75 (Very much)

The Shannon-Wiener Diversity Index (H ')

$$H' = -\sum_{i=1}^{S} pi \ln pi \quad (4)$$

With:
- H ' = Shannon-Wiener diversity index
- Pi = ni / N (comparison of the number of individuals (a type with all types)
- Ln = natural logarithm

IOP Conf. Series: Earth and Environmental Science **130** (2018) 012030 doi:10.1088/1755-1315/130/1/012030

Equitability Index (E)

$$E = \frac{H'}{H_{MAX}} \quad (5)$$

With: H ' = Shannon-Wiener's diversity index (H')
Hmax = maximum diversity index
E = Equitability Index

Dominance Index (D)

$$D = -\sum_{i=1}^{S} \frac{ni}{N}^{2} \quad (6)$$

With: D = Simpson's dominance index
Ni = the number of individuals of a kind
N = total number of individuals
S = genera

2.4. Water Quality Relationship with Phytoplankton

Environmental data such as temperature, penetration of light, light intensity, flow rate, pH, DO, BOD, COD were statistically tested and analyzed to observe its relationship with phytoplankton using SPSS correlation analysis Ver. 13 to see the current pollution level of the river.

3. Results and discussions

3.1. Belawan River Water Quality

The results obtained by the phytoplankton community as a bioindicator of Belawan river water quality are obtained as Table 1.

. Table 1. The average value of water quality in Belawan river

No	Parameter	Station I	Station II	Station III	Station IV	Station V
1.	Temperature (^{o} C)	28	29	28	28	29
2.	Light concentration (cm)	51	74	62	60	30
3.	Light Intensity (Cd)	244	267	228	573	143
4.	Flow Rate (sec / m)	24	71	24	42	17
5.	pH	7.86	7.93	7.50	7.60	7.63
6.	DO (mg/L)	8.4	6.8	6.8	4.8	4.7
7.	BOD_5 (mg/L)	1.40	0.70	1.70	1.05	1.05
8.	COD(mg/L)	15,488	9,152	10,342	18,304	24,800

It can be seen from Table 1 that the water temperature in the five stations ranges from 28-29 ^{0} C, with the highest temperature at station II and V which was situated near the habitation and estuary with overall the temperature is relatively the same. Penetration of light ranges from 30-74 cm with the highest amount of penetration of light at station II as because the area was more open with slightly overgrown plants, which could absorb light to easily penetrate into the water. The intensity of light ranges from 143-573 Candela with the highest light intensity in station IV, which was due to the ability of light to absorb was relatively high. The pH ranges from 7.60 - 7.93 with the highest pH at the second station II, which was near the industries, though the overall pH was almost the same

Dissolved Oxygen (DO) ranged from 4.7 - 8.4 mg/L with the highest dissolved oxygen at station I. This is caused by the environmental conditions that support photosynthesis well to contribute much oxygen in the waters. Biological Oxygen Demand (BOD) 5 ranged from 0.70 - 1.70 mg/L with

IOP Conf. Series: Earth and Environmental Science **130** (2018) 012030 doi:10.1088/1755-1315/130/1/012030

the highest BOD5 at station III which was a highly populated residential area that produced domestic waste in the form of organic materials in which the oxygen in the water was used by microorganisms to decompose the organic material. Chemical Oxygen Demand (COD) ranged from 9,152 – 24,800 mg/L with the highest COD at Station IV which was the estuary area where the substrate from upstream of the river that caused the area to have higher organic materials so that the oxygen needed to decompose those organic matters was also chemically high.

3.2. Abundance Value (A) (ind / m^2), Relative Abundance (RA) (%) and Presence Frequency (PF) (%) Phytoplankton in Belawan River

Results of the Abundance Value (A) (ind / m^2), Relative Abundance (RA) (%) and Presence Frequency (PF) (%) of Phytoplankton in Belawan River are shown in Table 2.

Table 2. Abundance Value (A) (ind / m^2), Relative Abundance (RA) (%) and Presence Frequency (PF) (%) Phytoplankton at Belawan River at Station I and II

No.	Taksa	Station I			Station II		
		A	RA	PF	A	RA	PF
I	Chlorophyta						
A	Chlorophyceae						
	1.Chaetosphaeridium						
	2. Closteriopsis						
	3. Gonatozygon	204.08	769.23	100.00	285.71	10.94	66.67
	4. Oocardium	326.53	12.31	66.67	326.53	12.50	33.33
	5. Pediastrum						
	6. Schroederia						
	7. Sphaeroplea						
	8. Stigeoclonium				163.27	6.25	66.67
	9. Trochiscia						
	10. Volvox	81.63	3.08	66.67			
	11. Zygnema	448.98	16.92	66.67	326.53	12.50	100.00
II	Chrysophyta						
B	Bacillariophyceae						
	12. Asterionella						
	13. Gomphonema				204.08	7.81	66.67
	14. Coscinodiscus						
	15. Cylindrotheca						
	16. Gyrosigma						
	17. Mastogloia				285.71	10.94	100.00
	18. Melosira						
	19. Navicula	244.90	9.23	100.00	367.35	14.06	66.67
	20. Nitzschia	367.35	13.85	66.67			
	21. Stauroneis	81.63	3.08	33.33	285.71	10.94	66.67
	22. Surirella						
	23. Synedra	204.08	7.69	33.33			
	24. Tabellaria						

III	Cyanophyta						
C	Cyanophyceae						
	25. Aphanizomenon	163.27	6.15	33.33			
	26. Coelosphaerium						
	27. Oscillatoria	530.61	20,00	100.00	367.35	14.06	66.67
	Total	**2653.06**	**100.00**		**2612.24**	**100.00**	

Table 3. Abundance Value (A) (ind / m^2), Relative Abundance (RA) (%) and Presence Frequency (PF) (%) of Phytoplankton at Belawan River at Station III, IV, and V

No.	**Taksa**	**Station III**			**Station IV**			**Station V**		
		A	**RA**	**PF**	**A**	**RA**	**PF**	**A**	**RA**	**PF**
I	Chlorophyta									
A	Chlorophyceae									
	1.Chaetosphaeridium							1836.73	10.34	100.00
	2. Closteriopsis							1020.41	5.75	66.67
	3. Gonatozygon	693.88	22.97	66.67						
	4. Oocardium									
	5. Pediastrum				367.35	8.91	66.67	1755.10	9.89	100.00
	6. Schroederia				163.27	3.96	33.33	285.71	1.61	33.33
	7. Sphaeroplea				204.08	4.95	33.33	2122.45	11.95	100.00
	8. Stigeoclonium									
	9. Trochiscia							367.35	2.07	33.33
	10. Volvox	285.71	9.46	66.67						
	11. Zygnema									
II	Chrysophyta									
B	Bacillariophyceae									
	12. Asterionella				612.24	14.85	100.00	2000.00	11.26	100.00
	13. Gomphonema	204.08	6.76	33.33	367.35	8.91	66.67	163.27	0.92	66.67
	14. Coscinodiscus				326.53	7.92	100.00	1510.20	8.51	100.00
	15. Cylindrotheca	163.27	5.41	100.00						
	16. Gyrosigma				81.63	1.98	33.33	489.80	2.76	66.67
	17. Mastogloia	367.35	12.16	100.00						
	18. Melosira				612.24	14.85	66.67	1755.10	9.89	66.67
	19. Navicula	81.63	2.70	33.33						
	20. Nitzschia	244.90	8.11	33.33						
	21. Stauroneis				326.53	7.92	66.67	163.27	0.92	33.33
	22. Surirella							163.27	0.92	66.67
	23. Synedra	448.98	14.86	66.67	163.27	3.96	33.33	244.90	1.38	100.00
	24. Tabellaria				204.08	4.95	33.33	1346.94	7.59	100.00
III	Cyanophyta									
C	Cyanophyceae									
	25. Aphanizomenon	367.35	12.16	100.00						
	26. Coelosphaerium							1673.47	9.43	100.00
	27. Oscillatoria	163.27	5.41	33.33	693.88	16.83	100.00	857.14	4.83	100.00
	Total	**3020.41**	**100.00**		**4122.45**	**100.00**		**17755.10**	**100.00**	

IOP Conf. Series: Earth and Environmental Science **130** (2018) 012030 doi:10.1088/1755-1315/130/1/012030

It can be seen from Table 2 and 3 that *Oscillatoria* at station I were found to have the highest Abundance value, Relative abundance and attendance frequency with 530.61 ind / m^2 (A), 20.00% (RA) and 100% (PF), while the lowest was found in *Volvox* and *Stauroneis*, with 81.63 (A), 3.08% (RA) and 33.33% (PF). *Navicula* and *Oscillatoria* were found in station II to have the highest Abundance value, Relative abundance and presence frequency were 367.35 ind / m^2 (A), 14.06% (RA) and 66.67% (PF), while the lowest were found from the genera of *Stigeoclonium* 163.27 (A), 6.25% (RA) and 66.67% (PF). *Gonatozygon* were found in station III to have the highest Abundance value, Relative abundance and attendance frequency, with 693.88 ind / m^2 (A), 22.97% (RA) and 66.67% (PF) and lowest in *Navicula* as equal to 163.27 (A), 6.25% (RA) and 66.67% (PF).

On the station IV, *Oscillatoria* were found to have the highest Abundance value, Relative abundance and attendance frequency, with 693.88 ind / m^2 (A), 16.83% (RA) and 100% (PF), and the lowest in *Gyrosigma* with 81.63 (A), 1.98% (RA) and 33.33% (PF). *Sphaeroplea* were found in station V to have the highest Abundance value, Relative abundance and presence frequency with 2122.45 ind / m^2 (A), 11.95% (RA) and 100% (PF), and lowest in *Gomphonema, Stauroneis* and *Surirella* of 163.27 (A), 0.92% (RA) and 66.67% (PF). Overall, *Sphaeroplea* has the highest abundance value, relative abundance, and presence frequency in the Belawan River as equal to 2122.45 ind / m^2 (A), 11.95% (RA) and 100% (PF) at station V, and while the lowest was found in *Volvox, Stauroneis* (Station I), *Navicula* (Station II), and *Gyrosigma* (Station IV) with 163.27 (A), 0.92% (RA) and 66.67% (PF).

3.3. *Value of Diversity (H') and Equitability (E) in Belawan River*

Based on the research conducted, the value of diversity (H') and equitability (E) in Belawan river were as follows:

Table 4. Diversity Values (H') and Equitability Index (E) of phytoplankton in Belawan River

	Station I	**Station II**	**Station III**	**Station IV**	**Station V**
H^I	**2.15**	**2.28**	**2.16**	**2.58**	**2.56**
E	0.18	0.19	0.23	0.21	0.21

It can be seen from Table 4 that the highest value of diversity (H') was found from station IV, with 2.58 while the lowest was from station I with 2.15. The diversity of phytoplankton in the 5 stations was low to moderate. Low diversity is when 0 < (H') <2,302, medium diversity is when 2,302 < (H') <6.907, and high diversity is when (H')> 6.907 [6]. The water in the river is said to be heavily polluted if (H') <1, moderately polluted if (H') is between 1.0 to 1, 5, while lightly contaminated when (H')> 2.0. The equitability (E) value ranges from 0.18 to 0.23 with the highest equitability found at station III, with the lowest at station I. Equitability value (E) ranges from 0 to 1, where value close to 0 means that equitability is low due to the dominant species [6] like *Sphaeroplea* and *Asterionella*. This means the number of individuals on the type is not equally and evenly distributed.

3.4. Correlation Analysis Value

Correlation value was obtained based on the measurement of the water quality parameters correlated with the value of diversity (Shannon-Wiener Diversity) as in Table 5 below:

Table 5. The Correlated Analysis Value obtained between water quality parameters and phytoplankton diversity

No	Parameters	Diversity (H')
1.	Temperature (^{o} C)	0.32
2.	Light concentration (cm)	-0.43
3.	Light Intensity (Cd)	0.41
4.	Flow Rate (sec / m)	0.02
5.	pH	-0.31
6.	DO (mg/L)	0.94
7.	BOD_5 (mg/L)	0.52
8.	COD(mg/L)	0.75

From Table 5 it can be seen that DO affects the diversity of phytoplankton as bioindicator. The obtained DO during the study ranged from 0.2 to 4.4 mg / L, this value of was far from the needs of aquatic organisms and partly enough to support the needs of phytoplankton [1], in which means that dissolved oxygen has a very strong relationship to the diversity of phytoplankton.

4. Conclusion

From research conducted on the phytoplankton community as a bioindicator of fertility in Belawan River, it can be concluded that:

1. Phytoplankton obtained consisted of 27 genera from 3 classes and 3 orders, with the highest abundance in the genera of *Sphaeroplea*, while lowest in genus *Volvox*, *Stauroneis*, *Navicula*, and *Gyrosigma*.
2. The diversity of phytoplankton in 5 stations was relatively low to moderate, while the level of pollution based on the diversity value of station I-V was classified as lightly polluted.

Acknowledgement
This research can be done with the help and contribution of various parties. To that end, the authors would like to thank all those who have contributed to this research, especially in the USU 2015 PNBP Research Program which has provided financial support for this research.

References

[1] Rudiyanti S (2009) Water Quality of Banger River Pekalongan Based on Biological Indicators. Journal of Saintek Fisheries. 4(**2**): 46-52.
[2] Edmonson WT (1963) Fresh Water Biology. Second Edition. Jhon Willey & Sons, inc., New York.pp. 274-285.
[3] Bold HC, and Wynne MJ (1985) Introduction to the Algae. Structure and Reproduction. Englewood Cliffs. New Jersey, Prentice-Hall
[4] Pennak RW (1978) Fresh-water Invertebrates of the United States. 2nd Edition. New York: John Wiley & Sons.
[5] Streble H, and Krauter, D (1988) Das Leben im Wassertropfen: Mikroflora und Mikrofauna des Süβwassers. Ein Bestimmungsbuch. Franckh-Kosmos Verlag GmbH, Stuttgart.
[6] Krebs, Charles J (1985) Ecology, the experimental analysis of distribution and abundance, 3rd edition. Harper & Row, New York.

Development of unidentified dna-specific hif 1α gene of lizard (*hemidactylus platyurus*) which plays a role in tissue regeneration process

T Novianti[1]*, M Sadikin[2], S Widia[2], V Juniantito[3], and E A Arida[4]

[1)]Biotechnology Department, Esa Unggul Universiy, Jl. Raya Arjuna Utara no. 9 Jakarta Barat 13
[2)]Biochemestry Department, Faculty of Health, University of Indonesia, Jl.raya salemba no 6 Jakarta Pusat
[3)]Patology and Anatomy Department, Faculty of Veteriner, Institute of Agriculture Bogor, Jl.Dramaga Bogor
[4)]Puslit Biologi, Lembaga ilmu Pengetahuan Indonesia (LIPI), Jl.Raya Cibinong, Bogor

*Email: titta@esaunggul.ac.id

Abstract. Development of unidentified specific gene is essential to analyze the availability these genes in biological process. Identification unidentified specific DNA of HIF 1α genes is important to analyze their contribution in tissue regeneration process in lizard tail (*Hemidactylus platyurus*). Bioinformatics and PCR techniques are relatively an easier method to identify an unidentified gene. The most widely used method is BLAST (Basic Local Alignment Sequence Tools) method for alignment the sequences from the other organism. BLAST technique is online software from website https://blast.ncbi.nlm.nih.gov/Blast.cgi that capable to generate the similar sequences from closest kinship to distant kindship. *Gecko japonicus* is a species that it has closest kinship with *H. platyurus.* Comparing HIF 1 α gene sequence of *G. japonicus* with the other species used multiple alignment methods from Mega7 software. Conserved base areas were identified using Clustal IX method. Primary DNA of HIF 1 α gene was design by Primer3 software. HIF 1α gene of lizard (*H. platyurus*) was successfully amplified using a real-time PCR machine by primary DNA that we had designed from *Gecko japonicus*. Identification unidentified gene of HIF 1a lizard has been done successfully with multiple alignment method. The study was conducted by analyzing during the growth of tail on day 1, 3, 5, 7, 10, 13 and 17 of lizard tail after autotomy. Process amplification of HIF 1α gene was described by CT value in real time PCR machine. HIF 1α expression of gene is quantified by Livak formula. Chi-square statistic test is 0.000 which means that there is a different expression of HIF 1 α gene in every growth day treatment.

Keywords: bioinformatics, alignment, hypoxia

1. Introduction

Tissue regeneration process occurs after the injury incidence. The first phase of tissue regeneration process is wound healing phase that some of cells undergo apoptosis process. Tissue regeneration begins fill in some cells that contribute to regeneration process. The process continued to de-differentiation some cells, regeneration bud formation (blastem), morphogenesis and tissue growth through mitotic cell division [1, 2].

Basically, the regeneration process is mediated by molecular process which some of genes play a role in the process. Regeneration in biology is based on the process of morphogenesis of multicellular organisms to repair the tissues or organs to maintain the physiological and morphological process [3]. In the wound healing phase, there is an increase of angiogenesis process that the growth of new blood vessels. Increased angiogenesis processes require the oxygen and nutrients necessary for tissue proliferation. In this phase can induce to acute hypoxia, which would be an important signal for the wound healing phase. Since hypoxia state, tissue would regulate cell proliferation, cell migration, and cell differentiation through induction by cytokines and cell signaling [1,3,4].

Relative hypoxia in injured tissue was caused by composition of oxygen only 1%. In hypoxic state, Hypoxia Inducible Factor (HIF) protein that was controlled by oxygen levels in the tissues will become stable. HIF-1α Protein will translocate to nucleus, join with subunit β and initiate to transcription process of gene expression [5,6]. The expression of several genes in a hypoxic state are Vascular endothelial growth factor (VEGF) gene, erythropoietin (EPO) gene, cytoglobin (cygb) gene, TGF α gene and other genes [7,8]. In hypoxic state, it can stimulate the production of extracellular matrix components and improve angiogenesis process. HIF 1α protein plays an important role in the expression of cytokines and growth factors such as cellular matrix during wound healing [1,9].

Knockout HIF 1α in muscle skeletal does not affect the severity of acute and regeneration of skeletal muscle, while HIF 1α knockout in myeloid cells yielded significant results to delay skeletal muscle regeneration [10, 11]. The stability of HIF 1α in the hypoxia state could initiate the expression of cygb suspected to act as a carrier of oxygen into the hypoxic tissues. The cygb protein is one of the hexacoordinate globin superfamily proteins that possess to bind oxygen ability strongly [12, 13].

The regeneration process begins with wound healing and de-differentiation process, regeneration of blastema formation, followed by cells differentiation, morphogenesis, and tissue growth [13.14]. Wound closure occurred because the skin epidermal cells at the wound edges proliferate by mitosis, then migrate to the inner surface of the wound, so that the entire surface of the wound is closed [15,16].

Lizard (*H. platyurus*) belongs to Reptile class, ordo *Squamata*, *Gekkonidae* family and Genus *Hemidactylus*, which has a high ability to regeneration the tissue of tail. Tissue regeneration of lizard's tail covers autotomy process to defend them. The lost tail will be replaced with a new tail within approximately in 7 days. The process of wound healing is an early phase in tissue regeneration involving inflammatory processes [11,14,17].

The development of specific unidentified gene is essential for analyze the availability these genes in a biological process. Tracking the specific DNA of HIF 1α genes is necessary to analyze contribution these genes in the regeneration process of lizard tail (*H. platyurus*). Approach with bioinformatics technique with multiple alignment methods is relatively easier method than the other and using PCR method to make sure that this gene can identification and amplification [18,19].

DNA sequences from any organism that has been identification could be search in NCBI website. NCBI as the gene bank center has some data, data for their closest kinship with alignment techniques and data about phylogenetic studies. Software on-line the most widely used alignment method is BLAST (Basic Local Alignment Sequence Tools) from the website blast.ncbi.nlm.nih.gov/Blast.cgi. The BLAST technique is capable to generate similar sequences from closest kinship to distant kinship. Mega Software is able to locate conserved base areas with Clustal IX method [18,19].

2. Methods

2.1 Gene tracking

HIF 1α gene of lizard is the target gene. The tracking of HIF 1α gene of lizard (*Hemidactylus Platyurus*) begins with the literature phylogenetic from the species closest kinship. *Gecko Japonicus* is the closest species with *Hemidactylus Platyurus*. HIF 1α genes of *Gecko Japonicus* were analyzed by the BLAST method to find some sequences HIF 1α gene from the other species. In the determination of conservation areas used Mega7 software with ClustalX technique. Some sequences do align each other with ClustalX Program to select the similarities base each other. We selected the base area which has the most similarity as the primary base for gene tracking to design primer DNA. We design the Primary DNA with Primer3 software and select high conserve areas manually in coding sequence. We design primary DNA with general parameters such as the number of nucleotides, GC content, the length of base sequence, the difference temperature between reverse and forward sequence, and there is no possibility of mutual complement between each otherthe bases inside [18].

2.2 RNA isolation process
The result of tissue regeneration tail lizard growth of days 1, 3, 5, 7, 10, 13, and 17 are the RNA isolation sample. The tissue was cut from the tail and the cell membrane was crushed by mechanically. The sample was added proteinase K 1 ul and cell lysis as 300 ul. The solution was incubated at 60 ° C for 15 minutes and homogenized every 5 minutes using a vortex. Then, add the MPC 15 ul to the solution and centrifuged at a temperature of -40^0 C for 10 min with a discharge of 10,000 rpm. The centrifugation of a supernatant is separated into a clean tube and an additional solution isopropanol 500 uL and homogenized. The supernatant solution was centrifuged for 10 minutes in speed 10,000 rpm. Then the total RNA will appear at the bottom of the tube. Discard the supernatant, wash the RNA with ethanol 70% in twice and saved in TE buffer 35 ul. RNA concentration 2μl was quantified and assessed purity using a varioscan quantodrop. The machine automatically count the purity of RNA from the ratio of absorbance at wavelength (λ) 260. Samples are stored in refrigerator temperatures -40^0 C.

2.3 QPCR gene expression
Dilated RNA isolation results with concentration ng/ul using TE buffer. Mix a solution of qPCR Kit from Kappa 10 mL for each sample composition are Kappa (2x) 5 mL, Forward primer DNA 0.2 mL, Reverse Primer DNA 0, 2 mL and Kappa (5 X) 0.2 mL, total RNA samples 2 μl and DEPF FW 2, 4 μl. Used control samples and housekeeping gene (18S) as a comparison of the results relative quantification of amplification produced. Amplification the genes used PCRmax eco48 real-time thermal cycle machine-Techne as 40 cycles. The results of circle amplification are captured by cyber green fluorescence, these results indicate the quantity amount of Double-stranded DNA formed. The data obtained from the real-time RT-PCR is a threshold cycle (CT) value. The value of ΔCT was calculated by reducing the CT 18S rRNA value with the CT value of the HIF 1 alpha gene. Intergroup ΔCT values were compared descriptively.

2.4 Statistic test
Statistical tests performed using SPSS method, to test the normality of the data and to test for differences in the data of daily treatment lizard tail growth

.3. Results and Discussion
3.1 Primary design DNA
The result of the nearest lizard kinship taxonomy study is Gecko Japonicus, because they comes from the same order, same class and same family, Geckonidae. Literature study from gene bank https://www.ncbi.nlm.nih.gov/ obtained data of HIF 1α gene of *G. japonicus*, has 4521 bp mRNA gene length and CDS region (coding sequence) at the base of 115 -2583 bp. The BLAST process of HIF 1 α Gecko gene sequence generates some sequences from a close family organism to a distant relative. The result of multiple alignment of HIF 1α gene (Figure 1) of some of these species contain some areas that have high conservation area, then it is assumed that sequences are in lizard too (*H. Platyurus).*

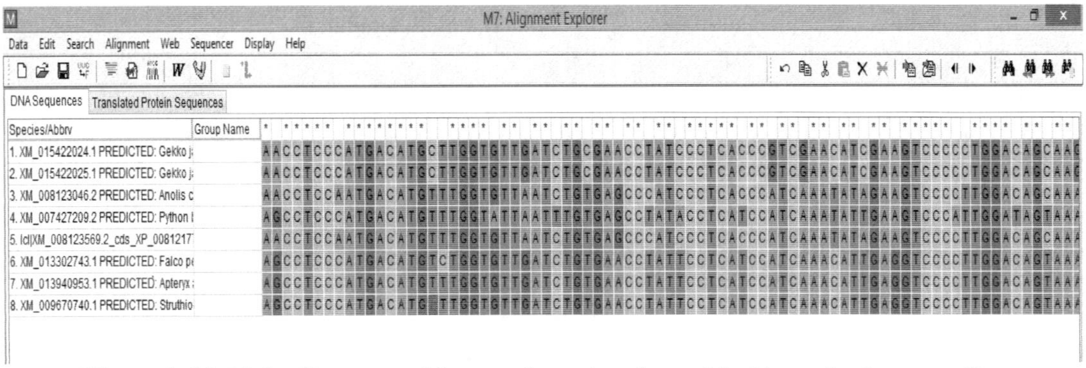

Figure 1. Multiple alignment with several species closest kinship each other was align by Mega7 software

```
ADDITIONAL OLIGOS
                        start  len      tm      gc%   any    3' seq

 1 LEFT PRIMER          1125   20   59.97   55.00  4.00  0.00
CGTGTGCGTGAACTACGTCT
   RIGHT PRIMER         1311   20   59.98   50.00  3.00  0.00
TAATGCGTCTGGCTCCTTCT
   PRODUCT SIZE: 187, PAIR ANY COMPL: 3.00, PAIR 3' COMPL: 0.00

 2 LEFT PRIMER          3524   20   60.01   50.00  6.00  0.00
CGACGCGTTTATATGTGGTG
   RIGHT PRIMER         3764   20   60.07   50.00  3.00  0.00
TATTCTGTCCCCCAGCAAAG
   PRODUCT SIZE: 241, PAIR ANY COMPL: 3.00, PAIR 3' COMPL: 2.00

 3 LEFT PRIMER           760   20   59.93   50.00  7.00  2.00
GAACCTCCCATGACATGCTT
   RIGHT PRIMER          966   20   60.02   55.00  3.00  0.00
CAGGGCGTGGTAGTATTCGT
   PRODUCT SIZE: 207, PAIR ANY COMPL: 5.00, PAIR 3' COMPL: 2.00

 4 LEFT PRIMER          3524   20   60.01   50.00  6.00  0.00
CGACGCGTTTATATGTGGTG
   RIGHT PRIMER         3766   20   59.93   50.00  3.00  1.00
GTTATTCTGTCCCCCAGCAA
   PRODUCT SIZE: 243, PAIR ANY COMPL: 3.00, PAIR 3' COMPL: 2.00
```

Figure 2. Several basic primers of nucleotide base using Primer3 software

Primary design results with primary software3 obtained some alternative primary DNA (figure 2). The primer3 design was selected (red colour), forward sequence starting at base 760 and reverse sequence starting at base 966. Base to 760-966 is a CDS region, and is present in high conservation areas according to multiple alignment results in Figure 1.

Primary DNA design for lizard (*Hemidactyulus Platyurus*) is obtained from *Gecko Japonicus* as the closest kinship with a unique base sequence and has high conservation areas, so it is expected that the primary designs obtained are most specific for the HIF gene 1 α lizard (*Hemidactyulus Platyurus*), with melting temperature (tm) 57^{0}C, GC value 55% and alkaline product length PCR 207 pb.

The result of design Primary DNA are :
Reverse : 5 'GAACCTCCCATGACATGCTT 3'
Forward : 5 'CAGGGCGTGGTAGTATTCGT 3'

3.2 Gene Expression test results

The expression of the HIF 1α gene mRNA in daily tail growth treatment after the autotomy process was measured quantitatively by the real-time PCR/ qPCR method. The result of real-time reaction - PCR is a threshold cycle (CT) value indicating the amplification where the fluorescence signal passes through the threshold marked by the increased number of amplicons. The increased amplicon shows that amplification of unidentified DNA band of HIF 1α genes was successfully from lizard samples by primary DNA bands from the result was design using the closest alignment and phylogenetic kinship method. The process of tracking DNA that has not been identified before, is very easy by using multiple alignment method.

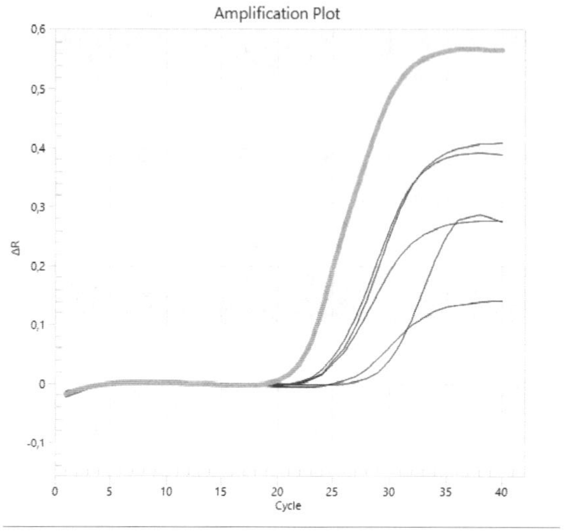

Figure 3. Graph of Amplification (Fluorescence Intensity Curve to Number of Cycles) HIF 1 α *H. platyurus* genes

At the graph of Figure 3, the increase of graph shows that tracking process and developing process of unidentified HIF-1 α gene from lizard (*Hemidactylus Platyurus*) has been identification successfully. The accumulation of PCR amplification product (amplicon) increases with increasing reaction cycle so the intensity of fluorescence will also be increased because CbyrGreen molecules bind to double-stranded DNA. The expression of HIF 1 α gene on daily treatment of lizard tail growth can be evaluated by comparing the mean value of the control group ΔCT with the mean ΔCT value of the treatment group shown in Table 1. The data was count by chi square method to see the different expression in daily treatment.

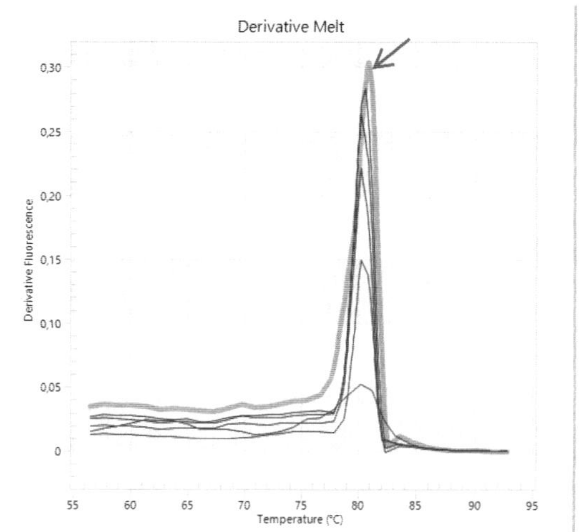

Figure 4. Graph of the Melting Lines of PCR Products showing the relationship between Fluorescence and Temperature Intensity in Melting-Curve Analysis

Melting Curve Analysis (MCA) is an analysis to estimate characteristics the double-stranded DNA denaturation becomes a single stranded DNA during the experience. The temperature when 50% of denatured DNA is known as a point melting (melting point). This temperature melting point value depends on the length DNA sequences and compositions of the bases of Guanin and Cytosin in DNA. Basically, primary specificity can be improved by optimization melting temperature of the DNA fragment. Optimization can be done with Melting Curve Analysis (MCA). The principle of MCA is melting temperature a DNA fragment is affected by the length and sequence of the fragment. If there is only one type of double stranded DNA fragment, it will appear one melting curve that is mean that DNA of amplification result is very specific [13]. The shape of the melt temperature curve can be seen in Figure 4.

Table 1. The meant of score ΔCT sample

No	sample	The meant of score ΔCT sample	The meant of score $^{-\Delta\Delta C}$
1	control	0,808	
2	1st day	0,7833	1,017
3	3rd day	-4.176	31,64
4	5th day	-6,9	209,09
5	7th day	5,573	0,0367
6	10th day	-1,861	6,359
7	13th day	-2,815	12,320
8	17th day	2,9475	0, 2269

Table 2. The results of chi square statistic test were significantly different between treatment group

	Observed N	Expected N	Residual
,0000	1	1,0	,0
,0367	1	1,0	,0
,3000	1	1,0	,0
1,0170	1	1,0	,0
6,3590	1	1,0	,0
12,3300	1	1,0	,0
31,6500	1	1,0	,0
209,0900	1	1,0	,0
Total	8		

Table 3. Chi- squaretest score

	delta_cv
Chi-Square(a)	,000
df	7
Asymp. Sig.	1,000

Statistic test results (Table 3) shows the value of chi-square 0.000, meaning that on every day growth shows a real difference of RNA expression. The ΔCT value of treatment growth day (Table 1)

was smaller (except on days 7 and 17) compared to the ΔCT in the control group. The change ΔCT values of PCR real time reaction showed the amount of cDNA amplicon from the mRNA template in the growth treatment group (excluding the 7th and 17th days) more than the control group. The value of ΔΔCT on the growth treatment of day-1, 3rd, 5th, 10th, and 13th-pointed tissue counts> 1.00, meaning that the HIF1α gene in the day group have a contribution in tissue regeneration process.

4. Conclusion

Develop and track unidentified gene HIF 1α-specific of lizard (*HemidactylusPlatyurus*) using bioinformatics with multiple alignment method and phylogenetic method has been done successfully. The results of primary design analysis from high conservation areas produce specific HIF 1αgecko genes, as evidenced by the occurrence of one peak at Amplified amplicon output chart. The result of chi square statistic test gives 0.000 result, which means there is real difference every day growth treatment.

Acknowledgments

We are grateful to the Ministry of Technology and Higher Education of the Republic of Indonesia who have provided scholarships for doctoral programs; Research Grant Program of University of Indonesia, Indonesian Institute of Sciences for giving permission to keep lizard; Dr. Marlina Ardiyani and Dr. Amir Hamidy as researchers from LIPI who have been give some idea for identification process of unidentified DNA; Eco Science Company Indonesia that had been give me a permission to use eco-realtime qPCR machine.

References

[1] Fausto N, Campbell JS, Riehlel KJ (2006) Liver Regeneration. *Hepatology*.**43** (2).

[2] Gauron C, Rampon C, Bouzaffour IM, Ipendey E, Teillon J, Volovitch M, Vriz S (2013) Sustained production of ROS triggers compensatory proliferation and irrequired. For regeneration to proceed *Scientific Reports*. **3**: 2084.

[3] Dharmayanti IP (2011) Metode Taksonomi Organisme Berdasarkan Sejarah Evolusi *Wartazoa*. **21**(1): 1-10.

[4] Maes C, Carmeliet G, Schipani E (2012) Hypoxia-driven pathways in bone development Regeneration, and diseases. *Nature Reviews Rheumatology*. **8**: 358-366.

[5] Komatsu DE., Bosch-Marce M, Semenza GL, Hadjiargyrou M (2007) Enhanced Bone Regeneration Associated With Decreased Apoptosisin Mice With Partial HIF-1α Deficiency. *J Bone Miner Res*. **22**(3): 366–374.

[6] Wan C, Gilbert SR, Wang Y, Cao X, Shen X, Ramaswamy G, Jacobsen GA, Alaql GS, Eberhardt AW, Gerstenfeld LC, Einhorn TA, Deng L, Clemens TL (2008) Activation of the hypoxia-inducible factor-1_accelerates bone regeneration. *PNAS*.**105**(2).

[7] Ke Q, Costa M (2006) Hypoxia-Inducible Factor-1 (HIF-1). *Mol Pharmacol*. **70**:1469-1480

[8] Love NR (2011) Genome-wide analysis of gene expression during Xenopustropicalis tadpole tail regeneration. *BMC Dev Biol*. **11**(70).

[9] Cramer T, Yamanishi Y, Clausen BE, Fo¨rster I, Pawlinski R, Mackman N, Haase VH, Jaenisch R, Corr M, Nizet N (2003) HIF-1alpha is essentialfor myeloid cell-mediated inflammation. *Cell* **112**: 645–657.

[10] Greijer AE ,Van der Wall E (2004) The role of Hypoxia inducible factor (Hif-1) in hypoxia induced apoptosis. *J Clin Pathol*. **54**: 1009-1014.

[11] Schremi S, Szeimmies, Pranti RM, Kareer L, Landthaler S, Babilas P (2010) Oxygen acute and chronic wound healing. *Br. J. dermato l*. **163**, 257-268.

[12] Singh S, Cansecoa DC, Manda SM, Sheltona JM, Chirumamilla RR, Goetscha SC, Yea Q Gerard RD, Schneidera JW, Richardson JA, Rothermela BA, Mammen PPA (2013) Cytoglobin modulates myogenic progenitoregeneration. *PNAS Published online*.**23**: E129–E138

[13] Soesilo NP (2009) Regenerasi Ekor Kadar (*Eutropis multifasciata* Kuhl) dan prospek aplikasinya. Pidato Pengukuhan Guru Besar. Dept. Biologi UGM,Yogyakarta

[14] Hutchins ED, Markov GJ, Eckalbar WL, George RM, King JM, Tokuyama MA, Fisher FE, Wade J, DeNardo DF, Rawls JA, Huentelman MJ, Wilson-Rawls J, Kusumi K (2014) Transcriptomic Analysis of Tail Regeneration in the Lizard Anolis Carolinensis Reveals Activation of Conserved Vertebrate Developmental and Repair. *PLOS ONE* . **9** (8).

[15] Mason SD, Howlett RA, Kim MJ, Olfert IM, Hogan MC, McNulty W, Hickey RP,Wagner JP, Kahn CR, Giordano FJ, Johnson RS (2004) Loss of skeletal muscle HIF-1alpha results in altered exercise endurance. *PloSBiol*. **2**: e288.

[16] Lukman A (2000) Mekanisme Regenerasi Anggota Tubuh Hewan *Biospecies*. **2**(2); 43 – 47.

[17] Nakatani Y, Kawakami A, Kudo A (2007) Cellular and molecular processes of regeneration, with special emphasis on fish fins *Dev Growth Differ* . **49**: 145–154.

[18] Santoso D (2001) Pengembangan pelacak DNAspesifik gen melalui bioinformatika: Indentifikasi gen penyandi protein biji 21 kDapada kakao UAH Indonesia Unit Penelitian Bioteknologi Perkebunan: 10-17.

[19] Wibowo SE, Muhammad, Djaelani A, Pancasakti HE (2013) Pelacakan Gen Sitokrom Oksidase Sub Unit I (COI) DNA Mitokondria Itik Tegal (*Anasdomesticus*) Menggunakan Primer Universal. *Bioma,* **15** (1) : 20-26

Phenotypic and molecular detection of ^{BLA}CTX-M gene extended-spectrum beta-lactamases in *escherichia coli* and *klebsiella pneumoniae* of north sumatera isolates

Mirzan Hasibuan[1,2], Dwi Suryanto[1*], and R Lia Kusumawati[2,3,4]

[1]Department of Biology, Faculty of Mathematics and Natural Sciences, University of Sumatera Utara Jln. Bioteknologi No. 1 Kampus USU, Medan 20155
[2]Clinical Research Unit, Clinical Microbiology Unit, University of Sumatera Utara Hospital, Jln. Dr. T Mansur No. 66 Kampus USU, Medan 20154
[3]Department of Microbiology, Faculty of Medicine, University of Sumatera Utara, Jln. Universitas No.1 Kampus USU, Medan 20155
[4]H. Adam Malik Hospital Medan, Indonesia.

*Email: dwisuryanto@usu.ac.id

Abstract. The application of antibiotics expanded-spectrum third-generation cephalosporin for the treatment of infectious diseases in hospitals is known contribute to increasing resistance due to the presence of the ^{bla}CTX-M gene in the bacteria producing ESBLs. This study was aimed to detect ESBLs, isolate phenotype and ^{bla}CTX-M genes on *Escherichia coli* and *Klebsiella pneumoniae* collected from H. Adam Malik Central Hospital. Phenotypes of the bacterial were detection using Vitek two compact, while the ^{bla}CTX-M genes were detection using polymerase chain reaction technique. The results showed that 85 (100%) isolates were ESBLs consisted of 41(48%) of *Escherichia coli*, and 44 (52%) of *Klebsiella pneumoniae*, respectively. ^{bla}CTX-M genes were detection in 62 (72.94%) of the isolates which 31 (36.47%) were *Escherichia coli*, and 31 (36.47%) of the isolates were *Klebsiella pneumoniae*, respectively. This study indicates the high prevalence of ^{bla}CTX-M genes in *Escherichia coli* and *Klebsiella pneumoniea* causing bacterial antibiotic resistance.

Keywords: antibiotic resistance, ESBLs, infections, molecular detection, polymerase chain reaction

1. Introduction

The presence of ^{bla}CTX-M gene in bacteria causes resistance to the broad spectrum of cephalosporin groups. ESBLs are enzymes derived from mutated beta-lactamase. This mutation leads to an increase in the enzymatic activity of beta-lactamase to penicillin, cephalosporins, and aztreonam. This resistance mechanism is widespread in the world, with reports of clinical isolates produce these beta-lactamase are from Europe, Africa, Asia, South America and North America [1,2]. ESBLs are most commonly produced by Enterobacteriaceae especially *Escherichia coli* and *Klebsiella pneumoniae* [3,4].

The use of beta-lactam group antibiotics, especially the third generation cephalosporin widely for the treatment of infections in hospitals, is a risk factor for infected by ESBLs producing bacteria

[5]. In addition to being resistant to beta-lactam antibiotics, ESBL-producing bacteria also often exhibit resistance to the use of quinolones. Excessive use of antibiotics, patients with severe disease, LOS (Length of Stay) and treated with invasive medical devices for a long time is also a high risk for being infected by ESBLs producing bacteria. Research on ESBLs is highly developed, coming from more than 30 different countries and reflects the distribution of ESBLs around the World[4]. Infection due to ESBL-producing bacteria are increasingly, and it is very difficult to treated since the choice of antibiotics becomes very limited and the emergence of new mutants [6], consequently this causes antibiotic failure against infectious bacteria [14,15].

Escherichia coli and *Klebsiella pneumoniae* are the primary causes of hospital-related infections and infections in the population (community) [6]. The worldwide increase in ESBL prevalence provides the basis for the importance of detecting ESBL phenotype and ^{bla}CTX-M genotypes in ESBLs-producing bacteria indispensable for improved therapeutic management and control of the bacterial transmission. This study aimed to detect ESBLs, and ^{bla}CTX-M gene in *Escherichia coli* and *Klebsiella pneumoniae* isolates from North Sumatra.

2. Methods

2.1 Bacterial isolates and phenotypic characterization

Sampling is done by purposive sampling based on the data of Germs Pattern of Haji Adam Malik Hospital Medan of the first semester of 2015. Bacterial isolates were collected from routine examination at Diagnostic Laboratory Installation, Microbiology Clinical Unit of Haji Adam Malik Hospital Medan. The phenotypic characterization was done using Vitek two Compact which included antimicrobial susceptibility test and ESBLs phenotyping.

2.2 DNA isolation

Isolates with ESBLs phenotype of *Escherichia coli* and *Klebsiella pneumonia* were subculture on Mac-Conkey agar and incubated 24 hours at 37°C. DNA isolation was conducted by the freeze-thaw cycling method, by inserting a bacterial colony into 20 µL aqua-bidest in a microtube and the vortex. The bacterial solution was freezer at -20°C for 10 minutes, followed by heating it at 90°C for 10 minutes using the dry block, freeze-thaw cycling done as six times. The freeze-thawed solution was spin at 13,000 rpm for 5 minutes. The supernatant was separated from cell, and subjected to DNA purification check.

2.3 Amplification and detection of CTX-M gene using PCR method

CTX-M gene amplification was performed by PCR technique using specific primer: ATG-TGC-AGY-ACC-AGT-AAR-GTK-ATG-GC forward and TGG-GTR-AAR-TAR-GTS-ACC-AGA-AYC-AGC-GG reverse [7]. Amplify the in PCR solution of 12.5µl master mix green go-Taq, 1µL of forward and reverse primer each, 8.5µL nuclease-free water, and 2µL bacterial DNA. Thermocycling reaction was conducted for denaturation at 94°C for 2 minutes, extended denaturation at 94°C for 1 minutes, annealing for 52°C for 30 sec, extension at 72°C at 45 sec, and extended extension at 72°C for 5 minutes, this reaction is 30 cycles. PCR product was visualized in mini gel electrophoresis and documented in UV Reader/Gel Documentation System.

3. Results and Discussion

3.1 Phenotypic characterization

Phenotypic characterization showed that all samples (n = 85) contained ESBLs. The ESBLs phenotype was determined from the results of the antimicrobial susceptibility test according to 2015 Clinical Laboratory Susceptibility Institute standard using Vitek 2 compact. The percentage of antimicrobial susceptibility is presented in Table 1.

Table 1. Percentage of Antimicrobial Susceptibility Test (AST)

Antimicrobial Agents		Antimicrobial Sensitivity (%)	
Group	Type of Antibiotics	Sensitive	Resistant
Penicillins	Amoxicillin	0	100
	Ampicillin	0	100
Cephalosporins	Cefotaxime	0	100
	Ceftriaxone	0	100
	Ceftazidime	0	100
	Cefepime	0	100
Monobactams	Aztereonam	0	100
Beta-lactamase Inhibitor	Amoxicillin/Clavulanic Acid	27	73.0
	Piperacillin/Tazobactam	43.5	56.4
	Cefoperazone/Sulbactam	63.5	36.4
Aminoglicosides	Gentamycin	35.3	64.7
	Amikacin	100	0
Carbapenem	Imipenem	85.8	14.2
	Meropenem	85.8	14.2
Fluoroquinolone	Ciprofloxacin	0	100
	Levofloxacin	25.8	74.2
Fosfomycin	Fosfomycin	7.0	93.0
Folate Pathway Inhibitor	Cotrimoxazole	0	100
Tigecil	Tigecycline	0	100

It seemed that all bacterial isolates were resistant to beta-lactam group antibiotics such as penicillin, cephalosporins, and monobactams, which indicated are carrying ESBLs. A number of isolates resistant to beta-lactamase inhibitors such as amoxicillin/ clavulanic, piperacillin/tazobactam, and cefoperazone/sulbactam acid were lower with the percentage of antimicrobial resistant of 73.0, of 56.4, and of 36.4%, respectively. Interestingly, all isolates were resistant to other classes of antibiotics such as tigecycline, cotrimoxazole, and ciprofloxacin. Saverin *et al.* [17] study in Surabaya, Indonesia, showed that ^{bla}CTX-M gene prevalence was found very high. 65 out of 73 isolates of ESBLs *Escherichia coli* bore ^{bla}CTX-M gene [17]. Another study reported that as much as 50.5% of ESBLs *Klebsiella pneumoniae* isolates in Tertiary-Referral Hospital, Bali, Indonesia was detected to have the ^{bla}CTX-M gene. This study showed the widespread of ^{bla}CTX-M ESBLs in Indonesia [18].

The antibiotic application can be pressure for bacterial cell. However, mutations allow them to survive. These bacteria pass through this trait to their offspring, leading to the evolution of resistance. Increased resistance due to mutation may facilitate by high bacterial cell number with short generation time[8]. Phenotypic resistance is expression by the presence of ESBL genes such as ^{bla}CTX-M. ESBL-producing bacteria are also often resistant to aminoglycosides, fluoroquinolones, tetracyclines, chloramphenicol, and sulfamethoxazole-trimethoprime [12].

3.2 Amplification of ^{bla}CTX-M gene

Amplification of ^{bla}CTX-M gene resulted in a 539 bp amplicon in both *Escherichia coli* and *Klebsiella pneumoniae* isolates (Figure 1) with the percentage of 72.94%.

IOP Conf. Series: Earth and Environmental Science **130** (2018) 012032 doi:10.1088/1755-1315/130/1/012032

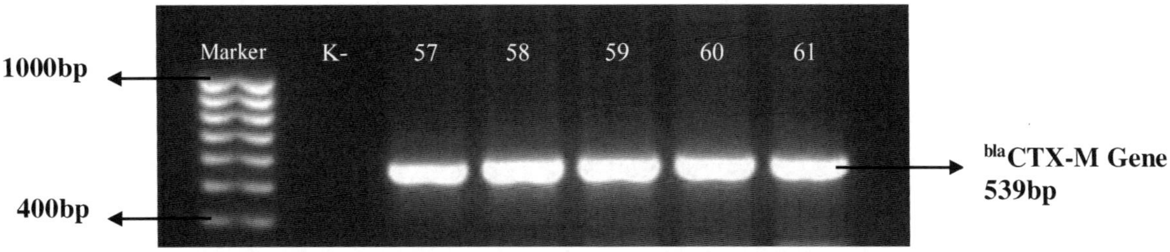

Figure 1. PCR profiles of molecular detection ^{bla}CTX-M gene of ESBLs
K- (Non-ESBLs), 57 (*E. coli* 11926), 58 (*K. pneumoniae*11688),
59 (*E. coli* 11779), 60 (*K. pneumoniae* 12006), 61 (*K. pneumoniae* 11901)

The primer used in this study amplified ^{bla}CTX-M gene in bacterial isolates resulted in the 539bp amplicon. It is primer could be used to amplify ^{bla}CTX-M gene. Hout et al. [7] using this primer showed that 50% of clinical isolates of the Cambodia Surgical Center collection was detected. Resistance to beta-lactam antibiotics may be due to inactivation of antibiotics by beta-lactamase enzyme, target modification of the penicillin binding protein (PBP), and failure penetration of drug disorders to PBP [15].

The presence of the ^{bla}CTX-M gene in bacteria is responsible for hydrolyzing antibiotics cefotaxime. Resistance may extend to other types of cephalosporin antibiotics [9]. Reported that unlike other countries, ^{bla}CTX-M genes were predominant in Europe. Several studies showed that ESBL types produce by *Escherichia coli* and *Klebsiella pneumoniae*. It was also reported that the ESBL prevalence increased with the emergence of ^{bla}CTX-M type genes [10]. Some researchers suggested that ^{bla}CTX-M gene found more common in ESBL isolates compared to that of SHV and TEM [16].

Today, the emergence of new variants on ESBLs producing microorganisms, especially ^{bla}CTX-M suggested the co-resistance involvement of other drug classes during endemics. The presence of the co-resistance is due to the different transmission of resistant gene types in the same clones [11]. Some studies showed that the ^{bla}CTX-M genotype is usually found in large plasmids often carrying other resistance genes of other antimicrobial agents including aminoglycosides, fluoroquinolones, chloramphenicol, tetracyclines mainly OXA and TEM [12].

It was find that 48% of *Escherichia coli* and 52% of *Klebsiella pneumoniae* isolates showed to have ESBLs, in which 36.47% of both bacteria was bearing ^{bla}CTX-M gene (Table 2). The other 27% of ESBL-producing bacteria might bearing another gene of beta-lactam resistance.

Table 2. Percentage of ESBL phenotype and ^{bla}CTX-M gene

Bacterial species	ESBLs phenotype	^{bla}CTX-M gene
Escherichia coli	41 (48%)	31 (36.47%)
Klebsiella pneumoniae	44 (52%)	31 (36.47%)
Total :	85 (100%)	62 (72.94%)

The ability of ESBL-producing bacteria to hydrolyze beta-lactam antibiotics most of the due to some mutations in several genes such as TEM and SHV. Such mutations generally occur in the active site of the enzyme causing enzyme activity increases [6]. High prevalence of ^{bla}CTX-M genes in *Escherichia coli* and *Klebsiella pneumoniae* bacteria may increase the risk of transfer of this resistance gene to an-ESBLs bacteria. This study was the first study in North Sumatra revealing the existence of ^{bla}CTX-M in ESBL *Escherichia coli* and *Klebsiella penumoniae*, therefore need for the hospital awareness to avoid cross infection through health facilities and paramedics.

IOP Conf. Series: Earth and Environmental Science **130** (2018) 012032 doi:10.1088/1755-1315/130/1/012032

4. Conclusions

This study showed that prevalence of ESBLs *Escherichia coli* and *Klebsiella pneumoniae* is quite high in North Sumatra, Indonesia. All isolates were ESBLs bearing. The molecular detection of the ^{bla}CTX-M gene showed that 72.94% of the isolates possessed the gene. However, the remains isolates of ESBLs may bear other genes but ^{bla}CTX-M gene.

Acknowledgments

The author gratefully acknowledges to Clinical Microbiology Unit of Diagnostic Laboratory Installation of H. Adam Malik Hospital Medan for providing bacterial isolates.

References

[1] Bonnet R (2004) Growing Group of Extended-Spectrum Beta-lactamases: The CTX-M Enzymes. Antimicrob Agent Chemother **48**: 1-14.

[2] Moland E, Black J, Hossain A, Hanson N, Thomson K (2003) Discovery CTX-M-like Extended Spectrum Lactamases in *Escherichia coli* isolates From Five U.S. States. Antimicrob Agent Chemother **47**: 2382–2383.

[3] Umadevi S, Kandhakumari G, Joseph M, Kumar S, Easow J, Stephen S, Singh U (2011) Prevalence and antimicrobial susceptibility pattern of ESBL producing Gram Negative Bacilli. J of Clin and Diagnostic Research **5**: 236-239.

[4] Yadav K, Adhikari N, Khadka R, Pant A, Shah B (2015) Multi Drug Resistant Enterobacteriaceae and Extended Spectrum Beta-lactamase Producing *Escherichia coli*: a Cross-sectional Study In National Kidney Center Nepal. Antimicrob Resist Infect Control **4**: 1-7.

[5] Tumbarello M, Spanu T, Bidino R, Marchetti M, Ruggeri M, Trecarichi E, Pascale G, Proli E, Cauda R, Cicchetti A, Fadda G (2010) Costs of Bloodstream Infections Caused by *Escherichia coli* and Influence of Extended Spectrum Beta-lactamase Production and Inadequate Initial Antibiotic Therapy. *Antimicrob Agents Chemotherapy* **54**: 4085–4091.

[6] Brooks GF, Carrol K, Butel J, Morse, S (2007) Medical Microbiology. 24. Mc Graw Hill.

[7] Hout B, Chamroeun O, Putheavy M, Vanvathanak V, Chonthida S, Vireak H, Agus R, Michael P, Steven N, Dustin H, Saqib N, James G, Ly T, Yong K, Gavin F (2015) Drug resistance in bacteria isolated from patients presenting with wounds at a non-profit Surgical Center in Phnom Penh, Cambodia from 2011–2013. *Tropical Diseases, Travel Medicine and Vaccines, Biomed Central* **1** (4): 1-11.

[8] Hawkey, P (2008) Molecular Epidemiology of Clinical Significant Antibiotic Resistance Genes. *Br. J. Pharmacol* 153: 406-13.

[9] Bradford P (2001) Extended-Spectrum Beta-Lactamases in 21st Century: Characterization, Epidemiology, and Detection of This Important Resistance Threat. *Clinical Microbiology Reviews* **14** (4): 933-951.

[10] Paterson D, Bonomo R (2005) Extended-Spectrum Beta-Lactamases: a Clinical Update. *Clinical Microbiology Journal* **18** (4): 658-686.

[11] Livermore DM, Canton R, Gniadkowski M (2007) CTX-M Changing the Face of ESBLs In Europe. *J Antimicrob Chemother* **59**: 165-174.

[12] Kanj SS, Corkill JE, Kanafani ZA, Araj GF, Hart R, Jaafar, Matar GM (2008) Molecular Characterization of Extended-Spectrum Beta-Lactamase Producing *Escherichia coli* and *Klebsiella spp* Isolates At a Tertiary-Care Centre In Lebanon. *Eur J Clin Microbiol Infect Disease* **14**: 501-504.

[13] Thenmozhi S, Moorthyl K, Suresh T, Suresh M (2014) Antibiotic Resistance Mechanism of ESBL Producing Enterobacteriaceae in Clinical Field: A Review. *Int. J. Pure App. Biosci*. **2** (3): 207-226.

[14] Pitout JD (2008) Multiresistant Enterobacteriaceae: the new threat of an old problem. Expert Rev. *Anti Infect, Ther.* **6** (5), 657-669.

[15] Tham J (2012) Extended Spectrum Beta-lactamase Producing Enterobacteriaceae: Epidemiology, Risk Factors, and Duration of Carriage. Department of Clinical Sciences, Lund University. Citation for published version (APA).

[16] Paterson D, Bonomo R (2005) Extended Spectrum Beta-lactamases: A Clinical Update. *Clinical Microbiology Journal* **18** (4): 658-686.

[17] Severin JA, Mertaniasih NM, Kuntaman K, Lestari ES, Purwanta M, Lemmens-Den T, Duerink DO, Hadi U, Van-Belkum A, Verbrugh HA, Goesens WH (2010) Molecular characterization of extended-spectrum beta-lactamases in clinical *Escherichia coli* and *Klebsiella pneumoniae* isolates from Surabaya, Indonesia. *Journal Antimicrob Chemotherapy*. Study Group Antimicrobial Resistance in Indonesia: Prevalence and Prevention' (AMRIN). **65** (3): 9-465.

[18] Fatmawati N, Tarini N, Budayanti N, Yuliandari P (2015) Molecular Characterization of Extended-Spectrum Lactamases-Producing *Klebsiella pneumonia* Isolated from Clinical Specimens at a Tertiary-Referral Hospital in Denpasar, Bali, Indonesia. *American Scientific Publishers.* **21** (2) : 219–221.

IOP Conf. Series: Earth and Environmental Science **130** (2018) 012033 doi:10.1088/1755-1315/130/1/012033

Production of litter and detritus related to the density of mangrove

Miswar Budi Mulya[1*] Arlen HJ[1]
[1]Department of Biology, Faculty of Mathematics and Natural Sciences, University of Sumatera Utara, Medan, Indonesia

*Email: miswarbm_ikl@yahoo.com

Abstract. Research about the production of leaf litter and detritus related to the density of mangrove trees has been done. The aims of this research are to know and analyze the amount of litter and detritus produced to the density of mangrove trees. The production and collection of leaf litter were carried out in five stations. Production of detritus and decomposition rate were calculated by measuring its dry weight. The density and level of mangrove trees were determined using transect quadratic method. The relationship between the leaf litter and detritus production ratio related to mangrove density were then analyzed. Results showed that mangrove trees with the density of 766.67 ind ha^{-1} ccould produce the amount of litter and detritus to about 28597.33 $gha^{-1}day^{-1}$and 1099.35 $gha^{-1}day^{-1}$ while mangrove trees with the density of 1300 ind ha^{-1} could produce the amount of litter and detritus to about 35093.33 g/ha/day and 1216.68 $gha^{-1}day^{-1}$ respectively. Data analysis showed that the increment of mangrove density is linearly related to the production increment of litter and detritus.

Keywords: Decomposition, organic material, nutrients, Karang Gading Wildlife Reserves

1.Introduction

Mangrove forest of Karang Gading Wildlife Reserve is located on the east coast of Sumatera Utara with the width of about 15765 Ha. Mangrove forest is very important to maintain the biological cycle balancing of mangrove trees shed. They store a great amount of organic matter in the form of leaf litter into tropical waters environments. Mangrove forest is a productive ecosystem with relatively high productivity. The high productivity is lineraly related to the rapid leaf litter production and efficient recycling of nutrients. Mangrove leaves play an important role as the largest contributor to the total production in the mangrove ecosystem [1]. The production of leaf litter plays an important role to the transfer of organic material. Only less than 10% of leaf litter degradation produced by mangrove trees tend to be consumed by living organisms [2].

Mangrove leaves litter is partially decomposed by bacteria and fungi into dissolved nutrients. Nutrients produced are useful for the growth of mangroves as well as a source of detritus which will then be exploited by aquatic organisms. Mangrove ecosystem is a habitat with abundant availability of food in the form of detritus [3]. Detritus is the basis of food network. There is a significant correlation between the production of litter and detritus with the density of mangrove trees. Information on the relationship between the production of litter and detritus with the density of mangrove trees in Karang Gading Wildlife Reserves, Sumatera Utara has not been well-studied.

2. Methods

The study was conducted in July - October 2016. Mangrove density was measured at the level of the tree, using the transect quadratic methods in five stations (Figure 1). Three plots with the size of 10 m x 10 m were created, and the leaf litter production was measured by collecting leaf litter from each station using 45 litter traps installed randomly under the tree canopy (1.5 m above ground level). The leaf litter which was collected in the traps were taken every 15 days for 3 months period. Total dry weight of the leaf litter shows the value of litter production. The litter decomposition rate is measured every 15 days up to 75 days by collecting and putting 10 grams of the wind-dried leaf litter from the traps into a litter bag tied at the base of the mangrove stem not to be drawn tidal currents.

The leaf litter decomposition rate is known by comparing the dry weight of the litter. Production of detritus known from the rest of the decomposition after the time of observation. The litter production, litter decomposition rate, and detritus production are analyzed using a formula [4], while the density of mangrove trees is also analyzed [5].

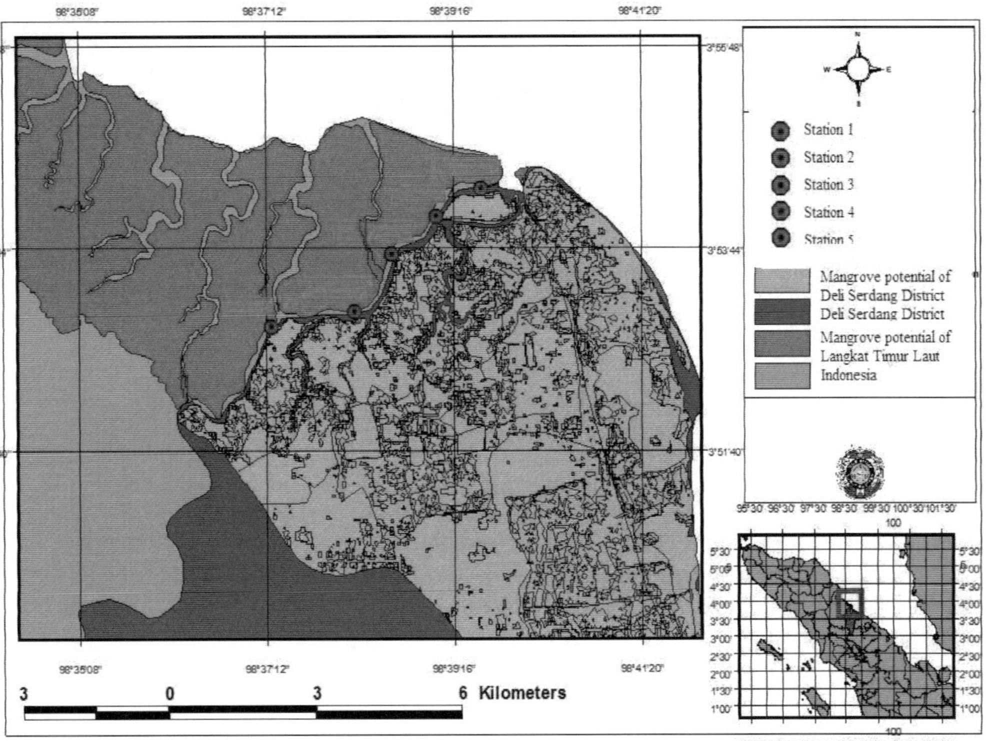

Figure 1. Map of Research

3. Results and Discussion

3.1. Mangrove Density

Twelve mangrove species from eight families were found during the study. The highest mangrove trees density at station 1 were from *Avicennia marina* (400 ind ha^{-1}), with lowest density from *Lumnitzera racemosa* (33.33 ind ha^{-1}). The highest density in station 2 was found from *Avicennia alba* (300 ind/ha), while the lowest mangrove trees density were found from *Bruguiera gymnorrhiza* (33.33 ind ha^{-1}). The overall mangrove density is presented in Table 1 below.

Table 1. Mangrove density (ind/ha) in each station

No	Spesies	Station				
		1	2	3	4	5
1	*Avicennia alba*	233.33	200.00	66.67	33.33	33.33
2	*Avicennia marina*	400.00	133.33	33.33	33.33	0.00
3	*Bruguiera gymnorrhiza*	66.67	33.33	100.00	66.67	33.33
4	*B. parviflora*	66.67	66.67	166.67	100.00	66.67
5	*Excoecaria agallocha*	66.67	66.67	66.67	100.00	66.67
6	*Hibiscus tiliaceus*	0.00	0.00	66.67	66.67	66.67
7	*Lumnitzera racemosa*	33.33	33.33	33.33	66.67	33.33
8	*Nypa fruticans*	0.00	0.00	0.00	0.00	233.33
9	*Rhizophora mucronata*	100.00	66.67	100.00	366.67	133.33
10	*R. apiculata*	100.00	100.00	100.00	300.00	66.67
11	*Sonneratia alba*	166.67	300.00	33.33	33.33	0.00
12	*Xylocarpus granatum*	66.67	66.67	33.33	33.33	33.33
	Total	1300	1066.67	800	1200	766.67

Hibiscus tiliaceus and *Nypa fruticans* were not found in station 1 and 2 due to the reason that both stations were located near the shore with higher salinity and pH, in which did not support the survival of both species. The highest mangrove tree density in station 3 was found from *B. parviflora* (166.67 ind/ha), while *A. marina, L. racemosa, S. alba,* and *X. granatum* were found to have the lowest density with 33.33 ind/ha.

The highest mangrove tree density in station 4 was found from *R. mucronata* (366.67 ind/ha) while the lowest was from *A. alba, A. marina, S. alba,* and *X. granatum* (33.33 ind/ha). The high density of *R. mucronata* was due to the sheltered location of station 4. Fresh water from mainland was also periodically supplied through the flow of the river together with the muddy substrate, in which they support growth. *N. fruticans* has the highest density at station 5 (233.33 ind/ha), while the lowest density was in *A. alba, B. gymnorrhiza, L. racemosa*, and *X. Granatum* with 33.33 ind/ha.Both *A marina* and *S.alba* were not found in the station 5, due to the location of the station that was situated near mainland with a muddy substrate, in which they barely support the growth of these species of mangrove. The highest density of mangroves was found in Station 1 (1300 ind/ha), followed by station 4 (1200 ind/ha), station 2 (1066.67 ind/ha), the station 3 (800 ind/ha), and station 5 (766.67 ind/ha). The low density of mangroves at stations 3 and 5 were due to the location of the stations which were situated near farm settlements, in which the possibility of degradation was higher, compared to the other three stations.

3.2. Litter Production

Litter production is an important part of the organic matter transfer process from mangrove vegetation to the soil. Production of mangrove leaves litter was different on each station. The measurement during the 75 days of observation is presented in Figure 2. Leaf litter production was the highest in station 1, with 35093.33 $gha^{-1}day^{-1}$, and the lowest was at station 5, with 28597.33 $gha^{-1}day^{-1}$. The high production of leaf litter in station 1 was due to highest density of mangrove trees in this station. Mangrove density has a positive correlation with the production of litter. Mangrove trees in natural conditions are capable of producing 5% more litter than the degraded mangrove [1]. Station 1 is located nearer to the shore. The rapid incase of leaf litter has a positive correlation with temperature and water salinity. Temperature and high salinity can cause leaf litter of mangrove to be easily abscised. The rising of temperature can reduce the humidity, in which will cause the mangrove transpiration to increase. In order to reduce it, mangrove leaves should be abscised. High temperature and salinity on the water surface (29.30%) accelerate litter fall by posing a physiological stress to mangrove species and to overcome this, mangrove leaves fall [6].

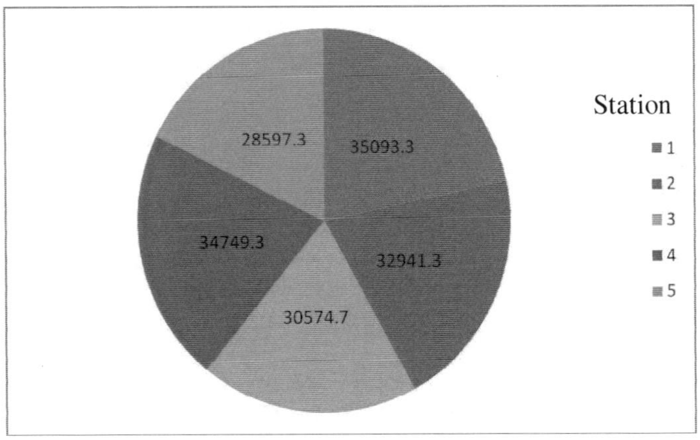

Figure 2. Litter production ($gha^{-1}day^{-1}$).

Station 4, which also has high litter production, was dominated by *R. apiculata* and *R. mucronata*. Both of these mangrove species have high litter production, due to the form of the leaves that are easily invalidated by wind and rain. Mangrove forests of Indian Sundarbans had produced the total litter production ranges from 3.312t ha^{-1} yr^{-1} to 5.874 t ha^{-1} yr^{-1} [6], showing a higher result compared to the litter production in mangrove forest of Karang Gading Wildlife North Sumatra.

3.3. Litter Decomposition

The first station has a higher litter decomposition rate than other stations (Table 2), typically because this station is an area that is inundated by a high or low tide, which caused it to have relatively higher salinity and temperature.

Table 2. Rate of litter decomposition (g/day) in each station

Station	Days to					Average	%
	15	30	45	60	75	($gday^{-1}$)	
1	0.0326	0.0239	0.0214	0.0347	0.0325	0.0290	24.60
2	0.0255	0.0215	0.0172	0.0214	0.0250	0.0221	18.77
3	0.0252	0.0205	0.0162	0.0208	0.0244	0.0214	18.18
4	0.0295	0.0229	0.0192	0.0257	0.0261	0.0247	20.96
5	0.0239	0.0200	0.0155	0.0204	0.0232	0.0206	17.48

The average rate of litter decomposition in each station was ranging from either 0.0206 $gday^{-1}$ to 0.0290 $gday^{-1}$ or 17.48% day^{-1} to 24.60% day^{-1}. Results showed that litter decomposition rate was higher in the first 15 days, and the value of percent residual litter was showing almost the same result.

Percentage of residual leaf litter during the decomposition process in each station (Figure 3), showed that the remaining litter ranges from 61.37% to 69.88% after 15 days of immersion, followed by a decrease after 30 days of immersion (48.89% to 54.81%), the remaining on day 45 was from 38.25% to 49.89%, while the remaining on day 60 and 75 were from 12.47% to 29.46%

and 8.75% to 17.55% respectively. This condition is because that leaf litter experiences a loss of organic material that was dissolved by rainwater, leaching, and decomposer. The degradation of mangrove litter in an aquatic environment is totally dependent on the microbial biomass and environmental conditions [7]. Furthermore, the litter decomposition occurrence caused by leaching of organic compounds should go through three phases: the oxidation of cellulose and lignin by microbes, the physics and then followed by biology fragmentation [8]. The presence of bacteria and fungi in mangrove leaf litter decomposition process could cause a rapid leaching on organic matter [9].

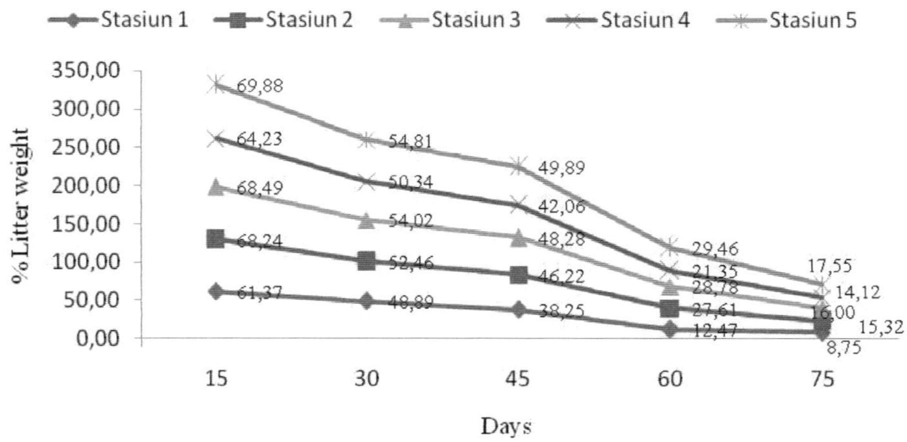

Figure 3. Percent residual litter in each station

3.4. Detritus Production

The highest detritus production was found at station 1 (0.122 $gm^{2-1}day^{-1}$ or 1216.68 $gha^{-1}day^{-1}$),while the lowest was found at station 5 (0.110 $gm^{2-1}day^{-1}$ or 1099. $gha^{-1}day^{-1}$)(Table 3).

Table 3. Detritus production in each station

Station	Days to					Average	Average
	15	30	45	60	75	$gm^{2-1}day^{-1}$	$gha^{-1}day^{-1}$
1	3.86	5.11	6.17	8.75	9.13	0.122	1216.68
2	3.18	4.75	5.38	7.24	8.47	0.113	1129.09
3	3.15	4.60	5.17	7.12	8.40	0.112	1119.94
4	3.58	4.97	5.79	7.86	8.59	0.115	1145.10
5	3.01	4.52	5.01	7.05	8.25	0.110	1099.35

The high production of detritus at Station 1 was due to the high litter production and decomposition rate compared to other stations. Results showed detritus production positively correlated with the production of litter. The higher production of litter, the higher the production of detritus will be.

3.5. Production of Litter and Detritus Related toMangrove Density

Results showed that litter and detritus production positively correlated with the density of the mangrove. The higher density of the mangrove, the higher litter and detritus will be produced (Table 4).

Table 4 . Production of litter and detritus relationship with Mangrove Density

Station	Mangrove density ind ha^{-1}	Litter production g ha^{-1} day^{-1}	Detritus production g ha^{-1} day^{-1}
1	1300	35093.33	1216.68
2	1067	32941.33	1129.09
3	800	30574.67	1119.94
4	1200	34749.33	1145.10
5	767	28597.33	1099.35

Station 1 with the density of mangrove 1067 ind ha^{-1} can produce as much litter 35093.33 g ha^{-1} day^{-1} and detritus 1216.68 g ha $^{-1}$ day^{-1}, dominated by *Avicennia marina*. Station 2 with a density of mangrove 1300 ind ha^{-1} can produce as much litter 32941.33 g ha^{-1} day^{-1} and detritus 1129.09 g ha^{-1} day^{-1}, dominated by *Sonneratia alba*. Station 3 with a density of mangrove 800 ind ha^{-1} can produce as much litter 30574.67 g ha^{-1} day^{-1} and detritus 1119.94 g ha^{-1} day^{-1}, dominated by *Bruguiera parviflora*. Station 4 with a density of mangrove 1200 ind ha^{-1} can produce as much litter 34749.33 g ha^{-1} day^{-1} and detritus 1145.10 g ha^{-1} day^{-1}, dominated by *Rhizophora mucronata*, and Station 5 with a density of mangrove 767 ind ha^{-1} can produce as much litter 28597.33 g ha^{-1} day^{-1} and detritus 1099.35 g ha^{-1} day^{-1}, dominated by *Nypa fruticans*.

4. Conclusion

1) Litter production in In Karang Gading Wildlife Reserves Sumatera Utara ranged from 28597.33 g ha^{-1} day^{-1} to 35093.33 g ha^{-1} day^{-1} while the detritus production ranged from 1099.35 g $ha^{-1}day^{-1}$ to 1216.68 g ha^{-1} day^{-1}.The density mangrove forest observed was from 766.67 ind ha^{-1} to 1300 ind ha^{-1}.
2) Mangrove trees with the density of 1300 ind ha^{-1} can produce as much litter of 35093.33 g ha^{-1} day^{-1}, with detritus production 1216.68 g ha^{-1} day^{-1}. Mangrove trees with the density of 1067 ind ha^{-1} can produce as much litter of 32941.33 g ha^{-1} day^{-1}, with detritus production 1129.09 g ha^{-1} day^{-1}. Mangrove trees with the density of 800 ind ha^{-1} can produce as much litter of 30574.67 g ha^{-1} day^{-1}, with detritus production 1119.94 g ha^{-1} day^{-1}. Mangrove trees with the density of 1200 ind ha^{-1} can produce as much litter of 34749./33 g ha^{-1} day^{-1}, with detritus production 1145.10 g ha^{-1} day^{-1}, and Mangrove trees with the density of 767 ind ha^{-1}, can produce as much litter of 28597.33 g ha^{-1} day^{-1}, with detritus production 1099.35 g ha^{-1} day^{-1}

Acknowledgments

The authors gratefully acknowledge the financial support by University of North Sumatra (BPPTN Research grant), Indonesia

References

[1] Zhila H, M Hossain and MZ Rozainah (2013) Litter Production in Both Natural and Degraded Mangrove Forest of Peninsular Malaysia. International Conference on Materials, Architecture and Engineering Technology (ICMAET), pp 6.
[2] Bosire JO, FD Guebas, JG. Kairo, J Kazungu, F Dehairs, N Koedam (2005) Litter degradation and CN dynamics in reforested mangrove plantations at Gazi Bay, Kenya. Biological Conservation 126: 287–295
[3] Reichard M, Jurajda, Simkova I, Matejusova (2002) Size-Related Habitat Use By Bitterling (Rhodeussericeus) in A Regulated Lowland River. Ecology of Freshwater Fish 11:112–122.
[4] Ashton E, PJ Hogarth R, Ormond (1999) Breakdown of Mangrove Leaf Litter in A Managed Mangrove Forest in Peninsular Malaysia. Hydrobiologia 413: 77-88.

[5] Bengen DG (2002) Introduction Technical Guidelines, and the Mangrove Ecosystem Management. Center for Coastal and Marine Resource Studies. Bogor Agricultural Institute. Bogor.

[6] Ghosh R and K Banerjee (2013) Inter-relationship between Physico-chemical Variables and Litter Production in Mangroves of Indian Sundarbans. Journal Marine Science Research and Development.1-4

[7] Mumby, PJ, AJ Edwards, Arias-J.E.Gonzalez, KC, Lindeman, PG. Blackwell A Gall, M Gorczynska, AR Harborne, CL Pescod, H Renken, CC Wabnitz and G Llewellyn (2004) Mangroves enhance the biomass of coral reef fish communities in the Caribbean Nature 427: 533-536.

[8] Kathiresan K, K Saravanakumar, R Anburaj, V Gomathi, G Abirami, SK Sahu and S Anandhan (2011) International Journal of Advanced Biotechnology and Research 2 (3): 382-389

[9] Chale FMM (1993) Degradation of Mangrove Leaf Litter Under Aerobic Conditions. Hydrobiologia. 257: 177 - 183.

Effect of *lawsonia innermis* (linn) leaves ethanolic extract on blood glucose and malondialdehyde level in alloxan-induced diabetic rats

Mutiara Indah Sari [1*], Syafruddin Ilyas[2], Tri Widyawati[3], Maya Anjelir Antika[4]

[1]Department of Biochemistry, Faculty of Medicine, Universitas Sumatera Utara Jl. Dr. Mansur Kampus USU Medan 20155, Indonesia
[2] Department of Biology, Faculty of Mathematics and Natural Science, Universitas Sumatera Utara Jl. Bioteknologi Kampus USU Medan 20155, Indonesia
[3]Department of Pharmacology and Therapeutic, Faculty of Medicine, Universitas Sumatera Utara Jl. Dr. Mansur Kampus USU Medan 20155, Indonesia
[4] Biomedical Master Program, Faculty of Medicine, Universitas Sumatera Utara Jl. Dr. Mansur Kampus USU Medan 20155, Indonesia

*Email: muti_dr@yahoo.com

Abstract The case of diabetes mellitus (DM) tends to increase worldwide. DM triggers the oxidative stress condition that caused by the increasing of free radical. The present study was conducted to evaluate the effect of giving ethanolic extract of *Lawsonia inermis* (Linn) leaves to the glucose and malondialdehyde (MDA) level in alloxan-induced diabetic Wistar male rats. The powder of dry leaves of L.inermis was macerated in ethanol 96% to obtain ethanolic extract (LLEE).Thirty five of rats were divided into five groups, ie. K (normal and given 0.9% NaCl solution), P1-P4 were induced using alloxan (120 mg/kg) intraperitoneally to get diabetic condition. Diabetic rats then were treated as follows: P1 (given 0.9% NaCl solution) P2 (LLEE (200 mg/kg BW), P3 (LLEE (400 mg/kg BW)), P4 (LLEE (600 mg/kg BW)). All groups were treated for 28 days. The fasting blood glucose levels were measured at day 1, 7, 14, 21, 28 whereas MDA levels were measured at the end of treatment. The result showed that LLEE improved blood glucose level (BGLs) of alloxan-induced diabetic rats significantly ($p< 0.001$) at 29th day. However, LLEE could not affect the MDA level ($p > 0.5$). The study concluded that LLEE have antihyperglycemic properties.

Keywords: Diabetes Mellitus, LLEE, BGLs, MDA

1. Introduction

Diabetes mellitus (DM) is a metabolic disorder in the body. The symptom of DM was elevated blood glucose levels because of abnormalities of metabolic pathways of carbohydrates, fats and proteins. It caused by insulin secretion abnormalities, insulin action, or both. Based on the data from the World Health Organization (WHO) the number of people with diabetes mellitus will increase continuously in the future. Around 415 million people were estimated suffering from diabetes mellitus in 2015. This number will keep continuously increase over the last three decades and will continuously be estimated to be 642 million people with diabetes mellitus by 2040 [1]

In 2015, 5 million people aged between 20-79 years died from diabetes mellitus. The aggregate accounts for 14.5% of total deaths in that group age due to various causes [2]. In Indonesia, Diabetes mellitus is increased from 2007 by 1.1% to 2.1% in 2013 [3]. Based on the pattern of population growth, it is estimated that by 2030 there will be 194 million people aged over 20 whose will become DM [4].

The condition of hyperglycemia in people with diabetes mellitus will lead to auto oxidative glucose reactions, protein glycation and activation of polyol metabolism pathways that will accelerate the formation of free radicals (oxidants). Oxidants were reactive compounds that can move electrons from other molecules and produce oxidation in the molecules [5]. Lipids that have double carbon chains can react with oxidants, this process was called lipid peroxidation. Lipid hydroperoxide was the main product of lipid peroxidation process . The lipid hydroperoxide structure was not stable and easy to be malondialdehyde (MDA), 4-hydroxy-2-nonenal (4-HNE), and several other aldehyde forms. The MDA profile in serum serves as a marker (marker) of cellular damage caused by oxidants [6]. Giving antioxidants can bond free radicals.

There were two antioxidants, namely; endogenous and exogenous antioxidants. Endogenous antioxidants come from within the body itself, consisting of Super Oxide Dismutase (SOD), Glutathione Peroxidase (GPx), Catalase, etc. Exogenous antioxidants were obtained from the outside through the foods that we consumed to help the body fight the excess free radicals in the body. Enhanced supply of antioxidants will help prevent the clinical complications of DM [7].

Exogenous antioxidants can be obtained from the plants which rich in antioxidants such as *Lawsonia inermis* (Linn). One part of *Lawsonia inermis* (Linn) plant that commonly used as medicine was its leaves. Phytochemical screening of its leaves demonstrated the present of a phenolic natural antioxidant compound (coumarin, flavonoid, naphthalene, lawsone, gallic acid derivatives) and glycosylated protein [8]. Research from Chikaraddy et al. [9], showed that the ethanolic extract *Lawsonia inermis* (Linn) at three doses i.e. 150, 200 and 400 mg/kg BW for three weeks administration able to decrease the blood glucose levels of diabetic rats. Thus, the decreasing of blood glucose levels will minimize the risk of oxidative stress in cells and tissues [10].

So far, no data has been reported about the correlation effect of *Lawsonia inermis* (Linn) ethanolic extract/ LLEE on blood glucose level and oxidative stress level compound MDA in alloxan-induced diabetic rats model. Therefore, the present study was conducted to investigate its activities on those markers.

2. Methods

2.1 *Research design*

The study was experimental analytical study using Post Test Control Group Design with the preclinical trial.

2.2 *Place of Research*

The study was conducted in Integrated Laboratory, Faculty of Medicine, Chemistry Laboratory of Faculty of Mathematics and Natural Sciences (FMIPA) and Biology Laboratory of FMIPA, Universitas Sumatera Utara (USU), in March 2017. The study was conducted after obtaining ethical clearance No 116/KEPH-FMIPA/2017 from the Ethics Committee of the Mathematics and Natural Sciences Faculty (MIPA) Biology USU Medan.

2.3 *Lawsonia inermis Ethanolic Extract preparation* [11].

The fresh leaves were obtained from Medan Denai area and were dried in temperatue room. The dried leaves then were blended in machine into powder. The extraction was done by maceration method. The powder was macerated with ethanol 96% (1:10 (w/v)). After 5 days the extract was filtered, the pulp was squeezed and washed with enough liquid to obtain 100 parts. Maserate was transferred into a closed vessel, left in a cool place that was protected from sunlight for 2 days. The concentrated extract was obtained using evaporator and then was kept in the freeze until used.

2.4 *Experimental animals*
The 35 males of healthy Wistar rats (*Rattus norvegicus*), ranging 150-200 g, and have never been used for any experiments previously.

The rats were individually stacked in the FMIPA Biology Laboratory and maintained at room temperature 25 ± 5 ° C. Giving sufficient food and beverage according to the treatment group with a 12 hour light cycle and 12 hours of dark cycle. After one week of acclimatization, the rats were divided into 5 groups randomly. Based on the sampling formula , the number of Wistar rat in each group was 5 rats. To avoid the risk of drop out of the Wistar rat, then 2 rats were added in each group [12], namely:

a Group I (K), negative control (normal) was not given any treatment, just like normal rats in general given feed and drink (*ad libitum*) in the cage.
b Group II (P1), Positive control (diabetic) were feed and drink (*ad libitum*) in the cage
c Group III (P2): LLEE 200 mg/kg BW, orally, for 28 days.
d Group IV (P3): LLEE 400 mg/kg BW, orally, for 28 days.
e Group V (P4): LLEE 600 mg/kg BW, orally for 28 days

The diabetic condition was obtained by injecting the rats using alloxan (120 mg/kg BW) in 0.1M citric buffer intraperitoneally (P1-P4) [13]. The blood glucose levels (BGL) were measured after 7 days using glucose meter (GlucoDr^{TM}). Rats with BGL of 250 mg/dl above were included in the study[14]

After the rats have BGL >250mg/dl , group P2-P4 were given extract ethanol of Lawsonia inermis Linn leaves in each 200 mg/kg BW, 400 mg/kg BW, 600 mg/kg BW [9][10]. The fasting blood was taken from rats tail were conducted by the using glucose test strips at day1, 7, 14, 21 and 28. After 28 days (4 weeks), the rats were sacrificed by surgical dislocation of the neck after light chloroform anesthesia. The 12 hours fasting blood of the rats was taken 3 ml from the heart and blood serum was stored at -80°C before examined.

.

2.5 *Determination of MDA Level*
The examination of blood MDA levels used Spectronic 21 with a commercial kit and read with at wavelength 532 nm.

2.6 *Process and Data Analysis*
All data were expressed as mean ± standard deviation (SD). Mean differences between the groups were statistically analyzed by one-way analysis of variance (ANOVA) at the alpha value of 0.05 followed by Post Hoc Test. Mean differences between pre-treatment and post-treatment were analyzed by paired *t*-test. Differences were considered significant at $p < 0.05$

3. Results and Discussion
The present study is evaluate the BGL of rats in several times i.e. before and after alloxan induction, at P1-P4 groups.

.

Table 1. The BGL fasting of rats before and after induced alloxan 120mg/kg BW

Group	Before (mg/dl)	After (mg/dl)	p*
P1	152±17.08	350±100.92	0.000
P2	119±14.57	440±21.22	
P3	112±14.50	487 ±101.79	
P4	105±13.21	426±149.46	

* *paired* t test

From the table 1 can be seen the BGL before and after induced alloxan 120mg/kg BW. The paired t-test of P1 –P4 showed the correlation of the BGL before and after alloxan induction ($p < 0.001$). Alloxan compounds were one of the most toxic diabetogenic substances, especially to pancreatic beta cells, and when given to experimental animals such as rats can cause diabetic to the rats. The alloxan toxicity mechanism begun with the entry of alloxan into the pancreatic β cells and the retrieval rate will determine the alloyan diabetogenic properties. The damaging of to β cells occurs through several processed simultaneously, through the oxidation of sulfhydryl groups and the formation of free radicals. The mechanism of action of the alloxan produces damage to pancreatic β cells primarily invading cellular compounds containing sulfhydryl groups, amino acids cysteine and proteins binding to SH groups (including enzymes containing SH groups). The alloxan reacts with two SH groups binding to the sides of proteins or amino acids to form disulfide bonds to inactivate proteins that result in impaired protein function [13].

Giving extract ethanol of Lawsonia inermis Linn leaves was done starting on the first day. The mean value of fasting blood glucose levels of animals experiment after given extract ethanol of Lawsonia inermis Linn leaves on day: 7, 14, 21, and 28 this time can be seen in table 2.

Table 2. The blood glucose level of alloxan-induced diabetic rats on day 7,14, 21, 28 after *Lawsonia inermis* (Linn) ethanolic extract administration

Group	The mean value of fasting blood glucose levels of rats (mg/dl)			
	Day 7	Day 14	Day 21	Day28
K	155± 9.16	81± 2.30	88 ± 19.00	112 ± 25.05
P1	379± 18.58	250±100.92	228 ± 64.08	309 ± 71.14
P2	421 ± 30.51	399±46.18	380.66 ± 55.89	450 ± 93.66
P3	402 ± 66.77	341± 41.52	253.66 ± 21.22	236 ± 66.57
P4	351 ± 2.30	359.66 ± 95.82	270.33 ± 31.64	256 ± 41.25

Table 2 showed that P1-treated group on day 7 still hyperglycemia. On day 14 and day 21 the BGLs of this group were decreased then tended to increase at day 28. Meanwhile, the decreasing of BGLs were demonstrated on P2-, P3-, and P4-treated group at day 7, 14, 21 and 28 measurements. Effective decreasing can be seen in giving the ethanolic extract of *Lawsonia inermis* (Linn) leaves 400mg/dl dose.
To determine the significance of the difference between BGLs on day 28 in all groups were done by ANOVA test with average blood glucose levels in each group that can be seen in table 3

Table 3. The mean blood glucose on all groups on day 28

Group	The Mean Blood Glucose (mg/dl)	p*
K	107.33	0.000
P1	196.00	
P2	119.67	
P3	112.33	
P4	105.67	

*ANOVA test

In this study there was an association between the extract ethanol of *Lawsonia inermis* (Linn) leaves with the decreasing of blood glucose levels in the group of experimental rat ($p<0.05$).
Lawsonia inermis (Linn) is native to tropical and subtropical plants such as Asia, South Africa, East Africa, North Africa, Northern Australia, which naturally grow also in tropical regions. Leaves are

small, greenish brown to the dull green, opposite in arrangement along the branches with short petioles, about 1.5-5cm long and 0.5cm wide to 2 cm. The main component of the leaf is a dye substance known as lawsone, 2-hydroxy-1,4-napthhoquinone-$C_{10}H_6O_3$ [16].

Based on phytochemical screening, the leaves of *Lawsonia inermis* (Linn) contained phenolic compounds (coumarins, flavonoids, naphthalene and gallic acid derivatives) glycosylation proteins. Flavonoids can reduce the activity of hydroxy radicals, superoxide anions and peroxide radicals by protecting lipid membranes against damaging oxidation reactions. Giving the antioxidants can increase the mass of pancreatic β cells and maintain insulin content in it [8].

Based on Choubey's et al. research [17], concluded that giving extract ethanol of Lawsonia inermis Linn leaves 500 mg/kg BW for 28 days effectively decreased blood glucose levels of rats. Research from Chikaraddy et al. [9], stated that the extract ethanol of *Lawsonia inermis* (Linn) leaves doses 150 mg/kgbw, 200 mg/ kg BW and 400 mg/kg BW for 3 weeks can decrease BGLs of albino rats. The decreasing blood glucose levels will decrease the risk of oxidative stress in cells and tissues [10].

Oxidative stress of rat's tissue in this study can be seen by measuring of rat's blood levels. In this study the ratio of blood MDA levels of rats in each group before and after giving extract ethanol of *Lawsonia inermis* (Linn) leaves can be seen in Table 4.

Table 4. The blood MDA level of rats on day 28

Group	MDA (μmol/gram)	P*
K	2.58	0.398
P1	2.74	
P2	2.43	
P3	2.31	
P4	2.40	

*Anova test

The increasing of MDA levels was found in alloxan-induced rats group at P1. The MDA level P2- was lower than K- and P1-treated groups. The P3-treated group showed the lowest level of MDA compared to the entire group. Normality test of MDA data of the present study showed p = 0.513, then one-way Anova analyzed result $p>0.05$ that concluded there was no association of giving LLEE to MDA level in the present study.This result was different from the previous study that the composition of flavonoid from the herbs able to decrease the level of MDA rats which was induced oxidative stress [18,19].

MDA is one of the product compounds of lipid peroxidation product used as a marker of oxidative stress. In high oxidative stress, there was a significant increase in serum MDA levels. The damaging to cells or tissues due to increased MDA in oxidative stress conditions can be slowed by the administration of antioxidants. Anti-oxidants are vital substances that can protect the body from free radical attack by neutralizing or reducing its negative effects. There are several antioxidant groups of endogenous antioxidants and exogenous antioxidants.

The flavonoid compounds in the leaves can not act as natural exogenous antioxidants with the ability to eliminate the free radical compounds in the body to inhibit the reaction of free radical in the lipid oxidation [8]. Thus, the mosaic blood level of alloxan-induced rats can be suppressed.

In the present study, although there was decreasing of MDA levels in the treatment group that received LLEE (P2-P4), statistically no significance. This result could be affected by the duration of LLEE administration that should be more than 28 days. The assume was supported by the decreasing of the BGLs that have not reached to normal level on the 28th days of the treatment.

4. Conclusions

Ethanolic extract of *Lawsonia inermis* (Linn) leaves at dose 400mg/kg BW able to decrease blood glucose level of alloxan-induced diabetic rats. MDA levels were not improved after LLEE administration.

Acknowledgments

The authors gratefully acknowledge that the present study is supported by Universitas of Sumatera Utara.

References

[1] World Health Organization (2016) Global Report on Diabetes, Isbn, vol. 978, p. 88.

[2] International Diabetes Federation (2015) IDF Diabetes Atlas, 7th edn. Available at: www.diabetesatlas.org.

[2] Riskesdas (2013) Kementerian kesehatan. Riset kesehatan Dasar.

[4] Perkumpulan Endokrinologi Indonesia (2015) Pengelolaan dan Pencegahan Diabetes Melitus Tipe 2 di Indonesia. PB. PERKENI.

[5] Ayala A, Munoz MF, Arguelles S (2014) Lipid peroxidation: production, metabolism and signaling mechanism of malondialdehyde and 4-hydroxy-2-nonenal. Oxid Med Cell Longev . 112: 21-8.

[6] Grotto D, Maria LS, Valentini J, Paniz C, Garcia GS, Pomblum VJ (2009) Importance of the lipid peroxidation biomarkers and methodological aspects for malondialdehyde quantification. Quimm Nova. 32(1): 169-74.

[7] Ganjifrockwala FA, Joseph JT and George G (2017) Decreased total antioxidant levels and increased oxidative stress in South African type 2 diabetes mellitus patients. Journal of Endocrinology, Metabolism and Diabetes of South Africa. 1(1):1–5

[8] Hsouna A, Ben T M, Culioli G, Blache Y, & Jaoua S (2011) Antioxidant constituents from Lawsonia inermis leaves: Isolation, structure elucidation and antioxidative capacity. Food Chemistry, 125(1), 193–200. https://doi.org/10.1016/j.foodchem.2010.08.060

[9] Chikaraddy A, Maniyar Y, & Mannapur B (2012) Hypoglycemic Activity of Ethanolic Extract of Lawsonia inermis Linn. (Henna) in Alloxan Induced Diabetic Albino Rats. International Journal of Pharmacy and Biological Sciences, 2(4), 287–292.

[10] Stangeland T, Remberg S F & Lye K A (2009) Total antioxidant activity in 35 Ugandan fruit and vegetables. Food Chemistry, 113(1), 85–91. https://doi.org/10.1016/j.foodchem.2008.07.026

[11] Chaudary G, Goyal S, Poonia P (2010) Lawsonia inermis Linnaeus: A Phytopharmacological Review. IJPSDR. 2(2):91-98

[12] Federer W(1991) Statistics and society: data collection and interpretation. 2nd ed. New York: Marcel Dekker

[13] Chougale AD, Panaskar SN, Gurao PM, Arvindeka AU (2007) Optimization of alloxan dose is essential to induce stable diabetes for prolong period. [cited 2016 August 10]. Available from: http://sciarlet.net/fulltext/?doi=ajb2007.402.408

[14] Nugroho AE (2006) Animal Models of Diabetes Mellitus : Pathology and Mechanism of Some Diabetogenics. Biodiversitas, Journal of Biological Diversity, 7(4), 378–382. https://doi.org/10.13057/biodiv/d070415

[15] Inawati, Syamsudin dan Winarno H (2007) Effect of henna leaves (Lawsonia inermis Linn) on decreased blood glucose, total cholesterol and trygliceride mice-induced alloxan. Jurnal Kimia Indonesia Vol. 2(1)

[16] Ojewunm O, Oshodi T, Ogundele O, Micah C, Adenekan S (2013) In vitro Antioxidant, Antihyperglycaemic and Antihyperlipidaemic Activities of Ethanol Extract of Lawsonia inermis Leaves. British Journal of Pharmaceutical Research, 4(3), 301–314.

[17] Choubey, A. et al. (2010). Hypoglycemic and Antihyperglycemic Effect of Ethanolic Extract of Whole. International Journal of Pharmaceutical Sciences and Research, 1(8), 74–77.

[18] Sari W M, Wahdaningsih S, Untari E K (2014) Efek Fraksi n-Heksana Kulit Hylocereus polyhizus Terhadap Kadar Malondialdehida Tikus Stres Oksidatif. Pharm Sci Res. Dec . Vol. 1 No. 3

[19] Gwarzo1 M Y, Ahmadu J H, Ahmad M B ,Dikko3 A U A (2014) Serum Glucose and Malondialdehyde Levels in Alloxan Induced Diabetic Rats Supplemented with Methanolic Extract of Tacazzea Apiculata. Int J Biomed S 236 ci vol. 10 no. 4

The Addition of Zeolite Adsorbents and Calcium Oxide on Purification of Bioethanol from Sugar Palm (*Arenga pinnata* Merr)

Netti Herlina*, **Ici Siska Dewi Harahap**

[1]Environmental Engineering, Faculty of Engineering, University of Sumatera Utara, Medan 20155, Indonesia

*Email: cecesrg@yahoo.com

Abstract. Bioethanol (C_2H_5OH) is a biochemical liquid produced by microorganisms through fermentation process on sugar molecules from carbohydrates. Bioethanol is a fuel of vegetable oil that has similar properties to premium. With its main product of palm juice, Sugar palm (*Arenga pinnata*) is a potential source of sugar and carbohydrate for bioethanol production. Production of palm juice can reach up to 12-14 liters/tree/day with total sugar content in palm juice ranges from 12-15%. The purpose of this research was to produce highly-concentrated bioethanol from palm juice through fermentation proccess to subtitude fossil fuel. This study was conducted with three stages of treatment, namely: the fermentation of palm juice, distillation of bioethanol, and purification of bioethanol with the addition of adsorbent zeolite and calcium oxide.

Keywords: Adsorption, fermentation, sugar

1. Introduction

The rapid growth on industrial and world population these days have resulted in a greater demand for energy source. Approximately 90% of the energy required is obtained from non-renewable fuel sources, such as fossil fuel [1]. Recent study reported that the amount of fossil fuel exploited continue to increase annually, and thus leading to the depleting amount of the fossil fuels itself [2]. As a result for the declining amount of fuel supply, the researchers worldwide are now intensively seeking for renewable and environmentally friendly energy sources. Some sources of renewable energy that we have discovered are solar, water, geothermal, and biomass [2].

Bioethanol (C_5H_2OH) is a biochemical liquid with the same properties as gasoline. Bioethanol itself is a product of sugar fermentation from carbohydrates with the help of microorganisms [3]. There are two forms of raw materials used to produce bioethanol: solid and liquid. The solid form includes corn, wheat, cellulose, algae and bagasse, while liquid form includes the palm juice from coconut (*Cocos nucifera*), nipa (*Nypa fructicans*), and sugar palm (*Arenga pinnata*). Between the two forms, liquid like palm juice has advantages over the solid form, mainly because that they can be directly fermented without initial processing, in which means that the time needed to produce the bioethanol will be shortened. Solid-formed materials, on the other hand, must go through a process of hydrolysis to convert starch polymers into simple sugars, whereas cellulose must be hydrolised in order to obtain the sugar molecules [4].

One of the plants that is potential to be utilized as raw material for bioethanol production is Sugar palm (*Arenga pinnata*). Palm trees are known to have a high amount of sugar, and they grow widely in 14 provinces in Indonesia such as Papua, Maluku, North Sumatra, North Sulawesi, and South Borneo. Data from the Directorate General of Plantation in 2004 showed that the land area dominated by sugar palm reached 60,482 Ha. The sugar content of palm sugar is approximately 12-15%, while the sugar content found in coconut palm and siwalan palm is 10.27% and 10.96% respectively. This information proved that Sugar Plam has great potency to be utilized as raw material in bioethanol production [5].

Preparation of bioethanol from sugar palm was conducted through three stages of the process, that is, fermentation, distillation, and bioethanol purification process.During the fermentation process, sucrose is broken down into glucose and fructose; the glucose molecules are then converted into bioethanol by *Saccharomyces cerevisiae* yeast. Looking into details, the sugar molecule as substrate is firstly converted to pyruvic acid through glycolysis. Afterwards, pyruvic acid is then converted into acetaldehyde and carbon dioxide with the help of pyruvate decarboxylase enzyme, which is then followed by the convertion of acetaldehyde into alcohol in the presence of alcohol dehydrogenase. The purification of bioethanol are going through the principle of distillation and adsorption by using porous media to absorb water contained in the bioethanol [6]. However, in order to be used as a renewable fuel, bioethanol must have at least > 99% concentration and is anhydrous so as not to be corrosive and damage the vehicle parts [7].

2. Methods

2.1 Experimental apparatus

To perform the study, an experimental apparatus has been designed and developed including the reactor for the fermentation process, as well as distillation Apparatus. After bioethanol is repeatedly distilled, bioethanol is redistilled using calcium oxide. To be purified using natural zeolite, bioethanol must first soaked using H_2SO_4 for 2 hours, and then soaked with natural zeolite for 5 hours.

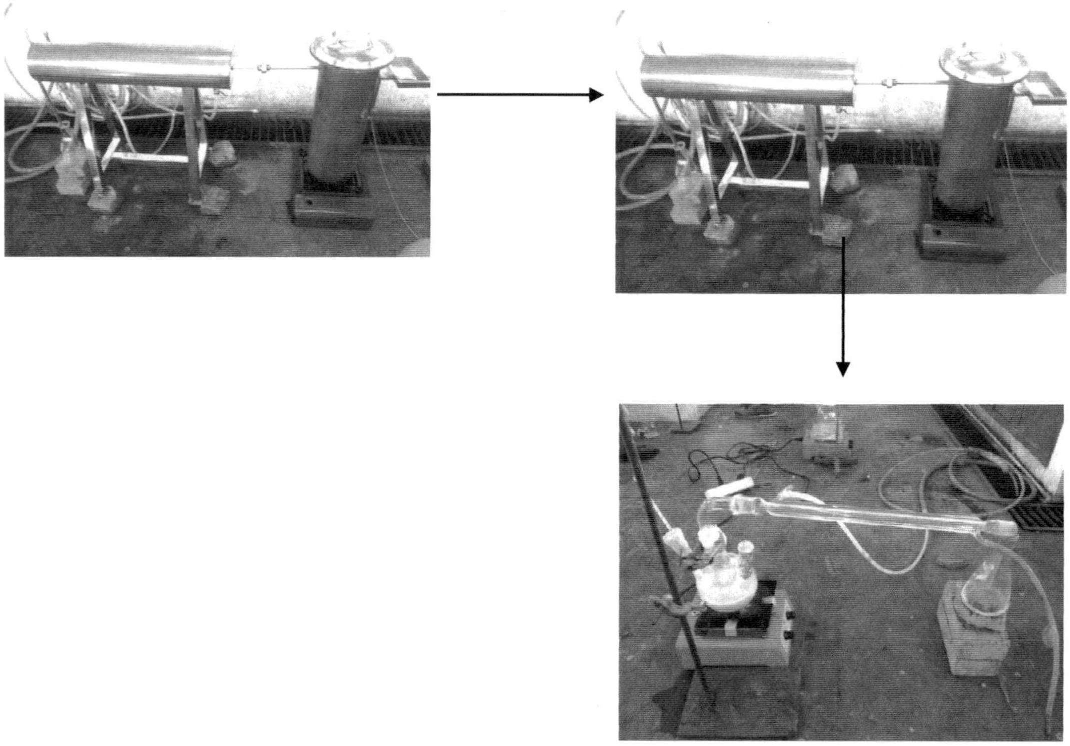

Figure 1. Distillation Apparatus for Calcium Oxide

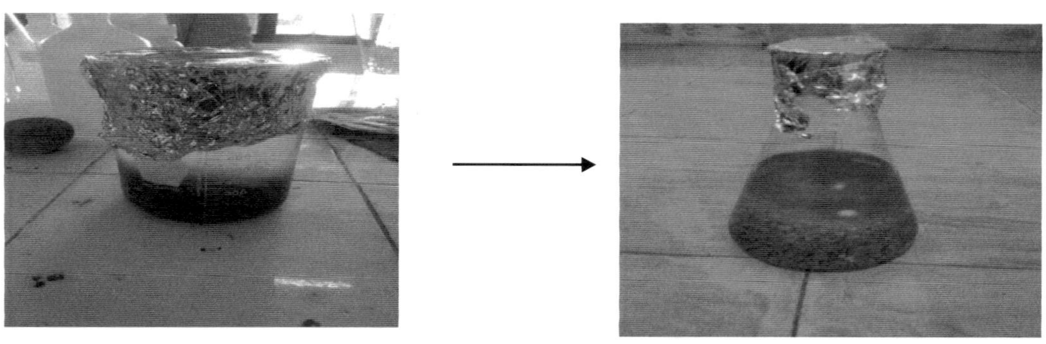

Figure 2. Experiment Apparatus for the purification of bioethanol with natural zeolite

2.2 Production of Bioethanol
Palm juice obtained from Sugar Palm (*Arenga pinnata* Merr) was inoculated with yeast *Saccharomyces cerevisiae* and left for up to 4 days to observe the optimum time for bioethanol production. The daily concentration of the bioethanol in the palm juice was measured afterwards.

2.3 Effect of Calcium Oxide Addition in Distillation Process
This process was done by adding calcium oxide powder during the distillation process of the bioethanol at 78 °C. The concentration of bioethanol obtained in this process was measured using Gas Cromatography method and then compared with the normal process without the addition of calcium oxide.

2.4 Effect of Natural Zeolite Addition in Purification Process
Natural zeolite used as adsorbent in this proess was activated by soaking the natural zeolite into 1M sulfuric acid to remove other particles from the zeolite while increasing its ability so absorb water from the ethanol. The natural zeolite was then heated in the oven at 200°C for 2.5 hours to activate the zeolite pores to absorb the water in ethanol. After the activation, ethanol was soaked using natural zeolite and kept for 5 hours.

2.5 Problem formulation
Concentration of bioethanol was measured using Gas Chromatography method, while the concentration calculation was done using the formula below:

$$C_{sample} = \frac{(\text{sample areas x sample concentration})}{\text{standard areas}}\% \qquad (1)$$

3. Results and Discussion
3.1 Effect of Fermentation Time on Bioethanol Produced
The effect of fermentation time (up to 4 days) on bioethanol concentration was measured and the result can be seen in Figure 1 below.

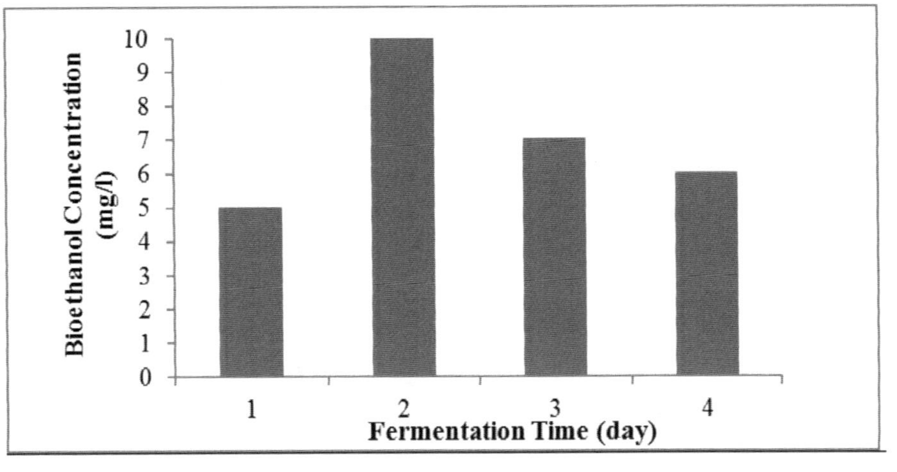

Figure 3. Effect of Fermentation Time on Bioethanol Produced

It can be seen from Figure 1 above that the concentration of bioethanol produced in the fermentation process increased from day 1 (5%) to day 2 (10%). However, the concentration dropped considerably

IOP Conf. Series: Earth and Environmental Science **130** (2018) 012035 doi:10.1088/1755-1315/130/1/012035

on the third and fourth day of the fermentation process, with 7% and 6% respectively. This means that the optimum time for *Saccharomyces cerevisiae* to multiply and produce bioethanol is approximately two days (48 hours). Subsequently, after reaching the optimum growth condition, the concentration of bioethanol in the medium will increase to about 10%, which will then start to become toxic for the cells and caused the activity of *Saccharomyces cerevisiae* to decrease [8].

The concentration of bioethanol beyond 10% is bad for yeast cell growth because they can be toxic to yeast cells, in which so that the maximum concentration of bioethanol allowed on yeast growth is 10%. This is due to that the enzymes in the yeast cells such as dehydrogenase and hexokinase are sensitive to high ethanol conditions, which then afterwards can interfere and become toxic to yeast growth activity [9].

To conclude, an appropriate fermentation time is required to produce high concentration of bioethanol [10]. The increasing fermentation time led to the further reaction of bioethanol, typically oxization of the bioethanol itself to acetic acid. In this case, the oxidation rate of bioethanol to acetic acid is higher than the rate of glucose converted to bioethanol, and thus causing the decrease amount of the bioethanol.

3.2 Effect of Addition of Calcium Oxide (CaO) in the Distillation Process

Distillate obtained was analyzed using GC (Gas Chromatography) to calculate the percentage of the ethanol produced after the purification process using distillation by the addition of CaO adsorbents. The observational data using GC (Gas Chromatography) can be seen in Table 1 below.

Table 1. Observational Data Using GC (Gas Chromatography)

No.	Name	Areas	Information
1	Ethanol Pa (1)	2.199.609	H = 5.861.90 T = 1.896
2	Ethanol Pa (2)	80.089.032	H = 31.317.514 T = 2.001
3	Ethanol Sample + CaO (1)	3.172.001	H = 8.546.98 T = 1.889
4	Ethanol Sample + CaO (2)	84.234.345	H = 30.665.978 T = 2.002

The GC measurement was conducted with two replication, in which then the concentration can be observed by measuring the average of both the sample and standard area using the calculation below:

$$\text{Sample Area} = \frac{3.172.001 + 84.234.345}{2}$$

$$= 43.703.173$$

$$\text{Standard Area} = \frac{2.199.609 + 80.089.032}{2}$$

$$= 41.144.320.5$$

Subsequently, the bioethanol concentration can be calculated by the following formula:

$$C_{sampel} = \frac{(\text{sample areas x sample concentration})}{\text{standard areas}} \%$$

$$= \frac{43.703.173 \times 94}{41.144.320.5}$$

$$= 99.85\%$$

It can be seen that the concentration of the bioethanol produced using calcium oxide (CaO) adsorbent was 99.85%. The addition of calcium oxide (CaO) to the distillation process resulted in the increment of the bioethanol concentration, from 94% for the initial purification to 99.85%. The excellent result in this study was typically caused by the ability of calcium oxide to bind the water molecules contained in the bioethanol, which then makes it able to reach > 99% concentration.

3.3 Effects of Natural Zeolite Adsorption in Purification of Bioethanol

Data of the concentration value of bioethanol through adsorption process using activated zeolite analyzed by using GC (Gas Chromatography) method can be seen on the Table 2 below.

Table 2. Observation Data Using GC (Gas Chromatography)

No.	Name	Areas	Information
1	Ethanol Pa (1)	2.199.609	H = 5.861.90 T = 1.896
2	Ethanol Pa (2)	80.089.032	H = 31.317.514 T = 2.001
3	Etanol Sample +Zeolite (1)	1.804.169	H = 8.546.98 T = 1.889
4	Ethanol Sample + Zeolite (2)	80.044.205	H = 30.665.978 T = 2.002

The GC measurement was conducted with two replication, in which then the concentration can be observed by measuring the average of both the sample and standard area using the calculation below:

$$\text{Sample Area} = \frac{1.804.169 + 80.044.205}{2}$$

$$= 40.924.187$$

$$\text{StandardArea} = \frac{2.199.609 + 80.089.032}{2}$$

$$= 41.144.320.5$$

Subsequently, the bioethanol concentration can be calculated by the following formula:

$$C_{sampel} = \frac{(\text{sample areas } \times \text{ sample concentration})}{\text{standard areas}} \%$$

$$= \frac{40.924{,}187 \times 94}{41.144.320{,}5}$$

$$= 95\ \%$$

It can be seen that the concentration of bioethanol obtained using the activated zeolite adsorbent was 95%, compared to the normal bioethanol purification which was 94%.

Research conducted by Novitasari and Kusumaningrum [6] showed that the variation of adsorption time of natural zeolite and 4A synthetic zeolite (8, 24, 32, 48 and 56 hours) showed different results at 84.34%, 81.92%, 74.49%, 72% and 74.92% respectively. This result indicated that the longer zeolite immersed in ethanol solution, the lower concentration of ethanol obtained from the adsorption process. This is due to that natural zeolite absorbs not only the water but also the ethanol produced itself. In addition, natural zeolites have varying pore sizes that more ethanol being absorbed could mean more water molecules released.

The 95% purity result obtained from this study is also in accordance with research conducted by Frita et al. [11], who purified 94% of ethanol using natural zeolite adsorbent with zeolite grain size of 50 mesh, 80 mesh, 100 mesh and 200 mesh showed increase to 95%, 95%, 95.5% and 95.5% respectively. The increment that only reached 95.5% from the initial concentration of 94% concluded that the higher concentration levels of the ethanol in the initial process to be adsorbed by zeolites, the less amount of concentration can be improved.

4. Conclusions

This study proved that the optimum fermentation time for *Saccharomyces cerevisiae* to produce bioethanol from palm sugar is approximately two days or 48 hours with 10% bioethanol concentration. The best bioethanol purification (up to 99.8% concentration) can be obtained by the addition of calcium oxide (CaO) during the process. The bioethanol produced in this research can be used as a fuel as its concentration is in line with the Indonesian Standard (SNI), which is > 99%. The success of this research proves that bioethanol obtained from palm juice can be utilized as an alternative substitute to fossil fuels.

Acknowledgments

The authors gratefully acknowledge that the present research is supported by Ministry of Research and Technology and Higher Education Republic of Indonesia. The support is under the research grant BP-PTN USU of Year 2017.

References

[1] Duruyurek M, Dusgun C, Fuat G, Mehmet, Zeliha S (2015) Production of Bioethanol from Waste Potato. Turkish Journal of Agriculture Food Science and Technology, 3(5): 331, 334, 2015.

[2] Wijaya I Made Anom Sutrisna, Arthawan I Gusti Ketut, Sari AN (2012) Potensi Nira Kelapa Sebagai Bahan Baku Bioetanol". Jurnal Bumi Lestari, Vol. 12, No.1.

[3] Gusti (2016) Pengaruh Suhu dan Volume Starter dalam Pembuatan Bioetanol dari Nira Aren (Arenga Pinnata Merr). Skripsi. Medan. Universitas Sumatera Utara.

[4] Manurung MM (2016) Pengaruh Volume Starter dan Agitasi dalam Pembuatan Bioetanol dari Nira Aren (*Arenga Pinnata* Merr). Undergraduate Thesis. Universitas Sumatera Utara.

[5] Akhir, YM, Chairul, Drastinawati (2015) *Pembuatan Bioetanol dari Fermentasi Nira Aren (Arenga Pinnata) Menggunakan yeast Saccharomyces cereviceae dengan Pengaruh Variasi Konsentrasi Nutrisi dan Waktu Fermentasi* JOM FTEKNIK 1(1).

[6] Novitasari DD, Kusumaningrum (2012) *Pemurnian Bioetanol Menggunakan Proses Adsorbs dan Distilasi Adsorbsi dengan Adsorben Zeolit*. Jurnal Teknologi Kimia dan Industri, 1(1): 534-539.

[7] Yuliana C, Silvia RY (2015) *Pemurnian Bioetanol Hasil Fermentasi Nira Nipah Menggunakan Proses Destilasi Adsorpsi pada Variasi Rasio Adsorben dengan Modifikasi* . JOM FTEKNIK 2(1).

[8] Kunaepah U (2008*) Pengaruh Lama Fermentasi dan Konsentrasi Glukosa terhadap Aktivitas Antibakteri, Polifenol Total dan Mutu Kimia Kefir Susu Kacang Merah*. Thesis. Semarang. Universitas Diponegoro.

[9] Neto, Pedro de Olivia, Claudia, Dorta, Ana F, Azevedo C, Valeria MGL, Douglas FS. (2013) The Brazilian Technology of Fuel Ethanol Fermentation-yeast Inhibotors

[10] Azizah NANA, S Mulyani (2012) *Pengaruh Lama Fermentasi Terhadap Kadar Alkohol, PH, dan Produksi Gas Pada Proses Fermentasi Bioetanol dari Whey Dengan Substitusi Kulit Nanas*. Jurnal Aplikasi Teknologi Pangan. Vol. 1. No. 2.

[11] Frita, Ornella, Ratnawulan, Gusnedi (2015) *Pengaruh Ukuran Bulir Zeolit Terhadap Kadar Bioetanol dari Tanaman Tebu (Saccharum officinarum)*. Pillar of Physics 5: 49-56

[12] Directorate General of Plantation (2004) The development of sugar palm in Indonesia. Proceedings of Seminar Nasional Aren. Research Institute for Coconut and Other Palms: 138-144.

IOP Conf. Series: Earth and Environmental Science **130** (2018) 012036 doi:10.1088/1755-1315/130/1/012036

The Effect of Paraquat Inhalation on Parkinsonism, Organ Morphology and Anatomy of Mice and Its Recovery with the Application of *Etliringea hemisphaerica* (Blume, R.M. Smith) Crude Extract

Choirul Muslim[1*], Santi Nurul Kamila[1]
Department of Biology, Universitas Bengkulu
Jl. W.R Supratman, Kandang Limun, Bengkulu 38371, Indonesia

*Email: choirulmuslim56@gmail.com

Abstract. This research is aimed to understand the effect of paraquat herbicide inhalation on Parkinsonism, morphology and anatomy change in mice, and its recovery with Etliringea hemisphaerica crude extract application. Sixty mice were placed into three following groups: group R0 were mice receiving standard food ransom, R1 were a group of mice receiving the regular food ransom plus inhalation of 1% paraquat, and R2 were a group of R1 plus obtaining 0,39mg/bw extract E. hemisphaerica (Bl.) R.M. Smith). After 2 X 7 days of sub-sequential application of both paraquat and "helani tulip" extract, we observed the effects. The examination included bradikinesia attitude, postural instability and rigidity, morphology and anatomy of brain, liver, blood, lung, and kidney. The data were tabulated and analyzed qualitative and semi quantitative description on the behavioural disorder, the alteration of morphology and anatomy, and their remedy based on Sander 2004, Junqueira and Carneiro, 2007. The results showed that the application of paraquat caused strong bradikinesia, postural instability and rigidity. The treatment of the extract was only resulting in the bradikinesia removal but was minor improving the consequence of postural instability and rigidity. Paraquat was not affecting the morphology of neural brain but was altering the morphology and anatomy of lung, liver, blood, and the kidney. In general, the negative impact of the paraquat was weakly eliminated by the treatment of "helani tulip" extract.

Keywords: Anatomy change, Etliringea hemisphaerica, herbicide, inhalation, Paraquat, parkinsonism, recovery

1. Introduction

Paraquat is a toxic herbicide which is freely sold at the market. It is the most applicable in agriculture to eradicate unwanted wild plant in the crop growing field, but it is also the most poisonous chemical product used in undeveloped farming [1], and may strongly trigger a neurodegenerative *Parkinsonism* [2]. The Parkinsonism might be appeared after someone contaminated with heavy metal ion (Mn, Hg, Pb, etc), inhaled pesticide and or herbicide (paraquat, rotenone, dieldrin, and lindane), MPTP (a by-product of synthetic opium), carbon monoxide (Anoxia), and post trauma syndrome [3]. The symptom of Parkinsonism is when anyone developed the abnormal tremor and shaky movement in hand, limbs, jaws, and facial muscle tissues [4]. Paraquat is one of the chemical compounds that may influence on the degeneration of neuron cells causes dementia and especially dopaminergic neuron-degeneration that might cause Parkinsonism and liver inflammation [5].

The study on herbal medication showed that some of the plant extract demonstrate the ability to recover and to prevent the Parkinsonism [6]. One of them is "helani tulip" or "honje hutan" (*Etlingera hemisphaerica* Blume, R.M. Smith) which contains an alkaloid, flavonoid, polyphenol, steroid, saponin, and atsiri oils. The extract may also include an antioxidant ascorbic acid to intoxicate the effect of the poisons [7].

Most of the Indonesian people are farmers and are still working disproportionately with the synthetic herbicide that might result in the occurrence of the Parkinsonism. Study with the application of the *Elitringea hemisphaerica* leaf extract has been able to recover Parkinsonism in mice that caused by mercury ion [8].

The injection application of paraquat on mice might have caused development of postural instability, rigidity and other parkinsonism phenomena and they are all displayed with the weak reflex movement avoiding the ravine, negative geotaxis reflex, reduction ability to swim [9]. The treatment with the Etliringea extract to the nauseous mice has been successfully rehabilitated the capability of smelling, seeing and hearing, and also the ability to lift up its body, although its improvement was slow.

As the inorganic farming system extensively uses paraquat that may cause the development of parkinsonism when it is inhaled via air respiration, we performed this investigation in order to understand whether the paraquat inhalation is affecting the development of parkinsonism in mice, and altering the morphology and anatomy of the lung, the liver, the blood, and the kidney, and whether the application of *Etliringea hemisphaerica* crude extract to the sick mice will remedy the nuisance.

2. Methods

We accomplished this research at our Laboratory during the period of June to September 2016. Preparation of the extract follows the work achieved by Gresinta, 2012 [10]. The leaf was finely ground, open drying and macerated on a shaker in the composition of one part of a fine crushed leaf and four parts of 96% ethanol. The maceration was taken place for a week. The macerate was separated with a filter so that we have a liquid crude extract. To get rid of the excess of ethanol, we evaporated it with a *rotary evaporator*, in a fifty degree Celsius of water bath. Finally, we have an extract in the form of pasta or a viscous liquid the "helani tulip" leaf crude extract.

We have sixty male mice placed in a proportional cage separated into three groups. The Group R0 were mice receiving standard pellet feed/ransom, R1 were a group of mice receiving the standard pellet ransom plus inhalation of 1% paraquat (*1,1-Dimetil-4,4-Bipyridium dichloride*) which has a brand product of proquat. The daily inhalation of paraquat has been carried on for seven days treatment respectively. The other groups were R2; these were a group of R1 receiving seven days of 1% paraquat and subsequently treated with obtaining 0,39 mg/gram of body weight "helani tulip" leaf extract (*Etliringea hemisphaera)* for another seven days of applications. After fourteen days of all application, we observed the effects. The examination included the development of bradikinesia, the postural instability and rigidity, the alteration of the morphology and the anatomy of brain, liver, blood, lung, and kidney as a consequence of the paraquat inhalation. We also compared and evaluated the effect of the "helani tulip extract" treatments to the recovery of the annoyance.

The data contain qualitative and semi quantitative performance of the behavioural disorder, the alteration of the morphology and the anatomy influenced by paraquat inhalation, and their recovery after the extract remedy. All the data were tabulated and analyzed descriptively. The comparative morphology and anatomy analysis uses Sander 2004, Junqueira and Carneiro, 2007.

The schemes of the treatments are as the following diagram:

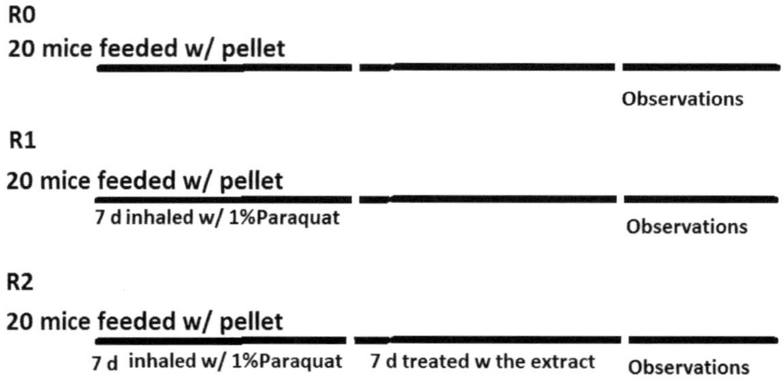

Figure 1. The scheme of the three group of treatments

3. Results and Discussion

3.1 Effect of Paraquat on appearance of Parkinsonism

In general, as indicated in the corresponding tables, we can demonstrate that yes 1 % paraquat that was being inhaled to white mice has a significant result to the manifestation of Parkinsonism that is the development of bradikinesia, postural instability and rigidity.

Table 1. The effect of 1% paraquat inhalation and "helani tulip" leaf extract on bradikinesia.

No	**Observation**	**The result (response to the treatment)**		
		Control (R0)	Paraquat (R1)	paraquat & extract (R2)
1.	Bradikinesia	Normal	Bradikinesia	Inhaled into normal

3.2 The effect on postural instability and rigidity

Table 2. The effect of 1% paraquat inhalation and the treatment of "helani tulip" leaf extract to mice muscle performance (agility to avoid the artificial gap/ravine), geotaxis reflex, and the skill to swim (direction, position, and activity of swimming

The variety of observation	**Treatment**		
	Control	Paraquat (R1)	Paraquat & extract
Reflex to avoid gap	Normal	Sharply decrease	Decrease
Reflex of Geotaxis	Normal	Sharply decrease	Decrease
Ability to swimming			
Swimming direction	Normal	Far Better	Far Better
Swimming Position	Normal	Normal	Normal
Swimming Activity	Normal	Far Better	Not sufficiently good

Table 3. The effect of 1% paraquat inhalation and the application of "helani tulip" leaf extract on the ability of smelling, seeing, and hearing, and lift up the body.

No	**Observation**	**Response to the treatments**		
		Control (R0)	Paraquat (R1)	paraquat and the extract
1.	Ability to smelling	Normal	Disappear /bad	Not fully recovered
2.	Ability to seeing	Normal	Disappear /bad	Not fully recovered
3.	Ability to hearing	Normal	Dropped sharply	Not fully recovered 100%
4.	Ability to lift up the body	Normal	Drop sharply	Not fully recovered 100%

Notes: Normal means having response 100% normal; Recovered to normal = having a positive response to the related challenge almost closed to normal (100%), dropped: positive response remain 50-99%, sharply dropped: positive response remains very low under 50%.

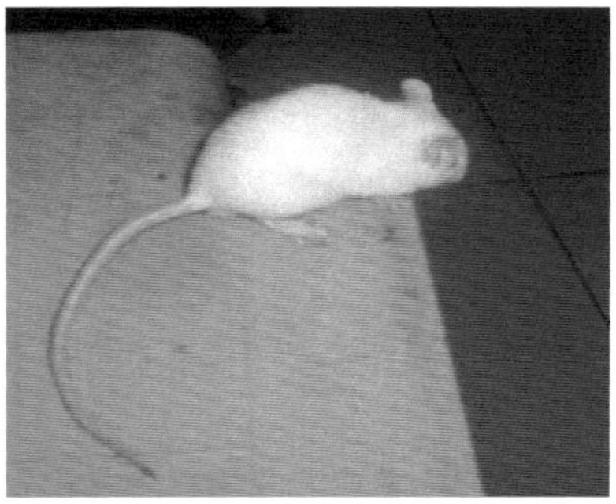

Figure 2. Reflex to avoid gap on control mice

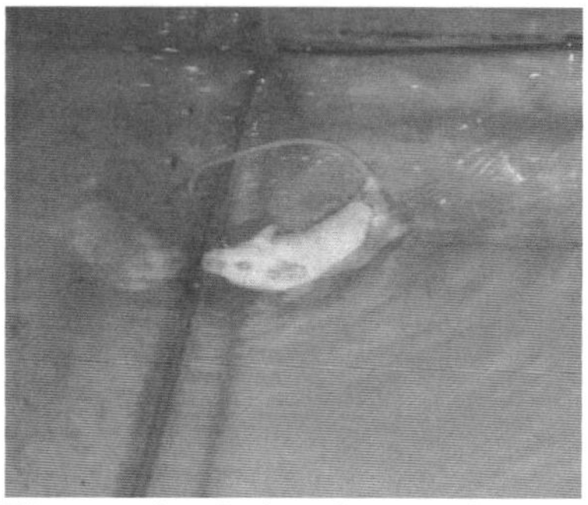

Figure 3. Swimming Reflex on control mice

Figure 3a. The behavior of lift up body and limbs on mice control

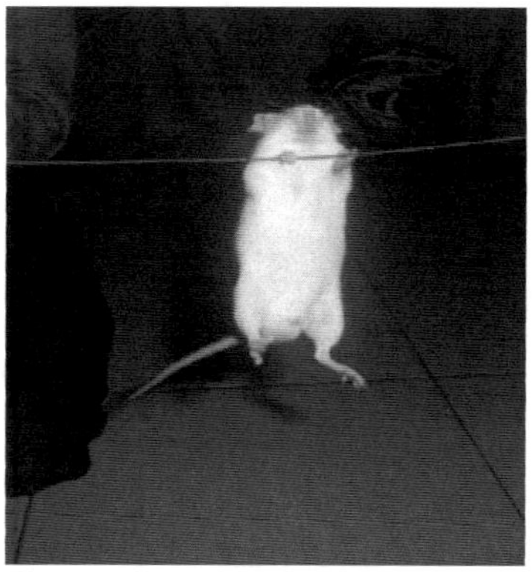

Figure 3b. Idem with figure 3a but on mice with paraquat (R1).

In general, the application of the extract was only resulting in the bradikinesia removal but was minor improving the consequence of postural instability and rigidity. The treatment of "helani tulip" leaf extract on Parkinsonism in mice has been able to fix or restored the bradikinesia in mice, but in general inadequately recovered the other bad appearance due to Parkinsonism. These are the ability to recover smelling, seeing and hearing capability, and also the capacity to lift up the body and its limbs back to the normal.

IOP Conf. Series: Earth and Environmental Science **130** (2018) 012036 doi:10.1088/1755-1315/130/1/012036

3.3 The effect of paraquat on the alteration of morphology and anatomy of selected organ, and its remedy affected by "the helani tulip" extract treatments

The following pictures showed the effects of Paraquat inhalation

R0. Control Treatment **R1. 1% Paraquat Inhalation** **R2. 1% Paraquat and Helani tulip leaf extract**

Figure 4. The effect of control (R0), 1 % paraquat (R1), and the paraquat and leaf "helani tulip" extract (R2) to the change of anatomy and morphology inner organ of mice.

The photo demonstrated the followings:
On mice organs control, the lungs display compactly pink in colour, whereas heart, liver, and kidney compactly old orange (R0); On mice organ of paraquat treatment, it is clear that its lungs showed compactly very pale pink, hearts have somehow swollen less densely orange, while the liver has exhibited barely pale as compare to the control and more enlarged and it seems due to inflammation (R1); On mice organ of paraquat treatment followed by leaf "helani tulip" extract it has shown that the lungs are very pale pink, the hearts still have inflammation with its colour remains a little bit recovered far close to the control, and the liver has recovered the colour, but it is still in the inflammation position (R2).

Paraquat was not affecting the morphology of neural brain but was strongly affecting on the morphology and anatomy of lung, liver, blood, and the kidney. In general, the applications of the "helani tulip" extract were weakly eliminating the negative influence of the paraquat.

Our result has yielded evidence on how the impact of paraquat treatment and its recovery with the application of the leaf "helani tulip" extract to the impairment effect of paraquat using mice as a model. In general, these results confirmed that the inhalation of 1% paraquat created the appearance of Parkinsonism. The behaviour test, mainly bradikinesia, postural instability, and tremor has indicated the above findings.

There is a significant effect of extract "helani tulip" leaf for the improvement of mice from bradikinesia caused by paraquat, but less significant impact for the repair of postural instability and tremor. Ones might expect that the recovery by the use of the extract needs to be longer in time periods.
In general, the treatment of 1 % paraquat was not affecting the morphological appearance in the mice brain. The organ has specific compact reddish yellow in structure and colour. In fact, the central neuron in brain has much protection mechanism from any toxic external compounds, providing that the influence must be under the threshold value. It is because there is a barrier between blood vessel and the brain that its function is to filter any entrance of foreign molecules to the central neuron system. This mechanism is

to prevent the entry of unwanted toxic compound, including antibiotic and bacterial cells to the central neuron system in the brain [11]. Only in the over whelming expose of the toxic compound, at the longer time exposure and followed by the weakness of immune system in our body, the possible access of the unwanted compound as paraquat and pathogen might penetrate the barrier. As a result, one might develop neurodegenerative syndrome as Parkinsonism appearance.

3.4 The effect of 1% paraquat inhalation to the morphological and anatomical lungs, hearts, liver, and kidney

As shown in the pictures, the lungs of mice control seem to have compactly pink or reddish pale in colour, while the heart, liver, and kidney have old orange in colour. When the 1 % inhalation of paraquat was applied, it results in the lungs change the colour into being paler in pink colour, while the hearts have altered into old reddish orange and a little bit enlarges but less condense due to the inflammation. The effect of paraquat inhalation to the livers and kidney showed to have paler and bigger as compared to the control in both livers, and kidneys.

The application of "helani tulip" extract for the recovery the mice organ from the impact paraquat inhalation is not fully accomplished. The lungs are still strongly pale, the hearts are still on enlarge because of inflammation, the colour of the liver has been close to normal, but it is yet in the swelling position.

These all are indications that the most in force of 1% paraquat inhalation are to the lungs organ. Histological structure of lungs that have the most influenced by 1 % paraquat inhalation is parts of alveoli. This is because of the breathing the air containing the paraquat directly to the lungs may results in the direct impact of paraquat to the alveoli in the lungs. While the outer epithelial layer of pleura and connective tissue is shown to be normal. An aspect of inflammation is able to be depicted in our pictures, in which there is an enlargement of the alveoli volume.

The paraquat is strongly affecting to the quality of blood coagulation. Our histology pictures have confirmed the strong coagulation on control but weaker on paraquat treatment and relatively regained coagulation after the application extract of leaf "helani tulip". The most probable of this occurrence is as a consequence of paraquat inhalation that preventing the blood coagulation.

3.5 The effect of 1% paraquat inhalation to the alteration of the kidney structure

In general, the weakness effect of 1% paraquat inhalation is to the inflammation of the kidney, and it is the least effect on structure kidney alteration.

We have observed the structure of cortex parts (and in part of medulla structures) of mice kidney. Investigation on both control and paraquat treatment suggested that the wall of the tissue of epithelium tubules are surrounding the parts of proximal tubules of the "contortus" cavity (P) and distal parts. The cavity of the P and D is brighter in control kidney rather than paraquat kidney. The structure of glomerulus (G) has shown to be spotted clustered, and the tissue of medulla parts is darker. This description is an indication that the structure of control kidney is more solid or compact, while the symptom of inflammation affected by paraquat is being demonstrated by the relative enlargement of P and D, and the structure of G is more protected.

3.6 The effect of paraquat inhalation to the alteration of liver histology

The effect of paraquat inhalation treatment to the histological structure of liver lobules of mice showed that the liver is experiencing an inflammation, which is relatively enlarged, and less condensed. The transparent hollow space is the ventral vena surrounded by hepatocyte tissue which has formed like elongated tendrils enclosed by the sinusoid branch of ventral vena. The liver Lobules from the control is relatively denser because of no inflammation, and therefore is more condensed and solid

4. Conclusions

We have come to the following conclusions:

1. The application of 1% paraquat inhalation has brought about the male mice to experiencing the appearance of Parkinsonism, in which the bradikinesia, the postural instability and the tremor are some of the indications. The treatment of leaf "helani tulip" extract to the Parkinsonism in mice affected the recovery of bradikinesia but did not effect on other behaviour of Parkinsonism. The remedial treatment of "helani tulip" extracts needs to be the longer application to restore the deleterious effect of paraquat.
2. The application of paraquat inhalation has made relatively obvious to the morphology and anatomy of lungs, hearts, livers, and kidneys. The poisonous effect of paraquat inhalation on mice has been being restored by the application of leaf "helani tulip" extract, and the restoration of morphology and anatomy of lungs, liver, hearts and kidney are some of the indications.
3. The application of paraquat inhalation did not result in the alteration of brains morphology, and the histology of central neural tissues of mice.
4. The application of paraquat inhalation, in general, decreased the ability of blood coagulation. While the application of leaf "helani tulip" extract relatively recovered the reduction coagulation ability caused by paraquat inhalation.

Acknowledgments

The support and fund for this research was provided by University of Bengkulu. The mice were available with the kindness of Denis at UNIB. We acknowledge and appreciate all their supports.

References

[1] Watts, M (2011) Paraquat Exposure And Parkinson's Disease. Pesticide Action Network Asia & The Pacific. Hamburg. Germany.

[2] Simon VA and A Taylor (1989) High Sensitivity HPLC Analysis Of Diquat And Paraquat With Confirmation. *Journal Chromatography Products* 479: 85-106

[3] Kang MJ, Suk JG, C Hyun (2009) Paraquat induces alternation of the dopamine catabolic pathways and glutathione levels in the substantia nigra of mice. *Journal Elsevier. Toxicology Letters*. Vol 188: 567-872.

[4] Nisticò R, Mehdawy B, Piccirilli S, Mercuri N (2011) Paraquat and Rotenone Induced Models of Parkinson's Disease. International Journal of Immunopathology and Pharmacology. 24(2): 315-324

[5] Muslim C (2014) The genetic and molecular biology based Parkinson disease. UNIB Press. Unversitas Bengkulu. Bengkulu.

[6] Viranda PM (2009) *Pengujian kandungan Fenol diukur dengan pengukuran kolorimetrik dengan larutan folin ciocalteu*. [Undergraduate Thesis]. University of Indonesia. Jakarta.

[7] Ruyani A, A Sundaryono, E Gresinta, Z Fakhrur, and D Samitra (2014) Potensial Assessment Of Leaf Ethanolic Extract "helani tulip" (*Etlingera hemisphaerica*) In Regulating Glucose And Tryglycerides On Mice (*Mus musculus*). *International Journal of Sciences.* Vol 3: 3-9

[8] Wardianti Y (2013) *Pengaruh Pemberian Ekstrak daun "helani tulip" (Etlingera hemisphaerica) Terhadap Pemulihan Perilaku Mus musculus Akibat Toksisitas Merkuri Dan Implementasinya Dalam Pembelajaran Fisiologi Hewan* [Thesis] University of Bengkulu. Bengkulu.
[9] Nababan NC, and C Muslim (2015) *Pengaruh Pemberian Ekstrak Daun "helani tulip" Hutan Etlingera Hemisphaerica (Blume) R.M.Sm Terhadap Gejala Parkinsonisme Pada Mencit Mus Musculus L. (1758) Swiss Webster Yang Telah Disuntik Paraquat.* Prosiding Seminar Nasional MIPA at University of Tanjung Pura Pontianak, 5-7 June 2015.
[10] Gresinta E (2012) *Uji Potensi Ekstrak Daun Etlingera hemisphaerica terhadap Jumlah Leukosit Mus musculus dan Implementasinya sebagai Modul Pembelajaran Sistem Imun* [Thesis] Bengkulu: University of Bengkulu
[11] Juncueira LC and J Carneiro (2007) *Histologi Dasar, Teks Dan Atlas*. Tenth Edition. Translated by J. Tambayong and Editor F. Dany. Penerbit Buku Kedokteran EGC. Jakarta.

Molecular genetics of craniosynostosis

Caterine[1], Elza Ibrahim Auerkari[2*]

[1]Orthodontic Residency Program, Faculty of Dentistry, University of Indonesia. Jl.Salemba Raya no.4, Jakarta, 10430. Indonesia.
[2]Department of Oral Biology, Faculty of Dentistry, University of Indonesia. Jl.Salemba Raya no.4, Jakarta, 10430. Indonesia.

[*]Email: eiauerkari@yahoo.com
Tel: +62213910344; Fax: +622131906289

Abstract. Tight regulation process and complex interplay occur along the osteogenic interfaces of the cranial sutures in normal growth and development of the skull. Cranial sutures serve as sites of bone growth while maintaining a state of patency to accommodate the developing brain. Cranial sutures are fibro-cellular structures that separate the rigid plates of the skull bones. Premature fusion of one or more cranial sutures leads to a condition known as craniosynostosis. Craniosynostosis is one of the most common craniofacial anomalies with a prevalence of 1 in 2,500 newborns. Several genes have been identified in the pathogenesis of craniosynostosis. Molecular signaling events and the intracellular signal transduction pathways implicated in the suture pathobiology will provide a useful approach for therapeutic targeting.

Keywords: cranial sutures, cranial development, signalling mechanism, sutures fusion,

1. Introduction

Tight regulation process and complex interplay occur along the osteogenic interfaces of the cranial sutures in normal growth and development. Cranial sutures serve as the center for the growth of bones of the skull. The signaling pathway and transcription factors that control the formation of cranial sutures have been identified [1]. The union of one or more cranial sutures before the proper timing is known as craniosynostosis.

Craniosynostosis is occurring in 1 in 2,000 – 2,500 birth in the general population. Classification of craniosynostosis may be based on suture involvement, such as single suture synostosis or multiple suture synostosis. Studies in the last decades have revealed several genes related to the mechanism of suture patency, such as FGFR2, FGFR3, and TWIST1. Here, we will give a brief review of the normal cranial development and further elaboration of craniosynostosis. Molecular biology aspect will be the focus of this review [2].

2. Normal cranial development

Growth may be defined as an increase in size by natural development through cellular proliferation and differentiation. Craniofacial growth is a complex process involving many interactions between different bones that make up the skull and between the hard and soft tissues. The processes that control craniofacial growth are not fully understood and are an area of extremely active research globally [3].

IOP Conf. Series: Earth and Environmental Science **130** (2018) 012037 doi:10.1088/1755-1315/130/1/012037

As this review will be focused on craniosynostosis problem, general bone growth mechanisms, the anatomy of the skull and normal growth of the cranial vault (which relates to the synostosis problem) in particular will be discussed briefly.

2.1. Anatomy of the Cranial Vault

The human skull is formed by two main parts, including the neurocranium and the viscerocranium. The neurocranium is the structures that enclose and protect the brain and sensory organs. The viscerocranium, on the other hand, includes the bones of the face and the palatal, temporal, pharyngeal, and auditory bones. Focus in this review would be around the neurocranium. The neurocranium comprises of five bones – a pair of frontal bones, a pair of parietal bones, and occipital bone that are connected by cranial sutures [4].

Cranial sutures are fibro-cellular structures that separate the rigid plates of the skull bones. Later on, sutures would be bridged by a dense network of collagen fibers and become increasingly interdigitated to unite the bone fronts [5]. The cranial sutures include metopic suture, coronal suture, sagittal suture and lambdoid suture. The metopic suture connects the frontal bones, the pair of parietal bones is connected by sagittal suture, coronal suture connects the frontal bones and parietal bones, and lastly, lambdoid suture connects the parietal and supraoccipital bones [2].

Ossification of the cranial vault primarily takes place via intramembranous ossification in which the mesenchymal cells, situated between the mesenchyme and dura mater, will undergo differentiation into osteoblasts. Osteoblasts would then form bone through secretion of an osteoid matrix. The mesenchymal cells in the middle of the suture remain undifferentiated during cranial vault development despite ossification of the cranial vault [2,4].

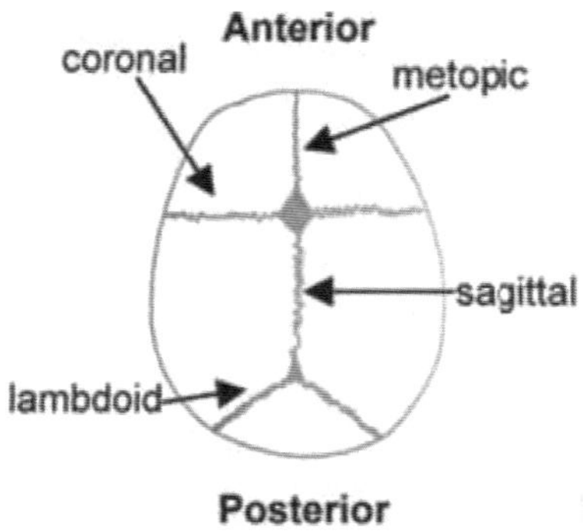

Figure 1. Cranial sutures (Source: Johnson D, Wilkie AOM, 2011) [6].

The principal function of cranial sutures is to allow the growth of the skull housing the rapidly developing brain, particularly during fetal and infant life. Also, cranial sutures assist in allowing distortion of the skull during delivery, absorb mastication load, and serve as absorbers of shock exerted by external forces [7]. Thus, suture patency is undeniably critical for normal craniofacial development.

2.2. Timing of craniofacial sutures fusion

The earliest suture to fuse is usually metopic suture, closing at three to nine months of age. Coronal suture, sagittal suture, and lambdoid suture fuse later in the third decade of life. Before the fusing of the suture, the equilibrium of sutural elasticity, calvarial osteogenesis, and the development of the brain maintains normal growth of the skull. Intrasutural mesenchymal tissue disruption by genetic mutation and environmental factors might initiate early union of one or more cranial sutures, causing substantial functional damage [8].

IOP Conf. Series: Earth and Environmental Science **130** (2018) 012037 doi:10.1088/1755-1315/130/1/012037

3. Craniosynostosis

Normal cranial and facial development is dependent on growth adjustment at the sutures as a response to the growth of the brain and facial soft tissue. Cranial sutures normally close when the children reach 2 to 3 years of age. Synostosis is the term used to describe the abnormal condition where cranial sutures close early in life. Thus, craniosynostosis is described as the fusion of one or more of the cranial sutures earlier than the usual expected time during fetal development. The premature fusion causes a lack of growth to the bones perpendicular to the fused suture. Conversely, as compensation for the lack of bone growth near the fused suture and developing brain, the bones at the non-fused sutures undergo compensatory overgrowth. This compensatory overgrowth causes a progressive distortion in the skull shape. Many complications might be associated with craniosynostosis because of the elevation in intracranial pressure, including sensory, respiratory, and neurological function, making craniosynostosis important to detect and treat [6,7,9].

3.1. Classification and Clinical Features

Classification of craniosynostosis may be based suture involvement, such as single suture synostosis or multiple suture synostosis. Single suture synostosis can be further classified according to the specific suture involvement, such as sagittal synostosis, coronal synostosis, metopic synostosis, and lambdoid synostosis. The sagittal synostosis is also called scaphocephaly, derived from the Greek word *scaphos* for boat and *cephalos* for head. Coronal synostosis can occur on one side of the skull or both side of the skull. Bilateral coronal synostosis caused a shortening of anteroposterior dimension and thus lead to a brachycephalic deformity. Metopic synostosis is connected to trigonocephaly in which compensatory growth take place at the posterior portion of the head. Lastly, lambdoid synostosis cause flattening of the occipital bone and contralateral growth in the parietal bone [10].

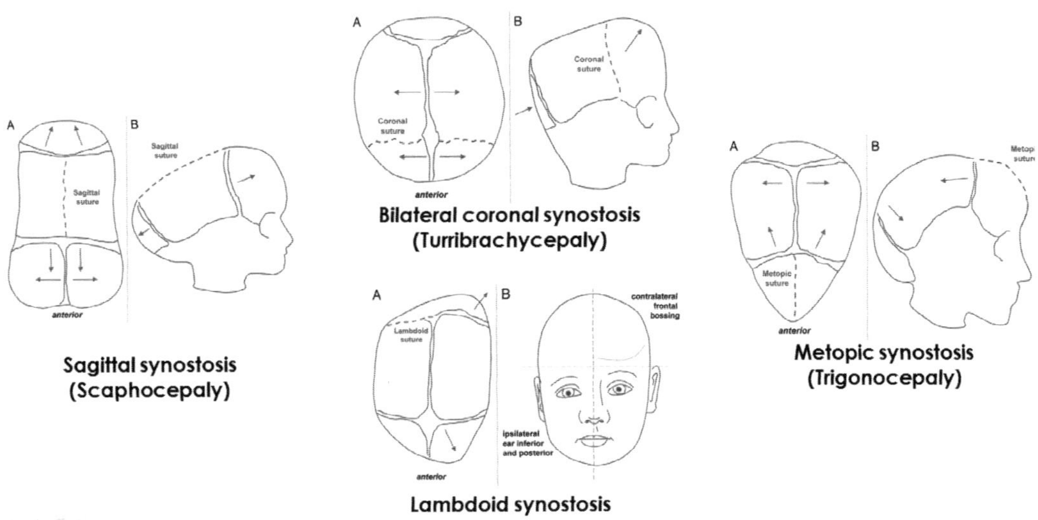

Figure 2. Classification of cranial synostosis according to involved suture (Source: Senarath-Yapa K, Chung MT, McArdle A, et al., 2012) [10].

3.2. Epidemiology

Craniosynostosis is occurring in 1 in 2,000 – 2,500 birth in the general population. Out of the reported cases, 25% cases are syndromic craniosynostosis. The remaining 75% account for nonsyndromic cases, that can be further subdivided according to the cranial suture involvement. Sagittal craniosynostosis is the most commonly seen type of nonsyndromic craniosynostosis, comprise of 40 – 50% of nonsyndromic cases, occuring in 1 in 5,000 births. Coronal synostosis occurred in 20 – 25% cases or 1 in every 1,000 births. Metopic synostosis and lambdoid synostosis account for 14% cases (1 in 15,000 births) and 3 – 5% cases (1 in 33,000 births) of nonsyndromic cases respectively [11]. In case of syndromic craniosynostosis, the most common syndromes are Pfeiffer syndrome affecting one every 25,000 births, Crouzon syndrome in about one every 100,000 births, and Apert syndrome in respective manner.

According to a retrospective case report by Anderson et al. in 2014, multiple sutures in nonsyndromic craniosynostosis is more prevalent in Asian community than in the white community. On the other hand, sagittal synostosis is significantly less common in Asian community than white community [12]. The prevalence rate of craniosynostosis in Indonesia is yet to be identified by the time this review was made. However, there is a first clinical and molecular report of Indonesian patients conveyed by Mundhofir et al. in 2013, published in Singapore Medical Journal [13].

3.3. Molecular Genetics of Craniosynostosis

Wilkie et al. (2010) did a 10-year prospective cohort of craniosynostosis on 326 children with craniosynostosis to identify the prevalence of single genes and chromosomal abnormalities associated with craniosynostosis. After a series of genetic analysis, those subjects are arranged into three genetic diagnosis categories: proven genetic, syndromic, and nonsyndromic. The proven genetic cases can be seen in 21% of the subjects with the major contribution of 86% are single gene disorders and the other 15% are chromosomal abnormalities (one patient had both). Mutations in FGFR2 are the most frequent cases account for 32% out of all the genetic cases, 25% of cases are FGFR3 mutations, followed by 19% account for TWIST1 mutations and 7% of EFNB1 mutations respectively [7,14].

Regulation of the formation and patency of cranial sutures by signaling pathways and transcription factors have been identified. The documentation of genes involved mostly is known by identification of mutagenic genes in syndromic forms of craniosynostosis [1]. Signaling molecules determined the fate of cells. The signaling molecules are chemical substances including hormones and neurotransmitters. These molecules are synthesized and secreted by cells to promote extracellular communication with other cells. The complex morphogenetic processes are arranged by FGF, TGF-β, BMP, and Wnt signaling pathway that respond to extracellular factors and environmental signals. Initiation of molecular events and activation of transcription factors take place when signaling molecules bind to their respective surface receptors. Cell behavior is highly dependent on the binding of transcription factors that would either express or repress the related genes [15,16].

FGF Signaling

Fibroblast Growth Factors (FGFs) belong to the family of polypeptides controlling several cellular processes. FGF signaling has an important role in prenatal and postnatal skeletal development. FGFs role in chondrogenesis and osteoblastogenesis control is through recruitment of the cells in the osteoblast lineage [17].

FGFs bind to high-affinity fibroblast growth factor receptors (FGFRs) that cause FGFR dimerization. These events lead to autophosphorylation on tyrosine residues initiating activation of transduction pathways as the signal transduction cascade is ligand-dependent. The activation of targets, later on, will transmit the signals to the nucleus in which cell differentiation or proliferation took place [4].

FGFRS comprised of 4 receptors of tyrosine kinase. The FGFRs structure consists of a tyrosine kinase domain, a hydrophobic transmembrane domain, and three immunoglobulin-like repeats (D1 – D3) of the extracellular domain [4,18]. Mutations either in FGF or their receptors (FGFR) may cause early sutural fusion, hence craniosynostosis, which indicates FGF signaling taking an active part in bone formation, including osteogenesis and chondrogenesis.

When FGFs bind to FGFRs, the intracellular receptor kinase comes into proximity. Hence phosphorylation and activation of the kinases occur. The activated kinase then activate intracellular substrates, such as phospholipase Cγ1 (PLC γ1) and FGFR substrate 2a (FRS2a). The activated FRS2a substrate causes an initiation of the RAS-MAPK pathway. The RAS-MAPK pathway affects cellular proliferation and differentiation. The phosphorylated extracellular signal-regulated kinase 1/2 (ERK 1/2) enter the nucleus and interact with transcription factors ERF, which is then exported from the nucleus. In the nucleus, ERF binds to RUNX2 and repress RUNX2 expression [4,17]. RUNX2 gene is essential for maturation of osteoblast.

FGFR mutations disturb the affinity of ligand binding site that affects the receptor activity, either decreasing the receptor specificity or increasing the receptor activity. In case of craniosynostosis, mutations of FGFRs result in overexpression of FGFRs thus trigger overproduction of ERK1/2. As a result, the number of ERF exported increased, and intranucleus ERF is not enough to suppress RUNX2 expression, causing osteogenesis to occur, hence, premature fusion of the suture (Figure 4) [4,17,19].

Several point mutations in FGFRs have been known and related to craniofacial malformations. Mutations in FGFR1 (Pro252Arg) are related to Pfeiffer syndrome. Mutations in FGFR2 (Ser252Trp and Pro253Arg) are linked to Apert, Pfeiffer, and Crouzon syndrome cases. Mutations in FGFR3 (Pro250Arg) are related to Muenke syndrome [4,13,18].

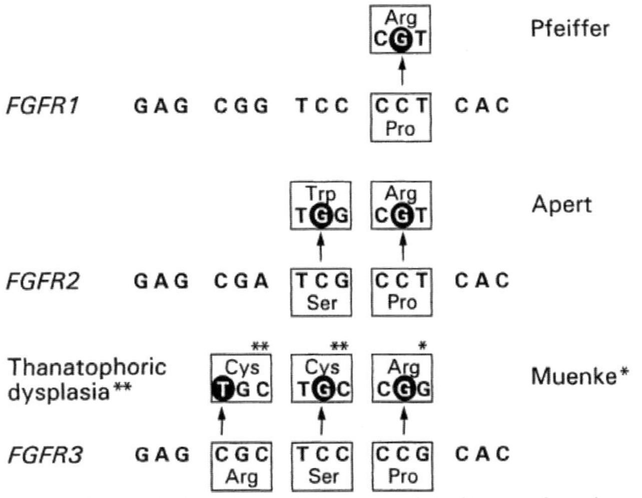

Figure 3. Common point mutations in FGFR gene related to syndromic craniosynostosis (Source: Muenke M, Wilkie A.O)

TGF-β / BMP Signaling

TGF-β family proteins consist of TGF-βs, BMPs, activins, and growth factors act through receptor kinases. TGF-β signaling is correlated with ERK1/2. Bone Morphogenic Proteins (BMPs) bind to type I/II transmembrane receptors. The activated receptors will cause signal initiation by phosphorylation of target proteins – Smad1, Smad5, Smad8 and Smad4 where Smad4 enter the nucleus and interact with RUNX2 and MSX2 gene expression. Overexpression of BMP signaling cause increased expression of RUNX2, thus lead to fusion of cranial suture. BMP signaling is regulated by extracellular modifiers such as noggin and glypicans. Noggin is BMP antagonist. Thus, overexpression of noggin relates to prevention of sutural fusion [4,16].

Wnt Signaling

The wingless-type integration sites (Wnts) molecules act through FZD (Frizzled 1–10) receptors to initiate intracellular signals and to trigger the accumulation of β –catenin that rule a variety of cell actions. β – catenin enter the nuclei and binds to T cell factor/lymphoid enhancer factor (TCF/LEF) transcription factors. Several reports have been made in relation to the importance of Wnt signaling for the proliferation of the neural crest mesenchyme. β – catenin is important in determining the differentiation of neural crest mesenchyme to osteoblast or chondrocytes [4,14].

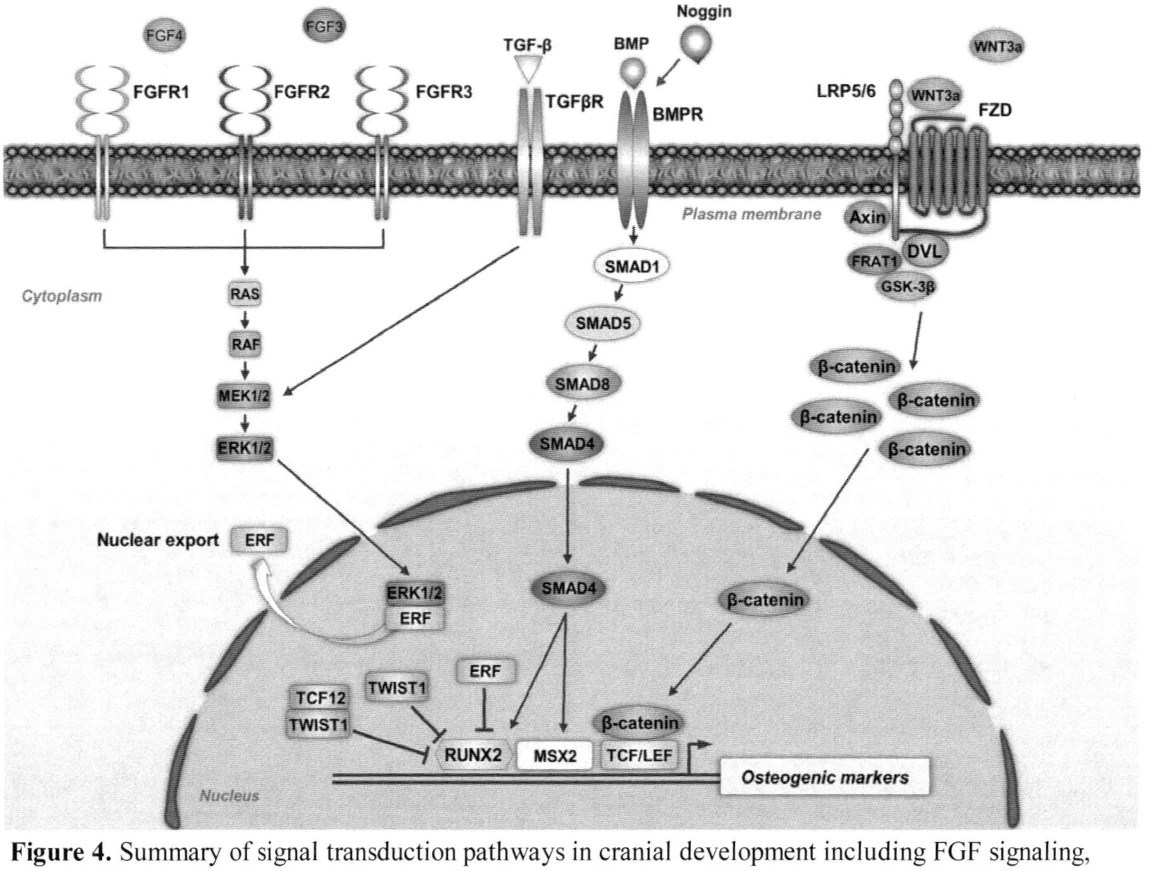

Figure 4. Summary of signal transduction pathways in cranial development including FGF signaling, TGF-β/BMP signaling, and Wnt signaling (Source: Katsianou MA, Adamopoulos C, 2016) [4].

3.4. Epigenetics of craniosynostosis

Nowadays, the term epigenetics referred to the study of changes in heritable gene function without changes in the sequence of the DNA. Environmental factors might bring an effect to epigenetic factors such as histone modification, DNA methylation, and post-transcriptional silencing by RNA interference. Epigenetics processes might cause gene expression or gene repression [20].

Dudacovic et al. (2015) conduct a study focusing on expression profiles of epigenetic regulators that influence osteogenesis. Molecular analysis revealed suppression of Enhancer of Zeste Homolog 2 (EZH2) during differentiation stage of adipose – tissue derived mesenchymal cells. EZH2 gene is located on the long arm (q) of chromosome 7, region 3, band 6, sub-band 1 (7q36.1). EZH2 gene is responsible for making histone methyltransferase. Histone methytransferase modify structural proteins that bind to DNA and shape chromosomes, called histone [21]

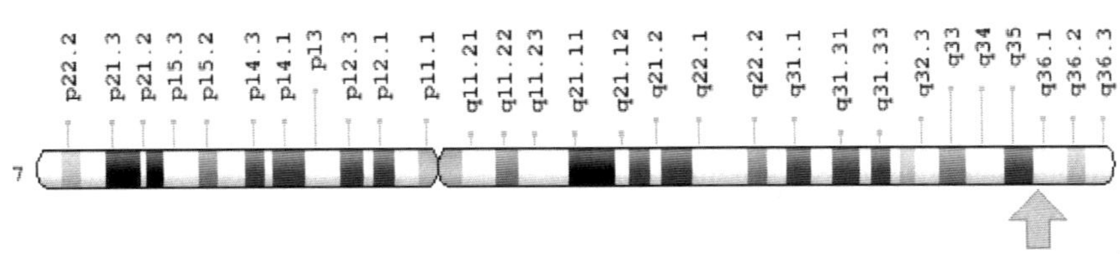

Figure 5. Location of EZH2 gene – 7q36.1 (Source: https://ghr.nlm.nih.gov/gene/EZH2)

EZH2 catalyzes tri–methylation of histone 3 lysine 27 (H3K27me3) that stop differentiation of osteogenic cells. EZH2 mediated H3K27 methylation assists chromatin condensation and heterochromatin formation. Heterochromatin limits the target DNA transcription. mRNA sequence analysis reveals that EZH2 gene inhibition encourages the production of bone extracellular matrix protein needed for a mineralized matrix. The study concluded that depletion of functional EZH2 in the mesenchyme leads to craniosynostosis [21].

MicroRNAs (miRNAs) are part of the non-coding RNAs, serve as regulators of gene expression after transcription process. MicroRNAs are involved in biological processes including cell differentiation, cell proliferation, and metabolism. FGF signaling pathway can be regulated by microRNAs. Loss of miRNAs regulation might cause disease progression. Regulation of FGF or FGFR expression by microRNAs can directly affect differentiation of a cell. For example, miR-338 in osteoblastic cells was found to unswervingly affect the regulation of 3' untranslated region (UTR) of FGFR2 that result in suppression of FGFR2 expression. Down regulation of miR-338 cause an increase in FGFR2 expression thus enhancing osteoblast differentiation [22].

3.5. Orthodonticperspective of craniosynostosis management

Craniosynostosis should be managed by a multispecialty team providing interdisciplinary care because of the complexity of the case. The interdisciplinary team comprises of professionals from several disciplines including: anesthesiologist, craniofacial surgeon, genetician, neurosurgeon, ophthalmologist, oral and maxillofacial surgeon, orthodontist, otolaryngologist, pediatricians, pedodontist, prosthodontist, psychologist, radiologist, and speech and language therapist [23].

Dental care for the patients with craniosynostosis yields special care because they have unique oral health and craniofacial growth problems. Common problems including malformed or missing teeth, delayed and ectopic eruptions, crowding, caries, and difficulties on performing oral hygiene procedure due to lack of motivation and limited physical ability. Orthodontic care facilitates better oral hygiene procedures by creating well-aligned and coordinated dental arches. Orthodontist serves as manager of growth data that plan the timing of jaw reconstruction and facial surgery [23].

The orthodontist might need to make dental casts, serial photographs, dental and cephalometric radiographs. Hand wrist radiographs, in some patients, might be needed to plan the interventions. The active role of the orthodontist in preparing the dentition at each stage is required, including: prediction tracings for surgery and provides treatment as indicated. The following table summarizes the dental, orthodontic, and surgical interventions according to child's age:

Table 1. Summary of dental, orthodontic, and surgical interventions [23].

Age	Intervention
Infancy onward	Pedodontist provides oral hygiene instruction, prophylactic and restorative treatment as needed
	Orthodontist takes serial photographs and radiographs as needed at regular intervals. Growth monitoring and coordination with pedodontist regarding dental extractions and with surgeons regarding midface and orthognathic procedures.
7 – 12 years	Phase 1 orthodontic treatment, including extractions
	Midface advancement planning

12 – 21 years	Phase 2 orthodontic treatment, planning for orthognathic surgery, records taking at regular interval, and detailing of orthodontic postsurgical.
21 years onward	Observation and records as needed

4. Conclusions

Craniosynostosis, the premature suture fusion of the skull, can be classified in many ways. It can involve one or more sutures. It can also be isolated nonsyndromic without other medical complications or syndromic if there are any other medical conditions involved. Studies in molecular biology of the cell suggest that a correct balance between cellular proliferation and differentiation is the key factor in the development of the skull. Regulation of cranial suture formation and patency by signaling pathways and transcription factors have been identified. The complex morphogenetic processes are arranged by FGF, TGF-β, BMP, and Wnt signaling pathway that respond to extracellular factors and environmental signals.

Many complications might be associated with craniosynostosis because of the increase in intracranial pressure, including sensory, respiratory, and neurological function, making craniosynostosis important to detect and treat. Craniosynostosis management involves a multidisciplinary team including professionals from many disciplines. Among them are a craniofacial surgeon, geneticist and orthodontist. Orthodontist's knowledge of craniofacial growth and the nature of the various craniosynostosis conditions is valuable in the timing and planning of jaw and facial surgery.

References

[1] Connerney JJ, Spicer DB (2011) Signal Transduction Pathways and their Impairment in Syndromic Craniosynostosis. In: *Craniosynostoses: Molecular Genetics, Principles of Diagnosis, and Treatment*. Basel: Karger 28-44.

[2] Beederman M, Farina EM, Reid RR (2014) Molecular basis of cranial suture biology and disease: Osteoblastic and osteoclastic perspectives. *Genes Dis* **1** (1) 120-125.

[3] Mitchell L (2013) *Introduction to Orthodontics*. 4th ed. London: Oxford University Press; 2013.

[4] Katsianou MA, Adamopoulos C, Vastardis H, Basdra EK (2016) Signaling mechanisms implicated in cranial sutures pathophysiology: Craniosynostosis. *BBA Clin* **6** 165-176.

[5] Khonsari RH, Olivier J, Vigneaux P, et al (2013) A mathematical model for mechanotransduction at the early steps of suture formation. *Proc Biol Sci* **280** (1759) 20122670.

[6] Johnson D, Wilkie AOM (2011) Craniosynostosis. *Eur J Hum Genet* **19** (4) 369-376.

[7] Twigg SRF, Wilkie AOM (2015) A Genetic-Pathophysiological Framework for Craniosynostosis. *Am J Hum Genet* **97** (3) 359-377.

[8] Cox TC, Horst JA, Hing A V, Cunningham ML (2015) Molecular Genetics and Biology of Craniofacial Craniosynostoses. *Stem Cell Biol Tissue Eng Dent Sci* **5** 499-520.

[9] Proffit WR, Fields HW, Sarver DM (2013) *Contemporary Orthodontics* **142** 5th ed. St.Louis; 2013.

[10] Senarath-Yapa K, Chung MT, McArdle A, et al (2012) Craniosynostosis. *Organogenesis* **8** (4) 103-113.

[11] Greenwood J, Flodman P, Osann K, Boyadjiev SA, Kimonis V (2014) Familial incidence and associated symptoms in a population of individuals with nonsyndromic craniosynostosis. *Genet Med* **16** (4) 302-310.

[12] Anderson IA, Goomany A, Bonthron DT, et al (2014) Does patient ethnicity affect site of craniosynostosis? *J Neurosurg Pediatr* **14** (6) 682-687.

[13] Faradz SMH, Mundhofir FEP, Sistermans EA, Hamel BCJ (2013) P.Ser252Trp and p.Pro253Arg mutations in FGFR2 gene causing apert syndrome: The first clinical and molecular report of indonesian patients. *Singapore Med J* **54** (3).

[14] Wilkie AOM, Byren JC, Hurst J a, et al. (2010) Prevalence and complications of single-gene and chromosomal disorders in craniosynostosis. *Pediatrics*.126(2):e391-e400.

[15] Kouskoura T, Fragou N, Alexiou M, et al (2011) The genetic basis of craniofacial and dental abnormalities. *Schweiz Monatsschr Zahnmed* **121** (7) 636-646.

[16] Gou Y, Zhang T, Xu J (2015) Transcription Factors in Craniofacial Development. From Receptor Signaling to Transcriptional and Epigenetic Regulation. *Current Topics in Developmental Biology* **115** 377-410.

[17] Marie PJ (2012) Fibroblast growth factor signaling controlling bone formation: An update. *Gene*. **498** (1) 1-4.

[18] Ciurea AV, Toader C (2009) Genetics of craniosynostosis: review of the literature. *J Med Life* **2** (1) 5-17. http://www.ncbi.nlm.nih.gov/pubmed/20108486.

[19] Fitzpatrick DR (2013) Filling in the gaps in cranial suture biology. *Nat Genet* **45** (3) 231-232.

[20] Armstrong L (2014) *Epigenetics*. New York: Garland Science, Taylor & Francis Group, LLC; 2014.

[21] Dudakovic A, Camilleri ET, Xu F, et al (2015) Epigenetic control of skeletal development by the histone methyltransferase Ezh2. *J Biol Chem* **290** (46) 27604-27617.

[22] Ornitz DM, Itoh N (2015) The fibroblast growth factor signaling pathway. *Wiley Interdiscip Rev Dev Biol* **4** (3) 215-266.

[23] Vargervik K, Rubin MS, Grayson BH, et al (2012) Parameters of care for craniosynostosis: Dental and orthodontic perspectives. *Am J Orthod Dentofac Orthop* **141** (4 SUPPL):S68-S73.

Medicinal plants used in the treatment of diabetes in karo ethnic, north sumatra, indonesia

Barita Raja Nasution[1*], T. Alief Aththorick[2], and Suci Rahayu[2]

[1]Graduate Study Programme of Biology, Faculty of Mathematics and Natural Sciences, University of Sumatera Utara, Jalan Bioteknologi Nomor 1, Campus USU. Medan City, North Sumatera, Indonesia.
[2]Department of Biology, Faculty of Mathematics and Natural Sciences, University of Sumatera Utara, Jalan Bioteknologi Nomor 1, Campus USU. Medan City, North Sumatera, Indonesia

[*]E-mail: baritarajanasution@gmail.com

Abstract. Medicinal plants derived from traditional medicines have played an important role in managing a variety of healthcare and diseases in Karo ethnic in North Sumatra, Indonesia. The study aimed to document the ethnobotanical information on medicinal plants used by traditional healers of Karo ethnic in the treatment of diabetes and to assess the crude extract of phytochemical constituents qualitatively from medicinal plant organ. The study was conducted on two Karo sub-ethnic living at the highland (Karo Gugung) in Karo Regency and the lowland (Karo Jahe) in Langkat Regency with the length of the study was eight months of observation. The survey was conducted using open-ended interviews among four traditional healers those who were selected by snowball sampling method; quantitative analysis of ethnobotanical data was performed by calculating the familiarity index (Fi). Fresh plant samples which were used for phytochemical analysis were collected using participatory method. The results showed that 15 plants were used to treat diabetes by Karo traditional healers. The plants are *Blumea balsamifera, Nypa fruticans, Bischofia javanica, Eleutherine americana, Allium cepa, A. sativum, Eugenia polyantha, Piper betle, P. nigrum, Citrus aurantiifolia, Boesenbergia pandurata, Curcuma longa, Kaempferia galanga, Zingiber montanum,* and *Z. officinale.* Familiarity index (Fi) value of each plant was 25 which explained that each of medicinal plant was used by only one traditional healer. The phytochemical screening showed that the crude plant extracts contained phenolic, terpenoid, steroid, and saponin.

Keywords: Etnobotany, phytochemistry, traditional medicine

1. Introduction

Use of ethnobotanical information in medicinal plant research has gained attention among the scientific community for the past few decade [1,2]. Herbal medicine showed an exponential growth and are gaining popularity in many countries in the present days because of their natural origin and fewer side effects [3]. Diabetes is the most current health problem in the world [4]. Diabetes is a chronic progressive metabolic disorder with life threatening complications including cardiovascular diseases, retinopathy, nephropathy, neuropathy which may eventually lead to death [5] and characterized by high blood glucose levels [6]. High blood glucose levels is either as a result of insufficient endogenous insulin production by the pancreatic beta cells (otherwise known as type-1 diabetes); or impaired insulin secretion and/or action (type-2 diabetes). Type-1 diabetes is an autoimmune disease characterized by T-cell mediated destruction of the pancreatic beta cells. In type-2 diabetes, there is a gradual development of insulin resistance and beta cell disfunction, strongly associated with obesity and a sedentary lifestyle [6,7]. Changes in human behavior and lifestyle over the last century have resulted in a dramatic increase on the incidence of diabetes worldwide [8]. The world prevalence of diabetes in the present days is 9% [4] and the incidence which was predicted among adults (aged 20-79) will increase from 6.4% in 2010 to 7.7% by 2030 and there will be 69% incrase in numbers of adults with diabetes in developing countries and a 20% increase in developed countries [9].

Insulin injection and hypoglycemic agents are essential and effective drugs for diabetes mellitus, but these compounds possess several adverse effects and have no effects on diabetes complications in the long term [10]. Complication is the major cause of morbidity and mortality in diabetes and management of diabetes without any side effects is still a challenge to the medical system [11]. In recent years, some studies have been conducted on the use of herbal/natural products in diabetes [4,12]. Most of the world population used traditional medicines for healthcare and medicine because of more effective, safe, and easy access [11]. Based on estimation opined by World Health Organization, more than three-quarters of population of developing nations use herbal medicine for their primary health care [13]; and various medicinal plants derived from traditional medicine have played important role in the treatment of various diseases in up to 80% of world population [10] and although many currently available drugs that have been derived from plants [11], there is no enough scientific information about some of them [10].

Various plants in nature are an extraordinary source of antidiabetic medicines [14]. Current anti-diabetic drugs as result that were gathered from research and development of anti-diabetic drugs suffer from lack of efficacy and also exert undesirable side effect, there by demanding new remedies [14,5]. Forest degradation as a consequence of forest conversion in many appropriate areas of Sumatra [15] has threatened the existence of medicinal plants used by local community [16]. Most plants used in traditional medicine were collected from the wild and only Few are cultivated [2]. In addition, only few traditional healers had apprentice [17]. Hence study of medicinal plants is needed.

Karo ethnic in North Sumatra, Indonesia has possessed a culture of using traditional medicine since a long time ago [18,19]. Karo community lived near the forest with various medicinal plants that played important roles in Karo's culture [19]. There are still few publications on medicinal plants in managing diabetes used by local community of Karo ethnic, North Sumatra [19,20,21].

Indonesia has come under scrutiny to reveal new information on diabetes remedies. Several parameters have been proposed such as study area and informant for interviews. The objective is to explore the medicinal plants used in the treatment of diabetes. The results are expected to supply the necessary information for the Indonesian Government to reveal new diabetes remedies in developing alternative solutions for managing and controlling diabetes.

2. Methods

2.1. Study area

The study was carried out in three districts consists of Berastagi District in Karo Regency, Merdeka District in Karo Regency, and Sei Bingai District in Langkat Regency, Province of North Sumatra, Indonesia. Geographical location of Karo Regancy and Langkat Regency are shown in Figure 1. Karo and Langkat Regency has a tropical climate with two seasons (Karo Regency has first rainy season from August until January, second rainy season at March until May, dry season at January, Juni, and July; Langkat Regency has first rainy season from August until February, second rainy season April and May, dry season at March, Juni, and July). Berastagi and Merdeka are two districts in Karo Regency, Berastagi District is geographically located at 98°28'-98°31'E and 03°10'-03°14'N at a distance of 67 km from Medan, the capital city of North Sumatra, with the elevation of 1.375 meters above sea level, with temperature ranging between 19°C-26°C. Merdeka District is at 98°27' - 98°31'

E and 03°10' - 03°14' N, with elevation of 1.162-1.453 meters above sea level, with temperature ranging between 16°C-17°C , at a distance of 65 km from Medan [22]. Whereas Sei Bingai is a district of Langkat Regency, which is located at 98°28'–98°30E and 03°11'–03°15'N at a distance of 60 km from Medan, with elevation of 106 meters above sea level [23].

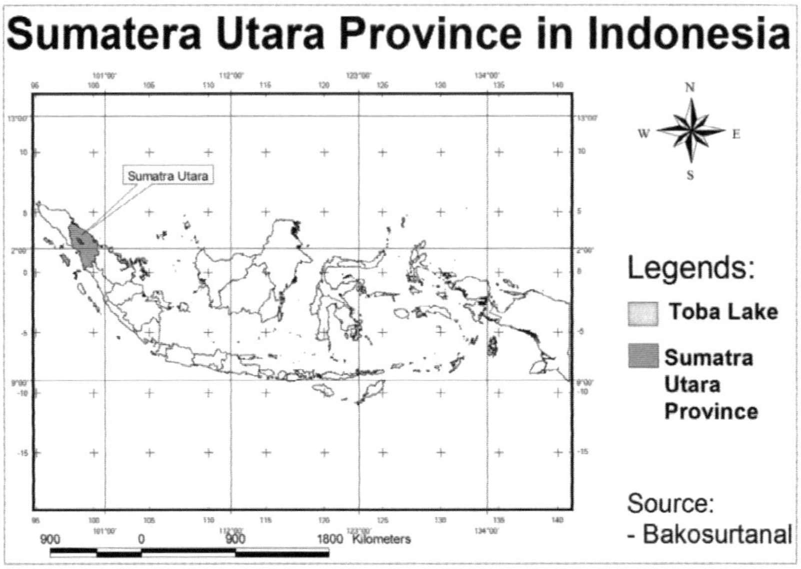

A

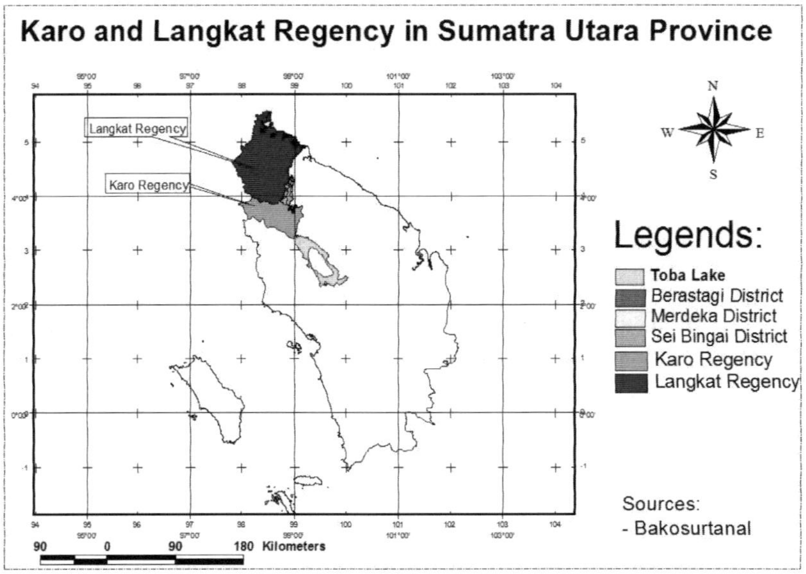

B

Figure 1. Study area location: (A) Sumatra Utara Province position in Indonesia, (B) Karo and Langkat Regency position in Sumatra Utara Province

Karo and Langkat Regency are cities in North Sumatra, Indonesia where Karo ethnic lived [18]. Primary forest in Leuser National Park (TNGL–Taman Nasional Gunung Leuser) encompass area from Province of Aceh until some part of North Sumatra in Karo and Langkat Regency as presented in

Figure 2. Karo people traditional life was connected to plants that were near to their settlement. Karo people's occupation is mainly farmer [18], but field observation during survey revealed that Karo life style slowly changing to be sedentary life style in recent years because of the alteration of their way to process their cultivation farm by hiring the immigrant.

2.2. Ethnobotanical data collection

Ethnobotanical data were collected using a method similar to Chinsembu (2016) [17]. Interviews were conducted among four traditional healers as respondents; snowball sampling method was used to obtain the respondents who were originated from two villages in Karo Regency representing Karo Gugung Sub ethnic (Karo people living at highland) and the rest two were from Langkat Regency who represented Karo Jahe sub ethnic (Karo people living at lowland). The first respondent of traditional healer was obtained based on information from local people and become key gatekeepers. Key gatekeepers were used to verify the presence of and access to next traditional healer in the study area.

Researcher introduced the team and explained the study to respondent before conducting interviews. Open ended interviews were performed individually with the traditional healer. Data of characteristics and information related to the medicinal use of plants for the management of diabetes were noted and recorded during interviews. Interviews were conducted in Indonesian language (*Bahasa Indonesia*).

During conversations with respondent of the traditional healer, information about the main symptoms of diabetes, traditional healing practice, and sources of ethnomedicinal knowledge were acquired. The data in relation to the plants were also recorded as follows: Local names, plant habits, plant parts used, the diabetes condition treated with the plants, and the modes of preparation and application of the plant remedies to the patient.

Participatory method was used to gather plants material sample for voucher specimens and crude extract for phytochemical screening. On site identification was conducted to assess familiar plant species. Voucher specimens of all plants were collected in herbarium plant presses and transported to the Herbarium MEDA, Department of Biology, Faculty of Mathematics and Natural Sciences, Universitas Sumatra Utara, Indonesia for identification, confirmation, and storage. IPNI (International Plant Name Index) was used to verify the botanical name of plants species. Plant material for crude extract phytochemical screening was weighed about 20 grams of fresh plant material. Phytochemical screening was conducted at Laboratory of Natural Product, Department of Chemistry, Faculty of Mathematics and Natural Sciences, Universitas Sumatra Utara, Indonesia.

2.3. Data analysis

Familiarity index (*Fi*) was used to acquire the quantitative analysis [17]. The *Fi* is a relative indicator of the familiarity of a plant species, which defined as the frequency of a given plant species mentioned as a medicine divided by the total number of traditional healers interviewed in the study. The *Fi* was calculated as follows:

$$Fi = \frac{N_a}{N_b} x100 \quad (1)$$

Where *Na* is the number of informants that mentions a species as a medicine, and *Nb* is the total number of respondents.

3. Results and Discussion

The result will be discussed in three subsections; they are traditional healers, ethnobotanical flora, and Phytochemical Screening of Medicinal Plants.

3.1. Traditional healer

Three out of four traditional healers had ever been engaged in some academical research as the respondent before the present study. The traditional healers were aged between 43 until 88 years old. Three out of four traditional healers acquired their medicinal plant knowledge by inheriting from their elder lineage (father or grandfather), whereas the rest received his medicinal plant knowledge from dreams. Only one traditional healer had an apprentice under his tutelage; the rest had none. Study on herbal medicines used in the treatment of diabetes in Arunachal Himalaya, Northeast Iran reported 76% of traditional healers were aged 40 years and above; 87% of traditional healer accepted their knowledge from their parents [11]. The traditional healers in this study received approximately 100-300 patients in one month. Field observations during survey revealed that the patients choosed to use traditional medicine despite modern medicine were because of the efficacy, low cost, and less side effect.

3.2. Ethnobotanical flora

Information on the plant's scientific names, local names, families, voucher numbers, habits, frequency indices, and parts used are presented in Table 1. Whereas information on traditional medicine remedies including diabetes severity treated, medicinal plants compositions, modes of preparation and application, other therapy used are presented in Table 2. Of a total fifteen plants were used by the traditional healer to treat diabetes in Karo and Langkat Regency, Sumatra Utara Province, Indonesia. The study revealed that the value of all plants were 25.

Table 1. Plants families, species, habits, part(s) used, and frequency index (*Fi*) of medicinal remedies in Karo ethnic, Sumatra Utara Province, Indonesia

No.	Family Species, local name, voucher number	Plant habit	Part(s) used	Frequency index
	Arecaceae			
1.	*Nypa fruticans* Wurmb, Ipah, brn02	Tree	Leaves	25
	Asteraceae			
2.	*Blumea balsamifera* DC., Galunggung, brn01	Shrub	Leaves	25
	Euphorbiaceae			
3.	*Bischofia javanica* Blume, Cingkam, brn03	Tree	Bark	25
	Iridaceae			
4.	*Eleutherine americana* Merr. ex K.Heyne, Pia-pia, brn04	Herb	Whole part	25
	Liliaceae			
5.	*Allium cepa* L., Bawang megara, brn05	Herb	Bulb	25
6.	*Allium sativum* L., Bawang mentar, brn06	Herb	Bulb	25
	Myrtaceae			
7.	*Eugenia polyantha* Barb.Rodr., Salam, brn07	Medium tree	Leaves	25
	Piperaceae			
8.	*Piper betle* L., Belo, brn09	Climbing shrub	Leaves	25
9.	*Piper nigrum* L., Lada, brn08	Climbing shrub	Fruits	25
	Rutaceae			
10.	*Citrus aurantiifolia* (Christm.) Swingle, Rimo acem, brn10	Small tree	Fruits	25
	Zingiberaceae			
11.	*Boesenbergia pandurata* Schltr., Temu-temu, brn11	Herb	Rhizome	25
12.	*Curcuma longa* L., Kuning gersing, brn12	Herb	Rhizome	25
13.	*Kaempferia galanga* L., Kaciwer, brn13	Herb	Rhizome	25
14.	*Zingiber montanum* Link ex A.Dietr., Kuning gajah, brn14	Herb	Rhizome	25
15.	*Zingiber officinale* Roscoe, bahing megara, brn15	Herb	Rhizome	25

All medicinal plants collected in this study were distributed into 9 families and 12 genera. The highest proportion of plant family was Zingiberaceae (5 species), followed by Liliaceae (2 species) and Piperaceae (2 species). The other plant family had only one species. The highest proportion of plants genera were *Allium, Piper,* and *Zingiber* with two species for each genera. Other genera had only one species. The study of the medicinal plants that collected by medicinal plants trader from Kabanjahe traditional market, Karo Regency, North Sumatra reported most of medicinal plants were species from Zingiberaceae family and that study reported seven species were used to treat diabetes [20] aside from the medicinal plants recorded in the present study. Of all plants species reported from that study, only one species (*Bischofia javanica* Blume, family Euphorbiaceae) was the same reported with in the present study. Another study about diabetes medicinal woody plants in Karo ethnic using diabetes suffering patients, farmers, local merchants, traditional medicinal processors, and the collectors reported 12 species distributed in 11 families [21]; only one species (*Eugenia polyantha* Barb.Rodr., family Myrtaceae) was the same species compared to the present study. Augmented amount of medicinal plant species reported in the present study with other two studies mentioned above revealed that total of thirty-two diabetes medicinal plants species in Karo ethnic.

Medicinal plants used to treat diabetes in the present study had various habits. 53% of total plants recorded in the present study were herbs, 27% trees, 13% climber, and 7% shrubs.The study of herbal medicines used in the treatment of diabetes mellitus in Arunachal Himalaya reported of 68% from 46 diabetes medicinal plants recorded were herbs and trees, and leaves were the majority part in the herbal formulations [11]. Medicinal plant parts used in diabetes remedies in the present study varies. 33% of plants parts used to treat diabetes in the present study were rhizomes, followed by leaves 27%, fruits 13%, bark 7% and whole parts 7%.

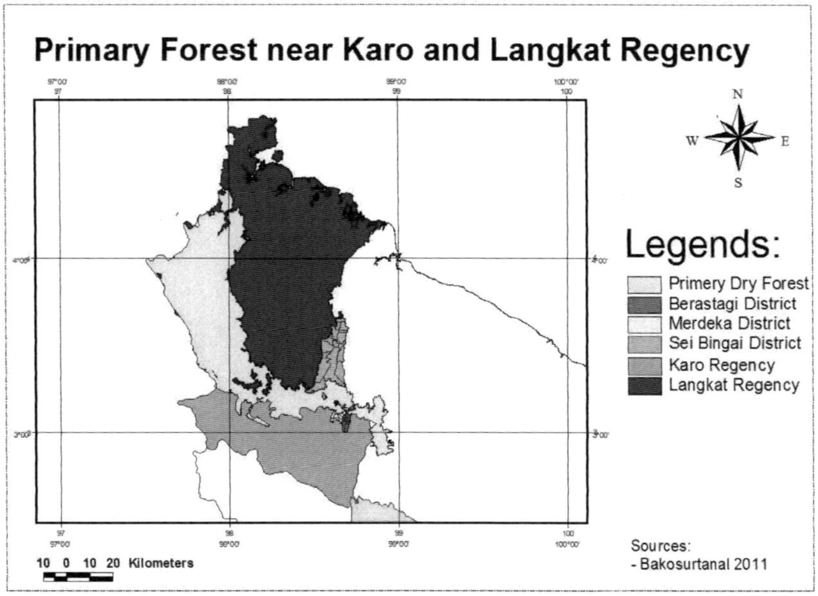

Figure 2. The primary forest near Karo and Langkat Regency, Sumatra Utara Province, Indonesia

The Familiarity index (*Fi*) value of 25 explained that each plant acquired from the study were used by only one traditional healer. The Familiarity index value of 25 indicated the high biodiversity of flora around the traditional healer settlement. The present study site which was located in the community settlement was near to primary forest as presented in Figure 2. Of all diabetes medicinal plants recorded in the present study, 27% were harvested form the wild. The results were contradictive compared to the data reported from another study that most of the medicinal plants species that were traded by medicinal plants trader in Kabanjahe traditional market, Karo Regency were harvested from the wild [20]. Meanwhile, the report on forest existence progress in Langkat Regency, North Sumatra reported the deforestation in Langkat Regency [24]. Hence the conservation of medicinal plant is needed. The present study recorded that most of the medicinal plants were cultivated because one of the traditional healer used the multi-component herbal formulations, one of which formulated of cultivated plants, the formulations had just being used to treat diabetes since last few months before the present study was conducted. The medicinal remedies which is used to treat diabetes was used to treat wounds and ulcers. The traditional healer has no knowledge on diabetes before; Therefore, the medicinal remedies showed good result in healing the diabetes patient within few months from severe ulcer for the first time when he used the medicinal remedies.

IOP Conf. Series: Earth and Environmental Science **130** (2018) 012038 doi:10.1088/1755-1315/130/1/012038

3.3. Ethnopharmacological analysis and medicinal plants crude extract phytochemistry

The traditional healers in the present study diagnosed diabetes by observing symptoms such as excessive and frequent urination, prolonged healing of wounds, polydipsia, polyphagia, weight loss, languid, difficult to concentrate, and blurred vision. Other than that, the traditional healer used the result of diagnosed with blood test laboratory. Hence two of the traditional healers in the present study made the medicinal formulations to treat diabetes based on the blood glucose levels as presented in Table 2. The study on diabetes herbal medicines from India reported that the traditional medicine diagnosed the diabetes patient by observing symptoms such as excessive urination, the presence of sugar in urine and swelling of legs [11].

Table 2. Traditional remedies used in the treatment of diabetes

No.	Diabetes with blood glucose level condition treated	Plants composition	Preparation and administration	Other therapy
1.	< 500 mg/dL	*Bischofia javanica* Blume	Crushed bark are fermented in nipa palm vinegar, drink concoction	-
2.	> 500 mg/dL	*Bischofia javanica* Blume	Crushed bark along with sweetened condensed milk and palm sugar are boiled in nipa palm vinegar, drink	-
3.	Blood glucose level is not determined, but used for diabetes patient with ulcer	*Allium cepa* L., *Allium sativum* L., *Boesenbergia pandurata* Schltr., *Citrus aurantiifolia* (Christm.) Swingle, *Curcuma longa* L., *Kaempferia galanga* L., *Piper nigrum* L., *Zingiber montanum* Link ex A.Dietr., and *Zingiber officinale* Roscoe	Pound all material except *Citrus aurantiifolia* (Christm.) Swingle at once, add squeezed of *Citrus aurantiifolia* (Christm.) Swingle fruit, pour into special vessel made from bamboo segment, externally used by rubbing concoction around ulcer, internally used by adding concoction into water and drink	-
4.	> 500 mg/dL	*Blumea balsamifera* DC.	Material is squeezed to drip out the fluid, drink	Reflexology
5.	< 500 mg/dL	*Eugenia polyantha* Barb.Rodr., *Nypa fruticans* Wurmb, *Piper betle* L.	Boil materials in water, drink decoction	Reflexology
6.	Blood glucose level is not determined	*Eleutherine americana* Merr. ex K.Heyne	Boil material in water, drink decoction	Param

The traditional healers in the present study used the single component recipeherbal formulations and also multi-component formulations as presented in Table 2. Herbal therapy with its richness and complex compounds in plants compositions has always been thought to act on multiple targets on the human body [14]. The modes of remedies processing used by the traditional healer in the present study were fermentation, boiling, pounding, and squeezing. Whereas, the administration of formulations which were applicated to the patient in the present study were orally administrated and rubbed to the certain body surface. The study on the usage of fermentation in the administration of herbal medicines reported that the fermentation improved the efficacy or reducing adverse effects of herbal medicines [5]. The traditional healers in the present study announced that the diabetes patients were cured without known side effects. The study on diabetes herbal medicines from India reported that the traditional healer announce that the diabetes patients were cured without any known serious side effect [11]. In addition to the traditional medicines to treat diabetes in this study, the other therapies were applicated such as reflexology and param. The traditional healer announced that the used of reflexology gave effect to heal and to stimulate the pancreas in producing insuline. The using of reflexology resulted to activate the body's own healing system to heal and prevent ill [25].

The phytochemical screening of medicinal plants in the present study revealed that crude extract contained phytochemical compounds such as phenolic, terpenoid, steroid, alkaloid, and saponin (as presented in Table 3). The crude extract that was screened from six medicinal formulations made by Karo traditional healers were phenolic, terpenoid, steroid, and saponin. In case of *P. betle* L crude extract, the data was different with the information reported in another study on the antioxidant effect of *P. betle* Linn. that major constituent of *P. betle* L. was phenol [26]. The absence of phenol constituent in *P. betle* L. that was shown in the present study because of plant organ sample of *belo* (*P. betle* L. local name in Karo ethnic) that was acquired randomly without specific kind as the informant (the traditional healer who used *P. betle* L. in his medicinal formulations) explained to the author. Several observations and discussions among Karo people in the study area revealed that there were three kinds of *P. betle* L. used among Karo people which consisted of *belo* (*sirih*), *beluntas*, and *sirih merah*. The three kinds of *P. betle* L. aforementioned were similar to each other. *Belo* (*sirih*) was the one used in typical medicinal formulations in Karo ethnic and was traded widely in the traditional market in Karo ethnic settlements; *beluntas* and *sirih merah* were usually used as ornamental besides as ingredients in medicinal formulations in Karo ethnic tradition.

Table 3. Phytochemical screening result of the medicinal plants used by Karo traditional healers in the treatment of diabetes

No.	Family Species, Karo name	Part(s) used	Phytochemical constituent	References
	Arecaceae			
1.	*Nypa fruticans* Wurmb, Ipah	Leaves	Phenolic, terpene, steroid, saponin	
	Asteraceae			
2.	*Blumea balsamifera* DC., Galunggung	Leaves	Phenolic, terpene, steroid	
	Euphorbiaceae			
3.	*Bischofia javanica* Blume, Cingkam	Stem bark	Phenolic, terpene, steroid, alkaloid, saponin	
		Root bark	Phenolic, terpen, steroid, alkaloid	
	Iridaceae			
4.	*Eleutherine americana* Merr. ex K.Heyne, Pia-pia	Root	Terpene, steroid	
		Bulb	Phenolic, terpene, steroid, alkaloid, saponin	
		leaves	Phenolic, terpene, steroid, alkaloid	
	Liliaceae			
5.	*Allium cepa* L., Bawang megara	Bulb	Phenolic, terpene, steroid, alkaloid	
6.	*Allium sativum* L., Bawang mentar	Bulb	Terpen, steroid, alkaloid	
	Myrtaceae			
7.	*Eugenia polyantha* Barb.Rodr., Salam	Leaves	Phenolic, terpene, steroid, alkaloid	
	Piperaceae			
8.	*Piper betle* L., Belo	Leaves	terpene, steroid, saponin	
9.	*Piper nigrum* L., Lada	Fruits	Phenolic, terpene, steroid, alkaloid	Gulcin, 2005; Nahak & Sahu, 2011
	Rutaceae			
10.	*Citrus aurantiifolia* (Dhristm.) Swingle, Rimo acem	Fruits	Phenolic, terpene, steroid, alkaloid	
	Zingiberaceae			
11.	*Bosenbergia pandurata* Schltr., Temu-temu	Rhizome	Phenolic, terpene, steroid, alkaloid	
		Root	Terpene, steroid	
12.	*Curcuma longa* L., Kuning gersing	Rhizome	Phenolic, terpene, steroid, alkaloid, saponin	
13.	*Kaempferia galanga* L., Kaciwer	Rhizome	Terpene, steroid, alkaloid	
14.	*Zingiber montanum* Link ex A.Dietr., Kuning gajah	Rhizome	terpene, steroid, alkaloid, saponin	
15.	*Zingiber officinale* Roscoe, Bahing megara	Rhizome	terpene, steroid, alkaloid, saponin	

4. Conclusions

The present study revealed that community in Karo and Langkat Regency still used the traditional medicine in the treatment of diabetes and other ailments or diseases. The traditional healer harvested the medicinal plants from the wild and the deforestation were two things that threated the existence of the medicinal plants and the traditional medicine itself. Hence the preserving of the traditional medicines knowledge and the cultivation of medicinal plants are important for the conservation of traditional medicines. The phytochemical screening result of diabetes herbal formulations made by Karo ethnic traditional healer in the treatment of diabetes were phenolic, terpenoid, steroid, and saponin. The efficacy and safety of all the reported plants with antidiabetic potential in the present study need to be evaluated for phytochemical and pharmacological studies in the further experimental investigations such as to determine the dosages, minimum inhibitory concentrations, bioactive compounds, and toxicities.

Acknowledgment
The authors gratefully acknowledge that the present research is supported by Ministry of Research and Technology and Higher Education Republic of Indonesia. The support is under the research grant BPPTN USU (*Bantuan Penelitian Perguruan Tinggi Negeri Universitas Sumatra Utara*) of the Year 2016 Contract Number 6049/UN5.1.R/PPM/2016.

References

[1] Heinrich M (2000) Ethnobotany and its Role in Drug Development. Phytother. Res. 14: 479-488.

[2] Cotton CM (1996) Ethnobotany: Principles and Applications. West Sussex: John Wiley and Sons Ltd, 13-14 p.

[3] Modak M, Dixit P, Londhe J, Ghaskadbi S, Devasagayam TPA (2007) Indian Herbs and Herbal Drugs Used for the Treatment of Diabetes. J Clin Biochem Nutr 40: 163-173.

[4] Tabatabaei-Malazy O, Larijani B, Abdollahi M (2013) A Novel Management of Diabetes by Means of Strong Antioxidants' Combination. Journal of Medical Hypotheses and Ideas 7:25-30. 10.1016/j.jmhi.2012.12.002

[5] Hussain A, Bose S, Wang J-H, Yadav MK, Mahajan GB, Kim H (2016) Fermentation, A Feasible Strategy For Enhancing Bioactivity of Herbal Medicines. Food Research International 81: 1-16. DOI:10.1016/j.foodres.2015.12.026

[6] Bilous R, Donelly R (2014) Buku Pegangan Diabetes. Jakarta: Bumi Medika, Indonesia, 3, 20 p.

[7] Nathan DM, Delahanty LM (2010) Menaklukkan Diabetes. Jakarta: PT. Bhuana Ilmu Populer, Indonesia, 7 p.

[8] Zimmet P, Alberti KGMM, Shaw J. (2001) Global and Societal Implications of the Diabetes Epidemic. Nature 414(13): 782-787. doi:10.1038/414782a

[9] Shaw JE, Sicree RA, Zimmet PZ (2010) Global Estimates of the Prevalence of Diabetes for 2010 and 2030. Diabetes Research and Clinical Practice 87: 4-14. DOI:10.1016/j.diabres.2009.10.007

[10] Bahmani M, Zargaran A, Rafieian-Kopaei M, Saki K (2014) Ethnobotanical Study of Medicinal Plants Used in The Management of Diabetes Mellitus in The Urmia, Northwest Iran. Asian Pac J Trop Med 7(Suppl 1): S348-S354. DOI: 10.1016/S1995-7645(14)60257-1

[11] Tag H, Kalita P, Dwivedi P, Das AK, Namsa ND (2012) Herbal Medicines Used in The Treatment of Diabetes Mellitus in Arunachal Himalaya, Northeast, India. Journal of Ethnopharmacology 141: 786-795. DOI:10.1016/j.jep.2012.03.007

[12] Sunaryati SS (2014) *14 Penyakit Paling Sering Menyerang dan Sangat Membahayakan*. Jogjakarta: Penerbit FlashBooks, Indonesia.

[13] Luqman S, Rizvi SI, Beer A-M, Khare SK, Atukeren P (2014) Efficacy of Herbal Drugs in Human Diseases and Disorders. Evidence-Based Complementary and Alternative Medicine 2014: (pp. 2) 273676. DOI: 10.1155/2014/273676

[14] Chang CLT, Lin Y, Bartolome AP, Chen Y-C, Chiu S-C, et al. (2013) Herbal Therapies for Type 2 Diabetes Mellitus: Chemistry, Biology, and Potential Application of Selected Plants and Compounds. Evidence-Based Complementary and Alternative Medicine 2013: pp. 378657, 33. DOI: 10.1155/2013/378657

[15] Setiawan Y, Lubis MI, Yusuf SM, Prasetyo LB (2015) Identifying Change Trajectory Over the Sumatra's Forestlands Using Moderate Image Resolution Imagery. Procedia Environmental Sciences 24(2015): 189-198. DOI: 10.1016/j.proenv.2015.03.025

[16] Shanley P, Luz L (2003) The Impacts of Forest Degradation on Medicinal Plant Use and Implications for Health Care in Eastern Amazonia. BioScience 53(6): 573-584.

[17] Chinsembu KC (2016) Ethnobotanical Study of Medicinal Flora Utilised By Traditional Healers in The Management of Sexually Transmitted Infections in Sesheke District, Western Province, Zambia. Brazilian Journal of Pharmacognosy 26(2016): 268-274. DOI: 10.1016/j.bjp.2015.07.030

[18] Prinst D (2011) Adat Karo. Medan: Bina Media Perintis, Indonesia, 65-70 p.

[19] Aththorick TA (2012) Etnoekologi dan Pengelolaan Sumberdaya Tumbuhan Masyarakat Karo di Cagar Biosfer Gunung Leuser Kabupaten Langkat Sumatera Utara. Bogor: IPB, Indonesia, 61, 141-142 p.

[20] Silalahi M, Nisyawati, Walujo EB, Supriatna J, Mangunwardoyo W (2015) The Local Knowledge of Medicinal Plants Trader and Diversity of Medicinal Plants in The Kabanjeh Traditional Market, North Sumatra, Indonesia. Journal of Ethnopharmacology 175: 432-443. DOI: 10.1016/j.jep.2015.09.009

[21] Situmorang ROP, Harianja AH, Silalahi J (2015) Karo's Local Wisdom: The Use of Woody Plants for Traditional Diabetic Medicines. Indonesian Journal of Forestry Research 2(2): 121-131.

[22] BPS Kabupaten Karo (2016) Karo dalam Angka 2016. Kabupaten Karo: BPS Kabupaten Karo, Indonesia.

[23] BPS Kabupaten Langkat (2016) Langkat dalam Angka 2016. Kabupaten Langkat: BPS Kabupaten Langkat.

[24] Siagian DR, Marbun T, Hermanto C, Alcantara AJ (2015) Landuse Conversion Impact Assessment on Landscape Provisioning Service for Rice Sufficiency in Langkat Regency, Indonesia. Procedia Environmental Sciences 24(2015): 3-14. DOI: 10.1016/j.proenv.2015.03.002

[25] Publication of Traditional and Complementary Medicine Division, Ministry of Health Malaysia (2011) Good Practice Guideline on Reflexology for Reflexologists Practicing in Malaysia. Kuala Lumpur: Traditional and Complementary Medicine Division, Ministry of Health Malaysia, Malaysia, 2 p.

[26] Srimani P, Mandal G, Ganguly S, Saha P, Sen R, et al. (2009) Antioxidant Effect of Ethanolic Extract of Piper betle Linn. (Paan) on Erythrocytes from Patients with HbE-beta Thalassemia. Indian Journal of Biochemistry & Biophysics 46: 241-246.

Genome sequence analysis of predicted polyprenol reductase gene from mangrove plant *kandelia obovata*

M Basyuni[1*], H Sagami[2], S Baba[3], and H Oku[4]

[1] Department of Forestry, Faculty of Forestry, University of Sumatera Utara, Jl. Tri Dharma Ujung No. 1 Medan, North Sumatera 20155, Indonesia
[2] Institute of Multidisciplinary Research for Advanced Material, Tohoku University, Sendai, 980-8577, Japan
[3] International Society for Mangrove Ecosystems, Faculty of Agriculture, University of the Ryukyus, 1 Senbaru, Nishihara, Okinawa 903-0213, Japan
[4] Molecular Biotechnology Group, Tropical Biosphere Research Center, University of the Ryukyus, 1 Senbaru, Nishihara, Okinawa 903-0213, Japan

*Email: m.basyuni@usu.ac.id

Abstract. It has been previously reported that dolichols but not polyprenols were predominated in mangrove leaves and roots. Therefore, the occurrence of larger amounts of dolichol in leaves of mangrove plants implies that polyprenol reductase is responsible for the conversion of polyprenol to dolichol may be active in mangrove leaves. Here we report the early assessment of probably polyprenol reductase gene from genome sequence of mangrove plant *Kandelia obovata*. The functional assignment of the gene was based on a homology search of the sequences against the non-redundant (*nr*) peptide database of NCBI using Blastx. The degree of sequence identity between DNA sequence and known polyprenol reductase was confirmed using the Blastx probability *E*-value, total score, and identity. The genome sequence data resulted in three partial sequences, termed c23157 (700 bp), c23901 (960 bp), and c24171 (531 bp). The c23157 gene showed the highest similarity (61%) to predicted polyprenol reductase 2-like from *Gossypium raimondii* with *E*-value $2e^{-100}$. The second gene was c23901 to exhibit high similarity (78%) to the steroid 5-alpha-reductase Det2 from *J. curcas* with *E*-value $2e^{-140}$. Furthermore, the c24171 gene depicted highest similarity (79%) to the polyprenol reductase 2 isoform X1 from *Jatropha curcas* with *E*- value $7e^{-21}$.The present study suggested that the c23157, c23901, and c24171, genes may encode predicted polyprenol reductase. The c23157, c23901, c24171 are therefore the new type of predicted polyprenol reductase from *K. obovata*.

Keywords: Biosynthetic pathway, dolichol, enzyme, leaf, polyisoprenoid.

1. Introduction

Mangrove forests have long been a rich source of phytochemical compounds producing polyisoprenoid alcohols [1,2] and other secondary metabolites [3,4]. Polyisoprenoid alcohols are classified to polyprenol and dolichol according to stereochemistry. A number of studies have been reported that the major polyisoprenoid in plant leaves were polyprenols rather than dolichols [5,6]. By contrast, in the animal world (especially in livers), dolichols were predominant, and few polyprenols occurred [7,8]

However, very recently, we are the first group to report that primary polyisoprenoid alcohols of Okinawan and Indonesian mangrove plants were not polyprenols but dolichols [1,2]. Therefore, the occurrence of larger amounts of dolichols even in the leaves of mangrove plants imply that polyprenol reductase is responsible for the polyprenol conversion to dolichol, may be active in mangrove plant leaves. In spite of the ever-present distribution of polyisoprenoid in the plant kingdom, their function plants world remain unclear, especially in mangrove plants. Two polyprenol reductase genes from *Arabidopsis thaliana* have been characterized [9]. However, information on the polyprenol reductase from mangrove species is limited. It, therefore, became necessary to understand the DNA sequence of polyprenol reductase in mangrove plants to obtain more insight into the physiological significance and biosynthetic pathway of these compounds. Here we report the early assessment of probably polyprenol reductase gene from the genome sequence of mangrove plant *Kandelia obovata*.

2. Materials and Methods

2.1. Sample collection

Fresh leaves of *K. obovata* were collected at Okukubi River, Okinawa, Japan. Total RNA was extracted from *K. obovata* leaves using cetyl trimethyl ammonium bromide (CTAB) procedure with minor modification to increase yields as previously described [10]. The total RNA was used for genome sequence.

2.2. Genome sequence analysis of probably polyprenol reductase

Genome sequence of *K. obovata* used shotgun sequencing for sequencing as previously reported [11]. The principle of genome sequencing consists of three steps, short fragment DNA from *K. obovata* were sequenced to a given length of coverage, then continue with genome assembly, and the last step was annotation [11]. Genome sequence of *K. obovata* using a BLAST search led to 3 nucleotide sequences that showed the highest similarity to polyprenol reductase.

2.3. Sequence and data analysis

Three nucleotide fragments named as c23157, c23901, and c24717 then were used for prediction of polyprenol reductase. The functional assignment of three DNA sequences was based on a homology search of the sequences against the non-redundant (*nr*) peptide database of NCBI (https://blast.ncbi.nlm.nih.gov/Blast.cgi) using BLASTX [12]. The degree of sequence identity producing significant alignments between DNA sequences and a known protein database was confirmed by the BLASTX probability score more than 50 bit or an E-value < 10^{-10} was considered to have significant similarity [13].

3. Results and Discussion

The three partial sequences of predicted polyprenol reductase from *K. obovata* termed c23157 (700 bp, coding for 228 amino acids), c23901 (960 bp, coding for 306 amino acids), and c24171 (531 bp, coding for 166 amino acids). As shown in Figures 1-3. The amino acid sequence of c23157 showed significant homology to known, predicted polyprenol reductase (Table 1). Among them, the highest homology of E-value ($2e^{-100}$), 61%, and 301 bits was depicted with predicted polyprenol reductase 2-like from *Gossypium raimondii*. Similarly, c23901 also showed significant to known, predicted polyprenol reductase as displayed in Table 2. The highest similarity of E-value ($2e^{-140}$), 82% identity and 403 bits was shown with Steroid 5-alpha-reductase DET2 from *Jatropha curcas*.

Furthermore, C24171 has the shortest DNA sequence; producing significant alignment to known, predicted polyprenol reductase from various plants (Table 3). Among them, the highest similarity of E-value ($7e^{-21}$), 79% identity, and 93.6 bits score was indicated to polyprenol reductase 2 isoform X1 from *J. curcas*. These results suggested that c23517, c23901, and c24171 encoded probably polyprenol reductase. The c23157, c23901, and c24171 are therefore the new type of predicted polyprenol reductases from *K. obovata*.

It has been reported that a common rule of two sequences are homologous if they are more than 30% matching over their whole lengths, the 30% condition misses many easily identified homologs [13]. In this context, three partial sequences showed more than 61% similarities to known polyprenol reductase in the plant kingdom. These results support previous results that primary polyisoprenoid alcohol in mangroves are dolichol, but not polyprenols where dolichols were found in examined mangrove samples [1-2]. Besides, the composition of polyisoprenoid may be an expression of the distribution of tissues in these mangrove plants [2].

```
        10        20        30        40        50        60        70        80
cttctgctagtgtagaagttcataattccgcagagattcttcatgcacttctacgttttgggtgtggtgtggaccacttt
 L  L  L  V  *  K  F  I  I  P  Q  R  F  F  M  H  F  Y  V  L  G  V  V  W  T  T  F

        90       100       110       120       130       140       150       160
cttgctggtcttaaattggcactatgcatataacatggctccattgggttttgactcatcggattacccctccgctgcta
  L  L  V  L  N  W  H  Y  A  Y  N  M  A  P  L  G  F  D  S  S  D  Y  P  S  A  A

       170       180       190       200       210       220       230       240
gtcacttgattgaaggttcagttacgttttccttgcacgagtcttatttacctgctgcaaagcatctttacaaggtttgg
S  H  L  I  E  G  S  V  T  F  S  L  H  E  S  Y  L  P  A  A  K  H  L  Y  K  V  W

       250       260       270       280       290       300       310       320
ctagctgtgttttcgcttttattaatggaaatgcatgtctcaaggcgcctttacgagacagtgcatgtctttaagtacag
 L  A  V  F  S  L  L  L  M  E  M  H  V  S  R  R  L  Y  E  T  V  H  V  F  K  Y  S

       330       340       350       360       370       380       390       400
ctcttcagctcgaatgcacattttggctatctgacaggcctttttttctatacagcagcacccctgtgcatttgtcact
  S  S  A  R  M  H  I  F  G  Y  L  T  G  L  F  F  Y  T  A  A  P  L  C  I  C  H

       410       420       430       440       450       460       470       480
tttgtgtattggatgctcttaattttgcagctgaccgagcaatgcagttgagtgttagaagtgaagagaagttgccatct
F  C  V  L  D  A  L  N  F  A  A  D  R  A  M  Q  L  S  V  R  S  E  E  K  L  P  S

       490       500       510       520       530       540       550       560
attgacttcaattggcaggcatacattaaacctatcacaaagcttaggtggtggtactgggctggtgctgctttattcac
 I  D  F  N  W  Q  A  Y  I  K  P  I  T  K  L  R  W  W  Y  W  A  G  A  A  L  F  T

       570       580       590       600       610       620       630       640
ttgggggtggctgcatcaatacatatgtcatgtgatttgggctctttgcgagaaaacaaggaaaaaactgatgaatatg
  W  G  W  L  H  Q  Y  I  C  H  V  I  L  G  S  L  R  E  N  K  E  K  T  D  E  Y

       650       660       670       680       690       700
tgattcctcatggtgactggtttgaacttgtttcctgtccacactacttggctgagatg
V  I  P  H  G  D  W  F  E  L  V  S  C  P  H  Y  L  A  E  M
```

Figure 1. DNA sequence of *K. obovata* predicted polyprenol (c23157) and its translation to amino acid sequence

Table 1. Distribution of the top 5 Blastx hits on c23157 sequence

Accession	Description	Identity (%)	Total score	*E*-value
XP_012457241.1	Predicted polyprenol reductase 2-like *Gossypium raimondii*	61	301	2e-100
XP_016724522.1	Predicted polyprenol reductase 2-like isoform X3 *G. hirsutum*	62	301	3e-100
XP_016724521.1	Predicted polyprenol reductase 2-like isoform X2 *G. hirsutum*	62	301	5e-100
XP_016724516.1	Predicted polyprenol reductase 2-like isoform X1 *G. hirsutum*	62	301	6e-100
XP_002531516.1	Predicted polyprenol reductase 2 isoform X1 *Ricinus communis*	62	298	2e-99

K. obovata green and yellow leaves have been shown to contain polyprenols $C_{75\text{-}100}$, $C_{45\text{-}95}$, respectively, while dolichol was $C_{70\text{-}100}$, $C_{75\text{-}100}$, respectively. In contrast, in the roots, no polyprenols detected, with 100% dolichols were $C_{80\text{-}105}$ [1]. This study indicated dolichol is more dominant than polyprenols in green leaves in the ratio of 76.3:23.7 and in the yellow leaves in proportions 42.9:57.1 [1]. In this context, the predominance of dolichol over polyprenol in *K. candel* and the majority of mangrove species suggested that existence of polyprenol reductases in mangrove leaves, which catalyze the conversion of polyprenol to dolichol corresponding to the SRD5A3 protein in animals [14], differ from those of other mangrove plants in reduction activity.

```
        10        20        30        40        50        60        70        80
tttttcctttttttttttttttatgaacactgaaacaccagttttcttttgcgattgtattccaagtaacatacatctgaa
 F  F  L  F  F  F  F  M  N  T  E  T  P  V  F  F  C  D  C  I  P  S  N  I  H  L  K

        90       100       110       120       130       140       150       160
aacatcagattacagactaaaaaaagaaattgcttcaatacaggaaaggtatcacagctttcctactcttgggataatct
  T  S  D  Y  R  L  K  K  E  I  A  S  I  Q  E  R  Y  H  S  F  P  T  L  G  I  I

       170       180       190       200       210       220       230       240
tccccaaacttctccagataccactgatgattggcacgtgccctaggaaccaagttggcacacgtgtagagaagaaaccc
F  P  K  L  L  Q  I  P  L  M  I  G  T  C  P  R  N  Q  V  G  T  R  V  E  K  K  P

       250       260       270       280       290       300       310       320
aaacccgaccccagaccacgtcatcagagcccaaccagtccattccacaatctccccaaagtagttggacagctcacca
 K  P  D  P  R  P  R  H  Q  S  P  T  S  P  F  H  N  L  P  K  V  V  G  T  A  H  Q

       330       340       350       360       370       380       390       400
actcaaaccatcctcctctcggcaccttgtacccaccacctcctctcttcaaggccaatatgaccttatcagcccaaata
  L  K  P  S  S  S  R  H  L  V  P  T  T  S  S  L  Q  G  Q  Y  D  L  I  S  P  N

       410       420       430       440       450       460       470       480
ttgatccacatgccgatcgcaaaaaccaacaacccaacaacaaacctatacgaaaaccacccatcgccgtcatagtcatc
I  D  P  H  A  D  R  K  N  Q  Q  P  N  N  K  P  I  R  K  P  P  I  A  V  I  V  I

       490       500       510       520       530       540       550       560
cttatagtgactaacccaccttgcctgcaagtacccattcaacagattgtacccaaaagccatgagggcaacgctgaccg
 L  I  V  T  N  P  P  C  L  Q  V  P  I  Q  Q  I  V  P  K  S  H  E  G  N  A  D  R

       570       580       590       600       610       620       630       640
gaaaaccaccgcctgttttggagatcttttgagtggtattccggtagagcaggagtgggtagaggcaagtgcgctggaaa
  K  T  T  A  C  F  G  D  L  L  S  G  I  P  V  E  Q  E  W  V  E  A  S  A  L  E

       650       660       670       680       690       700       710       720
tagtggatgagaaaaggggacatgagaatcagtgctttgggattggtgaagtgtcgacctatggggaaaggaggagtgt
I  V  D  E  K  R  G  H  E  N  Q  C  F  G  I  G  E  V  S  T  Y  G  E  K  E  E  C

       730       740       750       760       770       780       790       800
gagccagagggtggggctttccatgaggaaccaggccagtggcggcgccatggtggggccccatcctggacggctgtgct
 E  P  E  G  G  A  F  H  E  E  P  G  Q  W  R  R  H  G  G  A  P  S  W  T  A  V  L

       810       820       830       840       850       860       870       880
ttccgtacggtgcttgaaggaacctgagagagatgaaggttgggaatgctatgaggtagtgagaaaggagggcgtagtga
  S  V  R  C  L  K  E  P  E  R  D  E  G  W  E  C  Y  E  V  V  R  K  E  G  V  V

       890       900       910       920       930       940       950       960
aaaatcttctcatcggagacggccattttgatgttgattgaggggtgatctaacggtgagtggcaaaaacgtcttgt
K  N  L  L  I  G  D  G  H  F  D  V  D  *  G  G  D  L  T  V  S  G  K  N  V  L
```

Figure 2. c23901 predicted polyprenol reductase sequence and its amino acid sequence from *K. obovata*

IOP Conf. Series: Earth and Environmental Science **130** (2018) 012039 doi:10.1088/1755-1315/130/1/012039

Table 2. Distribution of the top 5 Blastx hits on c23901sequence

Accession	Description	Identity (%)	Total score	*E*-value
XP_012087922.1	Steroid 5-alpha-reductase DET2 *Jatropha curcas*	78	403	2e-140
OAY43658.1	Hypothetical protein MANES_08G087200 *Manihot esculenta*	78	402	3e-140
XP_008370052.1	Predicted steroid 5-alpha-reductase DET2 *Malus domestica*	76	385	1e-133
XP_002323554.1	3-oxo-5-alpha-steroid 4-dehydrogenases family protein *Populus trichocarpa*	79	385	6e-100
XP_008355710.1	Predicted steroid 5-alpha-reductase DET2-like *M. domestica*	62	298	2e-99

```
         10        20        30        40        50        60        70        80
gacatttataacctaaaatgtaagttgatacaaaaaataaaacaagacttgccacttgccactcgaaagccaaggcttaa
 D  I  Y  N  L  K  C  K  L  I  Q  K  I  K  Q  D  L  P  L  A  T  R  K  P  R  L  K

         90       100       110       120       130       140       150       160
acattataaagaaacctatttgcaagctgcaaaagatgaatacaatttaatcaatggtgtccggcatagatttcctcggc
  H  Y  K  E  T  Y  L  Q  A  A  K  D  E  Y  N  L  I  N  G  V  R  H  R  F  P  R

        170       180       190       200       210       220       230       240
ttctttacacgcacagaggcttgttaccatacatcaacttctgtacatcatggtcccagcccctaagagcaagaggtgca
L  L  Y  T  H  R  G  L  L  P  Y  I  N  F  C  T  S  W  S  Q  P  L  R  A  R  G  A

        250       260       270       280       290       300       310       320
gaactaattaaaaaacatagctacttgatctccagcatcataccactgcatttcactcaggcagacatgaaaattaacag
 E  L  I  K  K  H  S  Y  L  I  S  S  I  I  P  L  H  F  T  Q  A  D  M  K  I  N  R

        330       340       350       360       370       380       390       400
gtattcgaacttttgtcatcagttactgctccatcaagctgctgaatcgactaataaacaaaaggaaacacagtccgtcg
  Y  S  N  F  C  H  Q  L  L  L  H  Q  A  A  E  S  T  N  K  Q  K  E  T  Q  S  V

        410       420       430       440       450       460       470       480
atttcgtggataatcctcgaacttgtgcaggtaccacctctgtgtttctgctgctgcaaatgagagatttgccaccacaa
D  F  V  D  N  P  R  T  C  A  G  T  T  S  V  F  L  L  L  Q  M  R  D  L  P  P  Q

        490       500       510       520       530
atagaagaagaagccacatgaggatgtcactacctccactggcaactaa
 I  E  E  E  A  T  *  G  C  H  Y  L  H  W  Q  L
```

Figure 3. Nucleotide and amino acid sequence of predicted polyprenol reductase (c24171) from *K. obovata*

Table 3. Distribution of the top 5 Blastx hits on c24171 sequence

Accession	Description	Identity (%)	Total score	*E*-value
XP_012065467.1	Polyprenol reductase 2 isoform X1 *J. curcas*	79	93.6	7e-21
XP_002531516.1	Predicted polyprenol reductase 2 isoform X1 *R. communis*	81	92.0	3e-20
XP_009400472.1	Predicted polyprenol reducatse 1 isoform X2 *Musa acuminata* subsp. *malaccensis*	75	89.4	3e-19
XP_009400470.1	Predicted polyprenol reductase 1 isoform X1 *M. acuminata* subsp. *malaccensis*	75	89.0	2e-19
KVH95971.1	3-oxo-5-alpha-steroid 4-dehydrogenase C-teriminal *Cynara cardunculus* var. *scolymus*	75	87.8	7e-19

4. Conclusion

The current study suggested that the c23157, c23901, and c24171 genes may encode predicted polyprenol reductase from *K. obovata*. A detailed molecular cloning of the c23157, c23901, and c24171 genes are required further investigation to provide product confirmation of these genes and biosynthetic pathway of polyisoprenoid in mangrove plants.

Acknowledgment

A part of this study was supported by an International Research Collaboration and Scientific Publication Grant 2017 (No. 003/SP2H/LT/DRPM/IV/2017 to MB) from the Directorate for Research and Community Service, Ministry of Research, Technology and Higher Education, Republic of Indonesia.

References

[1] Basyuni M, Sagami H, Baba S, Iwasaki H, Oku H (2016) Diversity of polyisoprenoids in ten Okinawan mangroves *Dendrobiology* **75** 167–175.

[2] Basyuni M, Sagami H, Baba S, Oku H (2017) Distribution, occurrence, and cluster analysis of new polyprenyl acetones and other polyisoprenoids from North Sumatran mangroves *Dendrobiology* **78** 18–31.

[3] Bandaranayake WM (2002) Bioactivities, bioactive compounds and chemical constituents of mangrove plants *Wetl. Ecol. Manag.* **10** 421–452.

[4] Basyuni M, Oku H, Baba S, Takara K, Iwasaki H (2007) Isoprenoids of Okinawan mangroves as lipid input into estuarine ecosystem *J. Oceanography* **63**: 601–608.

[5] Swiezewska E, Sasak W, Mankowski T, Jankowski W, Vogtman T, Krajewska I, Hertel J, Skoczylas E, Chojnacki T (1994) The search for plant polyprenols *Acta Biochim. Pol.* **41**:221–260.

[6] Swiezewska E, Danikiewicz W (2005) Polyisoprenoids: Structure, biosynthesis, and function. *Prog. Lipid Res.* **44**: 235–258.

[7] Chojnacki T, Dallner G (1988) The biological role of dolichol. *Biochem. J.* **251**: 1–7.

[8] Sagami H, Kurisaki O, Oguro K, Chojnacki T (1992) Separation of dolichol from dehydrodolichol by a simple two-plate thin layer chromatography. *J. Lipid Res.* **33**: 1857–1862.

[9] Jozwiak A, Gutkowska M, Gawarecka K, Surmacz L, Buckzkowska A, Lichocka M, Nowakowska J, Swiezewska E (2015) Polyprenol reductase2 deficiency is lethal in *Arabidopsis* due to male sterility *Plant Cell* **27**: 3336–3353.

[10] Basyuni M, Baba S, Oku H (2017) Cloning and phylogenetic analysis of actin genes from mangrove family Rhizophoraceae *IOP Conf. Ser.: Mater. Sci. Eng.* **180**: 012247.

[11] Ekblom R, Wolf JBW (2014) A field guide to whole-genome sequencing, assembly and annotation *Evol. Appl.* **7**: 1026–1042.

[12] Altschul SF, Madden TL, Zhang JH, Zhang Z, Miller W, Lipman DJ (1997) Gapped BLAST and PSI-BLAST: a new generation of protein database search program. *Nucleic Acid Res.* **25**: 3389–3402.

[13] Pearson WR (2013) An introduction to sequence similarity ("homology") searching *Curr. Protoc. Bioinform.* **42**:3.1.1–3.1.8.

[14] Contagrel V, Lefeber DJ, Ng BG, Guan Z, Silhavy JL, Bielas SL, Lehle L, Hombauer H, Adamowicz M, Swiezewska E, De Brouwer AP, Blümel P, Sykut-Cegielska J, Houliston S, Swistun D, Ali BR, Dobyns WB, Babovic-Vuksanovic D, van Bokhoven H, Wevers RA, Raetz CR, Freeze HH, Morava E, Al-Gazali L, Gleeson JG (2010) SRD5A3 is required for converting polyprenol to dolichol and is mutated in a congenital glycosylation disorder. *Cell* **142**: 203–217.

Traditional risk factors as determinants of heart rate recovery in patients with diabetes mellitus type 2 without known coronary artery disease

Sanny Silaban[1,2*], A Afif Siregar[1,2], H Hasan[1,2], C Aryfa Andra[1,2]

[1]Department of Cardiology and Vascular Medicine, Medical Faculty, University of Sumatera Utara, Jl. Dr.T.Mansur No.5. Kampus USU, Medan 20155, Indonesia
[2]Haji Adam Malik General Hospital, Jl Bunga Lau No.17, Kemenangan Tani, Medan 20136, Indonesia

*Email : sannymarch.nss@gmail.com

Abstract. The impact of Traditional risk factors on heart rate recovery (HRR) has not been studied in patients Diabetes Mellitus type 2 without known coronary artery disease (CAD). For this reason, we sought to determine the association between HRR as cardiac autonomic dysfunction marker and traditonal risk factors. The study was conducted with a cross-sectional study involving 89 patients with Type 2 Diabetes Mellitus without known having coronary artery disease. The data was taken through anamnese and laboratory tests, and subjects who met the criteria were tested for a treadmill exercise to assess heart rate recovery in the first minute. In bivariate analysis Dyslipidemia, Hypertension, smoker, age, duration of DM$\geq$ 5 years, HbA1C $\geq$7.5, Peak Heart rate, functional capacity and ST depression ischemic have an association with heart rate recovery. In multivariate analysis patients with hyper triglyceride, smoker, overweight, duration of diabetes $\geq$ five years and HbA1C $\geq$ 7,5 have lower heart rate recovery significantly. Traditional risk factors are determinant factors for heart rate recovery in patients with Diabetes Mellitus type 2 without known coronary artery disease.

Keywords: Cardiac disease prevention, cardiac autonomic dysfunction, exercise test

1.Introduction

Diabetes Melitus is an important and independent risk factor for cardiovascular mortality and morbidity [1]. Most patients with DM develop autonomic neuropathy, which is one of the more common complications in this population. Fifteen studies that use different end points reported prevalence rates of 1% to 90% for its insidence [2]. Damage to autonomic nerve fibers that innervate the heart and blood vessels consequently lead to abnormalities in heart rate (HR) control and vascular dynamics [3]. Comparatively, the mortality of DM patients manifesting an autonomic neuropathy is markedly higher than DM patients without autonomic neuropathy [4,5].

Abnormal heart rate recovery (HRR) after exercise is an easy to measure marker of reduced parasympathetic activity. It has been found to be an independent predictor of mortality in submaximal as well as in symptom-limited exercise stress testing [6,7]. HRR is defined as the reduction in heart rate from the peak of strenuous exercise to1 min after termination of exercise. It is a new index clinically significant in evaluating the cardiac autonomic system, especially the parasympathetic

IOP Conf. Series: Earth and Environmental Science **130** (2018) 012040 doi:10.1088/1755-1315/130/1/012040

nervous function after exercise. Panzer et al reported that abnormal HRR was present among 42% of patients with impaired fasting glucose and 50% with diabetes [8]. HRR is considered as a strong predictor of increased cardiovascular disease and overall mortality, independently of other cardiac risk factors [9]. Several studies have also shown that abnormal HRR is a predictor of poor prognosis in patients with heart failure [10,11]. In addition, abnormal HRR is associated with severity of coronary artery disease [12].

However, patients with Type 2 Diabetes Mellitus (T2DM) have typically many shared complications, e.g., obesity and hyperlipidemia, which can be considered to be risk factors of T2DM rather than its consequences and which are known to affect in HRR in non-diabetic population [13,14]. Also reduced physical activity level and low physical fitness, which is associated with blunted HRR in a healthy population are common among patients with T2DM [13,15,16]. We, therefore hypothesized that HRR of T2DM patients might not be altered only because of the disease itself or its related comorbidities but also because of other earlier risk factors, e.g., obesity, hypertension, hyperlipidemia and lifestyle etc smoker, age, with low physical activity. In the present study, we examined the association between cardiovascular autonomic function as assessed by HR recovery after exercise and other potential factors, including demographic characteristics, metabolic variables that we conclude as traditional risks for coronary disease among patients with T2DM with unknown coronary artery disease (CAD).

2. Methods

The population of the present study consisted of 89 consecutive participants with T2DM who underwent treadmill exercise with Bruce protocol. The subjects for the study were male 67 (75%), and female 22 (25%), mean age 55±7,5 year, were volunteers for a routine health check-upon the Cardiac Centre of Haji Adam Malik Hospital, Medan, Indonesia. The subject's past medical histories were obtained through an anamnesis. Those who had history
of stroke, disability, COPD, established heart disease, and the other chronic heart disease were exlcluded. Subjects with hypertensive or on antihypertensive medication except for beta blocker dan Calcium channel blockers (CCB) non dyhidropiridine (DHP) were included in the study. We obtained written consents from all subjects,and the study was approved by Adam Malik Hospital and University of Sumatera Utara Medical School Institutional Review Board.

2.1 Demographic and Laboratory tests

Blood sampling for serum lipid profile and glucose was obtained after at least 14 hr of fasting. Post prandial blood glucose obtained after 2 hours meal. Diabetes Mellitus (symptom of diabetes and plasma glucose concentration ≥200 mg/dl or fasting plasma glucose ≥126 mg/dl or 2 hp >200 mg/dl) was diagnosed based on laboratory, patients information and medical history. Traditional risk defines as dyslipidemia, smoker, age >55 years, hypertension and obesity. As for Diabetes status, we include blod glucose control (HbA1C) and time duration the subject established with T2DM. Subjects were defined by their smoking habits as current or nonsmokers. Ex-smokers were considered those who quit smoking for at least 3 years. Body weight and height were measured in light clothing, and body mass index (BMI) was calculated. Hypertension was defined according to the current guidelines [17].

2.2 Exercise Protocol

Treadmill exercise stress test was done in all subjects using the Bruce protocol. Blood pressure, HR and the Borg scale of rating of perceived exertion (RPE) 20 were measured at 2 minute of each stage of the exercise [17]. The exercise was stopped when the subject demanded cessation of the treadmill due to exhaustion, or if the heart rate achieved was more than 85% of estimated maximal HR (220-age), or if the RPE was more than 18 . During the recovery phase, the subjects continued to walk for 30 second at the speed of 1.2 mph and then they sat down for 5 min with continued medical monitoring. HRR was calculated as the decrease of HR per minute between the peak exercise period and 1 min post exercise . Abnormal HRR define as the reduction of heart rate ≤ 12 b.p.m from peak heart rate to 1 minute after the termination of exercise. Those subjects who showed malignant arrhythmia, severe hypertension or hypotension, or unfinished exercise due to neuromuscular problem during exercise were excluded from this study. Positive myocardial ischemia was defined as horizontal or downsloping ST segment depression of ≥ 0.1 mV for 80 ms in 3 consecutive beats.

2.3 Statistical Analysis

Data with normal distribution expressed as a mean ± standard deviation and the others expressed as a median (lower-upper) value. Comparisons of means between the two groups were made by using a Student t-test. Bivariate analysis between variables was evaluated by Chi square. Regression logistic model was constructed with HRR as the dependent variable, and independent variables were those variables showing significant association with HR recovery upon uni and bivariate analysis. Back stepwise regression was used for the final model selection. Criteria for stepwise regression was 0.25 to included in multivariate analysis. Association between the independent variables was explored and insignificant association terms were removed from the final model. A p value <0.05 was considered statistically significant. SPSS 18 for was used for the analysis.

3. Results

Clinical Characteristic Demographic and laboratory characteristics of the population are shown in Table 1. The population study was around 55±7.39 years old. Half of the subjects had a sedentary lifestyle, with only 50.2 % doing regular exercise more than three times per week. Fifty-eight percent of the subjects reported that they were either hypertensive or on antihypertensive medications. Fifty-eight point one percent were smokers, and 63% subjects have dyslipidemia. There is significant different value in Lipid profile (HDL-c, LDL-c, TG) between subject with abnormal HRR (≤12 b.p.m) and normal HRR (> 12 b.p.m). In Diabetic profile, there are significant different value in HbA1C, fasting blood glucose and Duration of T2DM between normal and abnormal HRR. The BMI also show a significant between both of the group 27 (12.8-35) VS 24 (19-29), p = 0.001. As is shown in Table 2, these Treadmill test variables display a significant difference between those with abnormal and normal HRR proportion of peak heart rate, Mets, Chronotropic Incompetence, were different between the two groups

Table 1. Demographic Characteristic Patients

Characteristic	Heart Rate Recovery		All patients	P Value
	≤12 b.p.m N= 46	>12 b.p.m N=43		
Age (median), years	57±6.5	52±7.39	55±7.5	0.001
Male	34(38%)	33(37.5%)	67(75%)	0.760
Hypertension	35(38%)	17(19%)	52(58%)	0.001
Smoker	27(30%)	11(28.9%)	38(58.9%)	0.003
Dyslipidemia	35(38%)	23(25%)	58(63%)	0.01
HDL-c (mg/dl)	38.6 ±4.7	48 ±7.42	43±6.9	0.003
LDL-c (mg/dl)	178(92-265)	132(90-211)	156(90-265)	0.045
TG (mg/dl)	197±51	141±46	170±56	0.031
Total Cholesterol	241(156-376)	196(167-319)	212(156-376)	0.076
HbA1C	8.2(6.4-11.6)	6,5(6-10.2)	7.5(6-11.6)	0.001
Duration of T2DM	6 (3-10)	4(1-7)	4(1-10)	0.034
FPG (mg/dl)	147±46	119±28	135±51	0.02
2h post prandial glucose (mg/dl)	212(133-332)	178(121-245)	210 ± 51.9	0.001

BSA	<1.8(1.4-2.7)	1.8(1.4-2.1)	1.8(1.4-2.7)	0.275
BMI	27(21.8-35)	24(19-29)	26(19-35)	0.001
CCB	22(24%)	5(5%)	27(29%)	0.07
ACE-I/ARB	28(29.2%)	11(12%)	39(41.2)%	0.45
T2DM therapy	46 (51%)	43(49%)	89(100%)	
Oral	34(38%)	28(31%)	62(78%)	0.48
Insulin	12(13%)	15(16%)	27(22%)	0.373
Lipid Lowering Drug	21(23%)	29(32%)	50(55%)	0.054
Exercise	17 (19%)	28(31.2%)	45 (50.2%)	0.011

FPG: Fasting Plasma Glucose; BMI: Body Mass Index; TG: Triglycerides; HDL-c: High-Density Lipoprotein cholesterol; LDL-c: Low-Density lipoprotein Cholesterol; BMI: Body Mass Index; BSA: Body Surface Area; CCB: Calcium Chanel Blocker; ACE-I: Angiotensin Converting Enzym Inhibitor; ARB: Angiotensin II Receptor Blocker

Table 2. Characteristic Treadmill Test Result

Variable	Heart Rate Recovery		All patient	P Value
	≤12b.p.m	>12b.p.m		
Peak Heart Rate	136(76-187)	156(111-178)	148(76-178)	0.001
Peak Sistolic BP (mmHg)	179.2±28.11	179.6±21.34	179±24.9	0.92
Peak Diastolic BP (mmHg)	79(58-126)	77(65-118)	77(58-126)	0.86
Mets	7(1.8-10.2)	9.2(5-11)	7.6(1.8-11)	0.001
Duration Treadmil test (minute)	5.09(2.4-9.5)	6.57(3.4-10.49)	6.4(2.4-10.49)	0.002
Chronotropic Incompetence	25(28%)	7(7%)	32(35%)	0.001
ST depression positive	21(23%)	5(5.6%)	26(29%)	0.001

BP: Blood Pressure; Mets: Metabolic Equivalents

In Bivariat analysis as in Table 3 and 4, subjects who older than 55 years old [OR 4.81 (1.95-12.12)], hypertension [OR 4.81(1.954-12.12)], smoker [OR 4.13(1.667-10.19)], dyslipidemia [OR 4.76(1.12-6.8)], HDL-c <40 [OR 4.71(1.74-12.73)], LDL-c ≥130 [OR 2.7(1.12-6.58)], TG ≥150 [OR 8.31(3.13-21.60)], BMI≥25 [OR 4.24(1.35-8.01)] have higher probability to get abnormal heart rate recovery. HbA1C ≥7.5 has higher probability to abnormal heart rate recovery [OR 1.31 (4.6-36.4)], p=0.005 as seen in duration of diabetes ≥5 years [OR 6.86(2.46-18.30)], p=0.012.

By multivariate analysis in table 5, the presence of Triglyceride ≥150, smoker, BMI ≥25, HbA1C ≥7.5, Duration of DM ≥ 5 years were independently associated with slower HRR after an adjustment for the other variables. Peak heart rate and ischemic response in treadmill test also give a significant association with delay of HRR .

IOP Conf. Series: Earth and Environmental Science **130** (2018) 012040 doi:10.1088/1755-1315/130/1/012040

Table 3. Bivariate Analyses between HRR and Traditional Risk Factors

Variable	Heart Rate Recovery		P	OR	CI	
	≤12b.p.m	>12 b.p.m			Min	Max
Age (years)						
• ≥55	28 (68.3%)	13(31.7%)	0.04	**3.59**	1.48	8.54
• <55	18(37.5%)	30(62.5%)				
Gender						
• Male	34(50.7%)	33(49.3%)	0.76	0.89	0.32	2.25
• Female	12(54.5%)	10(45.4%)				
Hypertension						
• Yes	35(67.3%)	17(32.7%)	0.001	**4.81**	1.954	12.12
• No	11(29.7%)	26(70.3%)				
Smoker						
• Yes	27(71.1%)	11(28.9%)	0.003	**4.13**	1.667	10.19
• No	19(37.3%)	32(62.7%)				
Dyslipidemia						
• Yes	41(59.4%)	28(40.6%)	0.001	**4.76**	1.12	6.8
• No	5(25 %)	15(75 %)				
Lipid Profile						
• HDL-c ≤40	22(75.9%)	7(24.1%)	0.02	**4.71**	1.74	12.7
• HDL-c >40	24(40%)	36 (60%)				
• LDL-c ≥130	34(60.7%)	22(39.3%)	0.03	**2.7**	1.12	6.58
• LDL-c <130	12(36.4%)	21(63.6%)				
• TG ≥150	36(80%)	9(20%)	0.001	**8.31**	3.13	21.6
• TG <150	10(22.7%)	34(77.3%)				
• Total Cholesterol ≥200	34(68.8%)	16(32%)	0.04	**4.78**	1.93	11.7
• Total Cholesterol <200	12(30.8%)	37(69.2%)				
BMI						
• ≥25	32(68.1%)	15(31.9%)	0.03	**4.24**	1.35	8.01
• <25	14(33.3%)	28(66.7%)				

BMI: Body Mass Index; TG: Trygliceride; HDL: High-Density Lipoprotein cholesterol; LDL-c: Low-Density lipoprotein Cholesterol; BMI: Body Mass Index

Table 4. Bivariate Analyses between HRR and T2DM status

Variable	Heart Rate Recovery		P	OR	CI	
	≤12b.p.m	>12b.p.m			Min	Max
HbA1C						
• ≥7,5	33(80.5%)	17(19.5%)	0.005	**13.1**	4.6	36.4
• <7,5	13(26.5%)	26(73.5%)				
FPG						
• ≥126	12(26.6%)	33(73.3%)	0.06	4.01	0.27	2.87
• <126	34(77.3%)	10(22.7%)				
2h post prandial						
• ≥200	28(70%)	12(30%)	0.056	1.25	0.82	4.56
• <200	18(36.7%)	31(63.3%)				
Duration T2DM						
• ≥5 years	26(78%)	7(22%)	0.012	**6.86**	2.46	18.3
• <5 years	20(35.7%)	36(64.3%)				

FPG: Fasting Plasma Glucose; T2DM: Type 2 Diabetes Mellitus

Table 5. Logistic Regression with HRR as Dependent Variable

Variable	Coef	P	OR	CI 95%
Smoker	0.801	0.042	2.21	1.312-15.296
Triglyserides	1.27	0.015	3.567	1.378-20.23
BMI	0.871	0.024	2.388	2.594-23.31
Duration of T2DM	2.451	0.003	11.09	2.402-53.59
HBA1C	2.381	0.024	10.08	1.447-71.72
Peak Heart Rate	-1.574	0.015	0.265	0.006-0.913
ST Depression positive ischemia	1.412	0.184	2.21	1.83-3.813

4.Discussion

As shown in table 5, Triglyseride, BMI, Smoker, Duration of T2DM, HbA1C have independent association with the abnormality of heart rate recovery as a marker of subclinical cardiac diabetic neuropathy. Peak heart rate during exercise, and positive ST depression in precordial lead also showed significant association. This result shows that traditional risk factors can determine abnormal heart rate recovery as seen in coronary artery disease. Christina et al in 2001 also showed a similiar result with heart rate variability as its cardiac neuropathy marker. In a cross sectional study covered by 600 DM patients, the regresion analysis showed cardiac otonomic neuropaty incidence increase with positif corelation with hypertension, smoker, duration of diabetes, LDL, Cholesterol triglyseride and HbA1C. HbA1C value in this study was 8±1,8 with p <0.05 [18]. EURODIAB study also showed that there is a correlation between cardiac autonomic dysfunction with duration of DM (cut of point 8 years), smoker and HDL-c even though we have a different marker tool to established cardiac autonomic neuropathy [19]. It is necessary, however, to bear in mind that their patients had relatively well controlled glycemic state and that only a few of them had abnormal HR recovery according to the criteria mentioned.

However our limitation is the blood sample is taken only once to evaluate associaton between lipid profile and blood glucose status with heart rate recovery. Only one prior study has prospectively studied the relationship of slower HRR with change in risk factor levels over time, and similarly found no association between slower HRR and an increase in levels of glucose and insulin over time [20].

5.Conclusion

In conclusion, traditional risk factor is significantly associated with impaired vagal reactivation as showen in slower heart rate recovery. Therefore, evaluation of all cardiovascular traditional risk factors is necesarry to prevent cardiac autonomic dysfunction in Diabetes Mellitus patients.

Reference

[1] Wei M, Gaskill SP, Haffner SM, Stern MP (1998) Effects of diabetes and level of glycemia on all-cause and cardiovascular mortality. The San Antonio Heart Study. Diabetes Care 21(7): 1167-72.
[2] Vinik AI, Maser RE, Mitchel BD, Freeman R (2003) Diabetic autonomic neuropathy. Diabetes care 26: 1553-1579
[3] Schumer MPJS, Pfeifer MA. Cardiovascular autonomic neuropathy testing in patients with diabetes. Diabet Spectr 11:223-7.
[4] Ewing DJ, Borsey DQ, Bellavere F, Clarke BF (1981) Cardiac autonomic neuropathy in diabetes: comparison of measures of R-R interval variation. Diabetologia 21(1):18-24.
[5] Moţăţăianu A, Bălaşa R, Voidăzan S, Bajko Z (2013) Cardiovascular autonomic neuropathy in context of other complications of type 2 diabetes mellitus. Biomed Res Int 2013:507216.
[6] Cole CR, Foody JM, Blackstone EH, Lauer MS (2000) Heart rate recovery after submaximal exercise testing as a predictor of mortality in a cardiovascularly healthy cohort. Ann Intern Med 132:552–555
[7] Nishime EO, Cole CR, Blackstone EH, Pashkow FJ, Lauer MS (2000) Heart rate recovery and treadmill exercise score as predictors of mortality in patients referred for exercise ECG. JAMA 284:1392–1398
[8] Panzer C, Lauer MS, Brieke A, et al.(2002) Association of Fasting Plasma Glucose With Heart Rate Recovery in Healthy Adults. Diabetes 51: 803–807.
[9] Morshedi-Meibodi A, Larson MG, Levy D, et al.(2002) Heart rate recovery after treadmill exercise testing and risk of cardiovascular disease events (The Framingham Heart Study). Am J Cardiol 90: 848–852.
[10] Guazzi M, Myers J, Peberdy MA, et al.(2010) Heart rate recovery predicts sudden cardiac death in heart failure. Int J Cardiol 144: 121–123.
[11] Thomas DE, Exton SA, Yousef ZR (2010) Heart rate deceleration after exercise predicts patients most likely to respond to cardiac resynchronization therapy. Heart 96: 1385–1389.
[12] Deepak P, Vivekananthan, Eugene H, Claire E, Michael S (2003) Heart Rate Recovery After Exercise Is a Predictor of Mortality, Independent of the Angiographic Severity of Coronary Disease. J Am Coll Cardiol 42:831– 8.
[13] Georgoulias P, Orfanakis A, Demakopoulos N, Xaplanteris P, Mortzos G, Vardas P, Karkavitsas N (2003) Abnormal heart rate recovery immediately after treadmill testing: correlation with clinical, exercise testing, and myocardial perfusion parameters. J. Nucl. Cardiol 10 (5): 498–505.
[14] Kimura T , Matsumoto T , Akiyoshi M., Owa Y, Miyasaka N, Aso T, Moritani (2006) Body fat and blood lipids in postmenopausal women are related to resting autonomic nervous system activity. Eur J Appl Physiol 97 (5): 542–547.
[15] Carnethon MR., Jacobs Jr, Sidney S, Liu K. (2003) Influence of autonomic nervous system dysfunction on the development of type 2 diabetes: the CARDIA study. Diabetes Care 26 (11): 3035–3041.

[16] Carnethon MR, Jacobs Jr, Sidney S, Sternfeld B, Gidding SS, Shoushtari C, Liu K (2005) A longitudinal study of physical activity and heart rate recovery: CARDIA. Med Sci Sports Exerc 37 (4): 606–612.
[17] Lurbe E, Cifkova R, Cruickshank JK., Dillon MJ, Ferreira I et al.(2009) Management of high blood pressure in children and adolescents: Recommendations of the European Society of Hypertension. Journal of Hypertension 27: 1719−1742.
[18] Christina V, Michael P, Alexander K, Vasiliki A, Nicholas K, Nicholas T (2011) The association between cardiac autonomic neuropathy with metabolic and other factors in subjects with type 1 and type 2 diabetes. Journal of Diabetes and Its Complications 25: 159–167
[19] Kempler P, Tesfaye S, Chaturvedi N, Stevens LK, Webb DJ, Eaton S, Kerényi Z, Tamás G, Ward JD, Fuller JH (2002) Autonomic neuropathy is associated with increased cardiovascular risk factors: the EURODIAB IDDM Complications Study. Diabet Med 19: 900
[20] Carnethon MR, Jacobs DR Jr, Sidney S, Liu K (2003) Influence of autonomicnervous system dysfunction on the development of type 2 diabetes: the CARDIA study. Diabetes Care 26:3035–3041.

In vivo test of bitter *(andrographis paniculata* nees.) extract to ejaculated sperm quality

R Sumarmin[1]*, NK Huda[2], E Yuniarti[1], and Violita[3]

[1] Laboratory of Zoology, Department of Biology, Padang State University, Jl. Prof. Dr. HAMKA Air Tawar Padang, 25260
[2] Laboratory of Embryology, Department of Biology, Andalas University, Kampus Limau Manis, Padang 25321
[3]Laboratory of Plant Physiology, Department of Biology, Universitas Negeri Padang, Jl. Prof. Dr. HAMKA Air Tawar Padang, 25260

Email: ramadhan_unp@yahoo.com

Abstract. Sambiloto or Bitter (*Andrographis paniculata* Nees.), are often used to treat various diseases, such as influenza, cancer, anti-inflammation, anti-HIV, anti-mitotic and anti-fertility. This study aimed to determine the effects of the bitter (*Andrographis paniculata* Nees.) extract to ejaculated sperm mice quality (*Mus musculus* L. Swiss Webster). This research was conducted using Completely Randomized Design with 4 treatments, which are 0.0 g/b.w., (P0), 0.2 g/b.w., (P1), 0,4 g/b.w., (P3), or 0.6 g/b.w., (P4) bitter extract orally for 36 days. After treatment, the mice decapitated, dissected and collected the sperm from vas deferens. Then, the number of sperm counted by used the improved Neubauer and then stained by Eosin to count the abnormal sperm. Data analyzed by ANOVA (Analysis of Variance) then DNMRT. The results showed that the average numbers of sperm are 28.80 x 10^5 (P0), 19.50 x 10^5 (P1), 12.50 x10^5 (P2) and 9.50 x 10^5 (P3). The average abnormal sperm numbers are 18.33 x 10^5 (P0), 22.50 x 10^5 (P1), 31.50 x10^5 (P2) and 39.33 x 10^5 (P3). It showed that the effective treatment to decrease sperm number was 0.2 g/b.w., of bitter extract. It can conclude that the bitter (*Andrographis paniculata* Nees.) extract decreases the quality of the ejaculated sperm of mice (*Mus musculus* L.)

Keywords: andrographolide, bitter, ejaculated spermatozoa

1.Introduction

Family Planning Program has been announced by the government of the Republic of Indonesia as a national program. One of the efforts that have implemented in the program is the provision of contraceptives. Use of contraception in principle is to prevent fertilization or fusion of the sperm cell and an ovum cell. Means of contraception is more aimed at women, while men are still limited, so that the development of the male contraceptives is less than to the female contraceptives [1].

One of the plants that are expected to be anti-fertility is bitter *(Andrographis paniculata* Nees.). Bitter widely used as a medicinal plant that has many properties in various countries, among others, as an anti-oxidant, anti-diabetic, anti-fertility, anti-HIV, anti-flu, anti-malaria, anti-diarrhea and others [2].

Chemically this plant contains diterpenoid lactone. In the lactone, components are andrographolide, which is the main active ingredient of this plant [3]. Besides other content found in this plant are pani kulin and flavonoids. From various studies, the content of which are believed to fight disease is andrographolide. Besides, the bitter leaf contains saponins, alkaloids, and tannins [4].

Of the many chemical compounds found in bitter *(Andrographis paniculata* Nees.) Compound that acts as an anti-fertility is andrographolide. According to [5] andrographolide is the active substance contained in bitter that works to prevent cell division (cytokenesis). Cells can affect spermatogenesis occurs in the seminiferous tubules. Then [6] find that bitter extract still reveals the abnormality to development of mice fetuses.

According to Bardin [7], bitter extract in male mice given orally can decrease the number of epididymal spermatozoa. As the result of the declining number of spermatozoa found in the epididymis and disruption of the maturation of sperm cells, it can be expected to affect the ejaculated sperm contained in the vas deferens. To determine how it affects, then conducted a study entitled In Vivo Test of Bitter *(Andrographis paniculata* Nees.*)* Extract On the Quality of ejaculated spermatozoa of mice *(Mus musculus L.* Swiss Webster).

2. Materials and Methods

2.1 Materials

This study used a test animal mice *Mus musculus* L. Swiss Webster, 11-12 weeks, weighing 20-30 grams were obtaine from the Laboratory of Zoology Department of Biological Science Universitas Negeri Padang. Extracts of bitter *(Andrographis paniculata* Nees.) obtained from various processes performed at the Laboratory of Plant Physiology FMIPA Universitas Negeri Padang. The bitter plant obtained from the city of Padang, West Sumatra in conditions that are in bloom.

2.2 Methods

Mice were grouping into one control group and three treatment groups. Each group consisted of 6 mice. The control (P0) group was gave only the solvent of extract, and the food was the same as the treatment group. Group P1, P2, and P3 are the groups treated with the extract of bitter orally for 36 days with a dose of 0.2 g, 0.4 g and 0.6 g.

After the test of completion in each group, the mice were decapitated, dissected and cut the vas deferens of the left and right, place it in Petri dishes filled with 10 micron PBS *(Phosphate Buffered Saline*). To calculate the number of spermatozoa, it were counted using *the Improved Neubauer.* As for seeing and counting the abnormal number of sperm (the morphology of spermatozoa), using Eosin staining method. The data analyzed by ANOVA *(Analysis Of Variance)* and Duncan New Multiple Range Test.

3.Results and Discussion

From this research, it can be seen the effect of extracts of bitter *(Andrographis paniculata* Nees.) to the ejaculated spermatozoa number in mice *(Mus musculus* L.) is as Figure 1.

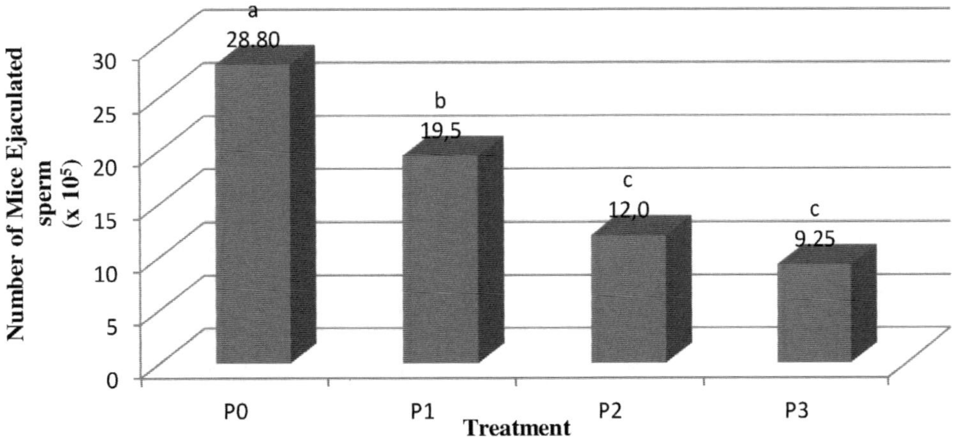

Figure 1. The number of ejaculated sperm of Mice on the Different Treatment of Extract Bitter. The number followed by different small letters, significantly $p<0.05$

After the extract bitter *(Andrographis paniculata* Nees.) With dosage 0.2 g, 0.4 g and 0.6 g in mice *(Mus musculus* L.) for 36 days there is decreases in the number of ejaculated spermatozoa. According to Pallav et al. [3], the number of spermatozoa produced is very dependent on the direct process that occurs during the process of spermatogenesis in the seminiferous tubules. When normal spermatogenesis takes place, it will produce normal sperm count as well. Conversely, if an interruption occurs during spermatogenesis process, the development will affect the number of spermatogonial cells formed spermatozoa. So with the division during spermatogenesis disorder, it is expected to lead to changes or damage to spermatozoa produced thus might change, and the damage will effect to decrease in the number of ejaculated spermatozoa.

Based on the test results can be seen that the use of 0.2 g dose was the most in reducing the number of ejaculated spermatozoa, and bias used as an anti-fertilizer dose. At a dose of 0.2 g is obtained an average of 12.00× sperm count 10^5 and has seen a significant decrease in the number of spermatozoa compared with controls who had 28.08×10^5.

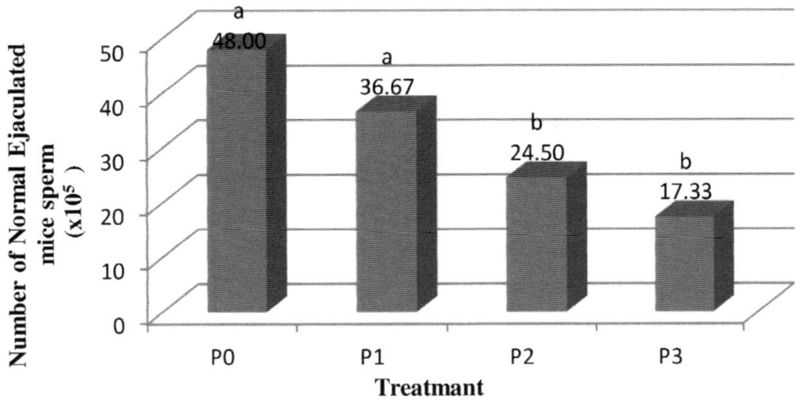

Figure 2. The number of normal spermatozoa of mice in various treatments of bitter extract. Histogram followed by superscript letters are different, significantly at $p<0.05$

The number of normal spermatozoa morphology examination (Figure 2) in the vas deferens mice *(Mus musculus* L.) obtained an average percentage of normal sperm morphology dose of 0.2 g/b.w., 0.4 g/b.w., and 0.6 g/b.w., lower than compared with the control. After further test was statistically, there are significant due to the extract of bitter *(Andrographis paniculata* Nees.). In this study, the higher dose given, the fewer number of normal spermatozoa. The dose that can using as an anti-fertilizer is on treatment with a dose 0.2 g/b.w., because it can inhibit or reduce the number of normal sperm.

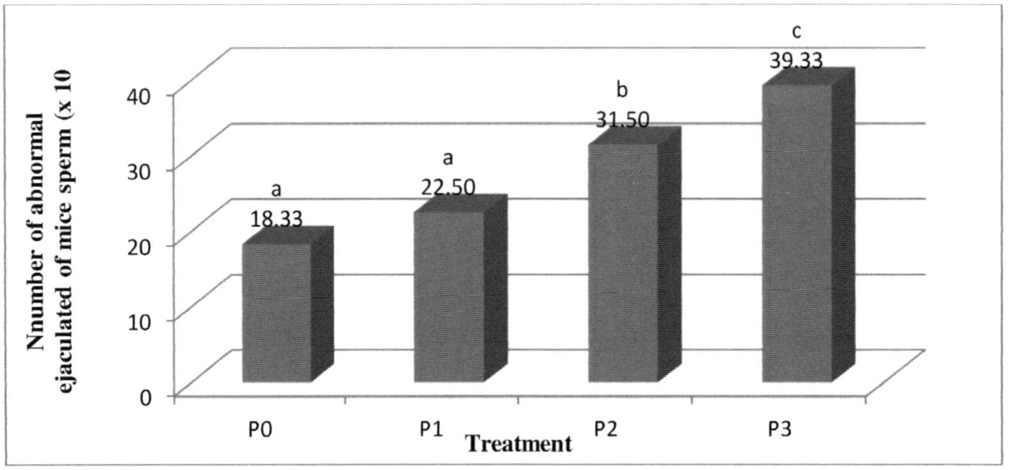

Figure 3. The number of ejaculated abnormal spermatozoa of mice in various treatments of bitter extract, significantly different at $p<0.05$

Based on Figure 3, it can be seen that the extract of bitter *(Andrographis paniculata* Nees.) also, cause abnormal spermatozoa. Abnormalities in spermatozoa of mice *(Mus musculus* L.) allegedly because of interference maturation phase of spermiogenesis spermatozoa in this case by the theory put forward in the seminiferous tubules, androgen function in controlling the process of spermatogenesis in meiosis and spermiogenesis process. In this study suspected antimitotic compounds contained in andrographolide has inhibited androgen produced, so that an interruption in the phase of spermiogenesis. So if the hormone androgen is a hormone that plays a role in spermatogenesis process when availability is a little, it will cause the process was interrupted and could result in abnormalities in sperm during spermiogenesis.

4.Conclusions

Based on the results obtained can be concluded that the extract of bitter *(Andrographis paniculata* Nees.) decreased the number of ejaculated spermatozoa of mice *(Mus musculus* L.) and decreased the number of normal sperm. In other words, the extracts of bitter *(Andrographis paniculata* Nees.) decreased the quality of the ejaculated spermatozoa of mice *(Mus musculus* L.).

Acknowledgments

The authors gratefully acknowledge that the present research is supported by Universitas Negeri Padang. The support is under the research grant BOP-PTN UNP of Year 2016 Contract Number NO. 171/SP2H/BOPTN/DP2M/VIII/2016.

References

[1] Kumar S, Dhanani T, and Shah S (2014) *Journal of Chromatograpic Science*. 52: 1043-1050.

[2] Wen WC, and Bi FL (2012) *Chinese Medicine*. 3: 136-1438 Bardin, CW. 1968 Testiscular Pituitary Axis in *Reproductive Endocrionology*. Philadhepia Saunders. pp. 618-646.

[3] Pallav KD, Anupam KP, and Ragini G (2014) *International Journal of Recent Advances in Multidisciplinary Research.* Vol. 01, Issue 04, pp. 105-1105 Elumalai, E., Banupriya R., Sangeetha T., and Madhumathi S. 2016. *International Journal of Pharma and Bio Science*. 7(4): 183-200.

[4] Dalimunthe, dan Aminah (2009) Interaction Paniculata *(Andrographis spaniculata)*. Department of Pharmacology. University of Northern Sumatra. 2013

[5] Joseph J and Solomon J (2014) *Med. Aromat Plants*. Vol. 3. Issue 4.

[6] Shivananda BG, Mohan M, and Khanan S (2013) *Journal of Pharmacognosy and Phytochemistr*y. Vol. 2 No. 1. pp: 342-348.

[7] Bardin CW (1968) Testiscular Pituitary Axis in *Reproductive Endocrionology.* Philadhepia Saunders. pp. 618-646.

Antihyperglycemic and antihyperlipidemic effects of pirdot (*saurauia vulcani* korth.) leaves extract in mice

Salomo Hutahaean[1*], Masitta Tanjung[1], Diah Puspita Sari[1], Vevy Elfia Ningsih[1]

[1]Department of Biology, Faculty of Mathematics and Natural Sciences, Universitas Sumatera Utara, Jl. Bioteknologi No. 1, Kampus USU, Medan 20155

*Email: salomo@usu.ac.id

Abstract. Approximately eighty percent of deaths in diabetic patients result from atherosclerosis, which is related to hyperlipidemia tendencies in diabetes. In North Sumatra, the use of plant-based ingredients as diabetes therapy has long been recognized. One of the local species which traditionally used was the pirdot plant (*Saurauia vulcani* Korth.). In this paper, we report the antihyperglycemic and antihyperlipidemic effects of the extract of pirdot leaves in model mice. In experiment I, twenty-five alloxan-induced diabetic mice was divided randomly into five groups of 5 mice, namely: control diabetic mice; diabetic mice + metformin; and three groups diabetic mice + pirdot leaves extract of 100, 200, or 300 mg/kg BW respectively. All the treatments were given daily for 21 days by oral gavage. In experiment II, another twenty-five mice were divided randomly into five groups of 5 mice. The treatments were as follows: a control group that did not receive any treatment; hyperlipidemic control (received quail yolk diet) for 30 days; and three groups of hyperlipidemic mice + orally treated pirdot leaves extract at a dose of 100, 200, or 300 mg/kg BW respectively. The result showed the pirdot leaves extract has the potential as antihyperglycemic. The effects obtained are equivalent to the effects of antidiabetic drug metformin. On the other hand, the antihyperlipidemic effect was not conclusive, because the extract lowered total cholesterol significantly, but no significant effect on triglyceride, marked reduced LDL, but significantly decreased the HDL level.

Keywords: Blood glucose level, HDL, LDL, *Saurauia vulcani*, total Cholesterol, tryglyceride

1. Introduction

Diabetes mellitus is a metabolic disease characterized by high blood sugar level caused by impaired insulin secretion, its action, or both. The predicted prevalence of world diabetes sufferers is 2.3% in 2000 to 4.4% by 2030, with the number of patients estimated to increase from 171 million to 368 million [1]. Therapies for the treatment of diabetes are the use of insulin and other drugs, such as sulfonylurea analogs, alpha-glucosidase inhibitors, and biguanides. These medications have certain side effects, such as causing hypoglycemia at higher doses, liver organ problems, lactic acidosis, and diarrhea [2]. The study of plant-derived antidiabetic agents is thus of great importance, including the scientific testing of plant material used in traditional medicine.

In North Sumatra, the use of plant-based ingredients as a diabetes drug has long been recognized. One of the local species of North Sumatra that has the potential is the pirdot plant (*Saurauia vulcani* Korth.). The Bataknese community traditionally used the pirdot leaf as a cure for various diseases, such as cancer, wounds, and diabetes [3,4]. Chronic diabetes is associated with long-term damage, dysfunction, and failure of different organs, such as the eyes, kidneys, nerves, heart, and blood vessels. Cardiovascular complications are the leading cause of death in diabetes mellitus. Eighty percent of deaths in diabetic patients result from atherosclerosis [5,6]. These cardiovascular complications are, among other things, related to hyperlipidemic tendencies in diabetes. In this paper, we report the antihyperglycemic and antihyperlipidemic effects of the extract of pirdot leaves in alloxan-induced diabetic mice.

2. Materials and Methods

2.1. Experimental Animals

The mice (Mus musculus L.), two-month old, weighing about 25-30 gram, were used in the study. They received from Faculty of Pharmacy, University of Sumatera Utara. Before the experiment, mice were kept two weeks in a cage for adjustment to new environment. During the study, animals were kept in 12 hours light / dark cycle at the temperature of 27-30 °C, and fed with standardize laboratory diet. Food and drinking water offered on an *ad-libitum* basis.

2.2. Preparation of Plant Extract

Pirdot leaves were obtained from the forest of Aek Nauli Parapat, North Sumatra. The leaves were air dried for approximately five days and then powdered with an electric blender. Extraction conducted with ethanol (70%; 40 hours) by maceration method. Macerate obtained was filtered and evaporated by using rotary evaporator at a temperature of 40°C [7].

2.3. Induction of Diabetes Mellitus

Diabetes mellitus experimentally induced in animals fasted for 18 hours by intraperitoneal injection of 150 mg/kg BW alloxan dissolved in NaCl solution (0.9%) [8]. Three days after injection, mice were screened for serum glucose levels. We considered the mice with blood glucose level above 200 mg/dL as diabetic and used for experiment I.

2.4. Induction of Hyperlipidemia

Hyperlipidemia experimentally induced by feeding the mouse with japanese Coturnix yolk. The yolk was separating from albumin and making emulsions by shaking it slowly. The Coturnix yolk diet was given 0.3 ml per day orally by gavage for 30 days. Animals received the food considered hyperlipidemic and used for experiment II.

2.5. Experimental Designs

2.5.1. Experiment I. The experimental research using 5 X 5 completely randomized design (CRD). Twenty-five mice were divided randomly into 5 groups of 5 mice with different treatment for each group, as follows: group K1 consisting of diabetic mice were orally administered an equivalent volume of vehicle (distilled water); group K2 was diabetic mice and orally given a standard antihyperglycemic drug (Metformin); group P1-P3 were diabetic mice and orally treated with extract of pirdot leaves at a dose level of 100, 200, or 300 mg/kg b.w. respectively. All treatments were given daily for 21 days by gastric intubation.

Blood glucose level were measured by using the EasyTouch glucometer test kit. The tail of mice was cut and the blood was allowed to drip onto glucose strip test, and after 10 seconds blood glucose level will appear on the screen. Blood glucose level measured at day 1, 7, 14, and 21 after alloxan injection. The body weight also measured on the same day.

2.5.2.Experiment II. The experimental research using 5 X 5 completely randomized design (CRD). Twenty-five mice were divided randomly into 5 groups of 5 mice with different treatment for each group, as follows: group K0 served as control and did not receive any treatment; group K1 served as hyperlipidemic control (received yolk diet) for 30 days; group P1-P3 were hyperlipidemic mice and orally treated with extract of pirdot leaves starting from day 11 to day 30 at a dose of 100, 200, or 300 mg/kg BW respectively.

Total cholesterol levels were measured by using the EasyTouch cholesterol test kit in the blood which taken from the tail of the mice. The tail of mice was cut, blood was dripped onto cholesterol strip test, and after 150 seconds cholesterol levels will appear on the screen. The measurement of total cholesterol performed on days 0, 10, 20 and 30. All animal dissected at the end of the experiment and 1 ml of blood was taken from the heart and placed into an EDTA tube. Blood was centrifuged at 4000 rpm for 15 minutes. Serum was taken and then used for measurement of triglycerides, HDL, and LDL [9]. Serum triglycerides estimated according to Colorimetric Enzymatic Test GPO (Glycerol-3-Phosphate Oxidase) method. HDL-Cholesterol and LDL-Cholesterol estimated according to the CHOD-PAP Enzymatic Colorimetric Test for Cholesterol With Lipid Clearing Factor (LCF) method using Mikrolab 300.

2.6. Statistic Analysis
All data analyzed with ANOVA and Duncan' test method. The differences between the means of the various groups were considered significant at $p<0.05$.

3. Results and Discussion
3.1. Experiment I. Antihyperglycemic Effect of Pirdot Leaves Extract
In the diabetic mice (K1), the blood glucose level increased from 226.8 ± 35.2 mg/dl to 292.4 ±74.9 mg/dl from day 4 after alloxan injection to day 24, while in metformin group (K2) BGL decreased from 230.2 ± 23.8 mg/dl to 135 ± 25.1 mg/dl. The extract of pirdot leaves at all dosages significantly lowered blood glucose level starting from day ten until the end of the experiment.

The antihyperglycemic effects of pirdot leaves extract at 100, 200, and 300 mg/kg were examined ten days after injection of the alloxan. As shown in Table 1, pirdot leaves extract tended to decrease BGL, but significant effect started at day 17. Furthermore, when the effect of the extract compared among the different groups of mice, BGLs were significantly lowered in all extract treatment compared to control (K1). No significant differences observed between all the extract treatment compared to the metformin group.

Table 1. Effect of Pirdot (*Saurauia vulcani* Korth.) leaves extract on blood sugar level in alloxan-induced diabetic mice.

Treatments	blood sugar level (mg/dL) on days-				
	0	**3**	**10**	**17**	**24**
K1	109.4 ± 36.7^{a}	226.8 ± 35.2^{a}	237.2 ± 43.9^{a}	267.8 ± 47^{b}	292.4 ± 74.9^{b}
K2	102.6 ± 22.8^{a}	230.2 ± 23.8^{a}	189 ± 30.3^{a}	154 ± 36.7^{a}	135 ± 25.1^{a}
P1	97.2 ± 14.5^{a}	301.4 ± 13^{a}	278.4 ± 97.4^{a}	192 ± 76.9^{a}	173 ± 74.8^{a}
P2	117 ± 38.3^{a}	225.2 ± 24.7^{a}	175.4 ± 24.9^{a}	134.2 ± 21.5^{a}	129.8 ± 23.4^{a}
P3	122.4 ± 15^{a}	279.8 ± 58.8^{a}	191.2 ± 41.5^{a}	156 ± 27.9^{a}	134.4 ± 41.7^{a}

Different superscripts mean statistically different ($P<0.05$).

3.2. Experiment II. Anti Hyperlipidemic Effect of Pirdot Leaves Extract
3.2.1. Total Cholesterol Level
The cholesterol level obtained every ten days. The result showed total cholesterol level increased in all groups after yolk diet administration (Figure 1).

The cholesterol level in control (K0) tended to stabilize between day 0 to 30, whereas in control yolk group the level increased starting from day 10 to the end of the experiment (day 30). Quail egg yolk contains high cholesterol content (220-250 mg per yolk) and an important saturated fat that can increase blood cholesterol [10].

In extract treatment groups, total cholesterol increased from day 0 to day 10, but decrease after extract administration. At the day 30, their level was not significantly different in comparison to control group (P>0.05). It seems that several compounds in this plant, especially flavonoids, saponins, and tannins [7], might have the potential to decrease the total cholesterol in mice blood.

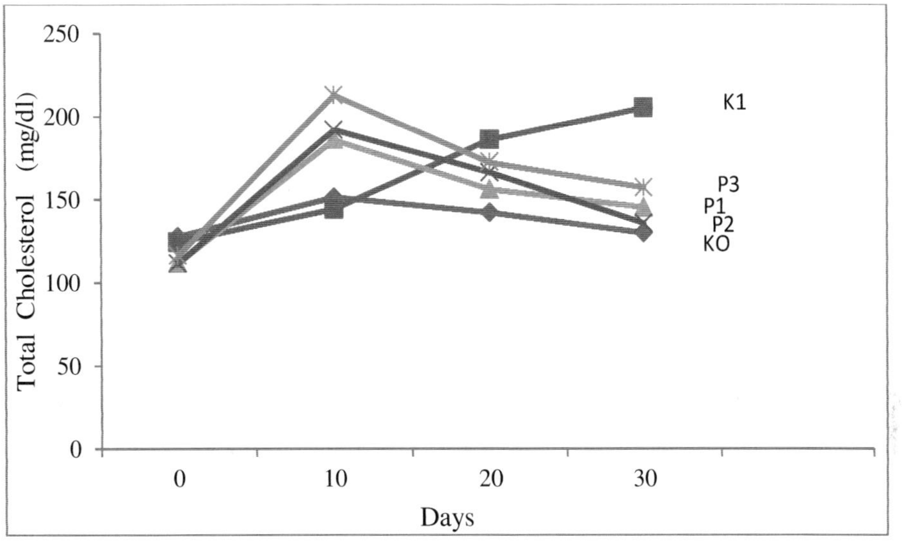

Figure 1. Total colesteol levels in mice after administration of high lipid diet and pirdot leaves extract at doses of 100 mg / kgBW, 200 mg / kgBW, 300 mg / kgBW for 30 days.

3.2.2. Serum Triglycerides

At the end of the experimental period, control yolk mice and all three leaves extract groups demonstrated a marked increase in serum triglycerides (Figure 2). Here, the yolk diet was able to increase serum triglycerides in animal test. According to Gani et al. [11], the rise and fall of blood triglyceride levels are influenced by the amount of fat consumed. In excessive consumption, fat will be stored in adipose tissue in the form of triglycerides. In animals with hypercholesterolemia an elevated triglyceride level occurs, due to visceral fat accumulation and decrease of lipoprotein lipase (LPL) enzyme activity [12].

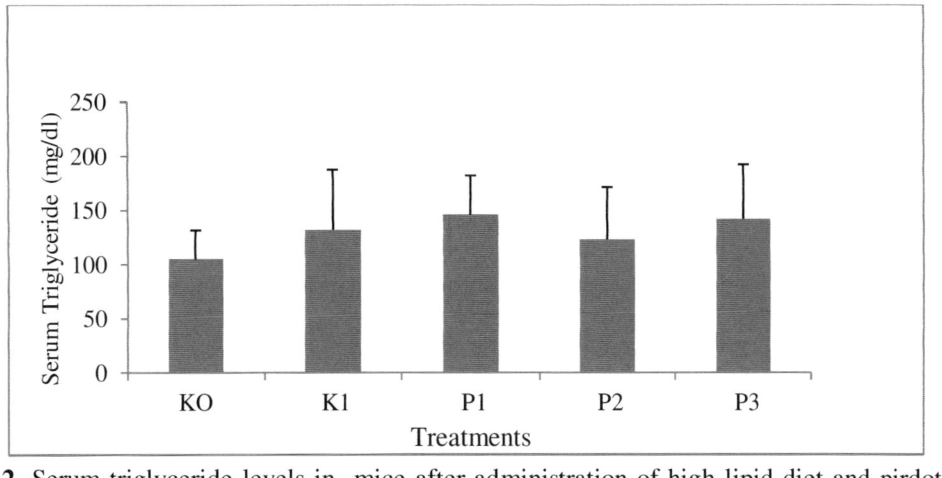

Figure 2. Serum triglyceride levels in mice after administration of high lipid diet and pirdot leaves extract at doses of 100, 200, or 300 mg/kg BW for 30 days.

We also found that statistically, the level of serum triglycerides of the control group (K0), control yolk group, and all three treatments group was not significantly different (p>0.05). It seemed that pirdot leaves extract did not decrease the level of serum triglycerides.

3.2.3. Low-Density Lipoprotein (LDL) Level

At the end period of the experimental, there was a marked higher LDL in yolk diet group and sharp decline in three extract groups (Figure 3), but statistically, there was no significant difference among all groups (P>0.05).

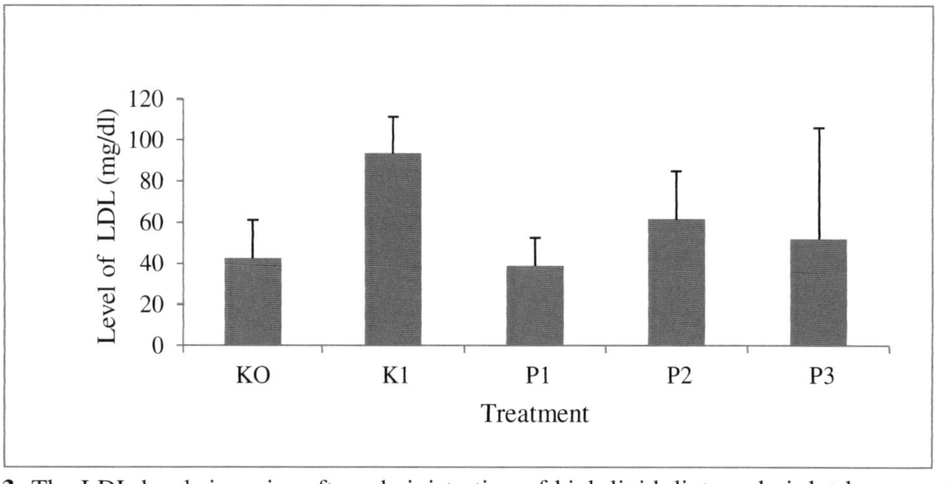

Figure 3. The LDL levels in mice after administration of high lipid diet and pirdot leaves extract at doses of 100, 200, or 300 mg/kg BW for 30 days

3.2.4. High-Density Lipoprotein (HDL) Level

Results showed that the highest HDL level found in group K1. Its level was significantly higher (P <0.05) compared to control and also to three extract groups. In Figure 4, the HDL level on the three extract groups is lower than the level in the yolk group. The levels of HDL obtained in three groups of extract treatment was not statistically different from control (P>0.05), and all were within normal range (40-60 mg/dl).

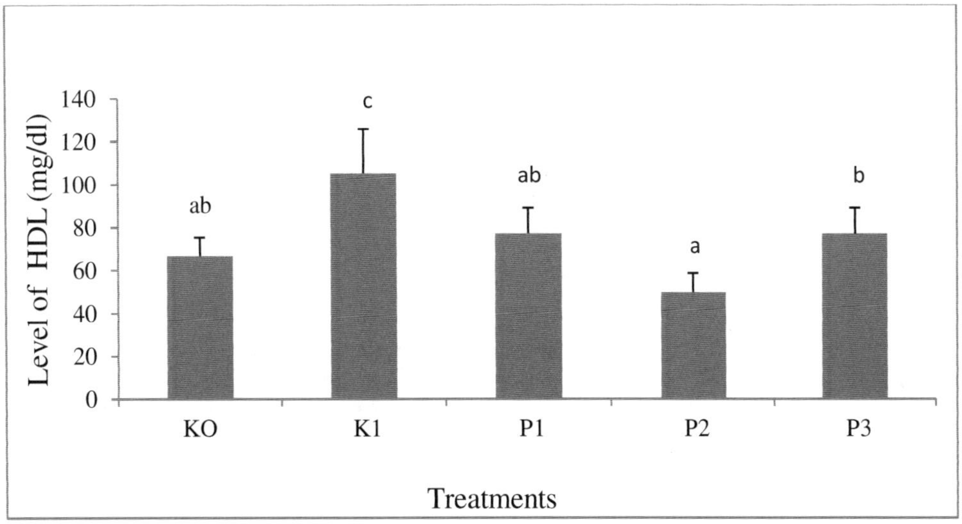

Figure 4. The HDL levels in mice after administration of high lipid diet and pirdot leaves extract at doses of 100 mg / kgBW, 200 mg / kgBW, 300 mg / kgBW for 30 days.

This result means that pirdot leaves extract decreased both LDL and HDL level. The decreased in both LDL and HDL have been discussed in pinto bean experiment [13], which reported that consumption of pinto beans showed decreases in both LDL- and HDL cholesterol which might be due to chemical composition had the effect on both plasma cholesterol.

In conclusion, the results obtained confirm the pirdot leaf extract has potential as antihyperglycemic. The effects obtained are equivalent to the effects of antidiabetic drug metformin. On the other hand, the antihyperlipidemic effect of the extract has not been conclusive, because the pirdot leaf extract lowered total cholesterol significantly, and also marked lowered LDL, but it also lowered the HDL level and no significant effect on triglyceride level.

Acknowledgement
We would like to acknowledge the support of DRPM Ministry of Research and Technology and the Higher Education Republic of Indonesia which has provided funding for this research through the PPT scheme (Contract Number: 003/SP2H/LT/DRPM/IV/2017, April 24th, 2017).

References

[1] Wild S, Mb Bchir, G Roglic, A Green, R Sicree, Mbbs, H King (2004) Global Prevalence Of Diabetes: Estimates For The Year 2000 And Projections For 2030. *Diabetes Care*. No. 27:1047–1053. DOI: 10.2337/diacare.27.5.1047.

[2] Fowler MJ (2007) Diabetes Treatment, Part 2: Oral Agents for Glycemic Management. *Clinical Diabetes*. vol. 25 no. 4: 131-134. https://*doi.org/10.2337/diaclin.25.4.131*

[3] Situmorang R, Harianja A, and Silalahi J (2015) Karo's Local Wisdom: The Use of Woody Plants for Traditional Diabetic Medicines. *Indonesian Journal of Forestry Research*, 2(2), 121-130. doi:http://dx.doi.org/10.20886/ijfr.2015.2.2.121-130.

[4] Silalahi M, J Supriatna, Eb Walujo, Nisyawati (2015) Local knowledge of medicinal plants in sub-ethnic Batak Simalungun of North Sumatra, Indonesia. *Biodiversitas* 16(1): 44-54. DOI: 10.13057/biodiv/d160106

[5] Aronson D, and Elliot J Rayfield (2002) How hyperglycemia promotes atherosclerosis: molecular mechanisms. *Cardiovascular Diabetology* 20021:1 DOI: 10.1186/1475-2840-1-1 ©Aronson and Rayfield; licensee BioMed Central Ltd.
[6] Luman A (2010) *Diabetes dan Penyakit Kardiovaskular*. Universitas Sumatera Utara.
[7] Sitorus P (2015) Characterization Simplisia and Ethanolic Extract of Pirdot (Saurauia Vulcani, Korth) Leaves and Study of Antidiabetic Effect in Alloxan Induced Diabetic Mice. *International Journal of ChemTech Research*, 8(6): 789-794.
[8] Szkudelski T (2001) The Mechanism Of Alloxan And Streptozotocin Action In β Cells of The Rat Pancreas, *Physiology Research*, 50: 536-54.
[9] Widyaningsih W, Prabowo A, and Sumiasih (2010) *Pengaruh Ekstrak Etanol Daging Bekicot (Achantina fulica) Terhadap Kadar Kolesterol Total, HDL, dan LDL Serum Darah Tikus Jantan Galur Wistar*. *Jurnal Sains dan Teknologi Farmasi.* 15(1): 1-10.
[10] Wiadnya IB, Zaetun S, and Sari WL (2014) Efektifitas Pemberian Filtrat Labu Siam (*Sechium edule*) terhadap Penurunan Kadar Kolesterol Total pada Hewan Coba Tikus Putih (*Rattus norvegicus*). *Media bina Ilmiah* 8(1): 50-55.
[11] Gani N, Momuat LI, and Pitoi MM (2013) *Profil Lipid Plasma Tikus Wistar yang Hiperkolesterolemia pada Pemberian Gedi merah (Abelmoschus manihot* L). *Jurnal MIPA UNSRAT online.* 2(1)*: 44-49.*
[12] Wresdiyati T, M Astawan and YH Lusia (2006) *Profil imunohistokimia Super Oksida Dismutase (SOD) pada jaringan hati tikus dengan kondisi hiperkolesterolemia*. *Hayati J Biosci.* 13: 85-89.
[13] Finley JW, Burrell JB and Reeves PG (2007) Pinto bean consumption changes SCFA profiles in fecal fermentation, bacterial populations of the lower bowel, and lipid profiles in blood of humans. *Journal of Nutrition*, 135: 2391-2398.

IOP Conf. Series: Earth and Environmental Science **130** (2018) 012043 doi:10.1088/1755-1315/130/1/012043

The test about blood serum capabilities in maintaining the quality of bull spermatozoa during storage in cep diluent at refrigerator temperature

Nur Ducha[1]

[1]Biology Department, Faculty of Mathematics and Science, Surabaya State University, Jalan Raya Ketintang Kampus Unesa Surabaya, Jawa Timur, Indonesia, 60231

email: nurducha@unesa.ac.id

Abstract. The storage of spermatozoa requires a protective material from cold shock events and the presence of free radicals.In CEP diluent contain BSA, that was used as spermatozoa protection. This study aim was to examine the ability of cow blood serum in replacing BSA as spermatozoa protective in CEP diluent. Fresh semen from Limousin bull was diluted with CEP diluent + BSA as control, in the treatment group were CEP without BSA, but replaced with 3%, 5%, and 7% serum from fresh blood. Spermatozoa quality tests included motility and viability. The motility of spermatozoa was observed by two people using a light microscope with 200 X magnification at temperature of 37°C. The method of viability observation was eosin nigrosin staining, and observed under a light microscope with 400 X magnification. The results showed that the replacement of cow blood serum with various concentrations gave different effects on the quality of spermatozoa. The best motility and viability of the treatment group was at serum concentrations of 5% after eight days storage and was not significantly different from the controls. The conclusion in this study was cow blood serum can replace BSA in CEP diluents.

Keywords: blood serum, CEP diluent, quality of bull spermatozoa

1. Introduction

The storage technology of spermatozoa becomes very important for the success of Artificial Insemination (AI). It requires media to increase the volume of semen and maintain the quality of spermatozoa. The medium is referred as semen diluent. A diluent is not only contains an energy source, but also a spermatozoa protective material from the cold shock incident as well as the presence of free radicals during the storage process [1,2].

The spermatozoa has the high content of polyunsaturated fatty acids (PUFAs) in the plasma membrane makes spermatozoa sensitive to damage caused by cold shock and lipid peroxidation by free radicals that affect motility, metabolism, ultrastructure, and fertility [3,4].

CEP diluent contained electrolyte ions, energy sources such as fructose, antibiotics, and there were also Bovine Serum Albumin (BSA) as macromolecule [5,6,7]. Bovine Serum Albumin (BSA) was often added in the diluent to maintain the quality of spermatozoa [8,9].

Blood is composed of blood plasma and shaped components. In the blood plasma there is a component of blood serum. The main component in both blood serum and BSA is albumin protein

[10,11]. BSA is the product of the isolation of blood plasma, the price is very expensive, while the blood serum in this study was obtained from fresh blood of cows through the process of centrifugation. Based on the same major constituent between BSA and blood plasma serum, the possibility of plasma serum from cow fresh blood may provide protection against spermatozoa during storage at low temperature. The aim of this study was to examine the serum plasma capability of cow fresh blood in CEP diluents on the quality of Limousin bull spermatozoa during storage at refrigerator temperature.

2. Methods

2.1 Making of diluent media

CEP extender contained NaCl 15 mmol/L; KCl 7.0 mmol/L; $CaCl_2(H_2O)_2$ 3.0 mmol/L; $MgCl_2(H_2O)_6$ 3.0 dmmol/L; NaHCO3 11.9 mmol/L; NaH_2PO_4 8.0 mmol/L; KH_2PO_4 20.0 mmol/L; fructose 55 mmol/L; sorbitol 1.0 gr/L; BSA 2.0 gr/L; Tris 133.7 mmol/L; penicillin 1000 IUI; streptomycin 1 gr; and citrate acid 42.6 mmol/L (Bioworld, USA) [5,6]. CEP with the addition of BSA was as a control group, and treatment group with different concentration from blood serum. All chemical were dissolved with deionized water, and sterilized with millipore membrane. Blood serum was obtained from fresh blood plasma that came healthy cow blood in slaughterhouses.

2.2 Preparation of fresh semen

The using of fresh semen was obtained from Limousin cattle at the Center for Artificial Insemination, Singosasari, Indonesia with the provision of eligible for the process of dilution and storage of motility $\geq$ 55%, viability $\geq$ 70%, morphological normality $\geq$ 75%. Fresh semen was diluted with CEP containing BSA for the control group, whereas in the treatment group was diluted with CEP which added blood serum from cow fresh blood 3%, 5%, and 7%. Semen dilution was determined 25 $x10^6$ spermatozoa/ml in accordance with Indonesian National Standard for storage of cow or goat spermatozoa at freezing temperatures. After the dilution process, the semen in the control group and the treatment was carried out in the refrigeratore at 4-5 ° C under dark conditions.

2.3 Observation of sperm motility

The motility of spermatozoa was observed using a light microscope. The semen of all treatments was taken using an glass needle, placed on a glass object, and then observed under a 200 X magnification of light microscope at 37 ° C, and the observations were made by two persons [10,11]. The motility of spermatozoa was observed daily until showed results below standard according to Indonesian national standards for storage or freezing process of not less than 40%.

2.4 Observation of sperm viability

Spermatozoa survival (viability) was observed using eosin negrosin staining [12,13]. Semen of all treatments and controls was taken using glass needles, dripped on a glass object, mixed with eosin nigrosin, and made smear preparations. The preparation were observed under a 400 X magnification of microscope, the percentage of living cells was calculated based on the counting of 100 cells. The observation of viability was performed daily to show results below the standard for storage at low temperatures was 50%.

3. Results and Discussion

3.1 Data of spermatozoa motility

The mean and repetitive standard of motility percentage of the Limousin bull spermatozoa in CEP diluent that containing and in the treatment group with BSA replacement using cow blood serum were shown in Table 1.

Table 1. Spermatozoa motility of Limousin bull daily in CEP + BSA (control) and CEP + cow blood serum (treatment group) diluents during storage at 4-5°C

Various of diluent	The mean motility percentage of the Limousin spermatozoa on daily								
	1	2	3	4	5	6	7	8	9
Control	55.00± 1.18^{b}	52.50± 1.60^{a}	51.00± 1.24^{a}	50.50± 1.13^{b}	48.50± 2.74^{a}	45.50± 1.94^{b}	45.50± 4.33^{a}	42.50± 1.92^{ab}	37.50± 1.56^{a}
Treatment A	53.50± 2.43^{ab}	53.50± 1.82^{a}	52.00± 2.61^{a}	50.50± 0.67^{b}	49.50± 1.31^{a}	46.00± 1.52^{b}	46.00± 3.13^{a}	41.50± 1.12^{ab}	39.00± 1.23^{a}
Treatment B	54.50± 1.68^{b}	51.50± 0.62^{a}	50.50± 0.69^{a}	48.00± 0.91^{a}	47.00± 0.47^{a}	47.00± 0.98^{b}	45.50± 3.39^{a}	43.50± 1.71^{b}	38.50± 1.39^{a}
Treatment C	51.50± 0.93^{a}	50.50± 1.02^{a}	49.50± 0.83^{a}	48.50± 0.84^{ab}	46.50± 3.22^{a}	44.50± 2.33^{a}	41.50± 2.64^{a}	38.50± 1.06^{a}	35.50± 1.41^{a}

Note : Control = CEPD + BSA ; treatment A = CEPD – BSA + blood plasma 3%; treatment B = CEPD – BSA + blood plasma 5%; treatment C = CEPD - BSA + blood plasma 7%.

Based on Table 1 showed a decrease in the percentage of motility during storage until the ninth days. Motility data that appropriate with Indonesian National Standards for artificial insemination (40%) after the storage process appeared on the 8th day of storage, storage on day nine was already below 40% in all treatments.

In the 8th day storage there was a difference of motility between treatments, the best motility showed in the 5% serum treatment, but also similar to the control group and the 3% serum treatment. 7% serum treatment showed the lowest mobility. However, on day 9, the percentage of motility showed no difference in all treatments.

3.2 Data of spermatozoa viability

The mean and deviation standard of viability percentage of the Limousin bull spermatozoa in CEP diluent + BSA and CEP no BSA but replaced with cow blood serum showed in Table 2.

Table 2. Spermatozoa viability of Limousin bull daily in CEP + BSA (control) and CEP + cow blood serum (treatment group) diluents during storage at 4-5°C

	The mean motility percentage of the Limousin spermatozoa on daily								
	1	2	3	4	5	6	7	8	9
Control	76.37^{b} ±2.45	74.15^{b} ± 1.88	72.42^{ab} ± 1.73	71.02^{b} ± 2.06	67.49^{a} ± 2.74	66.88^{a} ± 2.53	60.56^{a} ± 4.33	57.68^{b} ± 2.69	55.10^{b} ± 2.22
Treatment A	74.92^{ab} ±2.43ab	73.38^{b} ± 1.92	74.71^{b} ± 4.62	71.41^{b} ± 1.46	68.89^{a} ± 1.31	63.50^{a} ± 3.52	61.85^{a} ± 3.13	52.76^{ab} ± 2.12	49.18^{a} ± 1.90
Treatment B	72.60^{ab} ±0.68ab	72.40^{b} ± 0.82	70.89^{ab} ± 0.59	69.75^{ab} ± 0.71	68.43^{a} ± 0.47	65.66^{a} ± 0.88	62.01^{a} ± 3.39	56.97^{ab} ± 1.91	53.23^{ab} ± 1.59
Treatment C	71.30^{a} ±0.91a	69.30^{b} ± 0.62	68.64^{a} ± 0.52	67.94^{a} ± 0.44	64.26^{a} ± 4.22	61.98^{a} ± 3.23	59.18^{a} ± 2.64	52.12^{a} ± 1.02	49.45^{a} ± 1.81

Note: Control = CEPD + BSA ; treatment A = CEPD – BSA + blood plasma 3%; treatment B = CEPD – BSA + blood plasma 5%; treatment C = CEPD - BSA + blood plasma 7%.

Based on the data in Table 2 showed decreased viability of Limousin bull spermatozoa during storage. The percentage of spermatozoa viability showed a significant difference in various treatments on daily observation. The treatment group that approached the viability of the control group was at 5% treatment of serum.

Based on the results of the study showed that the replacement of BSA with cow blood serum in CEP diluents capable of maintaining the motility and viability of spermatozoa Limousin bull up to 8 days of storage. The best concentration of the treatment when compared with the control group was 5%.

Cow blood serum can protect against bull spermatozoa as well as BSA [14,15], this was because both have the same component of albumin. Blood serum is essentially blood plasma without fibrinogen and clotting factor [16,17]. Blood plasma is a light yellow liquid composed of 91% water, 7% protein and 2% other components such as ions, nutrients, gases, electrolytes, and hormones. The main proteins contained in it are albumin, globulin and fibrinogen. Albumin is the most numerous protein of 60% [18].

The albumin protein contained in BSA or serum becomes a transition link between Fe^{2+} and Cu^{+} metal ions, thereby minimizing the formation of OH^{-} radicals promoting peroxidation of lipid spermatozoa [19].

4. Conclusion
Bull blood serum capable to maintenance of spermatozoa quality in CEP diluent during storage at 4-5°C, so that capable to replace BSA in CEP diluent, and may be can use the other diluent.

Acknowledgments
The authors gratefully acknowledge that the present research was supported by Ministry of Research and Technology and Higher Education Republic of Indonesia. The support was under the research grant "Hibah Bersaing" of The Year 2015.

References
[1]. Cerolini S, Maldjian A, Surai P, Noble R (2000) Viability, susceptibility to peroxidation and fatty acid composition of boar semen during liquid storage. *Anim Reprod Sci.*58(1-2):99-111. doi:10.1016/S0378-4320(99)00035-4.

[2]. O'Hara L, Hanrahan JP, Richardson L, et al (2010) Effect of storage duration, storage temperature, and diluent on the viability and fertility of fresh ram sperm. *Theriogenology*.73(4):541-549. doi:10.1016/j.theriogenology.2009.10.009.

[3]. Sanocka D, Kurpisz M (2004) Reactive oxygen species and sperm cells. *Reprod Biol Endocrinol.* 2(1):12. doi:10.1186/1477-7827-2-12.

[4]. Guthrie HD, Welch GR (2012) Effects of reactive oxygen species on sperm function. *Theriogenology*. 78(8):1700-1708. doi:10.1016/j.theriogenology.2012.05.002.

[5]. Verberckmoes S, Van Soom A, Dewulf J, De Pauw I, De Kruif A (2004) Storage of fresh bovine semen in a diluent based on the ionic composition of cauda epididymal plasma. *Reprod Domest Anim.* 39(6):410-416. doi:10.1111/j.1439-0531.2004.00521.x.

[6]. Ducha N, Susilawati T, Aulanni'am, Wahyuningsih S, Pangestu M (2012) Ultrastructure and fertilizing ability of Limousin bull sperm after storage in CEP-2 extender with and without egg yolk. *Pakistan J Biol Sci.* 15(20). doi:10.3923/pjbs.2012.979.985.

[7]. Ducha N, Susilawati T, Aulanni'am, Wahyuningsih S (2008) Motilitas dan Viabilitas Spermatozoa Sapi Limousin Selama Penyimpanan pada Refrigerator dalam Pengencer CEP-2 dengan Suplementasi Kuning Telur. *Kedokt Hewan*. 7:2002-2005.

[8]. Uysal O, Bucak MN (2007) Effects of oxidized glutathione, bovine serum albumin, cysteine and lycopene on the quality of frozen-thawed ram semen. *Acta Vet Brno*. 76(3):383-390. doi:10.2754/avb200776030383.

[9]. Sariözkan S, Türk G, Cantürk F, Yay A, Eken A, Akçay A (2013) The effect of bovine serum albumin and fetal calf serum on sperm quality, DNA fragmentation and lipid peroxidation of the liquid stored rabbit semen. *Cryobiology*. 67(1):1-6. doi:10.1016/j.cryobiol.2013.04.002.

[10]. Mata-Campuzano M, Soleilhavoup C, Tsikis G, Martinez-Pastor F, de Graaf SP, Druart X (2015) Motility of liquid stored ram spermatozoa is altered by dilution rate independent of seminal plasma concentration. *Anim Reprod Sci.* 162:31-36. doi:10.1016/j.anireprosci.2015.09.004.

[11]. Thuwanut P, Chatdarong K, Techakumphu M, Axnér E (2008) The effect of antioxidants on motility, viability, acrosome integrity and DNA integrity of frozen-thawed epididymal cat spermatozoa. *Theriogenology.* 70(2):233-240. doi:10.1016/j.theriogenology.2008.04.005.

[12]. Chalah T, Brillard JP (1998) Comparison of assessment of fowl sperm viability by eosin-nigrosin and dual fluorescence (SYBR-14/PI). *Theriogenology.* 50(3):487-493. doi:10.1016/S0093-691X(98)00155-1.

[13]. Björndahl L, Söderlund I, Kvist U (2003) Evaluation of the one-step eosin-nigrosin staining technique for human sperm vitality assessment. *Hum Reprod.* 18(4):813-816. doi:10.1093/humrep/deg199.

[14]. Ashrafi I, Kohram H, Tayefi-Nasrabadi H (2013) Antioxidant effects of bovine serum albumin on kinetics, microscopic and oxidative characters of cryopreserved bull spermatozoa. *Spanish J Agric Res.* 11(3):695-701. doi:10.5424/sjar/2013113-3870.

[15]. Akhter S, Rakha BA, Iqbal R, Ansari MS (2014) Effect of bovine serum albumin on motility, plasmalemma, viability and chromatin integrity of buffalo bull spermatozoa. *Pak J Zool.* 46(1):115-120.

[16]. Martini & Nath J (2009) *Fundamentals of Anatomy and Physiology*. Vol 7th. doi:612.

[17]. Marieb EN, Hoehn K (2014) *Human Anatomy & Physiology.* Vol 7. doi:10.1038/nnano.2011.234.

[18]. Saladin K (2014) Anatomy & Physiology: The Unity of Form and Function. In: *Anatomy & Physiology: The Unity of Form and Function.* 1248. https://books.google.com/books?id=lmr8nQEACAAJ&pgis=1.

[19]. Perumal P, Nahak AK, Vupru K, Khate K, Balamurugan TC, and Prakash KR (2015) Effect of Addition of Bovine Serum Albumin on the Liquid Storage (5°C) of Mithun (Bos frontalis) Semen. *Journal of Cell and Tissue Research.* 15(1):4795-4800.

Density and White Shrimp Growth Pattern (*Penaeus merguiensis*) in Kampung Nipah Waters of Perbaungan North Sumatera

Sri Natalia Silaen[1*], and Miswar Budi Mulya[1]
[1]Department of Biology FMIPA University North Sumatera Jl. Dr. T. Mansyur No. 9, Padang Bulan Campus, Medan, 20155, North Sumatra

*Email: sri_silaen@yahoo.co.id

Abstract. The purpose of the study was to determine the density and pattern of growth of white shrimp (*Penaeus merguiensis)*of the village Nipah waters. The data collection was conducted by sampling using nets and fishing gear "Langge" (a tool) to determine the density of the three observation stations. The result showed that the distribution of white shrimp in the waters of the estuary and surrounding degraded over the past ten years. The highest density at station II is 0.56 and 5/m^2 and at least at the third station as much as 0.42 and 6/m^2. The correlation between the density of shrimp with depth as well as the fraction of the base substrate showed that only the depth of the waters who has any significant correlation with the density of shrimp, although the closeness of the relationship is small.

Keywords: aquatic estuary, density, white shrimp

1. Introduction

Indonesia is a country that has the potential of a most biggest source of sea shrimp. Shrimp is Indonesia's leading fishery commodity because of the high value or price of shrimp and the increasing demand for shrimp both in the domestic market and in International market. Shrimp has the largest export value of about 21% of world trade. For Indonesia alone shrimp are a mainstay export commodity and source of foreign exchange earnings considering about 50% of the total export of fishery products was sourced from this business. Shrimp exports to Indonesia over the last 25 years have an average growth rate of 27.8% [1].

Based on data from the Ministry of Marine Affairs and Fisheries (2011) for 5 years period from 2007 to 2011, the production of shrimp catch fisheries continues to fluctuate, in 2007 the production reached 258.976 tons, the year 2008 reached 236.922 tons, the year 2009 reached 236.870 tons, in 2010 as much as 227.326 tons [2]. In 2011 as many as 228.870 tons. The development of production for five years including decreased average -2.97%.

Indonesia currently ranks fourth in the world after China, Thailand, and Vietnam as the largest country regarding shrimp trade commodities. Based on the target of shrimp production in 2010, Indonesia has eight provinces that become largest shrimp producer, namely East Java, Central Java, DKI Jakarta, Lampung, Bali, North Sumatra, Southeast Sulawesi, and Sulawesi [3].

There are 11 types of penaeid shrimp that can categorized as having important commercial values in Indonesian water, consisting of two genera Penaeus and Metapenaeus [4]. In commercial terms, preferred are *Penaeus monodon* and *Penaeus merguiensis* (white shrimp), as both of these species can reach large sizes and good market for export.White shrimp (*Penaeus merguiensis*) was found in almost all Indonesian waters, ranging from estuaries of mangroves, coastal waters around mangroves such as estuaries, lagoons, and bays, to open waters [5].

According to North Sumatra Head of Fishery and Marine Affairs, Zonny Waldy said that the value of North Sumatera fish and shrimp exports moved up in early 2016 or to 38.975 million US dollars, the impact of fisherman catch and more selling price of up 0.85%. Head of the Central Bureau of Statistics (BPS) of North Sumatra, Wien Kusdiatmono said in January-February 2015, the export value of shrimp and shrimp was 37.753 million US dollars while in the same period 2016 was 38.975 million US dollars.

According to the fishermen in the area of white prawns sold at a price of about Rp 65.000,00 to Rp 75,000.00 per kg marketed in the area in Indonesia to Malaysia and also beneficial in the economy of white shrimp is also both consumed and good for the body. The increase in export value was driven by increasing exports and higher prices. Provincial Government of North Sumatra (2016) is continuously trying to improve the catch of fisherman ranging from helping fishing gear, environmental improvement around the sea to the provision of fishermen's insurance.

From some information above, data on white shrimp fluctuate every year and in Kampung Nipah itself is not known data related to the decrease and increase of white shrimp. Aquatic estuary in the village of Nipah, District Sei Nagalawan Perbaungan much overgrown with mangrove vegetation is very potential in supporting the life of white shrimp. To date, the abundance and growing data of the white shrimp population of *Penaeus merguiensis* in Nipah village estuary waters have not been obtained for that; it is necessary to conduct research.

2. Methods

2.1. Time and Place of Study

This research was conducted since August – September 2016 in Kampung Nipah Waters of Sei Nagalawan Subdistrict, Perbaungan Subdistrict, Serdang Bedagai Regency, North Sumatera. Geographically the estuary waters of Nipah village are at 03°35'32 "- 3°35'39.66"LUK and 99°5'33.00" - 99°6'18.06"BT.

2.2. Tools and materials

The tools used in the study are boat, buoy, fishing nets, nets, compass, refractometer, Hg thermometer, pH meter, sechidisk, analytical scale, strap rope, cool box, slider, roll meter, sample container, stationery, GPS, Meter cloth, net, digital camera, plastic, knife, paper label, wooden pillar, rubber band, raffia rope.The material used are white shrimp samples, water samples, and substrate.

2.3. Determination of Research Station

The research was conducted in 3 stations using purposive sampling. Station I is in coastal waters (3°35'32 "LU - 99°5'33.00" BT), station II is in the high seas (3°35'48,91"LU - 99°5'41.62 " BT) and station III is in the mouth of the estuary (3° 35'39.66"LU 99°6'18.06"BT).

2.4. Shrimp Sampling.

Shrimp sampling is done using a net with an area of 112.35 m^2 and "langge" at low tide in the morning noon. The net was spread to the surface of the water, then silenced for an hour at the bottom of the water and positioned in the direction of the fisherman's boat, after one hour the net was pulled and the shrimp snagged in the net catching of shrimp using "langge", done at night (18.00-20.00 WIB) only with 10.5 m^2 "langge". "Langge" was opened and inserted into the waters then left for 15 minutes and then lifted and taken the caught shrimp Measurement of Physical-Chemical Water and Substrate.

2.5. Parameters

Measurements of physical-chemical factors of water and substrate were carried out at each station by repeating the shrimp sampling period. Measurements of physical chemical parameters include water temperature, water crosses, water brightness, DO and BOD_5, pH, flow velocity, water depth, and substrate fraction.

2.6. Data analysis

The analysis of white shrimp data included abundance using the formula according to Brower et al. [6], growth pattern according to Sparre &Venema [7], white shrimp sex ratio using Campbell formula [8], and Pearson correlation analysis using Sugiyono formula [9].

3. Results and Discussion

The value of density obtained at station I is 0.52 ind/m^2, station II is 0.56 ind/ m^2 and station III is 0.42 ind/ m2 (Table 1). From the research that has been did shows that the most density value, found in station II followed the station I and the lowest at the station III.
Station II is an area of shrimp catching and at the station is also found a lot of white shrimp ripe gonad Station III was located at the base of the river estuary that has a lower salinity (28°/8) due to the flow of water entering the river into the waters, so that the white shrimp are also getting less.

Table 1. Density, and Sex Ratio of White Shrimp on each Station

No	Station	K (ind/m^2)	Male(tail)	Female(tail)	Sex Ratio
1	Station I	0.52	23	94	1:4 (4.086)
2	Station II	0.56	16	111	1:7 (6.93)
3	Station III	0.42	32	64	1:2 (2)

Based on the results from Table 1 it can also be seen that the ratio of white shrimp sex of *Penaeus merguiensis* at each station was dominated by female shrimp.
The results of this study indicated at the time of the study allegedly female shrimps spawning on the high seas. This was evidenced by a large number of female shrimps and has a maturity level 3 gonad (TKG 3). White shrimps obtained were at maximum size 51 g and length 19.6 cm and generally found in station II.

3.1. Physical Chemical Chemical Factor

Data on the measurement of physical, chemical factors of the waters during the research station can wa seen in Table 2 and Table 3.

Table 2. Physical Characterization of Aquatic Chemistry

Parameter	Unit	Station I	Station II	Station III
Temperature	°C	29.7	26.5	27.5
Salinity	°/∞	28	26	20
pH	-	5.6	7.8	7.2
DO	mg/L	6.36	7.04	792
BOD5	mg/L	1.3	0.7	0.7
Current velocity	m/s	0.9	2.8	0.7
Light intensity	Candela	1116	1034	1228
Penetration of light Meter	Meter	1	0.9	0.5

IOP Conf. Series: Earth and Environmental Science **130** (2018) 012044 doi:10.1088/1755-1315/130/1/012044

Table 3. Value of substrate from each station seeing

Fraction \ Station	I	II	III
Liat (%)	10	85	60
Sand (%)	70	10	25
Dust (%)	20	5	15

A. Temperature

From Table 2 it can be seen that the temperature values of the three stations ranged from 26.5°C to 29.7°C. The highest temperature at station I while the lowest temperature obtained at the station 8A II while for salinity at the station 3. The value of temperature and salinity obtained can still support the life of white shrimp. Mintarjo et al. (1984) state that temperatures affect shrimp growth and life where the rate of growth increases with the rise in temperature. [10] The range of aquatic environmental parameters is a factor that causes the distribution of white shrimp to spread evenly at different substrater depth and time intervals at 26°C-27°C. Barnes (1988) state that the criteria of a reasonable salinity range for aquatic organisms are °/8 - 30 °/8 [11]. Temperature and salinity were the most influential factors to shrimp abundance and survival [12]. Meanwhile, temperature is the most influential factor on survival and shrimp growth [13]. Temperature also has an important the population density and long relationship of shrimp weight. The increase in temperature increases the metabolism rate in the shrimp body to some extent and after that the temperature rise decreases [6].

B. pH

Based on the value of Table 2, the pH values ranged from 5.6 to 7.8 where the highest pH was at station II while the lowest was at the station I. The values obtained from the three observation stations supported the life of white shrimp. Life in water can still survive when the waters have a pH range of 5-9. [14] The differences in pH value in the water is due to the addition or loss of CO_2 through the process of photosynthesis that will cause changes in the pH in the water [5]

C. DO

The DO values in Table 2 obtained in the study ranged from 6.36-7.92 mg / L. DO is highest at station III whereas the lowest DO at station I. Optimum dissolved oxygen for aquatic biota life including white prawns ranging from 4.10 mg / L - 6.60 mg / L with a minimum tolerance limit is 4.00 mg / L. white shrimp can still live in waters that have a minimum oxygen content of 3.00 mg / L [16]. The high dissolved oxygen content in the waters was influenced by the temperature, pressure, and concentration of various dissolved ions in the waters. [15]

D. BOD_5

From Table 2, the BOD_5 value at each station ranged from 0.7-1.3 mg / L that the highest BOD_5 was at the station I and the lowest value was in station II and III. The BOD5 value is the amount of oxygen required by aerobic microorganisms in the process of decomposition of organic compounds, measured at certain temperature [2]. The difference of BOD_5 value at each station was caused by the number of different organic bodies at each research station. The high value of BOD is user as bioindicator about the number of organic material compounds that enter in water. The greater the BOD_5 value of water, the lower the dissolved oxygen content [17].

E. Current Speer

The current velocity at the observation station performed during the study ranged from 0.7 to 2.8 m/s, and the highest current velocity was in station II while the lowest current velocity had found at station III. Current velocity in estuary waters is affected by tidal water. The high concentrations of organic materials was caused by fresh water flow and water mixing due to tides [18]. Of these factors produce high productivity and make the white shrimp was found. The value of current velocity obtained at the three observation stations in Kampung Nipah Serdang Bedagai estuary still in the category of safe and normal and good in supporting the life of white shrimp.

F. Light intensity

Based on the observations in the field, in Table 2 above the results of the light intensity of each station ranges from 1.034-1228 Candelas where the highest light intensity is at the station III and the lowest at station II. The intensity of sunlight was reduced then the process of photosynthesis will be inhibited so that oxygen in water will also be reducer, where oxygen is needed aquatic organisms for metabolism. The results obtained from this research that a station the light intensity factor obtained is still in normal condition in the life of white shrimp in estuary waters of Kampung Nipah. [19]

G. Penetration of light

The value of light penetration was influenced by suspended particles dissolved in water thereby reducing the rate of photosynthesis [20]. The brightness of water is associated with suspended solids, the color of water and the light entering the water [21].

The results showed that the penetration value of light at each station ranged from 0.5-1 m, where the highest light penetration is at the station I and the lowest at station III, this was influenced by sunlight, suspended solids, measurement time and water color. A good brightness level ranges from 30-65 cm to support the productivity of aquatic organisms merguiensis de Man) in Mangrove Ecosystem Sei Tuan Sumatera Utara. The average value of light penetration in Nipah's estuary waters is still good for the survival of white shrimp (*Penaeus merguiensis*) and other organisms. [22]

H. Substrate Fraction

The station I sand fraction has the highest value of 70%. While at station 2 and station three clay fraction more Overall substrate fraction in Kampung Nipah Serdang Bedagai waters is general and the sandy fraction. The sandy clay substrate was favored by white. Water favored by white shrimp *Penaeus merguiensis* are water with sandy clay substrates [5]. Likewise, they prefer turbid water with clay substrates mixed with sand. [10]

3.2. White Shrimp Growth Pattern

Growth pattern is used to find out the relationship of a long weight of white shrimp at each station. The long-weight relationship of shrimp can be seen in Table 4.

Table 4. Connect below long-weight shrimp

No	Station	Average value b	Pattern of growth
1	I	1.924	Allometrik negative
2	II	3.578	Allometrik positive
3	III	2.355	Allometrik negative

From Table 4.1.3 it shows the growth pattern of white shrimp (*Penaeus merguiensis*) at station II is allometric positive where the growth of shrimp weight is more dominant compared with length, where as in station II and III allometric negative which indicates the addition of length is more dominant or Faster than weight. This is due to the white shrimp growth was influenced by physical-chemical factors of water and the availability of food. Growth are influenced by two factors: intrinsic factor (inside) and extrinsic factor (outside). Intrinsic factor is a factor that comes from within the white prawns are genetic, age, size of ability to use food and resistance of attack to disease [6].

IOP Conf. Series: Earth and Environmental Science **130** (2018) 012044 doi:10.1088/1755-1315/130/1/012044

3.3. Pearson Correlation Analysic

Pearson correlation analysis can was obtained by analyzing the relationship of white shrimp density with the physical-chemical factor in Nipah kampong estuary waters by using Pearson correlation analysis. The correlation (r) can was seen in Table 4.1.4 below this.

Table 5. Value of correlation result of white shrimp density with physical chemical factor of water

Parameters	Correlation Value
pH	0.996
Temperature	-0.859
Salinity	0.468
Penetration of light	-0.508
DO	0.531
BOD_5	-0.984
Current speed	0.580
Light intensity	-0.092

Table 5 showed that pH,temperature and BOD_5 played a very significant effect on the density of white shrimps. pH has a very positive effect on the density of white shrimp, which means that the density of white shrimps will be higher with increasing pH value. Temperature and BOD_5 have a very significant negative effect which means the density of white shrimp will be higher with the decrease in temperature value and BOD_5.

4. Conclusions

The value of white shrimp density of *P. merguiensis in* Nipah village waters ranged from 0.42-0.56 ind/m. Comparison of male and female shrimp in the three observation stations was more dominated by female shrimp. The pattern of white shrimp growth at stations I and III including allometric negative, while at station II is allometric positive. pH of water has a very positive effect on the density of white shrimp, while the temperature and BOD5 have a very significant negative effect on the density of white shrimp.

References

[1] Subagyo W (2005) Status of catching white shrimp (Penaeus Mergui Eensis De Man) In Cilacap waters and Surrounding and Proposed Management. Institut Pertanian Bogor: Bogor

[2] Forstner U (1990) Umweltschutztecechnik. Springer Verlag, Heidelberg Ministry of Marine Affairs and Fisheries (2011). Decree of the Minister of Marine Affairs and Fisheries of the Republik of Indonesia Number.KEP. 45 / MEN / 2011 concerning the Estimation of Fish Resource Potential in the Fisheries Management Area of the Republic of Indonesia. Ministry of Marine Affairs and Fisheries. Jakarta Minister of Energy and Mineral Resources, 2011, Handbook of Energy and Economic Statistics of Indonesia, 2011.

[3] Food and Agriculture Organization of the Uniniter Station (FAO) (2010) FAO Yearbook. Fishery and Aquaculture Statistics. Http://www.fao.org/fishery/publication / yearbook / e

[4] Nontji A (1993) Nusantara Sea. Publisher Djambatan. Jakarta

[5] Mulya, MB (2012) *Kajian Bioekologi UdangPutih* (*Penaeus merguiensis de Man*) di Ekosistem Mangrove Percut Sei Tuan Sumatera Utara. Disertasi PascasarjanaIPB. Bogor.

[6] Rahardjo MF, Sjafei DS, Affandi R, Hutabarat and Sulistiono (2011) Ichthyology. Bandung: Lubuk Agung. Page 111-113.

[7] Sparre P, and SC Venema (1999) Introduction to the study of tropical fish stock of manual books (Translation Edition). Food Organization Cooperation,United Nations with Center for Research and Development of Fisheries, Agency for Agricultural Research and Development. Jakarta. 438

IOP Conf. Series: Earth and Environmental Science **130** (2018) 012044 doi:10.1088/1755-1315/130/1/012044

[8] Campbell NA, JB Reece, and LG Mitchell (2003) Biology Third Edition. Jakarta: Erland, Jakarta

[9] Sugiyono (2005) Statistic for Research. Alfabeta, Bandung

[10] Naamin N (1984) Dynamics of Jerbung Shrimp Population (Penaeus Mergui Ensis de Man) in Arafura Waters and Alternative Management, Postgraduate Dissertation of IPB. Bogor

[11] Barnes RSK, and RN Hughes (1988) An introduction to Marine Ecology. Second. Edition. Blackwell Scientific Publication. London

[12] Martosudarmo B, and Ranoemihardjo BS (1980) Guidelinec for Paneid Shrimp Hatchery. Directorate General of Fisheries Ministry of Agriculture. Jepara Mukhtasor. 2007. Coastal and Marine Pollution. Jakarta:PT. Pradnya Paramita

[13] Rothschild BJ, Gulland JA (1982) Interim Report of the Workshop on the Scientific Basis for the Management of Penaeid Shrimp. NOAA Tech. Memorandum of NMFS-SEFC - 98. US Dept.Commerce

[14] Effendi H (2003) Telaah Kualitas Air bagi Pengelolaan Sumber Daya dan Lingkungan Perairan. Penerbit Kansisu, Yogyakarta

[15] Siagian S (2009) Diversity and Abundance of Fish and Its Linkage With Water Quality of Lake Siais Kabupaten Tapanuli Selatan. [Thesis]. Graduate program. Department of Biology. Medan: USU Repository

[16] Boyd CE, Fast AW (1992) Pond monitoring and management. Inside: Fast AW, Lester JL, editor. Marine Shrimp Culture-Principles and Practices. Elsevier.Amsterdam. Pp. 497-513

[17] Mukhtasor (2007) *Pencemaran Pesisir dan Laut*. Jakarta: PT. Pradnya Paramita

[18] Knox GA (1986) Estuarine Ecosystem : A System Approach. Vol. 1. CRC Press, Inc. United State

[19] Barus TA (2004) Pengantar Limnologi Studi Tentang Ekosistem Air Daratan. Medan: USU Press

[20] Affan JM, and Muhammadar (2011) Seed Breeder Technique Grouper (Epinephelus fuscoguttatuc In Procurement Needs of Grouper Seed Needs for Cultivators in Nangroe Aceh Darussalam. Aceh Development International Conference 2011. Kuala Lumpur.

[21] Odum EP (1994) *Dasar-Dasar Ekologi*. Edisi Ketiga. Yogyakarta: Universitas Gadjah Mada Press.

[22] Suwondo et al. (2004). "Kualitas biologi perairan sungai senapelan, sago dan sail Di kota pekanbaru Berdasarkan bioindikator plankton dan benthos".Jurnal Biogenesis. 1(1):15-20

Cucumber Plants (*Cucumis sativus* L.) Growth and Crop Yield of Chicken Manure Fertilized with Plant Spacing

Sri Pratiwi Aritonang[1*], Ernitha Panjaitan[2], Fetrus Parsaulian Tondang[3]

[1]Agriculture Faculty, Major Agrotechnology University of Methodist Indonesia
Jl. Harmonika Baru Pasar II Tanjung Sari Medan, 20132

*E-mail:pratiwiaritonang@gmail.com

Abstract. The research was conducted in Tanjung Sari, Kecamatan Medan Selayang Kotamadya Medan ± 32 meters above sea level. It started since July 2016 to September 2016. It was designed with randomization block design with two factorial experiments which are chicken manure and plant spacing. First factor was 4 doses of chicken manure, symbolized by K; K_0 = 1.5 kg/plot, K_1 = 2 kg/plot, K_2 = 2.5 kg/plot and K_3 = 3 kg/plot. Second was 4 different plant spacing, symbolized by J; J_0 = 30 cm x 60 cm, J_1 =: 35 cm x 60 cm, J_2 = 40 cm x 60 cm and J_3 = 45 cm x 60 cm. The result shows that giving 3kg/plot of chicken manure increases plant height to 162.15 cm with 22.44 number of leaves. Fresh fruitsper sample was weight 1121.88 g and per plot is 4.52 kg with 9.17 and 36.67 units of fruits per sample and plot respectively. With 45 cm x 60 cm (J_3) for plant spacing gives a plant with the height of 160.51 cm and 22.85 number of leaves. Fresh fruits obtained is 1216.67 g and 9.33 units per sample while per plot gives 4.90 kg and 7.33 units of fresh fruits. This plant spacing leads to a better output for the yield compared to narrower spacing. There are no interaction between chicken manure dosage and plant spacing towards plant height, number of leaves, fresh fruits weight and units per sample and plot.

Keywords: *Cucurbitaceae,* dose, Organic fertilizer, space

1. Introduction

Cucumber (*Cucumis sativus* L.) is a kind of plant in *Cucurbitaceae* (gourd) plant family largely consumed raw in Indonesia. As a vegetable, it is one of the good vitamin and mineral sources. Cucumber is commonly processed into fresh products like pickles, kimchi, and salad or as beverages like juice. Cucumber can also be used for beauty purpose as for body scrub and cleansing cream [1].

Cucumber plantation in Indonesia covered almost 61.000 hectares of land in 2010 and decreased to approximately 49.000 hectares in 2014 producing nearly 550.000 tonnes of cucumber each year. It is reported, the production to have slightly increase from 2010 - 2014 where the number dropped to 480.000 tonnes in 2014. Nevertheless, it could still suffice for cucumber consumption. Both of land and crop yield for cucumber should be taken into a concern to prevent import from taking place [2].

IOP Conf. Series: Earth and Environmental Science **130** (2018) 012045 doi:10.1088/1755-1315/130/1/012045

Fertilizer is used to increase soil fertility and the content of minerals to support plants growth and give optimum yield [3]. There are two fertilizers commonly used which are inorganic (chemical synthesized) and organic fertilizers, one of the examples is manure. Manure is natural and does not destroy the soil, provides macronutrients (nitrogen, phosphorous, potassium, calcium and sulfur) and micronutrients (iron, boron, zinc, cobalt, and molybdenum)[6].

Besides the dosage of manure, plant spacing is also one factor needs to be considered to create the right environment for crop growth and production [4].

Plant spacing has to be done in the right measurement. When space is wide, evaporation and weed growth are high[7]. On the contrary having not enough space leads to competitionbetween plants to get sunlight, nutrients and water. It experimented that space 30 cm x 60 cm is suitable for cucumber plants [5].

The usage of chicken manure in cucumber plants has not gained much information. Therefore, the author would like to research about growth and crop yield of chicken manure with plant spacing cucumber plants *(Cucumis sativus L).*

2. Methods

The research was conducted in Tanjung Sari, Kecamatan Medan Selayang Kotamadya Medan ± 32 meters above sea level and started in July to September 2016.

2.1. Experimental apparatus

Local cucumber seed, chicken manure, ultra gen fertilizer, soil, bamboo, plastic rope, paint, and water were used. Equipment like a pruning saw, hoe, knife, measurement, hand sprayer, weigher, rope, saw, triplex board, brushes, and stationery were needed.

2.2. Problem formulation

This research uses randomized block design with two factorial experiments which are chicken manure and plant spacing. The experiments are as follow:

Factor I is chicken manure symbolized by "K"; with four doses:

K_0:1.5 kg/plot; K_1:2 kg/plot; K_2: 2.5 kg/plot; K_3:3 kg/plot

Factor II is plant spacing to cucumber plants symbolized by "J"; with four different distances:

J_0:30 cm x 60 cm; J_1:35 cm x 60 cm; J_2:40 cm x 60 cm; J_3:45 cm x 60 cm

Therefore 16 different treatments were obtained as follow:

K_0J_0	K_1J_0	K_2J_0	K_3J_0
K_0J_1	K_1J_1	K_2J_1	K_3J_1
K_0J_2	K_1J_2	K_2J_2	K_3J_2
K_0J_3	K_1J_3	K_2J_3	K_3J_3

2.3. Experiment Procedural

2.3.1. Tillage

Before being used, land was measured, and weed was killed, and it was dug down 15 cm to 20 cm depth with a hoe.

2.3.2. Plot Plan

After plowing research plot was made with 120 cm x 100 cm measurement with the distance between each plot was 30 cm.

2.3.3. Chicken Manure Application

Chicken manure was embedded in the plot 1 week prior to planting. The application was done according with dose.

2.3.4. Planting
After tillage, cucumber seed was planted. With a dibber, 2 cm holes were made the soil to put the seed. Into every hole, as many as two seeds that used. Each hole was arranged to have 30 cm x 60 cm, 35 cm x 60 cm, 40 cm x 60 cm and 45 cm x 60 cm space which then was covered by plowed soil.

2.4. Plant Care
2.4.1. Watering
Watering was done twice a day; in the morning and afternoon but it was dependent on the condition the field.

2.4.2. Vining
Vining purpose was to make a trellis to improve care and as fruit support. Vining was done as early as ± 7 days after planting to avoid disturbing rooting. Trellis is made approximately 2 meters high.

2.4.3. Thinning
Thinning was done when plants were two weeks old by cutting one bad growing plant and leaving one good growing plant.

2.4.4. Binding
Raffia rope was used to bind plant onto trellis. It was aim to maintain trellis on vines in order support growth and care.

2.4.5. Weeding
Weeding plants to remove a weed was manually done by pulling out weed or using a hoe. This was done to plants after two and four weeks old.

2.4.6. Pest and Disease Control
Pathogen pest and disease control are one of the most important things in ongoing care for cucumber plants. It was done using active ingredients Deltamethrin (Decis 2.5 EC) with 2cc/liter concentration using a hand sprayer and was done accordingly to plant condition. Besides that, casting out livestock such as chickens and cucumber eaters from the planting area has to be done.

2.4.7. Harvesting
After cucumber is seven weeks old, harvesting is ready. Cucumber plants have to have harvest criteria as follow: at least 4 cm and 18 cm of diameter and length respectively, fruits which do not fulfill the required measurement can be harvested it there are defects on the peels either caused by nature or pests and deep green fruits with yellowish bottoms as indications of over ripeness. Harvesting was done every 2 days continuously by cutting the stem.

2.5. Observed Parameters
In this research, parameters analyzed were:
a. Plants Height (cm)
Plants height was measured from root neck to the highest point on plant samples. The measurement was done once a week since two weeks old until flowering took place.
b. Number of Leaves
Leaves counted were those perfectly opened on plant samples. Counting was done once a week from second to the fourth week after planting.

c. Weight of Fresh Fruits per Sample (g)
Fresh fruits per sample were weighted with a weigher. It is done after harvesting by weighing every fruits from each plant sample.
d. Weight of Fresh Fruits per Plot (kg)
Fresh fruit per plot is the result of all yield per plot from the harvest. Weighing was done at the end of harvesting by weighing fruit obtained.
e. Number of Fruits per Sample
It was counted at the end of research and after harvest by counting the number of fruits from each of plant.
f. Number of Fruits per Plot
A number of fruits per plot is the total of fruits obtained from all harvest session. The observation was done at the end of harvesting by summing up the total of fruits.

3. Results and Discussion

3.1. Plants Height

Table 1. Mean number of cucumber plant height at two, three and four weeks old with chicken manure application and plant spacing.

Treatment	Plants Height (cm)		
	2 weeks old	3 weeks old	4 weeks old
K_0	16.84	39.61a	154.39a
K_1	17.15	40.19ab	158.40b
K_2	18.09	41.23ab	159.70b
K_3	18.14	41.59b	162.15c
$BNJ_{0.05}$	-	1.71	2.34
J_0	16.61	39.30a	155.85a
J_1	17.94	40.88ab	158.24b
J_2	17.94	41.54b	160.02b
J_3	17.73	40.91b	160.51b
$BNJ_{0.05}$	-	1.71	2.34

Note: numbers followed by the same alphabets are based on BNJ test which means no difference from the test at 5% concentration

From Table 1 can be observed that the tallest plant obtained from treatment with chicken manure after three weeks planting, K_3 showed a significant height difference from those of K_0, but not from K_1 and K_2. At 4 weeks old, the tallest plant from K_3 showed significant difference from those of K_0, K_1, and K_3, Height from K_2 was far different from K_0, but not from K_1. Plant height in K_1 was significantly different from K_0.

The plant height influenced by chicken manure at four weeks old can be seen on Figure 1. From Figure 1 can be seen that cucumber plant growth increases with the dose of chicken manure which follows positive linear regression curve.

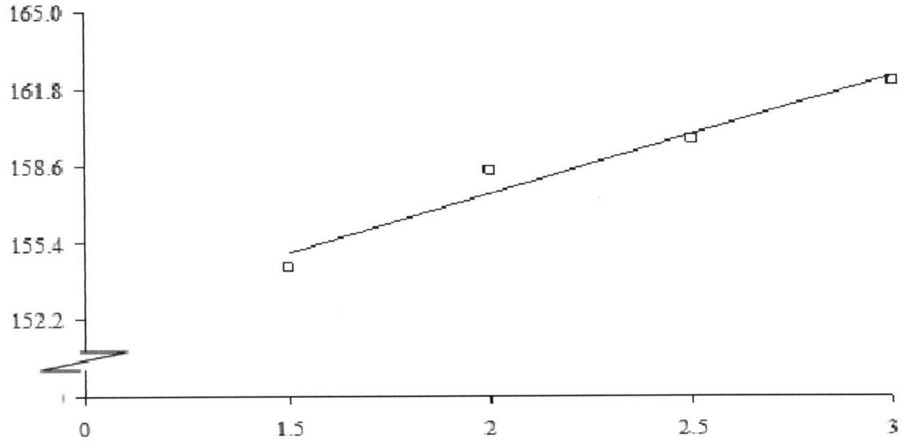

Doses of chicken manure (kg/plot)

Figure 1. Chicken manure dose to cucumber plant height at four weeks old.

3.2. Number of Leaves

A mean number of cucumber plants leaves at two, three and four weeks old with chicken manure application can be seen on Table 2. From Table 2 can be seen in that after three weeks with chicken manure application the plant with the most number of leaves is K_3 differ significantly from K_0, but not from K_1 and K_2. A number of leaves on plant K_2 was differed significantly from K_0, but not from K_1. The leaves on K_1 differ significantly from K_0. At four weeks, the plant with a most number of leaves is K_3 which shows the significant difference from those of plants K_0, K_1, and K_2. Plant K_2 has far greater number of leaves than K_0 but not than K_1. Plant K_1, however, has a significant difference number of leaves from K_0.

Table 2. Mean number of cucumber plant leaves at 2, 3 moreover 4 weeks old with chicken manure application and plant spacing.

Treatment	Number of Leaves		
	2 weeks old	3 weeks old	4 weeks old
K_0	4.96	10.00a	21.44a
K_1	5.23	10.38b	21.83ab
K_2	5.31	10.42b	22.00b
K_3	5.23	10.56b	22.44c
$BNJ_{0.05}$	-	0.37	0.40
J_0	5.00	9.90a	20.69a
J_1	5.17	10.19ab	21.75b
J_2	5.21	10.42b	22.42c
J_3	5.35	10.85c	22.85d
$BNJ_{0.05}$	-	0.37	0.40

The influence of chicken manure application to number of cucumber plant leaves at 4 weeks old can be seen on Figure 2. From Figure 2 can be seen that the number of leaves increases with the increase dose of chicken manure following the positive linear regression curve.

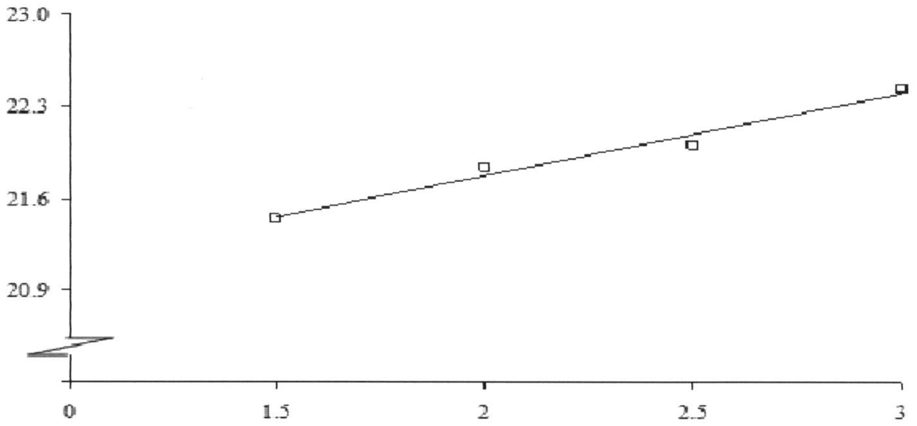

Figure 2. The influence of chicken manure dose to number of cucumber plant leaves at four weeks old.

3.3. Fresh Fruits Weight per Sample

The mean weight of fresh fruits per sample with chicken manure application and plant spacing can be seen in Table 3.

Table 3.Mean weight of fresh fruits per sample with chicken manure application and plant spacing (g).

Treatment	J_0	J_1	J_2	J_3	Mean
K_0	762.50	887.50	1008.33	1116.67	943.75a
K_1	904.17	979.17	1058.33	1304.17	1061.46ab
K_2	820.83	1104.17	1179.17	1187.50	1072.92b
K_3	1000.00	1050.00	1179.17	1258.33	1121.88b
Mean	871.88a	1005.21b	1106.25bc	1216.67c	
$BNJ_{0.05}$ = 127.39					

Note: numbers followed by the same alphabets are based on BNJ test which means no difference from the test at 5% concentration.

From Table 3 can be seen that at chicken manure application, the greatest weight for fresh fruits was from K_3 which differ significantly from K_0, but not from $K_1$and K_2. The weight of fresh fruits per sample in K_2 was far different from K_0 but not from K_1.Weight from K_1 does not differ far from K_0.

The influence of chicken manure application to fresh fruits weight per sample can be seen in Figure 3.

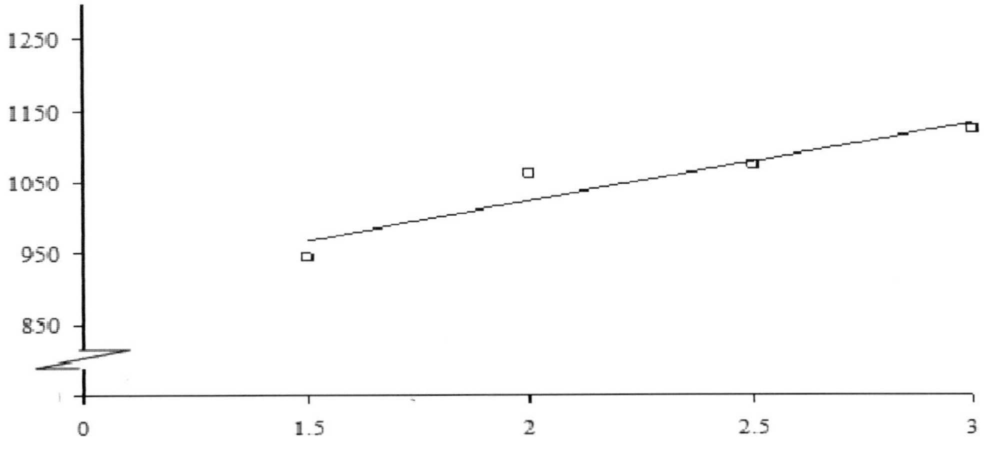

Doses of chicken manure (kg/plot)

Figure 3. The effect of chicken manure dose to fresh fruits weight per sample

3.4. Fresh Fruits Weight per Plot

Table 4. Mean weight of fresh fruits per plot at chicken manure application and plant spacing (kg).

Treatment	J_0	J_1	J_2	J_3	Mean
K_0	3.05	3.55	4.03	4.47	3.78a
K_1	3.62	3.92	4.23	5.22	4.25ab
K_2	3.28	4.42	4.72	4.75	4.29b
K_3	4.00	4.20	4.72	5.15	4.52b
Mean	3.49a	4.02b	4.43bc	4.90c	
$BNJ_{0.05} = 0.51$					

Note: a numbers followed by the same alphabets are based on BNJ test which means no difference from the test at 5% concentration.

From Table 4 can be seen that the heaviest fresh fruits obtained with chicken manure application are on plant K_3 which differ significantly from K_0, but not from K_1 and K_2. Fresh fruits weight, per plot at K_2 differ significantly from K_0 but not from K_1. The influence of chicken manure dose to fresh fruits weight per plot can be seen in Figure 4.

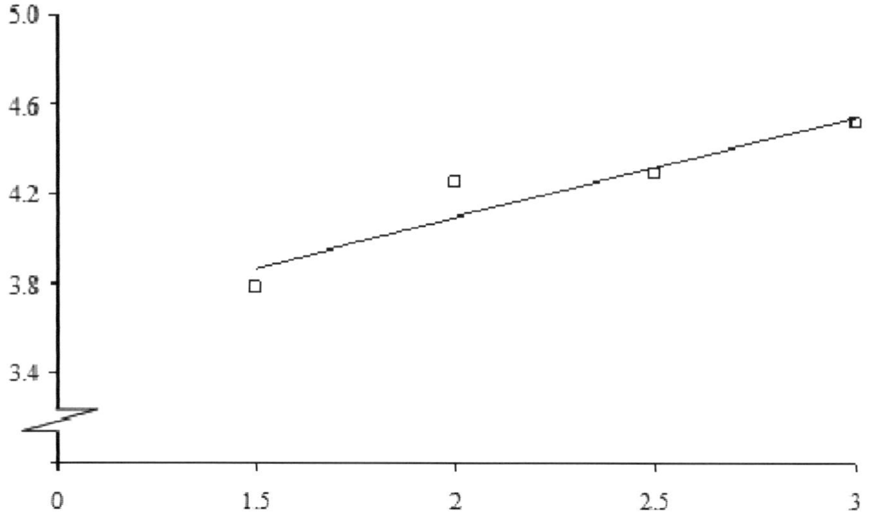

Figure 4. The influence of chicken manure dose on fresh fruits weight per plot.

3.5. Number of Fruits per Sample

Table 5. Mean a number of fresh fruits obtained per sample with chicken manure application and plant spacing.

Treatment	J_0	J_1	J_2	J_3	Mean
K_0	7.00	6.92	7..58	8.17	7.42a
K_1	7.33	7.00	8.42	9.00	7.94b
K_2	7.67	8.92	9.00	9.75	8.83c
K_3	8.08	8.67	9.50	10.42	9.17c
Mean	7.52a	7.88a	8.63b	9.33c	
$BNJ_{0.05}$ = 0.50					

Note: numbers followed by the same alphabets are based on BNJ test which means no difference from the test at 5% concentration.

From Table 5 can be seen that the biggest number of fresh fruits obtained from chicken manure application is at plant K_3 which differs significantly from K_0 and K_1, but not from K_2. The number of fruits per sample for plant K_2 differ significantly from K_0 and K_1. Number of fruits per sample from plant K_1 differ significantly from K_0.

The influence of chicken manure dose to a number of fruits per sample can be seen in Figure 5.From Figure 5 can be seen that the number of fruits increases with the increased dose of chicken manure following the positive linear regression curve.

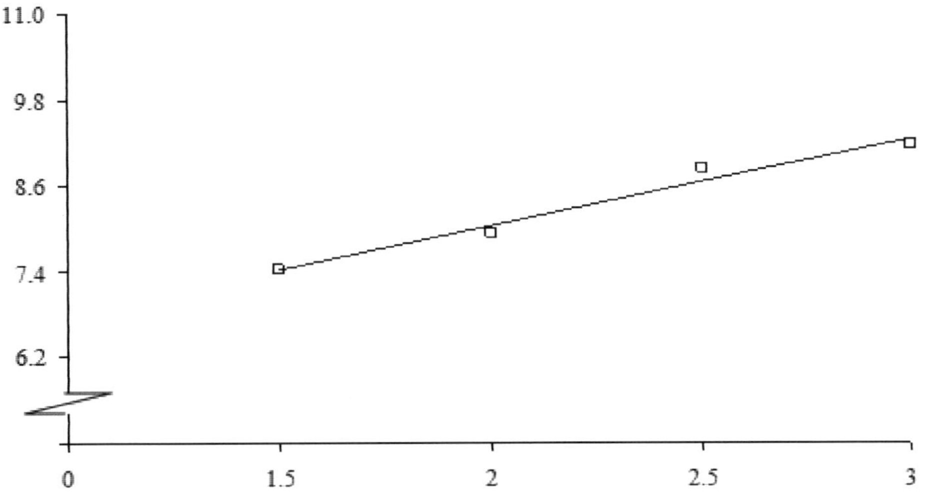

Doses of chicken manure (kg/plot)

Figure 5. The influence of chicken manure dose to the number of fruits per sample.

3.6. Number of Fruits per Plot

Table 6. The mean a number of fruits per plot with chicken manure application and plant spacing

Treatment	J_0	J_1	J_2	J_3	Mean
K_0	28.00	27.67	30.33	32.67	29.67a
K_1	29.33	28.00	33.67	36.00	31.75b
K_2	30.67	35.67	36.00	39.00	35.33c
K_3	32.33	34.67	38.00	41.67	36.67c
Mean	30.08a	31.50a	34.50b	37.33c	

$BNJ_{0.05}$ = 1.99

Note: numbers followed by the same alphabets are based on BNJ the test which means no differ from test at 5% concentration.

From Table 6 can be seen that the most fruits were obtained from plant K_3 which differs significantly from plants K_0 and K_1 but not from K_2. The number of fruits from plant K_2 was far different from K_0 and K_1. The number of fruits per plot in K_1 was far different from K_0.The influence of chicken manure dose to a number of fruits per plot can be seen in Figure 6.

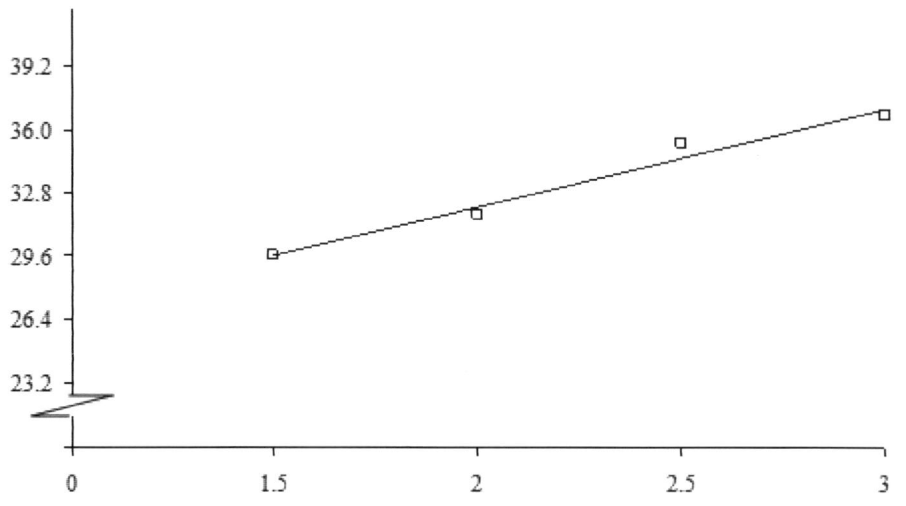

Figure 6. The influence of chicken manure dose to a number of fruits per plot.

3.7. The Influence of Chicken Manure Dose to Cucumber Plants Growth and Crop Yield

From ANOVA (analysis of variance) test shows that the application of chicken manure great influence on plant height, number of leaves, fresh fruits weight and number of fresh fruits per sample and plot.

Dosage of chicken manure up to3kg/plot can increase the growth of height and number of leaves. It is because chicken manure provides organic matters that influence physical, chemical and biological properties of soil [4]. That is also can affect like texture, water binding ability, humidity and nutrients in soil. Therefore, soil which has been given organic matter like chicken manure can bind water and nutrients which can be easily used by plants.

3.8. The Influence of Plant Spacing to Cucumber Plants Growth and Crop Yield

ANOVA test shows that plant spacing also great influence on plants' height, number of leaves, fresh fruits weight and number of fresh fruits per sample and plot.

Results show 45 cm x 60 cm (J_3) leads to the best cucumber plants growth and crop yield. If space is too tight being a competition between plants to get sunlight, nutrients, and water which then causes nutrients deficiency in half plants. Gardner, Pearce and Mitchell [2] stated that plant spacing is a factor influencing plant growth, because the absorption of solar energy leaves surface strongly influenced by this condition. When space is too close, it can affect vegetative growth and crop yield due todecreased rate of photosynthesis and leaves.

3.9. Interaction between Chicken Manure Dose and Plant Spacing for Cucumber Plants Growth and Crop Yield

ANOVA test shows chicken manure application, and plant spacing together do not affect plants' height, number of leaves, fresh fruits weight and a number of fresh fruits per sample and plot. It is because chicken manure application tends to improve soil physical properties for better cultivation, therefore supports better root growth. The four different spaces applied in the research are still optimum space for growing cucumber plants were no competition between plants for nutrients. Therefore the application of fertilizer in which chicken manure was used can result in better growth and crop yield with only slight differences in results with four different plant spacing.

IOP Conf. Series: Earth and Environmental Science **130** (2018) 012045 doi:10.1088/1755-1315/130/1/012045

4. Conclusions

1. 3kg/plot of chicken manure can increase plants height to 162.15 cm, with 22.44 number of leaves, giving 1121.88 g and 4.52 kg of fresh fruits per sample and plot respectively with 9.17 units and 36.67 units of fruit per sample and plot.
2. 45 cm x 60 cm plant spacing (J_3) gives 160.51 cm height and 22.85 number of leaves. Fresh fruits weight, 1216.67 g with 9.33 units of fruits and 4.90 kg with 7.33 units of fruits per sample and plot respectively. The result is a better compared to narrower plant spacing.
3. There is no real influence on height, number of leaves, fresh fruits weight and number of fruits per sample and plot between the interactions of chicken manure at certain dose with plant spacing.

References

[1] BadanPusatStatisik(2014) Kalimantan Tengah DalamAngka. KerjasamaBadanPerencanaan Pembangunan Daerah Propinsi Kalimantan Tengah denganBadanPusatStatistikPropinsi Kalimantan Tengah, Palangka Raya.

[2] Gardner FP, RB Pearce, and RL Mitchell(1991)Fisiologi Tanaman Budidaya. Universitas Indonesia Press, Jakarta.

[3] Guritno & SM Sitompul(2010) Analisis Pertumbuhan Tanaman. Gajah Mada, Yogyakarta.

[4] Hakim N, MY Nyakpa, AM Lubis, SG Nugroho, R Saul, MA Diha, Go BH, andHH Bailey(2005) Dasar-dasar Ilmu Tanah. Universitas Lampung, Bandar Lampung.

[5] Pangaribuan, DH, M Yasir and NK Utami(2012) Dampak Bokashi Kotoran Ternak dalam Pengurangan Pemakaian Pupuk Anorganik pada Budidaya Tanaman Tomat. *Jurnal. Agron. Indonesia* 40 (3):204-210.

[6] Samadi B(2010) Teknik Budidaya Mentimun. Kanisius, YogyakartaGilir. Penerbit Swadaya.

[7] Slamet Y et al.(2012) Pengaruh Pemangkasan Dan Pemberian Pupuk Organik Terhadap Produksi Tanaman Mentimun (*Cucumis Sativus* L). *Penelitian Agronomi Oktober 2012* 1(**2**): 107-114. ISSN: 2089-9858

[8] Tufaila M, and Dewi D(2014)Aplikasi Kompos Kotoran Ayam Untuk Meningkatkan Hasil Tanaman Mentimun (*Cucumis Sativus* L.) Di Tanah Masam. *Jurnal Agroteknos* 4(**2**): 119-126, ISSN: 2087-7706

[9] Walsen A(2015)Aplikasi Pupuk Subur Dengan Dosis dan Waktu Berbeda Pada Tanaman Ketimun (*Cucumis Sativus* L.). *Jurnal Budidaya Pertanian* 4(1): 29-37. [20 June 2016]

Characterization of protease from *bacillus* sp. on medium containing $FeCl_3$ exposed to magnetic field 0.2 mt

Sumardi[1*], Rochmah Agustrina[1], Christina Nugroho Ekowati[1] and Yovita Selvie Pasaribu[1]

[1]Department of Biology, Faculty mathematic and natural science, University of Lampung
Jalan Prof. Dr. Soemantri Brodjonegoro No. 1 Bandar Lampung, Lampung 35145
*Email: sumardi_bio@yahoo.co.id Telephone: +6285216391087 Fax : +62-721-704625

Abstract This purpose of this research is to determine the character of the protease enzymes from *Bacillus* sp. on media content of $FeCl_3$ exposed to 0.2 mT magnetic field. The data obtained were analyzed descriptively. The result showed that protease enzyme without Fe resulted in the highest activity at pH 8, temperature 30°C with the addition of activator Mn^{2+}, and V_{max} of 0.28 U/ml, and K_m of 4.60 U/ml. The protease enzyme on media without magnetic field exposure and containing Fe yielded the highest activity at pH 8, temperature 30°C with the addition of activator Mn^{2+}, and V_{max} of 0.33 U/ml, and K_m of 5.64 U/ml. The protease enzyme on medium with magnetic field exposure and use Fe as inductors have the highest activity at pH 9, the temperature of 55° C with the addition of activator Mn^{2+}, and V_{max} of 0.35 U/ml, and K_m 10.04 U/ml.

Keywords: determine, inductor

1. Introduction

The production of the enzyme with high activity can be achieved by optimizing of medium conditions such as pH, temperature, carbon and nitrogen sources, or by providing additional treatment such as exposure to a magnetic field of the medium [1]. Ion Fe is an essential nutrient for almost all microorganisms in growth because, of its role as a cofactor in some enzymes.

Iron or Fe is one of the ferromagnetic metal ions. Fe is an essential nutrient for optimal growth of a cell. Fe acts as a cofactor for several enzymes is required in the biochemical processes of cell, respiratory and photosynthesis reaction [2].

Fe in the form of Fe^{2+} is more easily utilized by bacteria as Fe^{2+} is soluble in water. If the Fe in the form of Fe^{3+} the user is first converted to Fe^{2+} with the rest of the energy that has been used for growth [3]. Magnetic field exposure on bacterial medium containing Fe metal will cause the magnetization in the medium to motion moment - the moment of the dipole into the direction of the magnetic field. The change in direction dipole moment Fe suspected that would cause changes in enzyme activity is controlled by the bacteria.

Based on the research Selfiana et al [4] demonstrated that exposure to 0.2 mT magnetic field for 10 minutes on Fe metal ions affect the activity of *Bacillus* sp. to produce a protease enzyme. The addition of the metal ions Fe in salt form $FeCl_3$ as much as a 0.01% exposed to the magnetic field 0.2 mT for

10 minutes yields the highest enzyme activity 0.06 U/ml.

So far, the protease enzyme results of these studies have not been characterized. Characterization needs to be done to determine the optimum pH and temperature, the metal ions that can increase the activity of the enzyme, as well as chemical kinetics optimal protease enzyme so that to obtain high enzyme activity. Therefore, characterization studies protease from Bacillus sp. on media containing $FeCl_3$ was exposed to a magnetic field of 0.2 mT necessary.

2. Method

2.1. The Culture of Bacillus sp.

Rejuvenation culture is done by scraping the ose culture of *Bacillus* sp., *Bacillus* sp which is a collection of the laboratory of microbiology FMIPA UNILA. Then grown on a modified medium composition comprising Mendel's with skim; 0.5 g; Yeast extract 0.35 g; Water Tryptone 0.35 g; NaCl 0.2 g; KH_2PO_4 0.245 g; $MgSO_4.7H_2O$ 0.035 g; $(NH4)_2SO_4$ 0.175 g in 100 mL of distilled water and agar 1.5 g as the compactor. Later, it was tilt incubated at 37°C in an incubator for two days.

Proteolytic test on a modified solid media Mendel's

$FeCl_3$ which has been exposed by the magnetic field 0.2 mT for 10 minutes is used as an inductor in the growth media. The proteolytic test consisted of three treatments as follows:

a. **Treatment of media without inductors ($FeCl_3$).**
   As a control treatment, a modified solid media Mendel has not given exposure to a magnetic field and are not given the inductor.
b. **Media treatment with the inductor ($FeCl_3$).**
   Treatment Negative is treated using a modified solid media Mendel's and given inductor $FeCl_3$ neither the media nor the inductors are being exposed to a magnetic field.
c. **Media treatment with exposure to a magnetic field in the inductor ($FeCl_3$).**
   Positive treatment is treated using a modified solid media Mendel's and given inductor $FeCl_3$. $FeCl_3$ was exposed to 0.2 mT magnetic field for 10 minutes before use.

Observations existence of a clear zone formed around colonies of bacteria do when the bacterial cultures were incubated for 18 hours at a temperature of 37°C. Colonies of bacteria and a clear zone formed around colonies of bacteria measured in diameter and further defined Proteolytic Index (PI):

$$PI = \frac{\text{the diameter of clear zone} - \text{the diameter of colony}}{\text{the diameter of the colony}} \quad (1)$$

2.2. Enzyme Production

Enzyme production is done by the respective treatment in the media without the addition of agar. Protease enzyme production is done by inoculating 5 ml starter *Bacillus* sp. in 45 ml liquid media Mendel's modified in Erlenmeyer 250 ml with the same treatment as the proteolytic test.

All treatments shake culture was incubated in an incubator at 120 rpm at a temperature of 40 ° C with a 24-hour long incubation time that is adjusted with previous studies. Extraction is done by centrifuging the protease enzyme of bacterial growth media at a speed of 10,000 rpm for 10 min at 4°C.

2.3. Protease Activity Test

0.1 ml protease sample of crude extract was added to the mixture using a 0.5 ml casein substrate as follows: A total of 0.5 ml of phosphate buffer pH 7 0:01 M was added and incubated at 39°C for 10 minutes. After wards plus 0.5 ml of 0.1 M TCA then incubated at 39°C for 10 minutes, and centrifuged at 10,000 rpm at 4°C for 10 minutes. Then 0.375 ml of the supernatant was taken and added with 1:25 ml of Na_2CO_3 0.4 M and 0:25 ml Folin reagent, incubated at 37°C for 20 minutes [5].

Observations were carried out by measuring the Optical Density (OD) at a wavelength of 578 nm. Blank value calculation is done in the same way, where the protease sample was replaced with distilled water. While the standard value calculation is done by replacing the protease samples with tyrosine 5 mM.

Protease activity was calculated in units of PU (protease units) per ml of the enzyme extract. One protease unit is defined as the amount of enzyme that can produce one µmol of tyrosine per minute on measurement conditions were measured using the following formula:

$$PU = \frac{Asp - Asbl}{Ast} \text{ X } \frac{1}{T} \quad (2)$$

Description:

PU : Unit Activity Protease (Units/ml)
Asp : Value Absorbance sample
Ast : Value Absorbance standard
Abl : Value Absorbance Blank
T : Time

2.4. Determination of optimum pH

Effect of pH on the enzyme activity is tested by reacting a solution of crude extract enzyme and 0.25% incubated casein substrate at 39°C in a buffer 0.05M different series, namely: pH 4.5 (citrate buffer), pH 6.7 (phosphate buffer), pH 8.9 (tris-HCl buffer), pH 10,11,12 (glycine-NaOH buffer).

2.5. Determination of Optimum Temperature

The effect of temperature on the enzyme activity is tested by reacting a solution of crude extract enzyme at optimum pH with temperature variation test 25, 30, 35, 40, 45, 50 55, 60, 65, and 70°C.

2.6. Effect of Metal Ion on Protease Activity

Effect metal ions Ca^{2+} Mn^{2+} Cu^{2+} Mg^{2+} Fe^{3+} in the form of a salt of each - each CaCl2, $MnCl_2$, $CuCl_2$, $MgCl_2$, and $FeCl_3$ as an activator. As an inhibitor of the protease activity used is metal ion chelating compounds, ethylene diamine tetraacetic acid (EDTA) by reacting a solution of crude extract of enzymes, enzyme substrates used as casein 0.25% with one mM of the metal ion. The enzyme is incubated with the metal ions and inhibitors for 10 minutes at a temperature and pH optimum then tested the activity of the enzyme.

2.7. Determination of K_m and V_{max}

Determination of K_m and V_{max} is done by testing the activity enzyme at its optimum temperature and pH variation casein different substrate concentrations of 0%, 0.5%, 1%, 1.5%, 2% and 2.5%. The results of the activity test then tabulated and made the curve relationship between concentration and specific activity of the enzyme casein. Once it is put in the Lineweaver-Burk linear equations. Km value and Vmax is obtained by the formula [6]:

IOP Conf. Series: Earth and Environmental Science **130** (2018) 012046 doi:10.1088/1755-1315/130/1/012046

$$\frac{1}{V_o} = \frac{k_m}{V_{maks}}\frac{1}{[S]} + \frac{1}{V_{maks}} \quad (3)$$

Specification:
V_o : initial velocity on the substrate concentration [S]
V_{max} : Maximum Velocity
K_m : Constanta of Michaelis-Menten enzyme for a particular substrate

3. Results and Discussions

3.1. Bacillus sp. protease production

Proteolytic activity *of Bacillus* sp. on solid media Mendel's modified evident from the formation of a clear zone around bacterial colonies growing. With the protease enzyme produced by *Bacillus* sp., casein in the *Bacillus* sp media. The hydrolyzed into smaller peptides and amino acids, their degradation process is characterized by the formation of a clear zone around the colony.

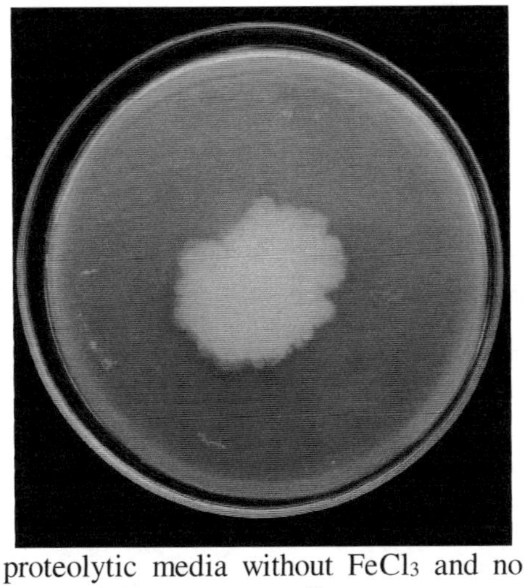

Figure 1. The test results of proteolytic media without $FeCl_3$ and no magnetic exposure. Isolate Bacillus sp. showed small the clear zone in protease medium

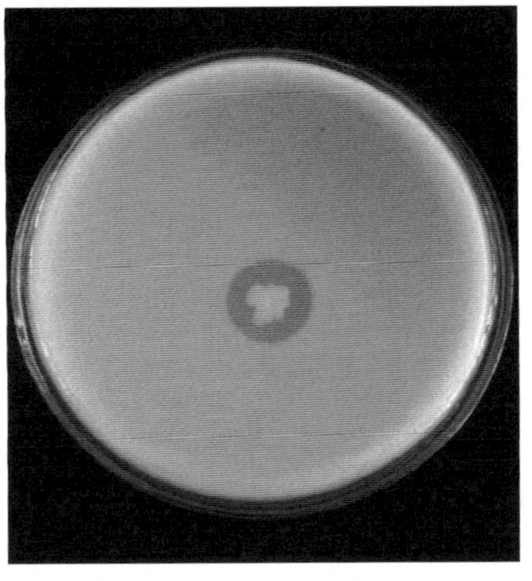

Figure 2. The test results on the media proteolytic treatment with Fe without exposure to magnetic fields. Isolate *Bacillus* sp. showed the clear zone in protease medium.

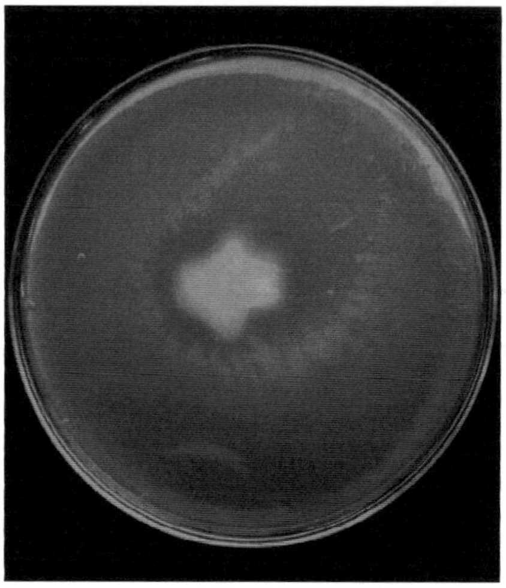

Figure 3. The test results on the media proteolytic treatment with Fe and exposure to magnetic fields. Isolate *Bacillus* sp. showed the clear zone in protease medium.

Proteolytic index value (PI) at the highest media *Bacillus* sp. after incubation for 18 hours of media treatment obtained by Fe outcome magnetic field exposure is equal to 5.62. The highest protease enzyme activity of 0.10 U/ml was obtained from the same media (Table 1).

Table 1. Comparison of proteolytic index and Protease Activity each treatment

Treatment media	Proteolytic Index (PI)	Protease Activity (U/ml)
Without Fe and no magnetic field exposure	0.54	0.08
without magnetic field exposure and containing Fe	2.65	0.07
with magnetic field exposure and containing Fe	5.62	0.10

The area that formed a clear zone indicates that the bacteria in a culture that has a high ability to change the substrate contained in a growth medium [7]. Proteolytic activity of bacteria is influenced by several factors given during treatment. The factors that influence proteolytic activity for each isolates are pH, substrate and enzyme concentration, temperature and the presence of activators and inhibitors. The high value of the proteolytic index on media treatment is given Fe results of magnetic field exposure is suspected because $FeCl_3$ which has been exposed by the magnetic field has induced a higher power and become more positive effect on the proteolytic activity of *Bacillus* sp.

Fe ions in salt $FeCl_3$ which is ferromagnetic have magnetic properties that the direction of the dipole moments of the movement direction of the external magnetic field. As a result of rectification movement can survive even if the external magnetic field has no [8].

If $FeCl_3$ which has been exposed by the magnetic field is added to the bacterial growth media that is stored on the magnetic properties of media expected to affect the growth of bacteria grown on these media. As reported by Sudarti et al [9] that the magnetic field can directly influence the metabolic activity of cells one to produce the enzyme. Farisna and Zulaikha [10] explains that *Bacillus* sp. able to grow on solid medium containing Fe as Fe metal ions required for the metabolism of bacterial cells even though in small amounts.

3.2. Determination of optimum pH

The optimum pH test results protease enzyme growth *of Bacillus* sp media prove that the treatment medium without Fe, pH optimum pH 8 to achieve the protease activity of 0.11 U/ml. While the optimum pH of protease enzyme in the treatment medium by Fe without exposure to magnetic fields reaches pH 8 with protease activity of 0.15 U/ml. As well as the optimum pH of the protease enzyme in the treatment by the medium Fe magnetic field exposure results reaches a pH of 9 with protease activity of 0.11 U/ml (Figure 4).

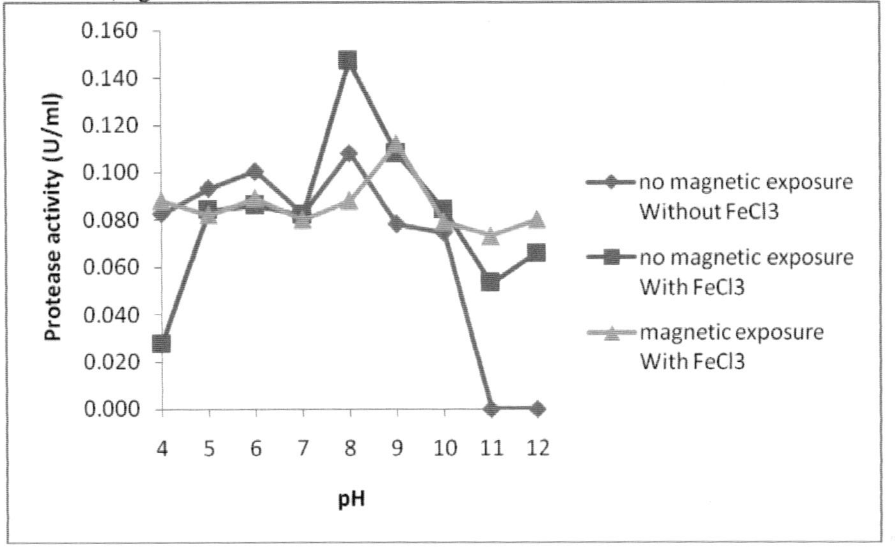

Figure 4. The activity of protease in difference pH

Bacillus sp isolates that it was inoculated in the three media treatments resulted in active proteases in the range of pH 8 and 9. Several studies have shown that the optimum activity of proteolytic enzymes varies from pH 8-10 [11]. Isolates of *Bacillus* sp. isolated from the digestive tract of chicken, protein degradation can grow at pH 7-9 [12]. Protease *.licheniformis B-05* LHSB also has an optimum pH 9. Some protease from genus *Bacillus* other has an optimum pH range 8-10 [13].

The optimum pH difference in the three treatment medium is thought to occur because of the influence of pH on the enzyme active site. The research result Ikehara et al [14], suggests that exposure to magnetic fields ELF has the effect of a reversible bond NH and CN of a peptide bond, and changing the secondary structure of the beta cells and alpha-helices in proteins, so easily have been the attachment of H^+ and OH^- on the side active enzymes.

At pH 4-7 protease enzyme activity tend to be smaller than the optimum pH for the number of H^+ ions that interfere with the enzyme active site, while at pH 10-12 decreased protease enzyme activity for enzyme functional groups disturbed by the many OH- that causes the enzyme denatured.

When $FeCl_3$ which has been exposed by the magnetic field is added in bacterial growth media will cause a difference in the number of H^+ and OH^- ions in the enzyme active site. Thus, there is a difference in the pH of the enzyme treated with magnetic field exposure are not given exposure [9].

3.3. Determination of optimum Temperature

Each enzyme has the maximum activity at a certain temperature; the enzyme activity will increase with increasing temperatures up to optimum temperature is reached. The optimum temperature of the protease enzyme in the treatment medium without Fe reaches a temperature of 30^0C with protease activity of 0.12 U/ml. Meanwhile, the optimum temperature of the protease enzyme treatment media by Fe without exposure to the magnetic field reaches a temperature of 30^0C with protease activity of 0.44 U/ml. As well, the optimum temperature of the protease enzyme treatment given media exposure results Fe magnetic field reaches a temperature of 55^0C with protease activity of 0.13 U/ml (Figure 5).

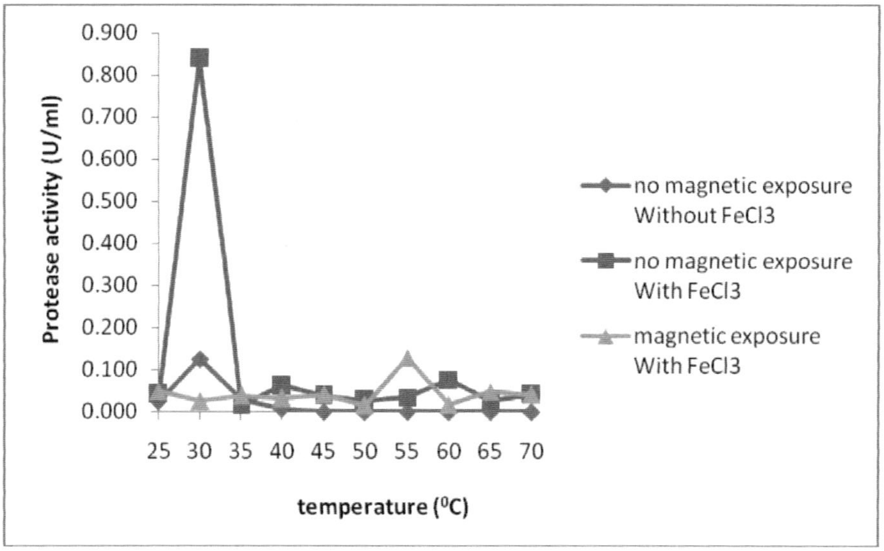

Figure 5. The activity of protease in difference temperature

Further increase in temperature will cause decreased enzyme activity. At temperatures lower than the optimum temperature, the enzyme activity is also low, due to lower activation energies available. The energy needed to create the conditions of an active complex level, both of the molecules of enzyme or substrate molecules.

Protease enzyme was produced on all three medium treatment has the optimum temperature 30 and 55°C. This is confirmed by the results of an intensive search Fitriani [15], where he reported isolates of protease *Bacillus* KUB B19 CC BPPT actively working at a temperature of 30 to 65°C with the optimum temperature at 60°C. Based on the optimum temperature of the protease is classified as thermophilic protease active means to work at high temperatures. Thermophilic proteins that are different characteristics in its structure compared to mesophilic proteins. Thermophilic proteins have more salt bridge structure coupled with the presence of the main chain hydrogen bonds are plentiful when compared to mesophilic proteins. Also, this protein has an amino acid residue more hydrophobic than mesophilic proteins. This will lead to increased activity of enzyme proteins at high temperatures [16]

The difference between the values of the activity of Fe media with media treatment without treatment by Fe without exposure to a magnetic field is thought to occur due to exposure to a magnetic field causes a change of Ca ion movement^{2+} and the increasing rate of movement of the ion. Such conditions lead to changes in transport in the cell membrane that affect cellular metabolism activity. It can have an impact on the process of cell growth. Increased bacterial growth will result in differences in enzyme activity value [17].

The temperature difference optimum media treatment by Fe the result of exposure to magnetic fields with the protease enzyme in media other treatments, allegedly caused by treatment $FeCl_3$ which is exposed to a magnetic field causes motion moments of the dipole direction of the magnetic field from the outside so that the influence of electromagnetic waves that pass through it. While the results of research Ikehara et al .[14], suggests that exposure to ELF magnetic fields have reversible effects on the NH and CN bond of peptide bonds, and changing the secondary structure of the beta cells and alpha-helical proteins. As a result is the release of bonds of constituent proteins and changes in the molecular structure.

3.4. *Effect of Metal Ion and Specific Inhibitor*

Addition of Mn^{2+} ions to a final concentration of 5 mM can raise the activity of the protease enzyme in the treatment medium without Fe into 0.27 U/ml. The addition of Mn^{2+} ions on the addition of Mn^{2+} ion concentration of 1 mM protease enzyme activity in the media treatment by Fe without exposure to a magnetic field, increasing to 0.18 U/ml. But with a final concentration of 5 mM was not able to raise the activity of the protease enzyme in media exposure treatment without magnetic field and added inductors Fe, The same thing happened on the media treatment of the protease enzyme by Fe result of exposure to magnetic fields. The addition of Mn^{2+} ions on the addition of Mn^{2+} ion concentration of 1 mM protease enzyme activity increased to positive treatment 0.30 U/ml. But, with a final concentration of 5 mM was not able to raise the activity of the protease enzyme in the treatment with magnetic field exposure and added inductor Fe (Figures 6 and 7).

The addition of EDTA inhibitor compounds with a final concentration of 5 mM greatly decreases the activity of the protease enzyme in the three media such treatment, until there is no activity at all three enzymes (Figure 6 and 7).

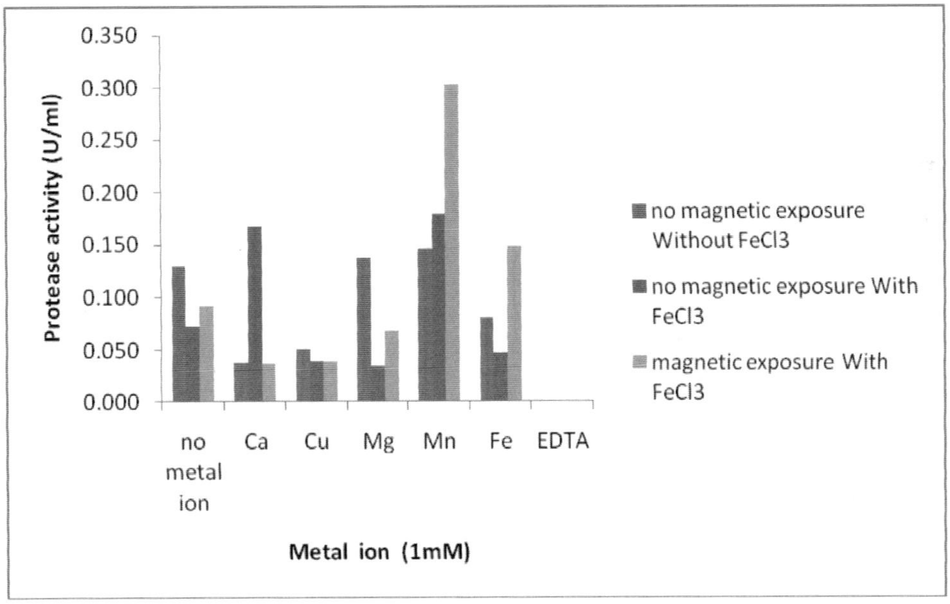

Figure 6. Effect of metal ions and specific inhibitor of the protease enzyme activity in a concentration of 1 mM

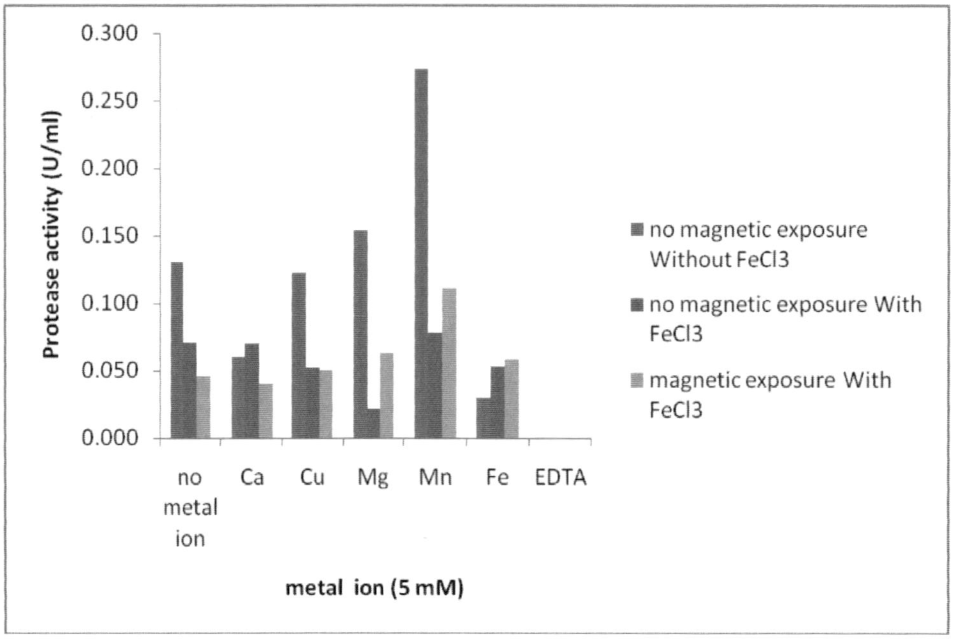

Figure 7. Effect of metal ions and specific inhibitor of the protease enzyme activity in concentrations of 5 mM

Test results show that a strong activator for the protease enzyme of isolates of *Bacillus* sp. the third was inoculated on media treatment is Mn^{2+} at a final concentration of 5 mM able to enhance the catalytic activity of the protease enzyme *of Bacillus* sp. The influence of the addition of metal ions from a few studies have been done, such as the activity measurement results with the addition of metal ions on the substrate casein, metal ion Mn^{2+} (5 mM) is a potent activator of protease *Bacillus epidermidis*, which can increase the activity of protease respectively 3 and 2 times folding of the protease control [18]. For *Bacillus* A1 strain Mn metal ion^{+2} as an activator can increase the enzyme activity according to the research it Adinarayana et al [19]. Inhibitors are compounds that can inhibit the enzyme activity. Extra activity inhibitor compound is one method to identify the enzyme. Differences in protease activity value on each concentration allegedly caused by the difference itself. Enzymes in treatment media which has the highest enzyme activity at a concentration of 1 mM decreased in concentration 5mM, allegedly can not grow optimum conditions of high metal concentrations, but the optimum conditions a low concentration.

Effect of protease inhibitors on the activity was shown in Figure 10, EDTA is a chelating metal (metalloprotease inhibitor) 5 mM slow down to up do not occur enzyme activity in it. The addition of 2 mM EDTA in protease of *Bacillus thermophilic* activity strain HS08 lowered relative to respectively 6% and 4% [20]. The results of analysis show that the protease activity of *Bacillus sp.* perfect inhibited by EDTA so that it can be concluded that these proteases are classified into metalloprotease protein [21]. Chantawannakul [22] reported a protease inhibitor that inhibits the fibrinolytic activity of strains *Bacillus* fermented soybeans 38 isolates from Thailand, namely 1.10 fenantrolin and identified as metalloprotease.

3.5. Protease Kinetics Chemistry

Determination of the value of K_m and V_{max} is useful to determine the enzyme kinetics that can be known how much bonding enzymes and enzyme substrate and how fast it can perform the activity. Determination of the value of V_{max} and K_m is done by measuring the initial velocity at various concentrations of protease enzyme substrate.

From the research that has been conducted shows that the increasing concentration of substrates will increase the reaction rate. However, at a certain concentration limits, no increases the reaction rate even if the substrate concentration is enlarged. This is because at a certain substrate concentration limit, all active parts have been met by the substrate or has been saturated with the substrate [23]. In this study, obtained the value of K_m and V_{max} of the enzyme protease in the treatment medium without Fe each - each amounted to 4.60 mM and 0.28 U/ml. It shows one ml of the enzyme can produce products as much as 0.28 units. As well, the value of K_m and V_{max} of the enzyme protease on media treatment by Fe without exposure to magnetic fields, respectively - each of 5.64 mM and 0.33 U/ml. Show one ml of the enzyme can produce products as much as 0.33 units. Meanwhile, the value of K_m and V_{max} of the enzyme protease media Fe treatment by exposure to magnetic fields result, each - each of 10.04 mM and 0.35 U/ml. This means that one ml of the enzyme can produce products as much as 0.35 units (Table 2).

Table 2. Comparison of Km and Vmax values between each treatments

Treatment media	Km (U/ml)	Vmaks (U/ml)
Without Fe and no magnetic field exposure	4.60	0.28
without magnetic field exposure and containing Fe	5.64	0.33
with magnetic field exposure and containing Fe	10.04	0.35

According to Kovacs et al [24], exposure to magnetic fields on a protein enzyme trypsin can produce different ranges in a redox reaction due to changes in the shape of the helix peptide bonds and hydrogen bonds, thus increasing the Km significantly. A value of K_m are high indicates a low affinity for the substrate and instead of.

In general, the crude extract of the enzyme will give a high value of K_m. It can be caused by the presence of contaminants other enzymes which also have an affinity for the substrate. The higher the value of K_m, the concentration required to achieve half of the maximum reaction rate is also higher. Maximum velocity (V_{max}) will be achieved when all of enzyme molecule has interacted with the substrate or enzyme has been saturated with the substrate, so that an increase in substrate concentration will not increase the rate of reaction [25].

4. Conclusion

From the discussion in this study, the character of the protease enzyme *of Bacillus* sp on media content of $FeCl_3$ in said the magnetic field of 0.2 mT. Where the characters are, among others: the optimum pH and temperature influence of metal ions and inhibitors, as well as the value of enzyme kinetics.

Acknowledgments

The authors gratefully acknowledge that the research is supported by University of Lampung of Indonesia (www.unila.ac.id). The support is under the research grant of Hibah Pascasarjana-UNILA of Year 2017 Contract Number 809/UN26.21/PP/2017.

References

[1] Mubarik NR (2001) Imobilisasi Protease *Bacillus subtilis* ATCC 6633 menggunakan Matriks Gel Poliakrilamida. *Jurnal Hayati* 8(1): 11-14.

[2] Shiying H, Y Feng, H Ren, and Y Zhang (2011) The Impact of Iron Oxide Magnetic Nanoparticle on the Soil Bacterial Community. *School of Biological Science and Medical Egineering.* 11: 1408 – 1417.

[3] Sutariningsih E (2012) *Potensi Pengguna Magnetic Pengguna Fe dan Hg*. Fakultas Biologi Universitas Gajah Mada. Yogjakarta.

[4] Selfiana I, Sumardi, Agustrina R (2016) The Effect of Metal Ions Fe and Zn Exposed to Magnetic Field 0.2 mT on The Production of Protease in *Bacillus* sp..*Prosiding USR International Seminar on Food Security*. 2016(1):73-74.

[5] Bergmeyer HV and Grassl (1983) *Method of Enzymatic Analisis 2*. Verlag Chemia, Weinhein.

[6] Wahyuntari B dan Hendrawati H (2012) Properties of an Extracellular Protease of *Bacillus megaterium* DSM 319 as Depilating Aid of Hides. *Microbiology Indonesia.* 6(2). doi : 10.5454/mi.6.2.4.

[7] Vijayaraghavan, Ponnuswamy and Samuel GPV (2013) A simple method for the detection of protease activity on agar plates using bromocresolgreen dye. *Journal Biochemistry Technology 4(3): 628-630.*

[8] Sipahutar WS (2015) *Efek Waktu Wet Milling dan Suhu Anneling Terhadap Sifat Fisis, Monostruktur dan Magnet Dari Flakes NdFeB*. Prosiding Fisika Universitas Sumatra Utara.

[9] Sudarti, Nurhayati, E Ruriani, and VT Hersa (2014) Prevalence of Salmonella Typhimurium on Gado-Gado Seasoning by Treatment of Extremely Low Frequency (ELF) Magnetic Field.*Artikel-ELF-Salmonella.* Jember University.

[10] Farisna ST and E Zulaikha (2015) *Resistensi Bacillus Endogenik Kalimas Surabaya terhadap Logam Besi (Fe)*. Jurusan Biologi FMIPA Institut Teknologi Sepuluh Nopember.

[11] Soeka YS and Sulistiani (2014) Karakterisasi Protease Bacillus subtilis A1 InaCC B398 Yang Diisolasi Dari Terasi Samarinda. *Berita Biologi Bidang Biologi. Puslit-LIPI.* 13(2).

[12] Nurhayati, Ninik and Sumaryanto (2012) Protease production by *Bacillus substilis* ATCC 1633. *Proceeding The Is' Conferece on Industrial Enzyme and Biotechnology.* 201-205.

[13] Johnvesl B, and Naik GR (2001) Study on production of thermostable alkaline protease from thermophilic and alkaliphilic *Bacillus* sp. JB99 in a chemically defined medium, *Journal. Process Biochemistry.*, 37: 139-144.

[14] Ikehara T, Yamaguchi H, Hosokawa K, Miyamoto H, and Aizawa K (2003) Effects of ELF Magnetic Fields on Membrane Protein Strucutre of Living HeLa Cells Studied by Fourier Transform Infrared Spectroscopy. *Bioelectomagnetics* 24(7): 457-464.

[15] Fitriani S (2013) Purifikasi Parsial dan Karakterisasi Enzim Protease dari Isolat B19 KUB BPPT CC.[*skripsi*]. Bogor(ID) :Institut Pertanian Bogor.

[16] Sadeghi A, Shahidi F, Mortazavi SA, and Mahalati N (2008) Evaluation of Different Parameters Effect on Maltodextrin Production by α-amilase Termamyl 2-x. *World Applied Sciences Journal*. 3 (1): 34-39.

[17] Kristinawati, Andika (2015) Pengaruh Lama Paparan Medan Magnet *Extremly Low Frequency* Terhadap pH dan Kadar Air Pada Proses Pembuatan Keju Jenis *Cream Cheese*.[*skripsi*]. Jember(ID) :Universitas Jember.

[18] Baehaki A (2012) Kolagenase *Bacillus licheniformis* F11 asal Palembang dan aplikasinya pada pembuatan peptida kolagen bioaktif.[*tesis*]. Bogor (ID) :Institut Pertanian Bogor.

[19] Adinarayana K, P Ellaiah and DS Prasad (2003) Purification and partial characterization of thermostable serine alkaline protease from a newly isolated *Bacillus subtilis* PE-1. *American Association of Pharmaceutical Scientists Pharmceutical Sciences Technology.* 4(4):440-448.

[20] Guang-rong H, Wei L, Dai DH, Wei-lian H (2011) Purification and characterization of a novel extracellular chitinase from thermophilic Bacillus sp. Hu1, *African. Journal. Biotechnology*. 10:2476-2485.

[21] Teufel P and Gotz F (1993) Characterization of an extracellular metalloprotease with elastase activity from *Staphylococcus epidermidis*. *Journal. Bacteriology. 175(13):4218-4224.*

[22] Chantawannakul P.(2001) Characterization of proteases of Bacillus strain 38 isolated from traditionally fermented soybean in Northen Thailand. *Science Asia* 28:241-248.

[23] Kumar D and Bhalla TC (2004) Purification and Characterization of a Small Size Protease from *Bacillus sp*. APR-4. *Journal Experiment Biology* 42: 515-517.

[24] Kovacs, Phillip E, Richard LV, and Pedro JJA (1997) The Effect of Static Magnetic Fields on Biological System : Implications for Enhanced Biodegradation. *Critical Reviews in Environment Science and Technology*. 27(4): 319-382.

[25] Trismilah and Lutfi A (2009) Membran polyethersulfone dan regenerated cellulose untuk ultrafiltrasi: Pengaruh pH terhadap proses ultrafiltrasi xilanase. *Jurnal Sains Dan Teknologi Indonesia*. 11(2): 76-83.

Quantity and quality of guinea pig *(cavia porcellus)* spermatozoa after administration of methanol extract of bitter melon (*momordica charantia*) seed and depot medroxy progesterone acetate (DMPA)

Syafruddin Ilyas[1,2]* Salomo Hutahaean[1], and Nursal[1]

[1]Biology Departement, Faculty of Mathematics and Natural Sciences, Universitas Sumatera Utara, Medan. Jl. Bioteknologi No. 1 Kampus USU, Padang Bulan, Medan, Indonesia.
[2]Featured Center for Science and Technology "Stem Cells", Universitas Sumatera Utara, Medan. Jalan Dr. Masyur No. 4, Kampus USU Padang Bulan, Medan, Indonesia.

*Email: syafruddinilyas2013@gmail.com

Abstract. The discovery of male contraceptive drugs continues to be pursued, due to the few participation of men associated with the lack of contraceptive options for men. The combination of bitter melon seed methanol extract and DMPA are the options that currently apply to men. Therefore, the use of guinea pigs as experimental animals conducted research using experimental methods with complete randomized design (CRD). There are 4 control groups and 4 treatment groups. The first group, control group of dimethyl sulphoxide (DMSO) for 0 week (K0), The second one, bitter melon seed extract of 50 mg/100g Body Weight/day for 0 week (P0), the third one, control group of dimethyl sulfoxide (DMSO) for 4 weeks (K1), the fourth one, bitter melon seed extract of 50 mg/100g BW/day for 4 weeks + Depot medroxy Progesterone Acetate (P1), the fifth one, control group of dimethyl sulfoxide (DMSO) for 8 weeks (K2), the sixth one, bitter melon seed extract of 50 mg/100g BW/day for 8 weeks + DMPA (P2), the seventh one, control group of dimethyl sulfoxide (DMSO) for 12 weeks (K3), the eighth one, bitter melon seed extract of 50 mg/100g BW/day for 12 weeks + DMPA (P3). Methanol extract of bitter melon seed to decrease the quantity and quality of guinea pig spermatozoa decreased significantly, i.e. viability and normal morphology of spermatozoa ($p<0.05$).

Keywords: antifertility, bitter melon *(Momordica charantia)*, guinea pig, sperm concentration

1. Introduction

According to the Survey of Indonesia Health Demographic at the 2002, family planning participation is still very low, only 4.4%, which include: condom use (0.9%), vasectomy / male surgery method (0.4%), intermittent intercourse (1.5%) and periodic abstinence (1.6%) [1]. The participation rate as a family planning acceptor still lower from Islamic countries, such as Bangladesh at 13.9% in 1997, and Malaysia at 16.8% in 1998.

Male was less interested in becoming "family planning" acceptor because there were not many contraceptive options available. Therefore it is necessary to develop from herbs one of the seeds of bitter melon (*Momordica charantia*) [2,3]. Combination of bitter melon seed with DMPA was better able to suppress spermatogenesis in mice [4,5,6,7] and rabbits to decrease the quantity and quality. The content of flavonoids in bitter melon seeds can lower serum testosterone levels so that it can cause decreased libido. Therefore, DMPA able to benefit as a source of testosterone in serum [8].

The workings of bitter melon seed extract and DMPA through the hormonal mechanism of hypothalamus-pituitary-testis [4]. Decreased intratesticular testosterone levels disrupt to spermatogenesis, so reduced spermatozoa or no one in the testis. Testosterone may enter the seminiferous tubule and bind by Androgen Binding Protein (ABP) so that it can be used for the growth and development of spermatogonia as its stem cell spermatozoa. Spermatogonia transformation become to spermatocytes, spermatids, and spermatozoa in the seminiferous tubules used to testosterone [9,10].

2. Methods

The study subjects used healthy male healthy guinea pigs aged 8-11 months (proven to have once-a-year offspring when mated with a female marmot) weighing 400-450g and placed in a clean cage. The study has received permission from the Research Ethics Committee of Health with no. 085 / KEPH-FMIPA / 2017.

2.1 Spermatozoa concentration

The spermatozoa suspension previously was homogenized by vibrating by hand or carefully stirred with a stirring glass. Removal of dilution liquid to 1.01 mark after sucking 0.005 mL spermatozoa suspension. Laying the spermatozoa suspension right on the edge of the cover glass to spread into the Neubauer Heumocytometer under a microscope with 400 times magnification. Then determine the concentration of spermatozoa in the field of view A, B, C, D. The total calculation of spermatozoa using the formula of determination of spermatozoa concentration in ml suspension cauda epididymis as follows: spermatozoa concentration = (N=1+2+3+4) x 50.000 sperm/mL. Calculating the total number of spermatozoa using the formula by computing boxes A, B, C, D [11].

2.2 Spermatozoa normal morphology

Calculation of spermatozoa morphology by taking ten μL of cement on top of object glass. Then fixation with ethanol 70% for 10-15 minutes and color with Giemsa for 15-20 minutes. Abnormalities or abnormal forms of spermatozoa seen primary and secondary deformities of 100-200 normal spermatozoa. The percentage of normal spermatozoa is the result of normal spermatozoa with the total number of normal and abnormal spermatozoa then multiplied 100% [11].

2.3 Spermatozoa viability

Checking of spermatozoa viability were performed with supravital painting, ten μL of semen added a 0.5% Eosin-Y solution then mixed on top of the object glass and covered with a glass cover. Spermatozoa that live colorless and colored dead. Then performed by microscope at 400x magnification and calculated against 100-200 spermatozoa. Express regarding percent of life gained from the results of live spermatozoa with the total number of live and dead spermatozoa multiplied by 100% [11].

3. Results and Discussion

Research was conducted in Medan, and various parameters was observed as follows: (1). Spermatozoa concentration, (2) Spermatozoa normal morphology, Spermatozoa normal morphology, and (3) Spermatozoa viability.

3.1. Spermatozoa concentration

The concentration of spermatozoa after giving of biter melon seed extract able to see Figure 1 below. There are different between treatment and control groups significantly ($p<0.05$) at the 12 and 16 weeks for administration of methanol extract of bitter melon seed. Testicles of guinea pig produced suppressing of spermatozoa concentrations.

The process of spermatogenesis or stages of spermatozoa cell formation ranging from spermatogonia and spermatocytes and spermatids greatly determine by concentration of spermatozoa. Provision of bitter melon seed extract allows decreasing testosterone levels in the testes (intratesticular testosterone) through repeated administration. Bitter melon seed extract contains β-sitosterol with a chemical structure similar to cholesterol. Cholesterol is the source of testosterone and enters the testicles so that the intratesticular testosterone is increased and consequently causes negative feedback to the hypothalamus and pituitary. The hypothalamus reduces the production of FSHRH (Follicle Stimulating Hormone Realizing Hormone) and LHRH (Luteinizing Hormone Realising Hormone) to affect the pituitary so that production of FSH and LH decreased. The reduction of FSH affects Sertoli cells in producing ABP (Androgen Binding Protein) or testosterone receptors. LH reduction suppresses Leydig cells to produce testosterone. These two ingredients are very important in the development and growth of spermatozoa cells. So that production interruption will reduce the production of spermatozoa cells in the testes.

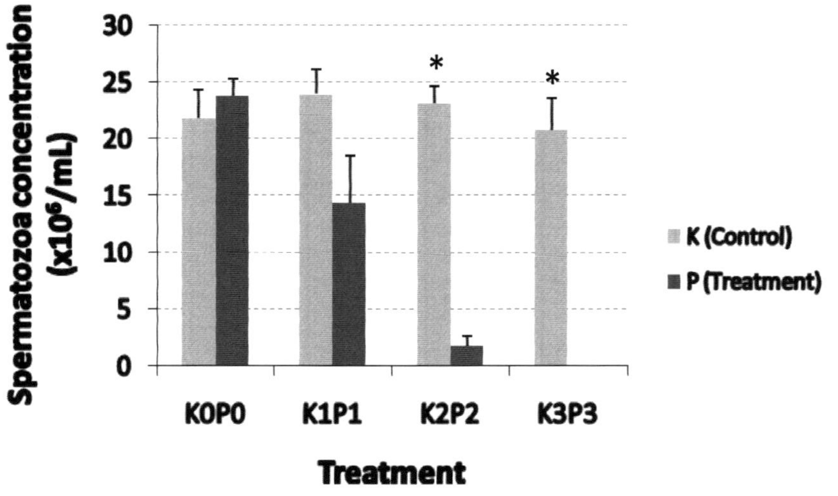

Figure 1. Bar graphic of spermatozoa concentration of guinea pig (x±SD). $p^{*}<0.05$=the same pair between control and treatment, K1P1=control and treatment on week 4th, K2P2=control and treatment on week 8th, K3P3=control and treatment on week 12th.

Beta sitosterol D-glucoside is a phytosterol contained in *Momordica charantia* Linn., and *Withania somnifera* (L.) Dunal and several other plants. Beta sitosterol D-glucoside has many pharmacological activities [12,13] such as androgenic, antiadenomic, anticancer [14,15] antiedemic, antiinflammatory [16]. Sitosterol has a structure such as cholesterol that later in the body can become the precursor of testosterone [2,17,18], so that if eating the extract of *Momordica charantia* seeds will cause increased testosterone levels to the highest culmination point and will eventually lead to decrease in testosterone serum body and testes [2,3]. DMPA replaces the decreased testosterone in the body, so there is no decrease in libido [4].

3.2. Spermatozoa normal morphology

In Figure 2 below we able to see the morphology of guinea normal spermatozoa after administration of methanol extract of bitter melon seed. Provision of bitter melon seed extract at weeks 8 (P2) and 12 (P3) showed significant results ($p < 0.05$) on morphology percentage of guinea pig spermatozoa. It is seen if we compare the research treatment with each control. This suggests that there are inhibitory effects of FSH and LH production due to negative feedback of testosterone on the hypothalamus (indirectly) or directly on the pituitary. Thus the functioning FSH in spermatogenic cell-form changes ranging from spermatogonium-spermatocytes-spermatids and spermatozoa are impaired. As a result, abnormal spermatozoa or defective spermatozoa (teratozoospermia) are present. According to Gordetsky [20], FSH levels have statistically significant associations with abnormal and morphological spermatozoa concentrations but are not related to the volume of semen (spermatozoa + liquid spermatozoa).

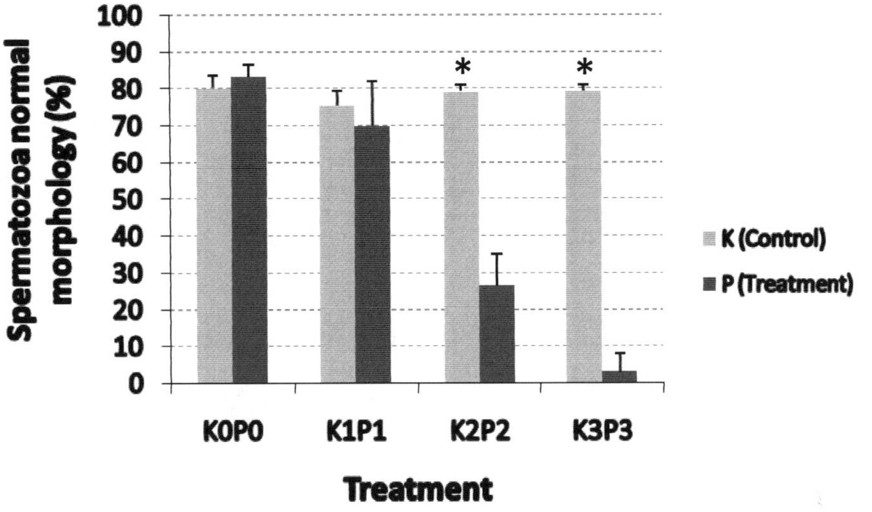

Figure 2. Bar graphic of spermatozoa normal morphology of guinea pig. $p^* < 0.05$=the same pair between control and treatment. K0P0=control and treatment on week 0, K1P1=control and treatment on week 4th, K2P2=control and treatment on week 8th, K3P3=control and treatment on week 12th.

3.3 Spermatozoa viability

Administration of methanol extract of bitter melon seed and DMPA in guinea pig, obtained by spermatozoa viability data as shown in the following Figure 3.

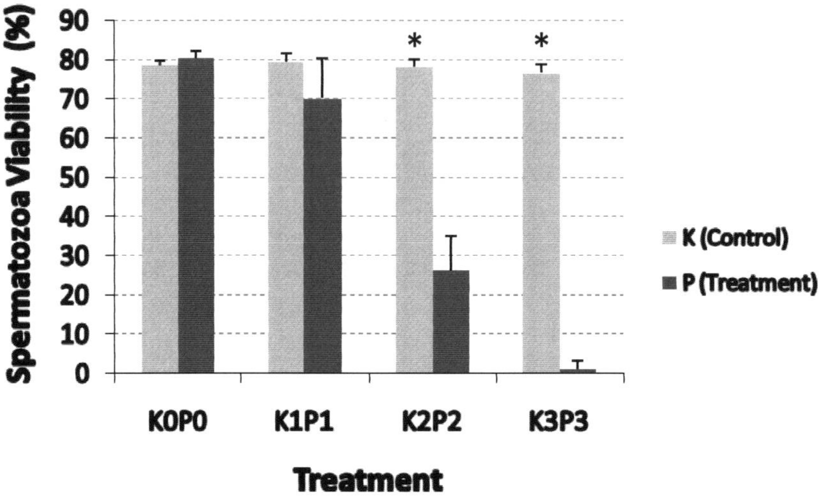

Figure 3. Bar graphic of spermatozoa viability of guinea pig. $p^* < 0.05$=the same pair between control and treatment, K1P1=control and treatment on week 4th, K2P2=control and treatment on week 8th, K3P3=control and treatment on week 12th.

Provision of bitter melon seed extract and DMPA at weeks zero and fourth did not affect significantly ($p > 0.05$), but at weeks 8 and 12 had a significant effect ($p < 0.05$). Bitter melon seed extract and DMPA caused by decreased testosterone levels in the testes due to at least LH levels. A small amount of tetosterone causes the production of LH by the pituitary to become to least level. Thus the production of testosterone from the Leydig cells decreases LH levels and causes a disruption to spermatogenesis, so that the production of spermatozoa viability also becomes less.

If the pituitary secrete LH and FSH slightly then the production of testosterone mice also slightly. So the concentration, morphology, and viability of spermatozoa become less. Testosterone is very important at the beginning and end of maturation of spermatozoa in the testes [20]. Lack of the hormone testosterone also triggers the occurrence of spermatogenic cell apoptosis [4,21,22] through the activation of caspase 3. Caspase 3 is the executor of spermatogenic cell death programmatically [23,24,25,26,27].

4. Conclusions

Methanol extract of bitter melon seed to decrease the quantity and quality of guinea pig spermatozoa significantly ($p < 0.05$) i.e. viability and normal morphology of spermatozoa. Therefore it was able to be candidate for herbal contraception.

Acknowledgments

The author would gratefully acknowledge the Indonesian Ministry of Research, Technology and Higher Education for providing the fund through Kemenristekdikti DRPM (second year of competition grant) program under the contract number 003/SP2H/LT/DRPM/IV/2017, April 20th 2017.

References

[1]. Suprihastuti (2000) Decision maker on the use of male contraceptives in Indonesia. Result analysis of SDKI - 1997. Jakarta.

[2]. Tumkiratiwong P, Ravicha Ploypattarapinyo, Urai Pongchairer, Wachiryah Thong-asa Iran. (2014) Original article Reproductive toxicity of *Momordica charantia* ethanol seed extracts in male rats. *J Reprod Med* **12**(10).

[3]. Yama OE, Francis ID, Ademola AO, Abraham AO, Cressie CN, Abayomi OO (2011) Sperm quotient in Sprague–Dawley rats fed graded doses of seed extract of *Momordica charantia. Middle East Fertility Society Journal.* **16**), 154–158.

[4]. Ilyas S, Hutahaean S, Nursal, (2016) Apoptosis overview of cerebellum Purkinje cell in mice (*Mus musculus* L.) after exposure to methanol extract of the seeds of bitter melon (*Momordica charantia*) and DMPA. *International Journal of PharmTech Research*. **9**(9), pp. 444-449.

[5]. Ilyas, S (2014a) Morphometric Analysis of Seminiferous Tubules in Rat Testes after Methanol Extract of Bitter Melon seed (*Momordica charantia*) and Depot Medroxyprogesterone Acetate (DMPA) Administration. *3rd Seminar of Pharmaceutical Science and Technology* (International Seminar).

[6]. Ilyas, S (2014b) Effect of Methanolic *Momordica Charantia* Seed Extract And Depot Medroxyprogesterone Acetate (DMPA) to Quantity and Quality of Rat Sperm. *International Journal of Pharmtech Research*. **6** (6) 1817-1823. 0974-4304.

[7]. Ilyas, S (2014c) Biology National Seminar, 2014 - FMIPA USU. Combination Effect of Methanol of Bitter Melon Seed Extract (*Momordica charantia*) and Progesterone Against Morphometry of Rat Leydig Cell (*Rattus* sp.). Univ. Sumatera Utara.

[8]. Ilyas S, Silvia W. Lestari, Nukman Moeloek, Asmarinah, Nurjati C. Siregar. (2013). Induction of Rat Germ Cell Apoptosis by Testosterone Undecanoate and Depot Medroxyprogesterone Acetate and Correlation of Apoptotic Cells with Sperm Concentration. **(45)**1.

[9]. Munell F, Carlos ASR, David MS, Oscar MT, and Jaume RS (2002) Androgen-Binding Protein and Reproduction: Where Do We Stand? *Journal of Andrology*, **23**(5).

[10]. Yi Ma, Hao-Zheng Yang, Long-Mei Xu, Yi-Ran Huang, Hui-Li Dai & Xiao-Nan Kang (2017) Testosterone regulates the autophagic clearance of androgen binding protein in rat Sertoli cells. Scientific Reports (**5**) 8894. Acces at August 16, 2017- www.nature.com/ scientific reports

[11]. World Health Organisation (WHO) (1999) *Laboratory manual for the examination of human semen and sperm-cervical mucus interaction*; 4th Ed. NY: Cambridge University Press.

[12]. Hasibuan R, Syafruddin Ilyas, Saleha Hanum (2015) Effect of leaf extract Haramonting (*Rhodomyrtus tomentosa*) to lower blood sugar levels in mice induced by alloxan. (88).6, pp 284-291.

[13]. Arianto A, Hakim Bangun, Urip Harahap, and Syafruddin Ilyas (2014) The Comparison of Swelling, Mucoadhesive, and Release of Ranitidine from Spherical Matrices of Alginate, Chitosan, Alginate Chitosan, and Calcium Alginate-Chitosan. (**6**)7, pp 2054-2063

[14]. Arianto A, Hakim Bangun, Urip Harahap, and Syafruddin Ilyas (2015) Effect of Alginate Chitosan Ratio on the Swelling, Mucoadhesive, and Release of Ranitidine from Spherical Matrices of Alginate Chitosan. **(8)** 4, pp 653-665

[15]. Masyithah C, Sumadio Hadisahputra, Syafruddin Ilyas (2015) Combinational effects of ethylacetate extract of *Zanthoxyluma canthopodium* DC. with doxorubicinon MCF7 breast cancer cells. (7)4, pp 651-653.

[16]. Satria D, Jansen Silalahi, Ginda Haro, Syafruddin Ilyas (2017) Poppy Anjelisa Z Hsb. Antioxidant and Antiproliferative Activities of an Ethylacetate Fraction of Picria Fel-Terrae Lour. Herbs. *Asian Pac J Cancer Prev*, (18)2, 399-403.

[17]. Taylor L (2002) Technical Data Report For Bitter Melon (*Momordica charantia*), Herbal Secrets of The Rainforest, 2nd Ed.: Sage Press Inc.

[18].Sen A, Poonam Dhavan, Kshitiz Kumar Shukla, Sanjay Singh, G. Tejovathi (2012) Analysis of IR, NMR and Antimicrobial Activity of β-Sitosterol Isolated from *Momordica charantia*. Science Secure Journal of Biotechnology (SSJBt) Volume 1 Issue 1 Page: 9-13.

[19].Supriya J, Tatke Pratima, Gabhe SY (2010) Marker Based Standardization Of Commercial Formulations And Extracts Containing Beta-Sitosterol D- Glucoside Using Hptlc. *International Journal Of Research In Ayurveda & Pharmacy, **1***(2), 616-623

[20].Gordetsky J, Edwin van Wijngaarden and Jeanne O ' Brien (2011) Redefi ning abnormal Follicle-Stimulating Hormone In The Male Infertility Population. BJU international 110, 568–572.

[21].Barrett KE, Barman SM, Boitano S and Brooks HL (2011) Ganong's review of medical physiology. 23rd edition,Tata Mc Graw Hill Education Private Limited. p. 404

[22].Masfria, Urip Hara hap, Maratua Pandapotan Nasutionl, Syafruddin Ilyas (2014) Cytotoxic Activity, Proliferation Inhibition and Apoptosis Induction of Rhaphidophora Pinnata (L.F.) Schott Chloroform *Fraction to MCF-7 Cell Line*. **(6)**4, pp 1327-1333.

[23].Ilyas, S (2015) Contraceptive Effect of Methanol Extracts of Pare Bitter Melon (Momordica charantia) and Depot Medroxyprogesterone Acetate (DMPA) on Rabbit (Oryctolagus Cuniculus). Report of Competitive Grant Research - Higher Education on The Third Year

[24].Shaha C, Rakshamani Tripathi and Durga Prasad Mishra (2010) Male germ cell apoptosis: regulation and biology. Phil. Trans. R. Soc. B ,365, 1501–1515

[25].Herwanto RY, Jenny Bashiruddin, Syafruddin Ilyas, M. Nadjib Dahlan Lubis (2015) Correlation of Noise Intensity to Heat Shock Response with Hsp 70, p53, Cytochrome C, Caspase 3 expressions and ultrastructure region of *Rattus norvegicus*'s cochlea. **(7)**1, pp 80-84

[26].Herwanto RY, Syafruddin Ilyas, Rr. Suzy Indharty (2016) HSP70 Gene Expression in Serum and Tissue of Rat Cochlear (*Rattus norvegicus*) Due to Noise Exposure and Heat. **(9)**11, pp 58-63.

[27].Moeloek N, Asmarinah Asmarinah, Nurjati C. Siregar, Syafruddin Ilyas (2008) Testosterone undecanoate and depo medroxyprogesterone acetate induced azoospermia through increased expression of spermatogenic cell caspase 3. **(17)**3

[28].Yusuf F, Syafruddin Ilyas, Harun AR, Damanik, Fatchiyah (2017) Microbiota Composition, HSP70 and Caspase-3 Expression as Marker for Colorectal Cancer Patients in Aceh, Indonesia. Acta Med Indones-Indones J Intern Med. **(48)**4.

Transforming Growth Factor Expression (TGF-β) Correlate with Serum Level of Malondialdehyde (MDA) after EVOO Administration in Preclinical Rat Models of Preeclampsia

Syafruddin Ilyas[1,2], Salomo Hutahaean[1], dan Evi Irianti[1]

[1]Departement of Biology, Faculty Mathematics and Natural Sciences, Universitas Sumatera Utara, Medan. Jl. Bioteknologi No. 1 Kampus USU, Padang Bulan, Medan, Indonesia.
[2]Featured Center for Science and Technology "Stem Cell," Universitas Sumatera Utara, Medan. Dr. Masyur Street, No. 4, Kampus USU, Padang Bulan, Medan, Indonesia.

Email: syafruddinilyas2013@gmail.com

Abstract. Preeclampsia can cause cell death either apoptosis or necrosis. One cause is the disturbance of the emergence of malondialdehyde (MDA). Very few reports on the role of Transforming Growth Factor Expression (TGF-β) in the remodeling process of placental cells and their association with serum MDA content. Research of true experiment with complete randomized design (CRD) with five treatment groups. The first group, preeclampsia negative control (T0). The second group, preeclamptic rats model (T1). The third group, preeclamptic rats model+EVOO 0.45g/kg-Body Weight/day (T2). The fourth group, preeclamptic rats model+EVOO 0.90g/kg-BW/day (T3). The fifth group, preeclamptic rats model+EVOO 1.8 g/kg-BW/day (T4). The results showed a significant effect of EVOO on TGF-β expression in preeclampsia rats, meaning that there was a role of TGF-β against pre-eclampsia placenta remodeling. There was a positive and strong relationship ($r=0.494$) as well as a very significant relationship ($p<0.01$) between TGF-β and the serum MDA.

Keywords: Apoptosis, Reactive Oxygene Species (ROS)

1. Introduction

Preeclampsia is a complication of pregnancy characterized by increased of blood pressure (hypertension) and signs of organ damage, such as kidney damage indicated by increased levels of protein in the urine (proteinuria). In humans, the symptoms of preeclampsia usually appear at the age of gestation entering the week 20 or later (most often between 24 and 26 weeks of pregnancy until shortly after birth) Preeclampsia is not realized by pregnant women can develop into eclampsia, a serious medical threat to maternal safety conditions pregnant and their fetus.

Preeclampsia is a medical complication of pregnancy and occurs in about 5-8% of pregnancies causing high maternal and fetal morbidity or mortality [1,2]. In 2013 hypertension in pregnancy is the second cause of AKI in Indonesia as much as 27.1% [3].

Caused of preeclampsia unclear because preeclampsia is "the disease of theories" [4]. The results of several factors that may cause preeclampsia include maternal adaptation imbalances that lead to systemic inflammatory reactions and endothelial dysfunction [5,6]. Endothelial dysfunction is a condition where there is no balance of mediatorial products by the endothelium to regulate vascular tone, platelet aggregation, coagulation, fibrinolysis. The occurrence of endothelial dysfunction is characterized by increased endothelin (ET-1) [7]. ET-1 is a peptide of 21 amino acids as a potent vasoconstrictor, and the main isoform produced from human endothelial cells causes vasoconstriction and causes hypertension and proteinuria which is a symptom of preeclampsia [8].

Until now there is no proper therapy to deal with preeclampsia, considering preeclampsia is a serious health problem to be addressed, so the need for prevention of this disease. One such prevention effort is the provision of antioxidants such as a combination of vitamin C and vitamin E supplementation. They were able to decreased the incidence of preeclampsia significantly [9, 10].

Extra virgin olive oil (EVOO), contains antioxidants that are tocopherol (vitamin E). The process of purifying olive oil produces some components, including hydroxytyrosol, and target tyrosols. Hydroxytirosol has been shown to be effective in increasing plasma antioxidant activity as well as a dilator to the oxidation of LDL (lipid density low). Tyrosol and other phenolic antioxidants can bind LDL so that can delay the process of atherosclerosis.

2. Methods

Animal Preparation Try Rats in adaptation to the environment for seven days with feeding in the form of basal rations in all rats. Plastic place of rats having a diameter of 80 cm with a height of 40 cm [11,15,13]. Basal ration composition is prepared based on standard AOAC (2005) that contains carbohydrates, proteins, fats, minerals, vitamins. An animal model of rat (*Rattus norvegicus*) from Universitas Gadjah Mada Yogyakarta with age of 10 weeks and body weight between 150-250 gram which has been approved by Ethical Research Commission of Animal by No. 068/KEPH-FMIPA/ 2017.

2.1 Research design

The study uscaous ofed a pure experimental method with Completely Randomized Design consisting of 5 treatment groups. (T0) of preeclampsia control, (T2) of preeclamptic rats + EVOO 0.45 g / kg BW / day, (T3) rats preeclampsia + EVOO 0.90 g / kg BW / day, T4) mice preeclampsia + EVOO 1.8 g / kg BW / day.

2.2 Measurement of MDA Serum Level

Malondialdehide content with TBARs method. Working procedures are carried out according to the standards set by the local laboratory. Mice blood samples as much as 2 ml, filled into a tube that already contains two drops of EDTA for subsequent centrifugation. After centrifugation at a speed of 3000 rpm, for 30 minutes taken plasmanya. Then prepare the necessary reagents, among others (1). SDS 200 μl, (2). EDTA 50 μl, (3). BHT 50 μl, and (4). Acetate 1500 μl. All the ingredients/solutions are inserted into the tube (which inside the tube already contains the plasma) and then vortexed for 10 seconds, then centrifuged at 300 rpm for 10 minutes. After that, the supernatant is taken away; the deposit is removed, then put into a tube that is already available, then add 1500 μl of TBA solution. In the meantime prepare six standard tubes (the contents therein are the four solutions mentioned above without mixing with the sample plasma), also prepare 1 blank tube (its contents aquabidest 700 μl). Then the standard tube and blanks are added TBA 700 μl, the two tubes are vortexed for 10 seconds. When finished, all the tubes are covered with cotton or gauze that is plastered, then inserted into the water bath at 940C for 1 hour. Then cool for 10 minutes, centrifuged for 10 minutes (except standard bloom) and reads the result on the "SmartSpecTMPlus" brand spectrophotometry from BioRad at 532nm wavelength. The way to read it is to take each sample with 1500 μl pipette then put it into a cuvette (try the cuvette glass when the reading is not blurry, always cleaned every time you replace the sample by rinsing with aquabidest 2 times, because the glassiness affects the reading result) to the spectrophotometry one by one from the sample.

2.3 Observation of TGF-β Expression with Immunohistochemistry (IHC)

This observation was performed on the placenta portion by immunohistochemical staining to determine the expression of TGF-β. The expression of TGF-β expression in placental cells was quantitatively observed using the Olympus BX51 microscope at 400 × magnification. The intensity of 4: 0 discordable expression when no expression on cells; 1 = when the staining is weak; 2 = when medium dyeing; 3 = when staining is strong. The pored-colored cell suspension: 0 when there is no expression on the cell; 1 = when <33% expression on cell; 2 = when 33-66% expression on cell; 3 =

when> 66% expression on cell. Calculated on 100 cells in hot spot area. Then summed up when the score 0-1 = negative, 2 = positive weak, 3-4 = moderate positive, 5-6 = strong positive [14,15,16,17].

2.4 Data analysis
This study used data analysis using quantitative analysis to look at placental histopathology and quantitative histopathology for TGF-β expression with ANOVA test to determine the difference between treatments followed by Post Hoc test to find out which treatments gave the best results.

3. Results And Discussion
Based on the results of research conducted, the results obtained as shown in Figure 1 below.

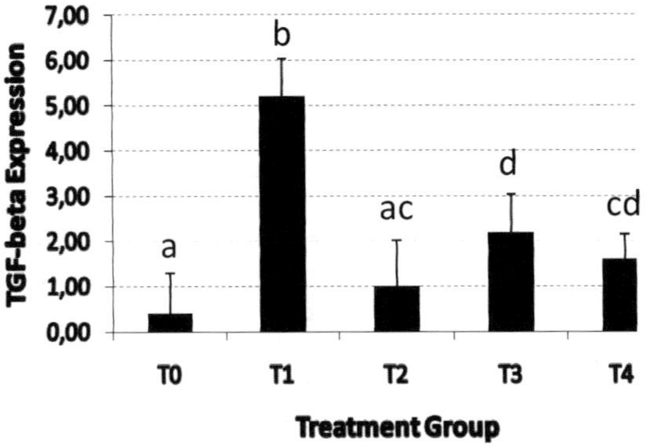

Figure 1. Graph Bar of TGF-β expression on the placenta of mouse model preeclampsia after treatment.

In rats, with preeclampsia, the presence of TGF-β expression was significantly greater ($p<0.05$) compared with the negative control and treatment of EVOO addition. This is due to an increase in oxidants or ROS (Reactive Oxygen Species) due to excessive NaCl content (above physiological/0.9%). The ROS shows the number of unpaired electrons so that they can lie in various cellular metabolic enzymes. As a result, the enzyme does not work and causes cell death either by necrosis or apoptosis [18,19,20,21,22]. Necrotizing cell death results in the release of proinflammatory (inflammatory) signals. As Iranloye [23] points out, that high salt levels can lead to increased Reactive Oxygene Species (ROS). Furthermore, Bertolino et al., [24] and Lebrin, [25] suggest that TGF-β is the main cytokine produced abundantly in vascular endothelial cells and trophoblasts. It plays a key role in various physiological processes, including embryonic growth and development, improvement of inflammation and angiogenesis.

EVOO levels in T2, T3 and T4 can significantly lower TGF-β expression in placental tissue ($p<0.05$) compared with T1 or preeclampsia mice without EVOO. This suggests a decrease in TGF-β activity as reduced cell damage or cell death (apoptosis or cellular necrosis). The content of EVOO more vitamin E content that serves as an antioxidant that can suppress oxidant activity (ROS) and cause cells to avoid the effects of oxidants or cell damage and cell death and the emergence of proinflamasi. As stated Khanduja et al., [26] that vitamin E can inhibit the activity of proinflammatory cytokines in the cells. High doses of vitamin E in the diet may play a role in reducing cytokine lipopolysaccharide and nitric oxide-induced proinflammation.

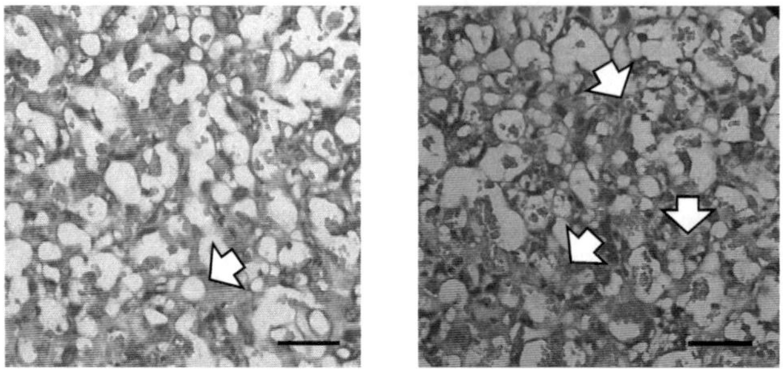

Figure 2. Expression of TGF-β in the placenta of preeclamptic model rat in treatment group (T) and control (T), —— = 100 μm

TGF-β expression is seen expressed in the cytoplasm and cytoplasmic membrane. The amount of expression, intensity and percentage of expression seen in the positive control group or preeclampsia without EVOO (Figure 1). This suggests that the activity of the TGF-β gene is more stimulated due to the physiological processes of cells that occur due to the administration of a high salt content (NaCl 6%). High salt levels disrupting cell membranes and lead to increased activity of death receptors such as DR (Death Receptor) or Tumor Necrosis Factor (TNFα). The receptor activity stimulates procaspase 8 to caspase 8 and results in an increase in caspase activity 3 as an apoptotic cell execution. Stimulant receptor stimuli may also increase Bax or Bid expression causing disruption to mitochondrial membranes. Damage to the mitochondrial membrane causes the outtake of the cell cytochrome and joins Apaf-1 and Pro-caspase 9 and forms apoptosome. Apoptosome triggers caspase 3 activity and leads to cell death by fragmentation (apoptosis). Also, cell death through necrosis can also occur due to increased levels of salt in the blood. Increased TGF-β can lead to cell death through apoptosis through a set of building information modeling (BIM) cells. As Ramesh et al. [27] find, there is a clear understanding of how TGF-β mediates apoptosis through regulatory mechanisms that may be controlled through the level of Bim expression (formerly known as Bcl2L 11).

Table 1. Correlation between TGF-beta and MDA-post after administration of EVOO on preeclampsia rat.

			MDA post	TGF-beta
Spearman's rho	MDA post	Correlation Coefficient	1.000	0.640**
		Sig. (2-tailed)	.	0.001
		N	25	25
	TGF-beta	Correlation Coefficient	0.640**	1.000
		Sig. (2-tailed)	0.001	.
		N	25	25

**. Correlation is significant at the 0.01 level (2-tailed).

Provision of EVOO to rats of preeclampsia may decrease MDA production produced by placental cells. MDA production signifies the presence of oxidant production by cell activity and as a sign of oxidative stress. By its role that MDA is often used as a biomarker of oxidative stress in clinical and non-clinical investigations. As Khoubnasabjafari et al. [28]) pointed out, MDA is used as a biomarker of oxidative stress in many diseases although it has wide variations in healthy people. According to the evidence gathered, most of the technical issues of MDA measurement have not been resolved and need further investigation, and the role of MDA biomarkers should be reevaluated by experts.

In this study we found a strong and significant relationship ($r=0.64$ and $p<0.05$) between TGF-β expression with post MDA levels in preeclamtic rats given EVOO (Figure 1).

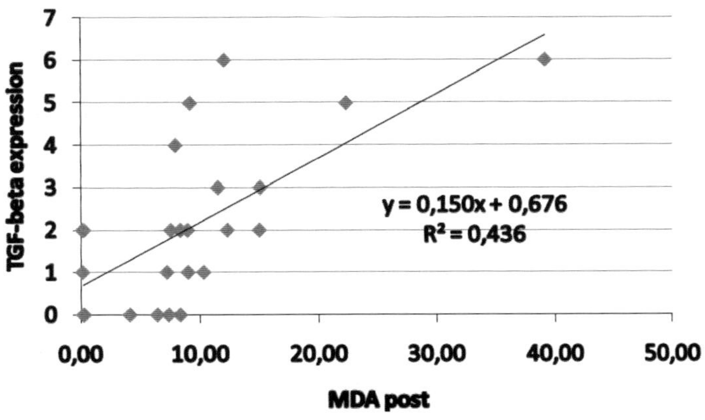

Figure 3. Linear graph of serum MDA relationship with TGF-β expression in placenta mouse model of preeclampsia.

This suggests the role of TGF-β in the process of regulation or the emergence of oxidative stress in rat placenta preeclampsia. By another study by Liu and Leen [29], that TGF-β is the most potent pro-fibrogenic cytokine and its expression is elevated in almost all fibrotic diseases. Although signaling via the Smad line is believed to play a central role in TGF-β fibrogenesis, emerging evidence suggests that reactive oxygen species (ROS) modulate TGF-β signals via different pathways including the Smad pathway. TGF-β1 increases the production of ROS [30] and suppresses antioxidant enzymes, causing a redox imbalance. ROS, in turn, induces TGF-β1 and mediates many of the TGF-β fibrogenic effects, forming a vicious circle (see graph flow chart on the right). Here, we review the current knowledge of feedback mechanisms between TGF-β1 and ROS in the development of fibrosis. Therapies targeting TGF-β-induced and ROS-dependent cellular signaling are a new approach in the treatment of fibrotic disorders.

4. Conclusions

Based on the result of the research, it can be concluded that there is a significant effect of EVOO on TGF-β ($p < 0{,}05$) placenta pre-eclampsia. There was a strong ($r = 0.494$) and significant relationship between TGF-β ($p < 0.01$) and post MDA levels.

Acknowledgments

The authors highly acknowledge and express our greatest gratitude to the Rector of University of Sumatera Utara that provides the research funding through TALENTA USU program under the contract number 5338/UN5.1.R/PPM/2017, May 22nd 2017, on the fiscal year 2017/2018.

Reference

[1]. Shah DA, and Khalil, RA (2015) Bioactive factors in uteroplacental and systemic circulation link placental ischemia to generalized vascular dysfunction in hypertensive pregnancy and preeclampsia. *Biochemical Pharmacology*, **95**(4), 211–226.

[2]. WHO (2011) Recommendations for Prevention and Treatment of Preeclampsia and Eclampsia, *WHO Department of Maternal and Child Health*, Geneva, Switzer-land

[3]. Ministry of Health RI (2014) Mother's Day's. InfoDATIN Center for Data and Health Information RI Leong x, Mustafa MR, Jaarin K. (n.d.). Nigella sativa and its protective role in oxidative stress and hypertension.

[4]. Cunningham LBHRS (2010) *Williams Obstetric.* 23 ed. United States of Ameri-ca: The McGraw-Hill Companies, Inc

[5]. Saucedo R, Valencia J, Manuel L, and Hern M (2014) Early Disturbed Placental Ischemia and Hypoxia Creates Immune Alteration and Vascular Disorder Caus-ing Preeclampsia, 45.

[6]. Yi KW, Jung SH, Cho GJ et al. (2014) Effects of sflt-1 and alpha 2-macroglobulin on vascular endothelial growth factor-induced endothelin-1 upregulation in human microvascular endothelial cell. **35**:64-69

[7]. Teixeira BC (2014) *Inflammatory markers, endothelial function and cardiovascular risk.* **13**(2):108–115 21
[8]. Salaets K, Scliessman, J Speiser R, Anh-Minh T, Wang E, Angerio A (2006) Role of endothelin - 1 in Atherosclerosis Georgetown. *Journal of health Sciences*, **3**(1):1-9
[9]. Takeuchi T (2001) Regulation of Platelet Aggregation in Vitro by Plasma Adenosine in Preeclampsia. Gynecol Obstet Invest **51**(1), 36-39.
[10].Satria D, Jansen Silalahi, Ginda Haro, Syafruddin Ilyas (2017) Poppy Anjelisa Z Hsb. Antioxidant and Antiproliferative Activities of an Ethylacetate Fraction of Picria Fel-Terrae Lour. Herbs. *Asian Pac J Cancer Prev*, **18**(2), 399-403.
[11]. Ilyas, S (2014) Effect of Methanolic *Momordica Charantia* Seed Extract And Depot Medroxyprogesterone Acetate (DMPA) to Quantity and Quality of Rat Sperm. *International Journal of Pharmtech Research*. **6** (6) 1817-1823. 0974-4304.
[12]. Hasibuan R, Syafruddin Ilyas, Saleha Hanum (2015) Effect of leaf extract Haramonting (*Rhodomyrtus tomentosa*) to lower blood sugar levels in mice induced by alloxan. **88**(6), pp 284-291.
[13]. Moeloek N, Asmarinah Asmarinah, Nurjati C. Siregar, Syafruddin Ilyas (2008) Testosterone undecanoate and depo medroxyprogesterone acetate induced azoospermia through increased expression of spermatogenic cell caspase 3. **17**(3)
[14]. Crowe M, Doetschman T, Greenhalgh DG (2000) Delayed Wound Healing in Immunodeficient TGF-β Knockout Mice. The Journal of Investigative Dermatology. **115**(1).
[15]. Das RK, Venkatraghavan V, Sheet D, Chakraborty C, Ray AK, Chatterjee J (2010) Evaluation of p63 Expression in Oral Submukus Fibrosis. Proceedings of 2010 International Conference on Systems in Medicine and Biology
[16]. Werner S, Grose R, (2003). Regulation of Wound Healing by Growth Factors and Cytokines. Physiol Rev, **83**: 835-870.
[17]. Yusuf F, Syafruddin Ilyas, Harun AR. Damanik, Fatchiyah (2017) Microbiota Composition, HSP70 and Caspase-3 Expression as Marker for Colorectal Cancer Patients in Aceh, Indonesia. Acta Med Indones-Indones J Intern Med. **48**(4).
[18]. Ilyas S, Hutahaean S, Nursal (2016) Apoptosis overview of cerebellum Purkinje cell in mice (*Mus musculus* L.) after exposure to methanol extract of the seeds of bitter melon (*Momordica charantia*) and DMPA. *International Journal of PharmTech Research*. **9**(9), pp. 444-449.
[19]. Masfria, Urip Hara hap, Maratua Pandapotan Nasutionl, Syafruddin Ilyas, (2014). Cytotoxic Activity, Proliferation Inhibition and Apoptosis Induction of Rhaphidophora Pinnata (L.F.) Schott Chloroform *Fraction to MCF-7 Cell Line*. **6**(4), pp 1327-1333.
[20]. Herwanto RY, Jenny Bashiruddin, Syafruddin Ilyas, M. Nadjib Dahlan Lubis (2015) Correlation of Noise Intensity to Heat Shock Response with Hsp 70, p53, Cytochrome C, Caspase 3 expressions and ultrastructure region of *Rattus norvegicus*'s cochlea. **7**(1), pp 80-84
[21]. Herwanto RY, Syafruddin Ilyas, Rr. Suzy Indharty (2016) HSP70 Gene Expression in Serum and Tissue of Rat Cochlear (*Rattus norvegicus*) Due to Noise Exposure and Heat. **9**(11), pp 58-63.
[22]. Masyithah C, Sumadio Hadisahputra, Syafruddin Ilyas (2015) Combinational effects of ethylacetate extract of *Zanthoxyluma canthopodium* DC. with doxorubicinon MCF7 breast cancer cells. **7**(4), pp 651-653.
[23]. Iranloye BO, Gabriel O. Oludare, Ayodele O. Morakinyo, Naomi A. Esume, and Lucy C. Ekeh (2013) Reproductive parameters and oxidative stress status of male rats fed with low and high salt diet. *J Hum Reprod Sci.* Oct-Dec; **6**(4): 267–272.
[24].Bertolino P, Deckers M, Lebrin F and ten Dijke P (2005) Transforming growth factor-beta signal transduction in angiogenesis and vascular disorders. Chest 128 (6 Suppl): 585S-590S.
[25]. Lebrin F, Deckers M, Bertolino P and Ten Dijke P (2005) TGF-beta receptor function in the endothelium. Cardiovasc Res 65: 599-608.
[26]. Khanduja KL, Pramod Kumar Avti Surender Kumar, Vandana Pathania, Chander Mohan Pathak (2005) Inhibitory effect of vitamin E on proinflammatory cytokines-and endotoxin-induced nitric oxide release in alveolar macrophages. **76**(23), 22 April 2005, Pages 2669-2680.
[27].Ramesh S, Gary M. Wildey, and Philip H. Howe (2009) Transforming growth factor β (TGFβ)-induced apoptosis: The rise & fall of Bim. Cell Cycle. **8**(1): 11–17.

[28]. Khoubnasabjafari M, Khalil Ansarin, and Abolghasem Jouyban (2015) Reliability of malondialdehyde as a biomarker of oxidative stress in psychological disorder. Bioimpacts. 2015; **5**(3): 123–127.

[29]. Liu RM and Leena PD (2015) Reciprocal regulation of TGF-β and reactive oxygen species: A perverse cycle for fibrosis. *Redox Biol.* **6**: 565–577

[30]. Ilyas S, Silvia W. Lestari, Nukman Moeloek, Asmarinah, Nurjati C. Siregar (2013) Induction of Rat Germ Cell Apoptosis by Testosterone Undecanoate and Depot Medroxyprogesterone Acetate and Correlation of Apoptotic Cells with Sperm Concentration. **45**(1).

IOP Conf. Series: Earth and Environmental Science **130** (2018) 012049 doi:10.1088/1755-1315/130/1/012049

Antibacterial activity of cinnamon ethanol extract (*cinnamomum burmannii*) and its application as a mouthwash to inhibit *streptococcus* growth

Syahdiana Waty[1,2], Dwi Suryanto[1*], Yurnaliza[1]
[1]Department of Biology, Faculty of Mathematics and Natural Sciences, University of Sumatera Utara, Jln. Bioteknologi No. 1, Kampus USU, Medan
[2]Medan Health Polytechnique, Ministry of Health, Jln. Jamin Ginting Km. 13.5, Lau Cih, Medan

Email: dwisuryanto@usu.ac.id

Abstract. Cinnamon bark has been commonly used as spicy and traditional medicine. It contains several antibacterial compounds such as flavonoids, saponins, and cinnamaldehyde. Several studies have been done to know the antibacterial effect on bacteria such as *Streptococcus* in vitro. This study aimed to examine the antibacterial activity of cinnamon ethanol extract against *Streptococcus* and its application as mouthwash to inhibit the bacteria. The cinnamon bark was macerated followed by extracted in 80% ethanol. Bacterial samples were isolated from dental plaque of patients visiting dental clinic drg. Syahdiana Waty in Medan, North Sumatra. The isolates were identified using Vitek 2 compact. Secondary metabolites were detected using previously described method. Antibacterial assay was done at extract concentration of 6.25%, 12.5%, and 25%. The result showed that alkaloids, flavonoids, saponins, and glycoside were detected in the extract. Nine bacterial species were identified as *Streptococcus mitis, S. sanguinis, S. salivarius, S. pluranimalium, S. pneumoniae, S. alactolyticus, Kocuria rosea, Kocuria kristinae,* and *Spingomonas paucimolis*. It showed that the extract of Cinnamon bark significantly inhibited *Streptococcus* growth, and it was effective as mouthwash.

Keywords: antibacterial activity, cinnamon bark, mouthwash, *Streptococcus*

1. Introduction

Cinnamon bark has been used for spicy and traditional medicine.Cinnamon bark and its leaves contain essential oils, saponins and flavonoids that have been widely used for healing various diseases [1]. The main compound of the essential oil is cinamaldehyde (60.72%), eugenol (17.62%) and coumarin (13.39%) which has antibacterial effect [2].

Prevalence of dental and oral diseases in Indonesia tends to increase in which caries is the most one. The dental caries and abnormalities of the dental begin with the formation of dental plaque [3]. The formation of plaque in the form of a thin layer attached to the tooth surface and sometimes also found in

IOP Conf. Series: Earth and Environmental Science **130** (2018) 012049 doi:10.1088/1755-1315/130/1/012049

the gums and tongue is caused by food debris used by pathogenic bacteria in the oral cavity [4]. Pathogenic bacteria present in the oral cavity include *Streptococcus mutans, S. viridans,S. pneumoniae, St. epidermidis,* and *St. aureus* [5].

There are various ways to reduce the accumulation of plaque in the oral cavity, among those are brushing teeth regularly, rinsing with antiseptic solution, cleaning interdental with dental floss, cleansing the tongue, chewing gum, and avoiding fermented carbohydrates [6]. The easiest way to remove plaque is by rinsing. Several studies have demonstrated the effectiveness and usefulness of an antiseptic mouthwash containing active compounds such as chlorhexidine and essential oils to prevent the formation of plaque and gingivitis [7].

One of the potential compounds in controlling plaque formation is in cinnamon bark. Puspita *et al.* (2013) showed that cinnamon extract has an effect on the growth of *S. mutans* which is carriogenic bacteria, but there is no information about the effectiveness of using cinnamon extract as a mouthwash to reduce dental plaque [2]. Therefore, in this study cinnamon ethanol extract applied as a mouthwash in inhibiting the growth of these bacteria was conducted.

2. Methods

2.1. Cinnamon bark maceration, extraction, and phytochemical test

Cinnamon bark was macerated followed by extraction in ethanol. A-2 kg of dried cinnamon bark was immersed in 20 L of distilled ethanol. Maserate was evaporated using a rotary evaporator at 60° C. The extract was fed into a water bath to remove remaining moisture. The extract was subjected to phytochemical tests including alkaloids, glycosides, anthraquinone glycosides, saponins, tannins, flavonoids, and steroids/triterpenoids.

2.2. Patients,plaque samples, and bacterial identification

Forty patients from Dental Clinic of Drg. Syahdiana Waty, Medan were selected based of Slovin's equation. Sample inclusion criteria were patients with dental caries of at least one tooth, aged 35-44 years, in good health and did not previously use antibiotic drug, willing to fulfill informed consent and willing not to brush their teeth before taking plaque samples. Samples were taken on buccal portion of 16, 26, 36 and 46 and labial of teeth 11 and 31. Samples were stored in containers and immediately taken to laboratory for bacterial culture. The samples were dissolved in sterilized water and subjected to be vortexed. Vortexed sample was streak on trypticase yeast cysteine and incubated at 37° C overnight. Single colony was identified using Vitek 2 compact.

2.3. Antibacterial assay of cinnamon bark extract and prepared mouthwash

Cinnamon bark extract was prepared in concentration of 6.25%, 12.5%, and 25% using dimethyl sulfoxide as solvent. Bacterial solution was prepared in 0.90% of NaCl and equalised to 0.5 Mc Farland. The bacterial solution was steak on entire surface of Muller Hinton agar using sterilized cotton bud. Paper disc immersed in each cinnamon extract concentration was put on the bacterial lawn in petri dish. The petri dish was incubated at 37° C overnight. Inhibition zone was measured as diameter of clearing zone around the paper disk subtracted to 0.5 cm. The extract with Minimum Inhibitory Concentration (MIC) was used as active compound in prepared mouthwash. pH and viscosity of prepared mouthwash were measured. Application of 10 ml mouthwash was conducted in patients for 30 seconds. The plaque sample and measuring bacterial cell number were conducted as previously described. Number of bacterial cell was measured as colony growth in total plate count agar.

2.4. Design and data analysis

This study was a pretest-post test one group design.Data was analyzed using Shapiro Wilk, Kolmogorov-Smirnov, Kruskal-Wallis, Wilcoxon and Man Whitney tests.

3. Results and Discussion

3.1. Cinnamon bark maceration, extraction, and phytochemical test

Maceration and extraction of 20 kg of the bark yielded 17.5 L of the brown macerate and 670 gram reddish brown extract, respectively. Phytochemical test showed that cinnamon bark ethanol extract contains secondary metabolite compounds such as alkaloid, saponin, flavonoid, and glycoside groups. Puspita *et al* (2013) showed that cinnamon bark extracts contained transinamaldehid, polyphenols, flavonoids, saponins, and tannins [2]. Awang *et al* (2013) reported that cinnamon bark essential oil contained cinnamaldehyde which is potential as an antimicrobial compound [8]. Alkaloid is one of the most organic compounds found in nature. It has prominent antimicrobial activity and has been widely used in antimicrobial treatment [9]. Saponin is a compound that has a working mechanism as antibacterial with its lipophilic properties capable of damaging cell membranes [10]. Flavonoid compounds disrupt bacteria by destroying the cytoplasmic membrane and causing leakage of important metabolites that inactivate bacterial enzyme systems [11].

3.2. Isolation of plaque bacteria

There were nine bacterial species found in the plaque samples, in which *Streptococcus* was the most common bacteria. There were six species of *Streptococcus* including *S. mitis, S. sanguinis, S. salivarius, S. alactolyticus, S. pneumoniae,* and *S. pluranimalium*. The other three bacterial species were *Kocuria rosea, K. kristinae*, and *Sphingomonas paucimolis* (Table 1).

Table 1. Frequency of *Streptococcus* and other bacterial species from dental plaque

No.	Bacteria species	Number of samples	Frequency (%)
1.	*S. mitis*	12	30
2.	*S. sanguinis*	8	20
3.	*S. salivarius*	4	10
4.	*S. alactolyticus*	3	7.5
5.	*S. pneumoniae*	1	2.5
6.	*S. pluranimalium*	3	7.5
7.	*K. rosea*	2	5
8.	*K. kristinae*	6	15
9.	*S. paucimolis*	1	2.5
	Total	40	100

Streptococcus mitis, S. sanguinis and *S. salivarius* were the most common bacteria in dental plaque. *Streptococcus mitis* is one of the bacteria of the mutans streptococci group (MSG). This group is known as cariogenic. It also included other species such as *S. mutans* and *S. sobrinus*. This group of bacteria is known as the pioneer bacteria involved in plaque formation and initiates dental caries [12].

Sterptococus sanguinis is one of species of *Viridans Group Streptococci* (VGS) which is most commonly found in dental plaque in the oral cavity. *S. sanguinis* and *MSG* have a strong association as *oral streptococci* and the most commonly found in dental plaque [12]. *Streptoccous salivarius* together with *S. sanguinis, S. mitis,*and *S. gordonii* known asthefirst colonizer the tooth surface, in contact with salivary glycoproteins on tooth surfaces via specific polymer capsules such as glucan and fructan. Furthermore, these bacteria play an important role in the process of dental caries together with other *Oral Streptococci* [13]. *Streptoccous pluranimalium* was a species of VGS that is rarely encountered in dental plaque. Dhotre *et al.* (2016) found three strains of *Streptococcus* that were very rare and unusual in dental plaque subgingival plaque such as*S. pluralimalium, S. thoraltensis* and *S. hyointestinalis* [14].

Bacteria *K. rosea* and *K. kristinae* are a gram-positive bacteria belonging to the family *Micrococcus*. Sabtie *et al* (2015) found 16 species of bacteria in dental supragingival plaque from 50 patients, in which two were *K. rosea* and *K. kristinae* [15]. The normal habitat of the *Kocuria* is on the skin but can also be found in the urinary tract in patients with urinary tract infections. This non-pathogenic bacteria arehuman normal flora. However, many studies showed that these bacteria might cause many infections of the urinary and gastrointestinal tract [16]. Many diseases in the gastrointestinal tract, urinary, and even other vital organ diseases are believed to be associated with the oral cavity diseases.

3.3. Antibacterial activity

The result of antibacterial test of cinnamon bark ethanol extract showed all concentration of the extract inhibited the bacterial growth (Figure 1.). Cinnamon bark extract of 6.25%, 12.5% and 25% showed inhibitory zone of 6.78, 9, and 11.68 mm, respectively (Table 2).

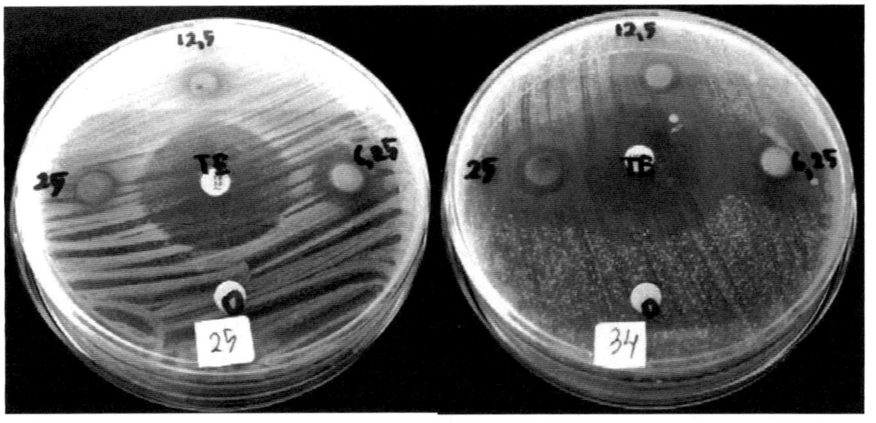

Figure 1. The bacterial inhibitory zone of cinnamon bark ethanol extract to *S. mitis* (25) and to *S. sanguinis* (34)

Table 2. Inhibitory zone of cinnamon ethanol extract to bacteria from dental plaque

Concentration of cinnamon extract	Diameters of inhibitory zone(mm)
6.25%	6.78
12.5%	9
25%	11.68
Tetracyclin	28.3

Data analysis showed that there was a significant difference in bacterial inhibition zone of the three concentrations. Furthermore, it was showneach concentration showed significantly different on inhibitory zone. The similar result showed in Puspita *et al.* (2013) to inhibit the growth of *S.mutans.* Al-Duboni *et al.* (2013) reported that cinnamon bark extract showed to have more antibacterial activity than that of ginger extract on *S. mutans* [17].

3.4. Examination of cinnamon extract mouthwash as antibacterial agent

It was shown that the prepared mouthwash had pH of 6.97 with viscosity of 1.01 cP.
The results of antibacterial activity found that population of bacteria were reduced (Figure 2.). It was shown that there was a significant difference to bacterial activity of the prepared mouthwash in the mouth before and after rinsing.

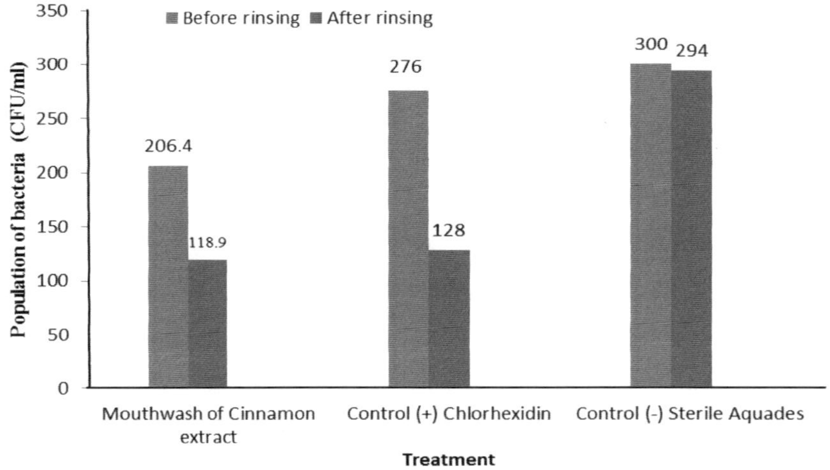

Figure 2. The average population of bacteria before and after rinsing with the mouthwash of Cinnamon extract, Chlorhexidin and sterile aquades.

From this study it was suggested to use this extract as an alternative in controlling pathogenic oral cavity bacteria. Many other plants such as starfruit leaves, gambir leaves, siwak, and lemon were reported to be used in mouthwash against bacteria in dental plaque in vivo. Fajriani and Mahrum (2015) reported that 40% lemon extract showed a significant effect in reducing bacterial number 30 minutes after rinsing [18].

4. Conclusions

Cinnamon bark ethanol extract contained phytochemical compounds such as alkaloid, flavonoid, saponin, and glikosid. The extract of each concentration showed antibacterial activity in vitro, and as mouthwash as well.

Acknowledgment

Many thanks are addressed to Ministry of Health, Republic of Indonesia for providing a scholarship and research fund.

References

[1] Pitojo S, and Zumiati (2016) *Tanaman Bumbu dan Pewarna Nabati*, Edisi. VIII. CV. Aneka Ilmu, Semarang. 64-66.

[2] Puspita A, Kholifa M, and Rochmanita, N (2013) Pengaruh Konsentrasi Ekstrak Kulit kayu manis (*Cinnamomum burmanni*) dalam Menurunkan Pertumbuhan *Streptococcus mutans* secara In Vitro. Fakultas Kedokteran Gigi UMS.

[3] Pratiwi R (2005) Perbedaan Daya Hambat terhadap *Streptococcus mutans* dari beberapa Pasta Gigi yang Mengandung Herbal. Majalah Kedokteran Gigi *(Dental Journal)*. 38.

[4] Chandrabhan D, Hemlata R, Renu B, and Pradeep V (2012) Isolation of Dental Caries Bacteria from Dental Plaque and Effect of Tooth Pastes on Acidogenic Bacteria. *Open Journal of Medical Microbiology*. 2. 65-69.

[5] Volk WA, and Wheeler (1993) *Mikrobiologi Dasar*, Edisi Kelima, Jilid I. Penerbit Erlangga, Jakarta.

[6] Ladytama, SR, Nurhapsari A, and Baehaqi M (2014) Efektivitas Larutan Ekstrak Jeruk Nipis (*Citrus Aurantifolia*) Sebagai Obat kumur Terhadap Penurunan Indeks Plak Pada Remaja Usia 12-15 Tahun- Studi di SMP Nurul Islami, Mijen, Semarang. *Odonto Dental Journal*. 1 (1).

[7] Endarti, Fauzia, and Zuliana E. (2006) Manfaat Berkumur dengan Larutan Siwak (*Salvadora persica). Majalah Kedokteran Nusantara*. 39(4) : 393-401.

[8] Awang AFIB, Susanti D, and Taher M (2013) Antimicrobial Activity and Synergic Effect of *Cinnamomum burmanii's* Essential Oil & its Isolated Compound (Cinnamaldehyde). *International Conference on Chemical, Agricultural and Medical Sciences*. 29-30.

[9] Chandukar PN, Tripathi N, Choudary A, and Murab T (2014) Antibacterial Properties of Cinnamon Stick Oil with Special References to *Streptococcus pyogenes* and *Pseudomonas aeruginosa*. *International Journal of Current Microbiology and Apllied Sciences*. 3 : 177-178.

[10] Liantari DS (2014) Effect of Wuluh Starfruit Leaf Extract for *Streptococcus mutans* Growth. *J Majority*. 3(7) : 27-33.

[11] Adzakiyah T, Lipoeta I, and Kusuma N (2015) Pengaruh Berkumur dengan Larutan Ekstrak Siwak (*Salvadora persica*) Terhadap pH Saliva Rongga Mulut. *Jurnal Sains Farmasi dan Klinis*. 2(1) : 74-77.

[12] Dziedzic A, Wojtyczka RD, and Kubina R (2015) Inhibition of Oral Streptococci Growth Induced by the Complementary Action of Berberine Chloride and Antibacterial Compounds. *Molecules*. 20 : 13705-13724.

[13] Maal KB, Majid, Boujari, and Zavareh FA (2010) Identification of *Streptococcus salivarius* Bacteriophage Isolated from Persian Gulf as a Potential Agent for Dental Caries Phage Therapy. *African Journal of Microbiology Research*. 4(20) : 2127-2132.

[14] Dhotre S, Suryawanshi N, Nagoba B, and Shelkar S (2016) Rare and Unusual Isolates of *Viridans Streptococci* From Human Oral Cavity. *Indian Journal Pathology and Microbiology*. 59(1) : 47-49.

[15] Sabtie HA, Ali IN, and Humudat YR (2015) Microbial Biofilm Formation as a Dental-fouling Microorganisme Attachment Phenomenon in Oral Cavity Environment. *Journal of Life Sciences*. 9 : 40-46.

[16] Ma ESK, Wong CLP, Lai KTW, Chan ECH, Yam WC, and Chan ACW (2005) Kocuria kristinae Infection Assosiated with Acute Choleocystitis. *BioMed Central Infectious Disease*. 5(60) : 1-3.

[17] Al-Duboni G, Osman MT, and Al-Naggar R (2013) Antimicrobial Effect of Aqueous Extract of Cinnamon and Ginger on Two Oral Pathogens Causing Dental Caries. *Research Journal of Pharmaceutical, Biological and Chemical Sciences*. 4(3) : 957-965.

[18] Fajriani and Mahrum (2015) Efectiveness of Lime (*Citrus aurantifolia*) Extract Solution in Inhibiting Bacteria *Streptococcus mutans* Case of Early Childhood Caries. *Donnish Journal*. 1(4) : 016-020.

RAPD Markers on Genetic Diversity in Three Populations of *Pisifera* Type of Oil Palm (*Elaeis guineensis*)

M Basyuni[1*], H Prayogi[1], L A P Putri[2], I Syahputra[3], E S Siregar[4], I Risnasari[1], R Wati[1], and D Arifiyanto[3]

[1] Department of Forestry, Faculty of Forestry, Universitas Sumatera Utara, Jl. Tri Dharma Ujung No. 1 Medan, North Sumatera 20155, Indonesia
[2] Department of Agroecotechnology, Faculty of Agriculture, Universitas Sumatera Utara, Medan, North Sumatera 20155, Indonesia
[3]Laboratory of Biomolecular of PT Socfin Indonesia, Martebing, Dolok Masihul, Tebing Tinggi 20991, North Sumatra, Indonesia
[4]Department of Biology, Faculty of Mathematics and Natural Sciences, Universitas Sumatera Utara, Medan 20155, Indonesia

*Email: m.basyuni@usu.ac.id

Abstract. Palm oil (*E. guineensis*) is one of the major commodity and contributing largely to non-petroleum oil of Indonesian foreign exchange. *E. guineensis* has three fruit types, dura (female), pisifera (male), and tenera —a hybrid between dura and pisifera. *Pisifera* plays an important function in the production of seed oil palm. The purpose of this research is to analyze genetic diversity of pisifera type of *E. guineensis* from three populations, Yangambi, Lame and Lame further cross in Bangun Bandar, North Sumatra, Indonesia. Eighteen samples for each population were analyzed using six RAPD markers. Results showed that RAPD markers were low polymorphic with 1.49, 1.39, and 1.00 average number alleles detected for Yagambi, Lame, and Lame further cross, respectively. The level of genetic diversity detected for each population was 0.28, 0.22, and 0.21 for Yagambi, Lame, and Lame further cross, respectively, indicating that the populations had little genetic variation. The highest of polymorphic information content (PIC) was found on the P11 primer of Yangambi (0.49) and P10 primer for lame further cross (0.49). By contrast, the lowest PIC belongs to P21 for Lame population (0.01). This data is likely to contributing oil palm breeding.

Keywords: Breeding, conservation, female sterile, genetic variation, North Sumatra

1. Introduction

Palm oil is one of the major commodities of Indonesia, and current world's largest producer and exporter of palm oil. The oil palm (*Elaeis guineensis* Jacq, Arecaceae), a tropical perennial plant originated from West Africa. *E. guineensis* categorized based on the presence or absence of the shell in their fruits as *dura* (thick-shelled/sh^+sh^+), *pisifera* (shell-less/sh^-sh^-), and *tenera* (thin-shelled/ (sh^+sh^-)—a cross between the *dura* and *pisifera* [1]. *Pisifera* (male parent) has an important function and high economic value in the production of oil palm seeds. In the oil palm seed industry, conservation of the *pisifera* pollen will be able to support a sustainable artificial pollination program in seed production and breeding. Therefore the population of *pisifera* is one important factor to be managed and developed [2].

Given the importance of *pisifera* as pollen sources are selected for commercial palm oil production of D×P planting materials. This economic important character of *pisifera* palms, which is female sterile due to premature rotting fruits thus unable to be fruits [3]. Information on genetic diversity of *pisifera,* therefore, is very important for survival and adaptation to environmental changes. However, a few studies on genetic diversity on the *pisifera* population in *E. guineensis* from North Sumatra. The aim of this research is to analyze genetic diversity of *pisifera* type of *E. guineensis* from three populations, Yangambi, Lame and Lame further cross in Bangun Bandar, North Sumatra, Indonesia.

2. Materials and Methods

2.1. Plant sample and DNA extraction

A total of 54 young leaves of individual *E. guineensis* were collected at Bangun Bandar Seed Production, Serdang Badagai, North Sumatra from three origins of *pisifera* as three populations: Lame (18 samples), Yangambi (18 samples), and Lame further cross (18 samples) and used for DNA extraction. Total genomic DNA was extracted from *E. guineensis* leaves using modified cetyl trimethyl ammonium bromide (CTAB) procedure [4]. The quality of the DNA was evaluated using 1% agarose gels and quantified using UV-Spectrophotometer. The DNA extraction was stored at -20 °C.

Six RAPD primers, namely P6 (5'-TACCACCCCG-3'), P10 (5'- GGACCCAACC -3'), P11 (5'-GTCGCCGTCA-3'), P15 (5'-TTGGCACGGG-3'), P19 (5'-AGCGCCATTG-3'), and P21 (5'-GGGGTGACGA-3') was used in this study as previously reported [5]. PCR amplification was performed using GoTaq green master mix 2X (Promega), primer 10 μM, DNA template, and nuclease-free water. PCR reaction was carried out in PCR PC-806 (Astec) for 1 cycle at 95 °C for 4 min; 40 cycles at 95 °C for 45 s, 35 °C for 45 s, 72 °C for 1 min and a final extension 72 °C for 7 min.

2.2. RAPD analysis

The amplicon was separated on 2% agarose gels using electrophoresis in 0.5X TBE buffer at 100 volts for 4-5 h. Gels were stained with GelRed (Biotium) and visualized by UV Gel Documentation system. The band pattern was analyzed by cluster analysis as previously described [5].

2.3. Data analysis

Polymorphic DNA banding patterns were generated by all the different primers were scored according to the presence (1) and absence (0) of the band of a particular molecular size to accumulate a binary matrix which was subjected to cluster analysis. Both faint and intense bands were scored if shown consistency in triplicate experiments.

Genetic variation overall populations were calculated using GenAlEX 6.5 [6] as a number of alleles detected (Na), effective numbers of alleles (Ne), Shannon index (I), unbiased genetic diversity (Uh), the percentage of polymorphism loci (PoP). Polymorphic information content (PIC) and H: heterozygosity was determined by [7].

3. Results and Discussion

Figures 1-2 show RAPD banding pattern using P10 and P11 primers (as representative) for population Yangambi and Lame further cross. The bands showing the different size of DNA from 150 to 4000 bp. The similar banding pattern each sample with different RAPD markers indicating no specific band present in this study.

RAPD markers were low polymorphic with 1.49, 1.39, and 1.00 average number alleles detected for Yagambi, Lame, and Lame further cross, respectively (Table 1). The level of genetic diversity detected for each population was 0.28, 0.22, and 0.21 for Yagambi, Lame, and Lame further cross, respectively, indicating that the populations had little genetic variation. The maximum heterozygosity for RAPD is 0.5 [7]. The Shannon index for each population was 0.42, 0.33, 0.30 for Yagambi, Lame, and Lame further cross, respectively.

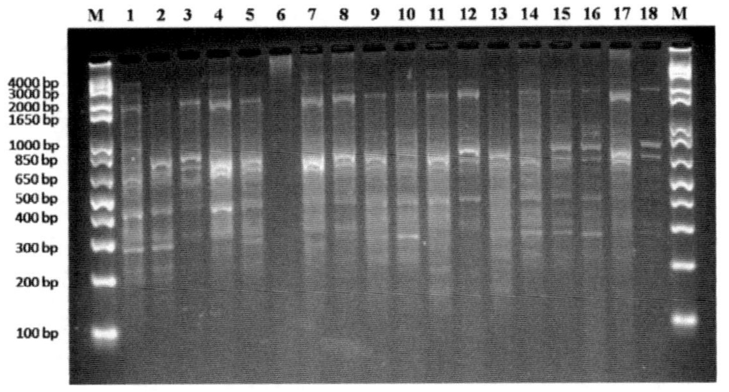

Figure 1. RAPD banding pattern using primer P10. M: marker, lane 1-18: samples from for Yangambi origin

Table 1. Genetic variation on oil palm of *pisifera*-type

Locus	N	Na	Ne	I	H	Uh	PoP (%)
Yangambi							
P6	18	1.85	1.60	0.52	0.35	0.37	92.31
P10	18	0.89	1.24	0.21	0.14	0.15	44.44
P11	18	1.50	1.42	0.38	0.25	0.26	75.00
P15	18	1.33	1.47	0.39	0.26	0.28	66.67
P19	18	1.47	1.52	0.43	0.29	0.31	73.33
P21	18	1.88	1.75	0.57	0.40	0.42	93.75
Mean	18	1.49	1.50	0.42	0.28	0.30	74.25
Lame							
P6	18	0.92	1.13	0.15	0.09	0.10	46.15
P10	18	1.56	1.52	0.45	0.31	0.32	77.78
P11	18	1.62	1.60	0.48	0.33	0.35	81.25
P15	18	2.00	1.54	0.52	0.34	0.36	100.00
P19	18	2.00	1.30	0.37	0.22	0.23	100.00
P21	18	0.25	1.01	0.03	0.01	0.01	12.50
Mean	18	1.39	1.35	0.33	0.22	0.23	69.61
Lame further cross							
P6	18	0.15	1.07	0.05	0.04	0.04	7.69
P10	18	1.44	1.55	0.45	0.31	0.33	72.22
P11	18	1.50	1.54	0.45	0.31	0.32	75.00
P15	18	1.00	1.38	0.29	0.20	0.22	50.00
P19	18	0.67	1.30	0.22	0.16	0.17	33.33
P21	18	1.25	1.42	0.36	0.25	0.26	62.50
Mean	18	1.00	1.38	0.30	0.21	0.22	50.12

(N: number of individuals, Na: number of alleles detected, Ne: effective number of alleles, I: Shannon index, H: heterozygosity, Uh: unbiased genetic diversity, PoP: percentage of polymorphism loci)

IOP Conf. Series: Earth and Environmental Science **130** (2018) 012050 doi:10.1088/1755-1315/130/1/012050

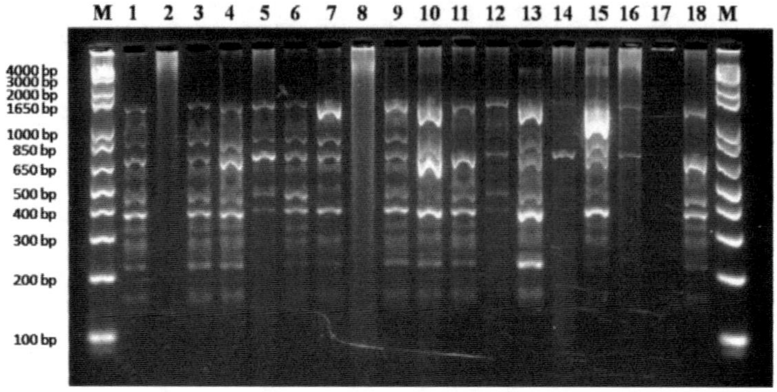

Figure 2. RAPD banding pattern using primer P11. M: marker, lanes 1-18: samples from Lame further cross origin

Table 2. Locus, size band, frequency of band pattern, and polymorphic information content (PIC) on *pisifera* type of *E. guineensis*

Locus	Size band (bp)	Frequency of band	Polymorphic information content (PIC)
Yangambi origin			
P6	300-3000	0.30	0.41
P10	220-4000	0.26	0.38
P11	170-3500	0.53	0.49
P15	200-3000	0.33	0.44
P19	150-4000	0.25	0.37
P21	200-4000	0.39	0.47
Lame origin			
P6	300-3000	0.05	0.10
P10	220-4000	0.22	0.34
P11	170-3500	0.27	0.39
P15	200-3000	0.24	0.36
P19	150-4000	0.13	0.23
P21	200-4000	0.01	0.01
Lame further cross			
P6	300-3000	0.05	0.08
P10	220-4000	0.44	0.49
P11	170-3500	0.41	0.48
P15	200-3000	0.22	0.34
P19	150-4000	0.14	0.24
P21	200-4000	0.27	0.39
Total		4.53	6.01
Mean		0.25	0.33

The population of Yangambi had a high percentage of polymorphism loci (74.25%) compared to other populations, where, Lame further cross contained the lowest one (50.12%). This result parallels with a number of alleles found. The highest of polymorphic information content (PIC) was found on the P11 primer of Yangambi (0.49) and P10 primer for lame further cross (0.49). By contrast, the lowest PIC belongs to P21 for Lame population (0.01). PIC is important for genetic studies in relation to molecular markers. In this context, the P11 primer is suitable for all populations with PIC value 0.39-0.49. RAPD is a dominant marker, maximum PIC value is 0.5 due to two alleles per locus and affected by a number of alleles [7].

This study demonstrates one way in which DNA markers (RAPD) might be used to assess genetic variation and assist in the development of oil palm breeding [5]. RAPD markers may provide fast and cheap methods for identifying oil palm type, in addition to two-dimensional thin layer chromatography, an alternative approach to identify fruit type of oil palm [8].

4. Conclusions

The level of genetic diversity detected for three populations of *pisifera*: Yagambi, Lame, and Lame further cross was little genetic variation. The conservation of *pisifera* is needed for securing genetic materials. This finding is likely to contributing oil palm breeding.

Acknowledgment

This study was partly supported by Grant-in Aids for PUPT (Penelitian Unggulan Perguruan Tinggi) 2014 (No. 4808/UN5.1.R/KEU/2014 to MB) and Riset Pengembangan Ilmu Pengetahuan dan Teknologi (IPTEK) 2016 (No. 017/SP2H/LT/DRPM/II/2016 to MB), and PUPT (Penelitian Unggulan Perguruan Tinggi) 2014 (No. No. 003/SP2H/LT/DRPM/IV/2017 to MB) from the Directorate for Research and Community Service, Ministry of Research, Technology and Higher Education, Republic of Indonesia.

References

[1] Basyuni M, Amri N, Putri LAP, Syahputra I, Arifiyanto D (2017) Characterization of fresh bunch yield and the physicochemical qualities of palm oil during storage in North Sumatra, Indonesia *Indones. J. Chem.* **18** 182–190.

[2] Barcelos E, Rios SDA, Cunha RNV, Lopes R, Motoike SY, Babiychuk E, Skirycz A, Kushnir S (2015) Oil palm natural diversity and the potential for yield improvement *Front Plant Sci.* **6** 190.

[3] Arolu IW, Rafii M Y, Marjuni M, Hanafi MM, Sulaiman Z, Rahim HA, Kolapo OK, Abidin MIZ, Amiruddin MD, Din AK, Nookiah R (2016) Genetic variability analysis and selection of pisifera palms for commercial production of high yielding and dwarf oil palm planting materials *Ind. Crops Prod.* **90** 135–141.

[4] Basyuni M, Baba S, Oku H (2017) Microsatellite analysis on genetic variation in two populations of red mangrove *Rhizophora mangle* L. (Rhizophoraceae) and its implication to conservation *IOP Conf. Ser.: Mater. Sci. Eng.* **180** 012243.

[5] Suthish DK, Mohankumar C (2007) RAPD markers for identifying oil palm (*Elaeis guineensis* Jacq.) parental varieties (*dura & pisifera*) and the hybrid *tenera Indian J. Biotechnol.* **6** 354–358.

[6] Peakell R, Smouse PE (2012) GenAlEx 6.5: genetic analysis in Excel. Population genetic software for teaching and research–an update *Bioinformatics* **28** 2537–2539.

[7] Nagy S, Poczai P, Cernák I, Gorji AM, Hegedűs G, Taller J (2012) PICcalc: an online program to calculate polymorphic information content for molecular genetic studies *Biochem. Genet.* **50** 670–672.

[8] Arifiyanto D, Basyuni M, Sumardi, Putri LAP, Siregar ES, Risnasari I, Syahputra I (2017) Short Communication: Occurrence and cluster analysis of palm oil (*Elaeis guineensis*) fruit type using two-dimensional thin layer chromatography *Biodiversitas* **18** 1587–1492.

Species of leafy liverworts in protected forest of simancik 1, regency of deli serdang, north sumatera

Nursahara Pasaribu[1*], Etti Sartina Siregar[1], and Wardatur Rahmi[1]

[1]Department of Biology, Faculty of Mathematics and Natural Sciences, University of Sumatera Utara, Jl. Bioteknologi No. 1 Campus USU, Medan 20155, Indonesia

[*]Email: nursahara@usu.ac.id

Abstract. Data and information of leafy liverworts diversity in Sumatera, especially North Sumatera, are still less reported. Research is conducted in order to invent and describe the morphological character of leafy liverworts in Protected Forest of Simancik I, regency of Deli Serdang, North Sumatera. Samples were collected along the hiking tracks of study site. Twenty two spesies belonging to 12 genera and 6 families were found. They were only one species of *Frullaniaceae*, eleven species of *Lejeuneaceae*, two species of *Lepidoziaceae*, two species of *Lophocoleaceae*, four species of *Plagiochilaceae* and two species of *Radulaceae*. Those species were found as epiphyte on the tree truck and some on the decaying wood. The highest diversity of leafy liverworts was from *Lejeuneaceae* which consisted of two subfamilies: *Lejeuneoideaea* and *Ptychanthoideae* and seven genera which are *Lejeunea, Archilejeunea, Lopholejeunea, Dendrolejeunea, Ptychantus, Mastigolejeunea* and *Thysananthus* while the lowest diversity was from *Frullaniaceae*.

Keywords: Bryophyte, *Lejeuneaceae*, *Lepidoziaceae*, Liverwort, *Plagiochilaceae*, *Radulaceae*, Sumatera

1. Introduction

Bryophytes are non-vascular plants which belongs to the second largest group after Angiospermae and has a very wide geographic spread. There are about 15,000-25,000 species of bryophytes in the world [1,2]. The most diverse bryophytes are found in subtropical and tropical regions especially in tropical rainforest areas [3,4]. Bryophytes are divided into three groups namely liverworts, mosses, and hornworts [4]. Bryophytes generally grow attached to various types of substrates i.e trees, decayed wood, litter, soil and even rocks with humid environment conditions and adequate sunlight.

Liverworts may grow as epiphytes, erect upward, dangle down, attach or lay down on the surface of the substrate [5], epiphytes in trunks and tree branches while some may also grow by attaching to the leaves epiphyll in lowland rainforests [6]. Bryophytes have an important role in the forest ecosystem, which maintain moisture and contribute in nutrient cycles, especially the carbon cycle [7]. Bryophytes also have potential utilization in the health field, since certain species contained a number of active ingredients, including anti-fungal, antibacterial, antitumor [8,9] and potential source of antioxidants in cosmetics, health, and food supplement industries [10]. Study of bryophytes in Indonesia is still less reported.

Data of liverworts biodiversity is still focused on Java origins, while in Sumatra, especially in North Sumatra is still scarce. Protected Forest of Simancik I or *Taman Hutan Raya* (TAHURA) Bukit

Barisan is located in Deli Serdang, District of North Sumatra. We assumed that the forest might be suitable habitat to find various species of liverworts. Until now, collections and information of leafy liverworts in Simancik 1 have not ever been reported yet, based on specimen collections in Herbarium MEDANESE without any records on bryophytes especially leafy liverworts. Therefore, the study will be the first attempt in collecting and describing leafy liverworts species based on morphological characters that inhabit Simancik Forest.

2. Methods

2.1. *Study Site Description*

The field study was conducted during five months of field observation at Protected Forest of Simancik 1 (098° 38' E, 03° 05' N, 250-1100 m asl), Taman Hutan Raya (TAHURA) Bukit Barisan, North Sumatera which covers an area of 9.800 ha. The area has a relative flat topography. The average range of annual rainfall, humidity and temperature were 2000-2500 mm, 80-90%, and 13-29 °C respectively. Commonly found vegetation were species such as: *Pinus mercusii, Altingia exelsa, Podocarpus imbricatus, Toona sureni, Casuarina, Eucalyptus, Cupressus, Agathis* and members of *Dipterocarpaceae, Moraceae, Arecaceae, Melastomataceae and Lauraceae*.

Collected samples were brought and identified at Laboratory of Plant Systematics, Faculty of Mathematics and Natural Sciences, University of North Sumatra, Medan. Forest area administratively located at Rambai Village, Sinembah sub-district of Tanjung Muda Hilir, Regency of Deli Serdang, North Sumatera, Indonesia.

2.2. *Field Methodology*

Leafy liverworts were sampled by surveying through accessible tracks. Each species found during survey is assessed by looking for its spot characters. For species growing as epiphytes on trees or trunks, a pocket knife is used to dissect samples from its substrate. Samples were then stored in paper envelope and written down local information such as: coordinates, altitude, temperature, humidity, light intensity and soil pH using standard analytical instruments.

2.3. *Laboratory Identification*

All specimens were air-dried to prevent excessive humid and fungal contamination. Specimens were identified using following guides to bryophyte and liverworts: A Revision of Japanese *Lejeuneaceae. The journal of the Hattori Botanical Laboratory* [11], The Genus *Plagiochilla* (Dum.) Dum. In Southeast Asia [12], Systematic Botany Monographs [13], Guide to the Bryophyte of Tropical Amerika, Vol 86 [1], Mengenal *Bryophyta* (Lumut) Taman Nasional Gunung Gede Pangrango Volume 1 [6], Guide to the Liverwort and Hornworts of Java [7]. Morphological characters were examined by using light-compound microscope. Essential characters following species identification are: length, width, lobes, lobulles, and perianths.

3. Results and Discussions

The study found twenty-two species of leafy liverworts belonging to twelve genera and six families: *Frullaniaceae* (1 species), *Lejeuneaceae* (11 species), *Lepidoziaceae* (2 species), *Lophocoleaceae* (2 species), *Plagiochillaceae* (4 species), and *Radullaceae* (2 species). Data are shown in Table 1.

Table 1. Species of Leafy Liverworts at Protected Forest of Simancik 1

N	Families	Species	Substrates
1.	*Frullaniaceae*	*Frullania apiculata*	Trunk
2.	*Lejeuneaceae*	*Lejeunea rotundistipula*	Trunk
3.		*Archilejeunea planiuscula*	Trunk
4.		*Dendrolejeunea fruticosa*	Trunk
5.		*Lopholejeunea acutifolia*	Trunk
6.		*Lopholejeunea eulopha*	Trunk dan Wood log
7.		*Lopholejeunea nigricans*	Trunks

8.		*Lopholejeunea subfusca*	Trunks
9.		*Mastigolejeunea replata*	Trunks
10.		*Mastigolejeunea virens*	Trunks
11.		*Ptychantus stratus*	Trunks
12.		*Thysananthus spathulistipus*	Trunks
13.	*Lepidoziaceae*	*Bazzania japonica*	Trunks
14.		*Bazzania vittata*	Trunk and Twig
15.	*Lophocoleaceae*	*Heteroscyphus argutus*	Trunk and Wood log
16.		*Heteroscyphus coalitus*	Trunk and Wood log
17.	*Plagiochilaceae*	*Plagiochila gracilis*	Trunk and Twig
18.		*Plagiochila teysmannii*	Twig
19.		*Plagiochila spathulifolia*	Twig
20.		*Pedinophyllum interruptum*	Trunk and Twig
21.	*Radulaceae*	*Radula javanica*	Trunk and Wood log
22.		*Radula retroflexa*	Trunk

Total leafy liverworts species found in this study are considered lower than previous reports at Sibayak Forest, North Sumatera. Previous study found 163 species of liverworts which belonged to 53 genera and 22 families. Numerous species of liverworts found in Sibayak were due to the wide spread of bryophytes across habitat [14]. The forest was known to ranging from sub-alpine to low alpine forest while in our study, Simancik was only known for its lowland-type forest (600 m asl). Intensity of survey, sampling period and macro-/microclimate variation may also cause difference in finding bryophytes in each area [15].

Members of *Lejeuneaceae* were mostly found in our study. *Lejeuneaceae* was the speciose of leafy liverworts in the tropics and estimated about 90 genera with 1600 species. The family was commonly found in montane forest in contrary with our results from from lowland forest [1,16]. Only a small portion of species from *Lejeuneaceae* that were revealed in our study. Previous studies have reported a number of species from *Lejeuneaceae* in Java (160 species) and North Sumatera (49 species) [7,14,17].

Measurement of physico-chemical parameters at Simancik 1 was conducted although data not shown. The average temperature, humidity and light intensity was 25-29 ^{o}C, 48-50%, and 13-123 lux respectively which was assumed to be suitable for growth of certain bryophyte species [4,18,19]. Elevation is also a limiting factor of finding bryophytes. An increase of elevation will cause a drop of temperature, especially humidity and wind flow [2,20,21]. Other factors that support bryophyte existence in habitat are their growth substrates. Species of leafy liverworts were dominantly found attached to tree trunks followed by twigs and wood logs in our study. The specific term is given as *Corticolous* (epiphyte at trunks) and *Ramicolous* (epiphyte at twigs) [3,22].

4. Conclusions

Leafy liverworts found in this study are still in a small portion of total species in the tropics. Our results are the first report in revealing species richness of leafy liverworts in lowland forest of Simancik 1, North Sumatera. Attempt to collect more specimens is needed to give a precise number of leafy liverworts from North Sumatera. Alternative study site and sampling intensity must be considered in future study.

References

[1] Gradstein SR, SP Churchill and N Salazar-Allen (2001) *Guide to the Bryophytes of Tropical America*. New York: The New York Botanical Garden Comp.

[2] Glime JM (2007) *Bryophyte Ecology Volume 1. Physiologycal Ecology*. Michigan Technological University and the International Association of Bryologists.

[3] Gradstein SR and T Pocs (1989) *Bryophyte*. Book Chapter: Lietth, H and MJA Werger. 1989. *Tropical a Forest Ecosystems*. Amsterdam. Elsevier Science Publisher B.V. Pages 314-315.
[4] Sporn SG, MM Bos, Hoffstatter-Muncheberg, M Kessler and SR Gradstein (2009) Microclimate Determines Community Composition but not Richness of Epiphytic Understorey Bryophytes of Rainforest and Cacao Agroforests Indonesia. *Funct Plant Biol.* **36(2)**:171-179.
[5] Damayanti L (2006) Koleksi Bryophyta Taman Lumut Kebun Raya Cibodas. Volume II No.4. UPT Balai Konservasi Tumbuhan Kebun Raya Cibodas. Sindanglaya. Cianjur.
[6] Hasan M and Ariyanti N (2004) Mengenal Bryophyta (Lumut) di Taman Nasional Gunung Gede Pangrango Volume 1. Cetakan pertama. Taman Nasional Gunung Gede Pangrango.
[7] Gradstein SR (2011) Guide to the Liverworts and Hornworts of Java. Bogor. Seameo Biotrop.
[8] Shi YQ, CJ Zhu, BQ Yuan, BQ Li, J Gao XJ Qu, B Sun, YN Cheng, S Li, X Li and HX Lou (2009) Marchantin C, a Novel Microtubule Inhibitor from Liverwort with Anti-Tumor Activity Both *In Vivo* and *In Vitro*. *Cancer Lett.* **276(2)**: 160-170.
[9] Veljić M, A Ćirić, M Soković, P Janaćković and PD Marin (2010) Antibacterial and Antifungal Activity of the Liverwort (*Ptilidium Pulcherrimum*) Methanol Extract. *Arch Biol Sci Belgrade.* **62(2)**: 381-395.
[10] Dey A, and JN De (2012) Antioxidative Potential of Bryophytes: Streaa Tolerance and Commercial Perspective: A review. *Pharmacologia.* **3(6)**: 151-159.
[11] Mizutani M (1961) A revision of Japanese Lejeuneaceae. *Journal of the Hattori Botanical Laboratory* **24**: 116-180.
[12] Inoue H (1989) The genus *Plagiochila* (Dum.) Dum. In Southeast Asia. Tokyo. Academia Scientific Book Inc.
[13] So ML (2001) Systematic Botany Monographs of *Plagiochilla* (*Hepaticeae, Plagiochillaceae*) in China. *Journal of the Hattori Botanical Laboratory* **60**: 207.
[14] Siregar ES, NS Ariyanti, and SS Tjitrossoedirdjo (2014) *Lejeuneaceae* Anak Suku *Ptychanthoideae* di Hutan Sibayak Sumatera Utara. *Floribunda* **4(8)**: 218-225.
[15] Gradstein SR, NM Nadkarni, T Kromer, I Holz, and N Noske. (2003) A Protocol and Representative Sampling of Vascular and Non-vascular Epiphyte Diversity of Tropical Rain Forests. *Selbyana* **24(1)**:105-111.
[16] Gradstein SR, and H Culmsee (2010) Bryophyte Diversity on Tree trunk in Montane Forest of Central Sulawesi, Indonesia. *Tropical Bryology* **31**:95–105.
[17] Söderström L, SR Gradstein, and A Hagborg (2010) Checklist of the Hornworts and Liverworts of Java. *J Phytotaxa* **9**:53–149.
[18] Friedel A, GV Oheimb, J Dengler, and W Härdtle (2006) Species Diversity and Species Composition of Epiphytic Bryophytes and Lichens A Comparison of Managed and Unmanaged Beech Forests in NE Germany. *Feddes Repert.* **117**:1-2, 172-185
[19] Ariyanti NS, MB Merijin, K Kuswata, ST Sri, E Gunahardja, and SR Gradstein (2008) Bryophytes on Tree Trunks in Natural Forest, Selectively Logged Forest and Cacao Agroforests in Central Sulawesi, Indonesia. *Biol conserv.* **141**: 2516-2527.
[20] Whitmore TC (1984) *Tropical Rain Forest of the Far East.* Oxford (DE). Clarendon press.
[21] Enroth J (1990) Altitudinal Zonation of Bryophytes on the Huon Peninsula, Papua New Guinea, A Floristic Approach, with Phytogeographic Considerations. *Trop Bryol.* **2(2)**:61-90.
[22] Gonzales-Mancebo JM, A Losada-Lima and S McAlister (2004) Host Specificity of Epiphytic Bryophyte Communities of Laurel Forest in Tenerife (Canary Island, Spain). *The Bryologist.* **106(3)**; 383-384.

Predicting size limit of wild blood python (*python brongersmai* stull, 1938) harvesting in north sumatera

Markus Mangantar Pardamean Sianturi[1,4*], Erni Jumilawaty[2], Delvian[3], and Adrian Hartanto[2]

[1]Graduate School, Management of Natural Resources and Environment, University of Sumatera Utara, Jl. Prof. T. Ma'as, Campus USU
[2]Department of Biology, Faculty of Mathematics and Natural Sciences, University of Sumatera Utara, Jl. Bioteknologi No. 1, Campus USU
[3]Faculty of Forestry, University of Sumatera Utara, Jl. Tridharma Ujung No 1. Campus USU, Medan City, North Sumatera 20155, Indonesia
[4]North Sumatera Conservation Agency (BBKSDA–SU), Directorate General of Natural Resource and Ecosystem Conservation, Ministry of Environment and Forestry, Jl. Sisingamangaraja KM 5,5 No. 14, Marendal, Medan City.

*Email: sianturi.markus@gmail.com

Abstract. Blood python (*Python brongersmai* Stull, 1938) is one of heavily exploited wildlife in Indonesia. The high demands on its skin trade have made its harvesting regulated under quota-based setting by the government to prevent over-harvesting. To gain understanding on the sustainability of *P. brongersmai* in the wild, biological characters of wild-caught specimens were studied. Samples were collected from two slaughterhouses from Rantau Prapat and Langkat. Parameters measured were morphological (Snout-vent length (SVL), body mass, abdomen width) and anatomical characters (Fat classes). Total samples of *P. brongersmai* in this research were 541 with 269 male and 272 female snakes. Female snakes had the highest proportion of individuals with the best quality of abdominal fat reserves (Class 3). Linear models are built and tested for its significance in relation between fat classes as anatomical characters and morphological characters. All tested morphological characters were significant in female snakes. By using linear equation models, we generate size limit to prioritize harvesting in the future. We suggest the use of SVL and stomach width ranging between 139,7 – 141,5 cm and 24,72 – 25,71 cm respectively to achieve sustainability of *P. brongersmai* in the wild.

Keywords: Anatomy, conservation, morphology, *Pythonidae*, sex ratio, slaughterhouses

1. Introduction

Skin trade has become a major supplier for global leather industry to achieve optimum industrial outputs. Majority of global worldwide supply on reptilian skin is provided from South East Asia, comprising countries such as Malaysia, Viet Nam, Lao People's Democratic Republic, and Indonesia. Indonesia is known to supply skin from five species, viz. the water monitor (*Varanus salvator*), two species of python (the reticulated python, *Python reticulatus* and the blood python *Python curtus* complex), the rat snake (*Ptyas mucosa*), and the spitting cobra (*Naja sputatrix*) [1]. Pythons have been heavily exploited in Indonesia since rising demands on commercial leather industry with globally known and well-studied species, reticulated pythons or *Python reticulatus* [2,3]. Other members of *Pythonidae* like Burmese python and Short-tailed python are currently facing serious threats from exploitation in the wild and their trade are still managed and approved in Appendices of Convention on International Trade in Endangered Species of Wild Fauna and Flora [4].

As formerly described as subspecies of *P. curtus* complex, *Python brongersmai* now become one of the most exported Short-tailed python species in Indonesia with central harvesting area located in North Sumatera [5,6]. While many comprehensive studies have been conducted to assess population of reticulated pythons in Indonesia, it is still detrimental to *Python brongersmai* as second most exploited species. In Indonesia and Malaysia, sustainable harvests are hoped to be achieved by using quota-based settings. However, study case on reticulated pythons showed many negative effects on applying quota which then recommended placing by an alternative strategy which is by promoting size limit of future catchment [3]. Size limit is one of biological attributes that may be used as predictor along with other attributes: like population demography and reproductive output [7]. Predicting size limit may be an alternative way to manage harvest by giving governmental force into local gatherers [8]. A practical size limit need to be pointed out to local hunters to achieve sustainable population of *P. brongersmai* in the wild. Here we provided current biological characteristics of wild-caught *P. brongersmai* in order to construct predictive unit to become a size limit approach for future sustainable harvesting in North Sumatera.

2. Methods

2.1 Study Sites

The research was conducted at rural area located in Rantau Prapat and Langkat, districts of North Sumatera. Based on this study, wild *Python brongersmai* were collected by local hunters around oil palm plantation. Surveys were conducted once per month for three months to ensure high catchment volume due to stockpiling and fluctuative catchment rate or gain of snakes in the wild. Snakes were gathered at slaughterhouse and for confidential issue, specific location was not declared in this research. Wild-caught snakes were carried using plastic sacks and delivered alive until being slaughtered by the local gatherers.

2.2 Measurement of morphological and anatomical characteristics

Snakes were directly counted (N) during three months of observation. Specimens were measured immediately after they were killed. Sex assessment was done by observing protruding hemipenis in male but not in female. Morphology and anatomy as biological characteristics were measured such as: Snout-vent length (SVL; cm), body mass (g), stomach width (cm) and abdominal fat reserves. Measurement was carried out using calliper with accuracy of 0.1 centimeter. Abdominal fat reserves were scored using arbitrary unit with following categorization: Class 0 (zero) being the lowest quality of fat reserve with no sign of fat deposits; Class 1 (one) with diameter between 1 – 2 cm, Class 2 with diameter between 2 – 3 cm, Class 3 being the best quality of fat reserve with diameter above 3 cm (Figure 1).

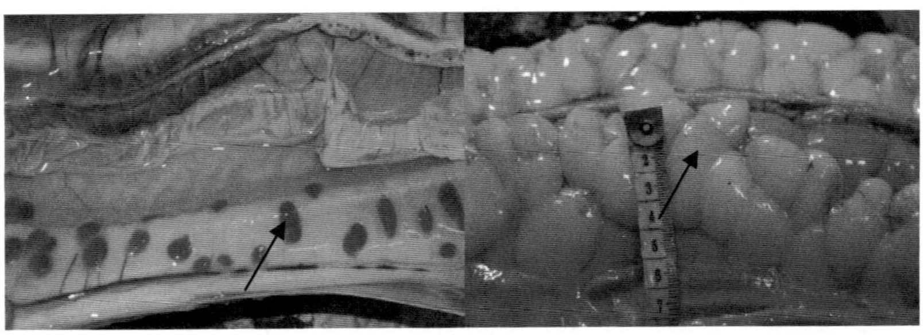

Figure 1. Representative images of Abdominal Fat Reserves: Left (Class 0, Poor); Right (Class 3, Best)

2.3 Confirmation on reproductive characters
Dissection on freshly-skinned bodies was made to ensure their sexes by visual inspection as well as categorizing their maturity state. Determination of male snakes was based on presence of male gonads, testes; with categorization of adult male having turgid testes with opaque and thickened efferent ducts. Determination of female snakes was based on presence of female gonads such as: oviductal eggs, vitellogenin ovarian follicles, and corpora lutea in adult females.

2.4 Data analysis
All numerical characters including morphological and anatomical characters were analyzed descriptively (mean values and standard deviations/ SDs) as showed in tables and graphs. Relationship between morphological and anatomical characters were subjected to linear regression model and significance was tested with the aid of software Minitab ver. 16.0.

3. Results and Discussion
We measured 541 snakes, comprising 269 male and 272 female *Python brongersmai* during three months of sampling period. Among two study areas, Rantau Prapat district had higher collection of P. brongersmai (N = 511) than Langkat district (N = 30). Specimens data were pooled (N = 541) and presented (Table 1).

Table 1. Biological attributes of *Python brongersmai* collected from North Sumatera

Trait	Male	Female	Total
N	269	272	541
Snout-vent length (cm)	135,03 (9,46)	140,18 (11,03)	137,41 (11,86)
Stomach width (cm)	23,69 (2,05)	24,91 (2,35)	24,3 (2,28)
Body mass (kg)	3,62 (0,75)	4,19 (0,95)	3,9 (0,91)

Note: Table shows mean values (SD in parentheses)

During study, we found that slaughterhouses had coordination with specific small home-scale leather industry or local gatherers to accommodate supply of raw materials on snake skins. Fluctuating catchment rate was known to explain different collection between areas in different districts and in different time [9]. Even though, we only studied blood python caught in the slaughterhouses, reticulated python (*P. reticulatus*) was also found in this study and not considered in our data. Processing techniques applied to the killed snakes were considered same as reported previously [2]. By looking at our data of nearly 1 : 1 (269 : 272) ratio of male versus female snakes, we may indicate that harvesting occured in random and non-selective meaning. Both male and female snakes had opportunity to be caught due to same habitat characteristic and movement in the wild with most natural habitat being converted into oil palm plantation. Future research will have to reveal significance of anthropogenic cause of converted natural habitat to movement ecology of this species. Survey on *P. brongersmai* in 1999 showed more males than females as well as greater morphological characters.

Morphological characters showed a diverse variation of samples during survey. Based on survey in 1999, the smallest size was 85 cm, which was not found in this study [2]. It may indicate that species with mentioned size did not exist anymore in the wild but accurate field study needs to be conducted to strenghthen our assumption. Trappers or Hunters also had their personal judgment for not catching snakes with small sizes due to economical reasons [6]. Differences between male and female snakes or sexual size dimorphism were found slightly different in this study. Female snakes tend to show bigger size than male due to investment on reproductive output [10]. This study showed domination in size of female snakes to almost all morphological characters. The bigger female will ensure successful reproduction in the wild and tend to positively correlated with survivorship of offsprings and fecundity [11]. Abdominal fat reserves were dominantly found at class 3 in female and class 2 in male snakes (Figure 2).

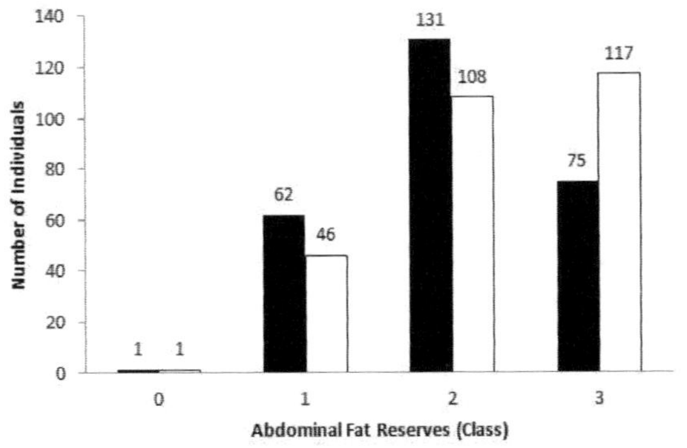

Figure 2. Proportion of *P. brongersmai* individuals with different classes of abdominal fat reserves: Male (Black bars), Female (White bars)

The differences showed a higher proportion of reproductively female snakes in the wild by considering indirect clue from proportion of individuals with a classes of abdominal fat reserves. By observing ovarian follicles, it was found that majority of class 3 female showed connection to total individual that mature and reproductively active. We later tested whether distribution of fat classes were related to morphological characters (Figure 3). Based on statistical test, relationship between fat classes distribution and other traits was found significant in female snakes. Male snakes only showed significant relatioship between fat classes and stomach width. Hence, SVL, body mass and stomach width variables may be used to predict classes of abdominal fat reserves in female snakes (Table 2).

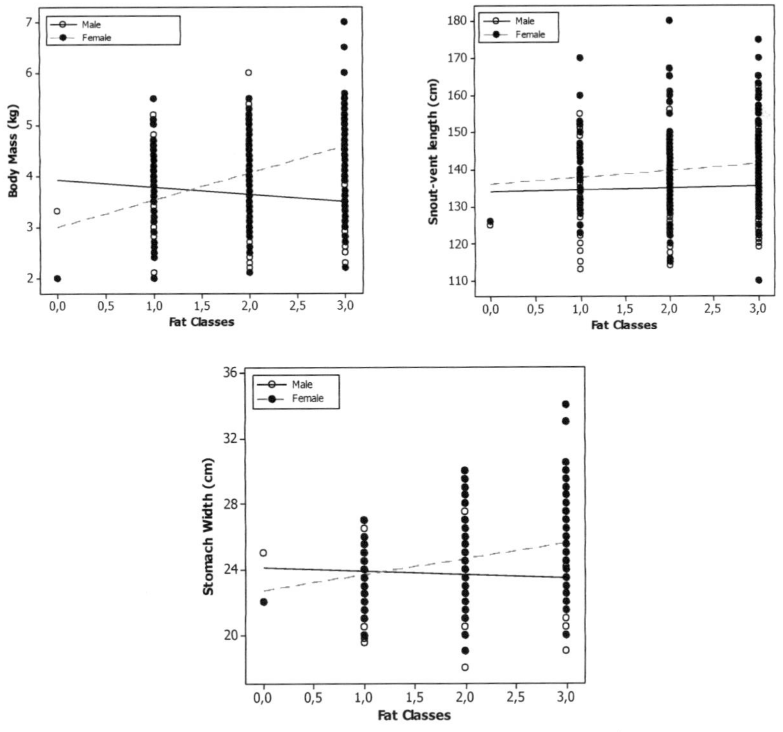

Figure 3. Fat classes relative to morphological characters: SVL, stomach width and body mass

Table 2. Relationship between fat classes and other traits of male and female *P. brongersmai*

Fat Classes	Traits	Statistical test
	Snout-vent length (cm)	R-sq = 0,00%; r = 0,04; $P = 0,55$
Male	Stomach width (cm)	R-sq = 0,15%; r = -0,07; $P = 0,24$
	Body mass (kg)	R-sq = 1,44%; r = -0,13; $P < 0,05$
	Snout-vent length (cm)	R-sq = 1,11%; r = 0,12; $P < 0,05$
Female	Stomach width (cm)	R-sq = 8,83%; r = 0,30; $P < 0,05$
	Body mass (kg)	R-sq = 16,89%; r = 0,41; $P < 0,05$

The result also supports the fact that female snakes tend to invest more energy input that may affect their morphology which also explained for the sexual dimorphism [15]. To assess sustainable harvest in the future, several acts must be carried out, one of which by promoting practical size limit of catchment. Based on significant relationships, we can predict optimum morphological characters of female snakes since it showed a positive correlation between stated traits. Before considering our finding on linear equation models on snakes traits, many studies focused on body condition scores or indices to predict anatomical character based on morphology. Since body indices were still a long debate and argued to provide an indirect measure on reproductive condition in snakes [12,13,14], here we provide basic empirical evidence by directly relating all morpho-anatomy variables using simple linear regression models as supported by statistical analysis. Female with fat class categorizing between 2 and 3 must be considered as priority harvest since it still appeared in large volume of catchment that may related to their numerous individuals in the wild. By applying linear equation models, several size limit are suggested with fat class (2 and 3) as predictors (Table 3).

Table 3. Size prediction of female *P. brongersmai* in relation with morphological characters

Trait	Linear equation model	Size limit
Snout-vent length (cm)	$y = 136,1 + 1,801x$	139,70 – 141, 50
Stomach width (cm)	$y = 22,75 + 0,98$	24,72 – 25,71
Body mass (kg)	$y = 2,99 + 0,53x$	4,06 – 4,58

In field application, we assume that measuring SVL and stomach width characters will be more applicable rather than weighting each snakes by using mass analyzer. Therefore, we hope to affect our local hunters to specify their priorities by following instruction based on limited attributes. Since the models were formulated on female snakes, more efforts have to be made in the field to differentiate sex correctly. We also hope to inform local gatherers so they are aware of excluding snake with the size outside of suggested ranges back to their natural habitat. Future study is needed to test whether this model may enhance survivorship of lower fat classes by evaluating caught snakes in the same area.

4. Conclusions

We proposed size limit restriction on harvesting *Python brongersmai*. The size limit are determined by relating number of individuals exposing large anatomical representatives (fat classes) with their morphological characters. Restriction are made for each of morphological characters but we suggest the use of SVL and stomach width as size ranges to follow up ease of application in the field.

Acknowledgments
We would like to thank Mr. Sudirman for his support by permitting us to conduct measurement of caught snakes in Rantau Prapat. We would also like to express our highest gratitude to North Sumatera Conservation Agency (BBKSDA–SU) regarding the official permit and informations.

References

[1] Iskandar DT, Erdelen WR (2006) Conservation of amphibians and reptiles in Indonesia:issue and problems. *Amphibian and Reptile Conservation* **4**(1):60-87.

[2] Shine R, Ambariyanto, Harlow PS, Mumpuni (1999) Ecological Attributes of Two Cormecially-Harvested Python Species in Northern Sumatra. *Journal of Herpetology* **33** (2):249-257.

[3] Natusch DJD, Lyons JA, Mumpuni, Riyanto A, Khadiejah S, Mustapha, Badiah N, Ratnaningsih S (2016) Manajemen Perdagangan Berkelanjutan Kulit Ular Sanca Batik di Indonesia dan Malaysia. Occasional Paper of the IUCN Species Survival Commission No. 61. Gland, Switzerland: IUCN.

[4] Convention on International Trade in Endangered Species of Wild Fauna and Flora (CITES). 2017 accessed from https://cites.org/sites/default/files/notif/E-Notif-2016-064-A.pdf

[5] Keogh JS, Barker DG, Shine R (2001) Heavily exploited but poorly known: systematics and biogeography of commercially harvested pythons (*Python curtus* group) in Southeast Asia. *Biological Journal of Linnean Society* **73**:113-129.

[6] Siregar J (2012) Upaya Pelestarian Pemanfaatan Ular Sanca Batik (*Python reticulatus*) dan Ular Sanca Darah Merah (*Python brongersmai*) Ditinjau Dari Aspek Penangkapan dan Pemasarannya di Provinsi Sumatera Utara [Thesis]. Bogor Agricultural Institute. Bogor [Indonesian]

[7] Kasterine A, Arbeid R, Caillabet O, Natusch D (2012) The Trade in South-East Asian Python Skins. International Trade Centre (ITC), Geneva.

[8] Natusch DJD, Lyons JA, Mumpuni , Riyanto A, Shine R (2016) Jungle Giants: Assessing Sustainable Harvesting in a Difficult-to-Survey Species (*Python reticulatus*). *PLoS ONE* **11** (7):e0158397.

[9] Nainggolan K (2015) Karakteristik Panenan Ular Sanca Batik *Python reticulatus* Di Sumatera Utara [Thesis]. Bogor Agricultural Institute. Bogor [Indonesian]

[10] Lillywhite HB (2014) How Snakes Work:Structure, Function, and Behavior of the World's Snakes. Oxford University Press, USA.

[11] Rivas JA, Burghardt GM (2001) Understanding sexual size dimorphism in snakes:wearing the snake's shoes. *Animal Behaviour* **62**:F1-F6.

[12] Weatherhead PJ, Brown GP (1995) Measurement versus estimation of condition in snakes. *Canadian Journal of Zoology* **74**:1617-1621.

[13] Waye HL, Mason RT (2007) A combination of body condition measurements is more informative than conventional condition indices: Temporal variation in body condition and corticosterone in brown tree snakes (*Boiga irregularis*). *General and Comparative Endocrinology* **155** (2008): 607-612.

[14] Labocha MK, Schutz H, Hayes JP (2013) Which body condition index is best? *Oikos* **000**:001-009.

[15] Aubret F, Bonnet X, Shine R, Lourdais O (2002) Fat is sexy for females but not males: The influence of body reserves on reproduction in snakes (*Vispera aspis*). *Hormones and Behavior* **42**: 135-147.